Biomass: Sustainable Energy Resource

Biomass:
Sustainable Energy Resource

Edited by **Hannah Seabrook**

SYRAWOOD
PUBLISHING HOUSE
New York

Published by Syrawood Publishing House,
750 Third Avenue, 9th Floor,
New York, NY 10017, USA
www.syrawoodpublishinghouse.com

Biomass: Sustainable Energy Resource
Edited by Hannah Seabrook

International Standard Book Number: 978-1-68286-359-6 (Hardback)

Printed in China.

Contents

Preface

The world is advancing at a fast pace like never before. Therefore, the need is to keep up with the latest developments. This book was an idea that came to fruition when the specialists in the area realized the need to coordinate together and document essential themes in the subject. That's when I was requested to be the editor. Editing this book has been an honour as it brings together diverse authors researching on different streams of the field. The book collates essential materials contributed by veterans in the area which can be utilized by students and researchers alike.

Biomass has emerged as one of the most prominent sustainable energy resource in the past few decades. It refers to the waste produced by plant or animal sources, and is a renewable source for fuel production. Included in this book, is a detailed explanation of the various concepts such as methods of growing energy crops, biomass consumption technology, and biofuel production and conversion. This book traces the progress of this field and highlights some of its key concepts and applications. It aims to serve as a resource guide for students and experts alike and contribute to the growth of the discipline.

Each chapter is a sole-standing publication that reflects each author´s interpretation. Thus, the book displays a multi-facetted picture of our current understanding of applications and diverse aspects of the field. I would like to thank the contributors of this book and my family for their endless support.

Editor

Changes in woody species composition following establishing exclosures on grazing lands in the lowlands of Northern Ethiopia

Wolde Mekuria[1]* and Mastewal Yami[2]

[1]International Water Management Institute (IWMI), P. O. Box 4199, Ban Nongviengkham, Xaythany District, Vientiane, Laos People's Democratic Republic.
[2]International Water Management Institute (IWMI), P. O. Box 5689, Addis Ababa, Ethiopia.

Restoring vegetation in low rainfall areas is difficult and urges the need to design an effective and low-cost method of vegetation restoration. This study was undertaken in the lowlands of northern Ethiopia to: (1) investigate how exclosure age affects restoration of degraded native plant species richness, diversity and aboveground standing biomass, and (2) identify soil characteristics, which affect effectiveness of exclosures to restore degraded native vegetation. Replicated (n = 3) 5-, 10- and 15-year-old exclosures were selected and each exclosure was paired with an adjacent grazing land to detect changes in vegetation variables following establishing exclosures on communal grazing lands. All exclosures displayed higher species richness, diversity and aboveground biomass when compared to the adjacent grazing lands. Results on vegetation composition indicate that all exclosures are at early stage of succession. In all exclosures and grazing lands, vegetation variables displayed significant ($p < 0.05$) correlations with soil variables indicating that consideration of soil fertility will help enhance natural regeneration in exclosures. Our study indicates that the establishment of exclosures on degraded communal grazing lands can be effective in restoring degraded native vegetations, and with time, exclosures may obtain an important role as source of seeds of indigenous woody species.

Key words: Grazing pressure, land degradation, land use conversion, native vegetation, soil variables, vegetation restoration.

INTRODUCTION

Degradation of vegetation has eventually led to desertification in Africa than any other continent (FAO, 2001). A total of 340 million ha of woody vegetation in dryland zones of Africa have become degraded through human activities (FAO, 2001). The dryland vegetation in Ethiopia resides in arid and semi-arid lands, which occupy 50% of the land area (Jama and Zeila, 2005) and comprise the largest vegetation resource of the country (Mengistu et al., 2005). At present, this dryland vegetation is facing intense degradation as a consequence of deforestation, agricultural land expansion and overgrazing (Lemenih et al., 2005; Mengistu et al., 2005). This in turn affects the productivity and diversity of forest, woodland and bushland resources. Counteracting the degradation of vegetation by rehabilitating degraded dryland forests is one of the central concerns of the government and local communities in Ethiopia (Girma, 2001; Kebrom, 2001).

In Tigray, northern Ethiopia, communities and local authorities started to establish exclosures to rehabilitate degraded communal grazing lands about three decades ago. Exclosures are areas closed from the interference of human and domestic animals with the goal of promoting natural regeneration of plants and reducing land degrada-

*Corresponding author. E-mail: wolde_mekuria@yahoo.com.

tion of formerly degraded communal grazing lands (Mekuria et al., 2011). Exclosures are usually established in steep, eroded and degraded areas which have been used for grazing in the past (Descheemaeker et al., 2006). The inception of exclosures dates back to the 1980's and coincided with the large-scale land rehabilitation and soil and water conservation programs in Ethiopia (Nedessa et al., 2005). Exclosure management and protection have proven to be effective because of the participation of the local communities in the establishment of exclosures and the responsibility of local authorities for management and protection of the area (Descheemaeker et al., 2006). Following establishment, the vegetation recovery process in exclosures consistently starts with the rapid recovery of herbaceous species (that is, including grasses). After three to five years, shrub and tree species gain importance and suppress the abundance of herbaceous species (Assefa et al., 2003).

Several case studies conducted in the central and northern highlands (e.g. Mengistu et al., 2005; Yami et al., 2006; Berhane et al., 2007; Yayneshet et al., 2009) and southern lowlands (Angassa and Oba, 2010) of Ethiopia have shown that exclosures can be effective in enhancing composition, diversity, and density of vegetation. However, exclosures can also have negative impacts such as causing shortage of fuel wood and increasing grazing pressure on the remaining communal grazing areas (Mekuria et al., 2011). Therefore, it is important to make a careful evaluation of the benefits and costs of exclosure establishment (Descheemaeker et al., 2006). One important aspect in evaluating plans for establishing additional exclosures on communal grazing lands is the potential of exclosures for vegetation restoration.

Although exclosures have been established in the northern lowlands of Ethiopia for about three decades, empirical data on the effectiveness of these protected areas in restoring vegetation are lacking. Particularly, predictive relationships among vegetation composition, exclosure age, and soil factors that control the restoration of native vegetation have not been explored well. This information is critical, however, for evaluation of existing exclosures in restoring degraded native vegetation and for deciding whether additional exclosures should be established. In addition, land managers working in the northern lowlands of Ethiopia at local level are facing challenges to base their decision on scientific information due to lack of studies. This in turn urges the need to explore the effectiveness of exclosures in restoring degraded native vegetation and the controlling factors under existing biophysical and socio-economic conditions of the northern lowland of Ethiopia, and support local land managers in their decision by generating empirical data.

The study was carried out in the northern lowlands of Ethiopia with the following objectives: (1) to investigate how exclosure age affects restoration of degraded native plant species richness, diversity and aboveground standing biomass, and (2) to identify soil characteristics influencing the effectiveness of exclosures in restoring degraded native vegetation. Based on a review of more than 15 existing case studies (e.g., Smit et al., 1999; Assefa et al., 2003; Osem et al., 2002, 2004; Mengistu et al., 2005; Abebe et al., 2006; Yami et al., 2006; Berhane et al., 2007; Muchiru et al., 2009; Yayneshet et al., 2009; Hosseinzadeh et al., 2010; Verdoodt et al., 2010) and their findings related to changes in vegetation composition following establishing exclosures and the controlling factors, we formulated the following hypotheses: (1) after establishing exclosures on communal grazing lands, recovery of plant species richness will be moderated by soil fertility, where we expect a higher species richness and diversity in relatively fertile sites as our study sites are degraded, (2) the composition of naturally regenerated plant species will be dominated by herbaceous species in the younger exclosures and shrub and tree species in the older exclosures, (3) aboveground biomass will be improved after establishing exclosures on communal grazing lands, and (4) the older exclosures will have comparable species composition with the surrounding church forests. The analyzed plant species richness, diversity, density, and aboveground biomass from exclosures were compared with those of adjacent communal grazing lands. The trends in the changes were then used as indicators of exclosure effectiveness to restore degraded vegetation.

METHODS

Study area

The study was carried out in three districts of Tigray (12° - 15° N latitude and 36° 30' - 40° 30' E longitude), the northernmost region of Ethiopia (Figure 1, Table 1). All sites have a tropical, semi-arid climate. Mean annual rainfall (for the years 2000 to 2006) varied between 488 and 645 mm year^{-1} with an average of 562 mm year^{-1}. Mean minimum temperature ranged from 11 to 17°C and mean maximum temperature ranged from 26 to 34°C (Ethiopian Meteorological Service Agency, 2007). The rainy season usually occurs between June and September with the growing period of 60 to 90 days. According to the district Agricultural and Rural Development Office, average farm size ranges from 0.75 to 2.5 ha per household.

Soils of the study sites were classified into three major groups: Leptosol, Regosols, and Cambisols (WRB, 2006; Table 1). The most common woody vegetation species included *Acacia seyal*, *Acacia melifera, Acacia etbaica,* and *Balanites aegyptiaca.* Diverse assemblages of grasses and herbs dominated the under-storey vegetation.

According to the districts Agricultural and Rural Development offices, major land uses at the study sites included cultivated lands (between 24 and 62% of the area), forest lands (0 to 14%), exclosures (6 to 21%), grazing lands (2 to 18%) and others (1 to 30%). Mixed crop-livestock farming is the backbone of households' livelihoods in all of the study sites. Major cultivated crops include *Zea mays* (maize), *Eragrostis teff* (teff), *Sorghum bicolour* (sorghum), and *Arachis hypogea* (peanut).

In the study areas, the first exclosures were established two decades ago and accordingly we selected replicated (n = 3) exclosures of 5, 10 and 15 years which are paired with adjacent

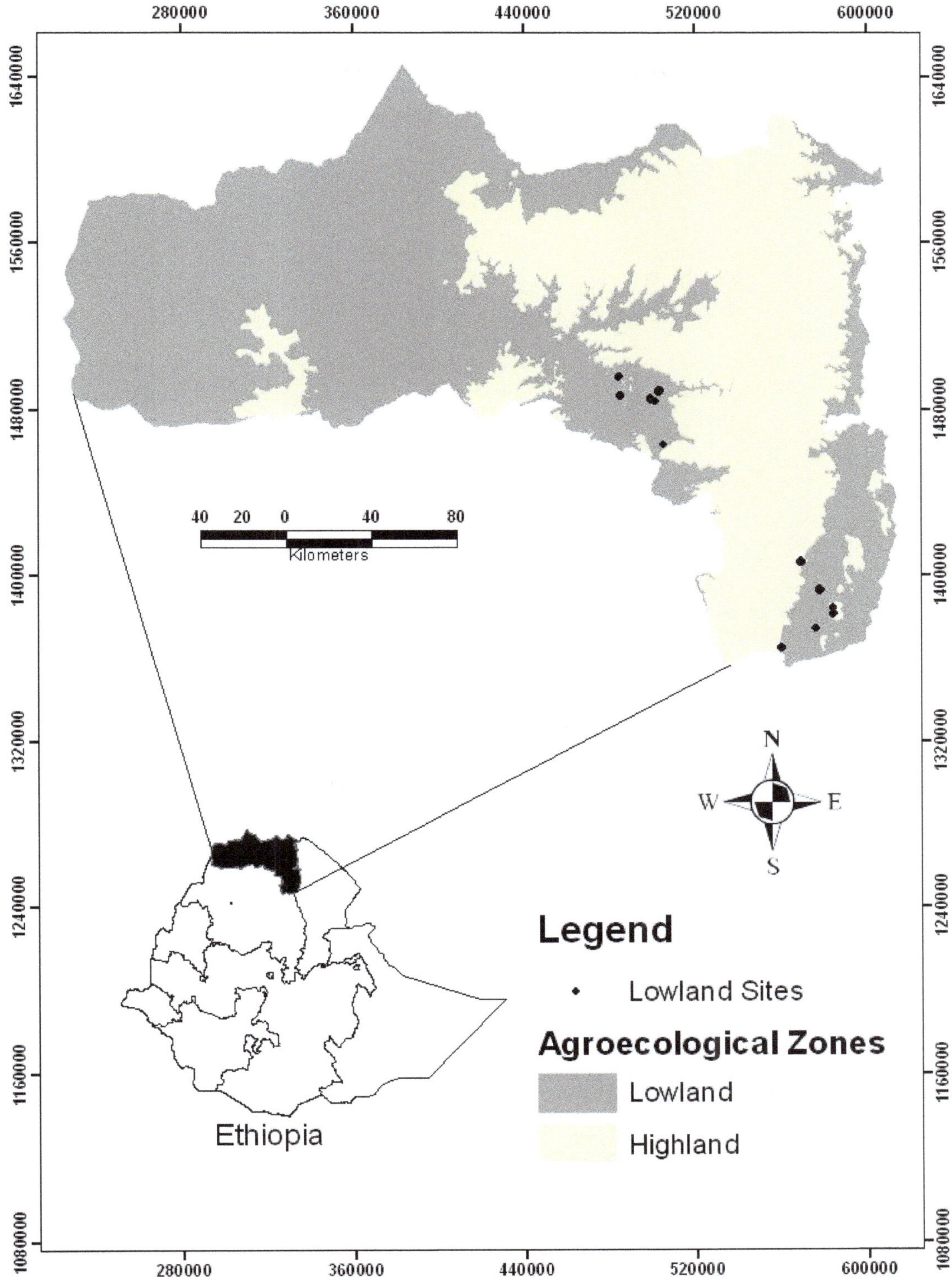

Figure 1. Location of the study area in the lowlands of Tigray (ranging from 500 to 1500 m a. s. l.), Northern Ethiopia with the distribution of specific sites indicated by (•).

communal grazing lands. In total, we selected nine paired exclosures and grazing lands that are distributed throughout the lowlands of Tigray (Figure 1). The area of the selected exclosures ranged from 7.1 to 85 ha whereas that of the adjacent communal grazing lands ranged from 2.8 to 38.3 ha (Table 1). Natural features such as large gullies and man-made features such as roads usually

Table 1. Area, location and major soil types of exclosures and adjacent grazing lands.

Variable	Replicate	Site					
		5-Year exclosure	Adjacent grazing land	10-Year exclosure	Adjacent grazing land	15-Year exclosure	Adjacent grazing land
Area (ha)	I	18.2	5.5	28.9	38.3	19.7	14.0
	II	28.6	14.1	10.4	3.7	85.0	2.8
	III	7.1	11.2	14.5	12.4	10.8	3.7
District	I	Endamehoni		Endamehoni		Tanqua-Abergele	
	II	Alemata		Endamehoni		Tanqua-Abergele	
	III	Tanqua-Abergele		Tanqua-Abergele		Tanqua-Abergele	
Major soil type	I	Regosol	Regosol	Cambisol	Cambisol	Leptosol	Leptosol
	II	Regosol	Regosol	Regosol	Regosol	Leptosol	Leptosol
	III	Leptosol	Leptosol	Leptosol	Leptosol	Leptosol	Leptosol

demarcate the boundaries of exclosures and the adjacent grazing lands. Based on the fact that exclosures are not fenced, guards are hired by the local administration on a food-for-work basis (Yayneshet et al., 2009). In exclosures, grazing is not allowed, but grass harvesting (using a cut-and-carry system) is permitted once a year, typically after seeding stage, starting three to five years after exclosure establishment (Descheemaeker et al., 2006). The main reason to restrict grass harvesting is to restore the soil seed bank.

In the adjacent communal grazing lands, unlimited access for free grazing is practiced. Although, there is variation in managing the communal grazing lands across the lowlands of Tigray, the communal grazing lands are under heavy grazing pressure and severe land degradation as the communal grazing lands are grazed throughout the year and most of the people depend on fuel wood sourced from the communal grazing lands to meet their energy demand (Zenebe, 2007). In both exclosures and communal grazing lands, soil and water conservation structures such as hillside terraces, stone bunds, and micro-basins have been constructed. However, the soil and water conservation structures are in better conditions in exclosures (that is, not damaged by livestock or human) because livestock and human activities are restricted (Mekuria et al., 2011).

In addition to exclosures and the adjacent grazing lands, we selected three isolated forest fragments that remain around churches, also called "church forests". In northern Ethiopia, the only areas where patches of forests can be found are areas surrounding the Ethiopian Orthodox Tewahido churches and these church forests have survived as a result of the traditional conservation system and protective patronage of the churches (Wassie et al., 2009). In the studied church forests, livestock grazing, collection of fuel wood and construction materials were allowed depending on the needs of the local community after thorough discussion among church administration and local communities on the need of specific use of forest resources. The area of the church forests in our study varied between 5.9 and 15.3 ha.

Experimental design

We used space-for-time substitution approach (Pickett, 1989; Fukami and Wardle, 2005) to detect changes in vegetation composition, diversity and aboveground biomass after conversion of communal grazing lands to exclosures with ages of 5, 10 and 15 years. This approach, which has a long tradition in ecology, is a method that is used to extrapolate temporal dynamics of vegetation succession by comparing multiple sites of different ages since disturbance or land use change in a region. In using space-for-time substitution approach, the experimenter does not establish the perturbation but instead selects sites where the perturbation is already running or has run.

The implicit assumption of this approach is that each paired communal grazing land and exclosure age should have comparable initial conditions such that changes in vegetation composition, diversity and aboveground biomass are a consequence of the land use change (that is, exclosure establishment). Thus, we selected sites with a pair of communal grazing land and exclosure adjacent to each other. Prior to establishment, exclosures and the adjacent communal grazing lands had similar conditions because the exclosures were established on part of the degraded communal grazing lands, which were used for livestock grazing (Mekuria et al., 2011). To cover the variability in soil and topography, we selected three replicates for each paired communal grazing land and exclosure age throughout the lowlands of Tigray (Figure 1). We also selected replicated (n = 3) church forests to detect whether exclosures had reached species richness and diversity level of these forests. At each exclosure, grazing land and church forest, we randomly established two to four transects spaced at a minimum distance of 75 m. The length of transects varied between 170 and 1340 m in exclosures, from 140 to 859 m in adjacent grazing lands and from 250 to 1120 m in church forests, depending on area of the site. The number of transects per site was based on vegetation density, spatial heterogeneity of vegetation, and area of the site (Mengistu et al., 2005). Transects were parallel to each other and to the topography of the landscape. Along each transect, we delineated three landscape positions (upper slope, mid slope and foot slope), and in each landscape position we established a sampling plot of 20 m × 20 m. In the entire study, we studied 153 sample plots of which 75 were in exclosures, 54 in adjacent communal grazing lands and 24 in church forests.

Vegetation inventory, analyses and estimation of aboveground standing biomass

In each 20 × 20 m plot, we measured the following individual plant variables: diameter at breast height (DBH), or for smaller and multi-stemmed shrub, diameter at stump height or at the height of 30 cm (d_{30}) from the ground, crown diameter and total height. DBH and d_{30} were measured using calipers. Crown diameter was measured using a measuring tape. Total height was measured using either a clinometer or a measuring tape depending on tree height. We also identified the species identity of plants encountered in each plot by the support of local people. For plants, which could not be identified in the field, herbarium specimens were collected, properly dried in a plant press, and identified at the National Herbarium in Addis Abeba University, where voucher specimens were deposited.

The average total density of woody plant species per hectare was derived from the total number of individuals recorded in the sample plots at each exclosure age (ranged from 24 to 27 plots, or 0.96 to 1.08 ha) and the respective adjacent communal grazing lands (18 plots, or 0.72 ha) as well as in the 24 plots (0.96 ha) at church forests. To determine the species similarities between exclosures, adjacent communal grazing lands and church forests, we used Sorensen's similarity index (Sorensen, 1948):

$$Ks = \frac{2c}{a+b} \times 100$$

Where, Ks = Sorensen's similarity coefficient; c = number of species common to both sites (that is, an exclosure age and the adjacent communal grazing land or an exclosure age and church forest); a, b = number of species found in site one and two (that is, in exclosure or in adjacent communal grazing land; in exclosure or church forest), respectively. To determine native (indigenous) plant species diversity in exclosures, adjacent communal grazing lands and church forests, and to conduct comparisons in species diversity between exclosures and adjacent communal grazing lands, we used the Shannon-Wiener index of diversity (*H'*) (Shannon, 1948):

$$H' = -\sum_{i=1}^{s} P_i \ln(P_i),$$

Where, P_i is the proportion of individuals of the abundance of the ith species as expressed as a proportion of the total. We used only the Shannon-Wiener index of diversity because similar to Simpson's and other index it takes into account species richness and proportion of each species within a community (that is, within exclosures, adjacent communal grazing lands and church forests). To determine the similarity in abundance of different species within a community, we used the Shannon-Wiener index of evenness (J):

$$J = \frac{H'}{H'_{\max}} = -\sum_{i=1}^{s} P_i \ln\left(\frac{P_i}{\ln(s)}\right)$$

Furthermore, the type of plant species found in exclosures, adjacent communal grazing lands and church forests were described in terms of richness, plant family and life forms such as herbaceous-, shrub-, and tree-species. In our study, differences in species richness are equivalent to the value of beta diversity (that is, species turnover) provided that we did not encounter a species unique to the communal grazing lands.

To estimate aboveground biomass, we identified dominant woody species using our data on vegetation inventory. The dominant woody species were determined based on the relative importance value (that is, the sum of relative basal area, relative frequency and relative density) in each exclosure, adjacent communal grazing land and church forest. This approach ensures that species that are few in number but productive (that is, large biomass) are not excluded. The number of woody species selected for biomass estimation varied between 2 and 10 in exclosures, between 1 and 8 in adjacent communal grazing lands, and between three and 10 in church forests.

We used the methods of Hoff et al. (2002) and Snowdon et al. (2002) for measuring aboveground woody biomass. The dominant woody species were grouped into three diameter classes in order to minimize errors that can arise from variable sizes of individuals.

In exclosures and the adjacent communal grazing lands, we harvested and weighed selected individuals representing the dominant species. Altogether we harvested and weighed 270 trees and shrubs (162 from exclosures and 108 from communal grazing lands). In the church forests, the partial harvest method (Snowdon et al., 2002) was employed because tree felling is strictly forbidden; biomass was collected form 24 individuals. Fresh mass of aboveground vegetation was adjusted to dry mass using the measured moisture contents, determined by oven-drying sub samples of stems, branches, and leaves at 65°C until constant mass was attained (about 78 h).

Statistical analyses

We conducted paired t-tests to test for significant differences in density, plant species richness, diversity, evenness, and above-ground standing biomass of the naturally regenerated plant species between paired exclosures and adjacent communal grazing lands. We checked the equality of the variances in both groups before starting the interpretation of the statistical results from the t-tests. Pearson correlation tests were conducted to examine the relationships between plant species richness, diversity and above-ground biomass and independent variables using the mean values of the three landscape position for each land use type (n = 12).

RESULTS

Vegetation composition in exclosures and adjacent communal grazing lands

In exclosures, we recorded between 37 and 46 plant species representing 23 to 29 families, whereas we recorded between seven and 30 plant species representing six to 22 families in the adjacent communal grazing lands (Table 2). All exclosure ages contained equivalent number of plant families that were represented by two or more species. In all exclosures, Fabaceae/Leguminosae and Poaceae contributed the greatest number of species. Exclosures contained more herbaceous, shrub and tree species than the adjacent communal grazing lands (Table 3). Moreover, the number of shrub and tree species increased with the increase in exclosure age. However, the proportion of herbaceous and shrub species was considerable (ranging from 73% in the 5-year-old exclosure to 61% in the 15-year-old exclosure). The similarity of vegetation between an exclosure and the adjacent communal grazing land varied between 61 and 29 % and decreased with the increase in exclosure age (Figure 2).

Species richness, diversity and aboveground standing biomass in exclosures and adjacent communal grazing lands

All exclosures displayed higher plant species richness, diversity and aboveground standing biomass when compared to the adjacent communal grazing lands (Table 3). The 15 year-old exclosure displayed the highest proportional increase in species richness following establishment (that is, 80.4% increase compared to the adja-

Table 2. Total number of native plant species recorded in the entire sampled plots in exclosures, grazing lands and church forests*.

Variable	5-ex	AGL	10-ex	AGL	15-ex	AGL	Church forest
Total number of sampled plots	24	18	27	18	24	18	24
Area of the total sampled plots (ha)	0.96	0.72	1.08	0.72	0.96	0.72	0.96
Total number of species recorded	46	30	37	22	44	7	49
Plant families (number)	28	22	23	14	29	6	35
Families represented by two or more species (number)	7	4	7	4	6	1	8

*5-ex, 10-ex and 15-ex refer to 5, 10 and 15 years exclosures respectively. AGL, Adjacent grazing lands.

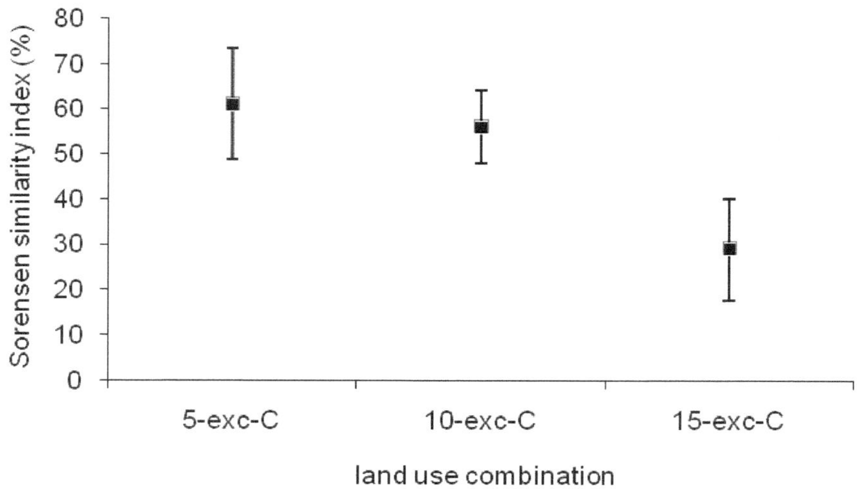

Figure 2. Sorenson similarities index (%) between an exclosure and adjacent communal grazing land. C refers to the adjacent communal grazing land for each respective exclosure age.

cent grazing land), while we observed the highest proportional increase in aboveground biomass in the 5 year-old exclosure (Table 3). The influence of exclosure age on plant species richness and diversity was not consistent, as the increases in plant species richness and diversity following the establishment of exclosures on communal grazing lands was significant ($p < 0.05$) only between 15-year-old exclosure and the adjacent communal grazing land (Table 3). However, the increase in aboveground biomass following the establishment of exclosures was significant ($p < 0.05$) between all paired exclosures and communal grazing lands (Table 3). Moreover, we did not detect significant differences ($p > 0.05$) in Shannon Wiener evenness index values between any of the exclosure ages and the adjacent communal grazing lands.

Relationships between soil properties and vegetation variables

In exclosures, plant species diversity, richness and aboveground standing biomass were positively and significantly ($p < 0.05$) correlated with % OM, % N, available P and clay content, while negatively correlated with sand and silt content (Table 4). In communal gra-zing lands, plant species diversity was positively and significantly ($p < 0.05$) correlated with % OM and silt content. Plant species richness was positively correlated with available P and sand content whereas negatively correlated with % OM and clay content. Aboveground biomass was positively correlated with % OM and clay content whereas negatively correlated with available P and sand content.

Vegetation variables in church forests

The total number of plant species recorded in the entire church forests was comparable with the total number of species recorded in exclosures (Table 2). Trees accounted for 33%, shrubs accounted for 31%, and herbaceous species, which includes grasses, climbers, forbs and epiphytes accounted for 36%. The average height of plant species found in the church forests was 4.2 (± 0.6) m, whereas the average total density was 954 (± 249) stem ha^{-1}. The Shannon-Wiener diversity value of plant species found in the church forests was 1.76.

Table 3. Average (± SE, n = 3) of plant species richness, diversity, evenness, aboveground standing biomass, vegetation cover, and herbaceous, shrub and tree species in exclosures and communal grazing lands and their differences.

Variable	Exclosure age (year)	Exclosure	Adjacent grazing land	Difference a	*t-value*
Richness (number)	5	19.6 (4.3)	12.0 (5.9)	7.6 (5.8)	1.05
	10	16.0 (2.1)	13.0 (5.2)	3.0 (3.2)	0.54
	15	24.0 (6.7)	4.7 (0.9)	19.3 (6.9)*	2.88
Diversity (Shannon Wiener index)	5	1.1 (0.6)	0.7 (0.5)	0.4 (0.22)	0.99
	10	1.3 (0.3)	1.0 (0.4)	0.3 (0.12)	1.04
	15	1.6 (0.3)	0.7 (0.3)	0.9 (0.25)**	3.71
Evenness (Shannon Wiener index)	5	0.48 (0.1)	0.3 (0.2)	0.18 (0.08)	1.28
	10	0.68 (0.09)	0.61 (0.16)	0.07 (0.09)	0.64
	15	0.64 (0.08)	0.59 (0.18)	0.05 (0.09)	0.49
Aboveground biomass (mg ha^{-1})	5	9.9 (1.20)	1.4 (0.4)	8.5 (0.6)**	17.7
	10	13.0 (2.9)	7.1 (2.5)	5.9 (0.9)*	2.65
	15	29.0 (8.4)	16.3 (7.3)	12.7 (1.5)**	7.31
Herbaceous species (number)	5	8.0 (0.6)	1.7 (1.2)	6.3 (1.7)**	4.75
	10	5.0 (0.5)	3.0 (0.6)	2.0 (0.6)	2.44
	15	7.0 (2.1)	0.3 (0.3)	6.7 (2.3)*	3.16
Shrub species (number)	5	6.3 (2.8)	6.7 (3.5)	-0.3 (3.8)	-0.07
	10	6.7 (0.9)	5.7 (2.6)	1.0 (1.7)	0.36
	15	7.7 (1.5)	1.3 (0.3)	6.3 (1.8)**	4.24
Tree species (number)	5	5.3 (1.2)	3.7 (1.3)	1.7 (0.3)	0.93
	10	4.3 (1.2)	4.3 (2.0)	0.0 (1.5)	0.00
	15	9.3 (3.2)	3.0 (0.6)	6.3 (2.9)	1.96
Total density (stem ha^{-1})	5	1260 (420)	699 (192)	561 (238)	1.21
	10	858 (102)	536 (55)	321 (46)*	2.76
	15	1203 (84)	428 (225)	775 (232)*	3.21
Height (m)	5	1.84 (0.13)	1.19 (0.44)	0.65 (0.53)	1.43
	10	2.48 (0.49)	2.31 (0.39)	0.17 (0.19)	0.28
	15	2.78 (0.41)	2.66 (0.53)	0.12 (0.26)	0.19
Canopy cover (%)	5	38.3 (5.1)	12.8 (3.2)	25.5 (7.3)**	9.26
	10	45.0 (6.8)	36.3 (5.5)	8.8 (6.1)*	3.39
	15	74.2 (5.4)	25.5 (7.6)	48.7 (9.0)**	13.2
Under-canopy cover (%)*	5	60.9 (4.7)	6.3 (0.7)	54.5 (4.7)**	10.9
	10	55.9 (6.1)	28.3 (4.4)	27.6 (5.5)**	7.91
	15	43.9 (6.8)	7.0 (3.6)	36.8 (8.2)**	35.7

*Under canopy cover refers mainly of grass cover. [a]Differences (calculated as: exclosure - adjacent grazing land) were significant at *$p \leq 0.05$, **$p \leq 0.01$ after paired t test.

DISCUSSION

Comparability of exclosures and the adjacent communal grazing lands

The inherent assumption of space-for-time substitution approach is that exclosures and the adjacent communal grazing lands had similar conditions before exclosure establishment. This assumption is confirmed by the fact that exclosures are established in the northern lowlands of Ethiopia on part of communal grazing lands that have been used for grazing in the past (Descheemaeker et al., 2006). In addition, we tested this assumption using varia-

Table 4. Pearson correlation coefficients (r; *n* = 12) between soil characteristics and plant species diversity, richness and aboveground standing biomass within exclosures and communal grazing lands.

Soil characteristic	Exclosure			Communal grazing land		
	Diversity	Richness	Biomass	Diversity	Richness	Biomass
% OM	0.44*	0.04	0.61*	0.49*	-0.62*	0.60*
% N	0.51*	0.23	0.70*	0.34	-0.19	0.09
P (ppm)	0.56*	0.06	0.53*	0.43	0.81*	-0.88*
% Sand	-0.62*	-0.30	-0.84*	-0.24	0.77*	-0.87*
% Silt	-0.12	-0.66*	-0.18	0.73*	-0.27	0.39
% Clay	0.64*	0.84*	0.88*	-0.39	-0.87*	0.90*

*Significant at $p < 0.05$.

bles measured in the paired exclosures and communal grazing lands that are less dependent on land use (e.g. soil texture) and detected no difference in soil texture between any of the paired exclosures and adjacent grazing lands (Mekuria et al., 2011). This indicates that the paired sites were comparable and differences in native plant species richness, diversity and aboveground standing biomass measured between the paired exclosures and adjacent communal grazing lands were mainly caused by land-use change (that is, exclosure establishment) and not by inherent site variability.

Aboveground biomass, plant species richness and diversity in communal grazing lands

In communal grazing lands, the low aboveground standing biomass, richness and diversity of plant species is a result of free grazing practices and human interference, resulting in severe overgrazing, which affects regeneration and growth of plant species negatively. A study in Ethiopia (Taddese et al., 2002) demonstrated that free grazing resulted in lower species richness and biomass when compared to ungrazed plots due to its impact on soil physical properties. Similarly, a study in Inner Mongolia, China also demonstrated that overgrazing resulted in severe damage to vegetation (Zhao et al., 2011). The degradation of plant species in the communal grazing lands that are included in our study is critical, as more than 90% of local communities depend on bio-energy sources (wood, dung) that are obtained from grazing lands to meet their household energy demand (Zenebe, 2007), illustrating the urgent need to restore these useful shrub and tree species.

Changes in aboveground biomass, species richness and diversity in relation to exclosure age

The higher plant species richness, diversity and aboveground standing biomass in all exclosures compared to the adjacent communal grazing lands (Table 3) illustrates that rehabilitation of the degraded grazing lands occurs in relatively short period of time after restricting human and livestock interference. Other case studies conducted in Ethiopia have also shown that exclosures provide favorable microhabitats for plants, which can be an effective method to improve aboveground biomass, density, composition and diversity of naturally regenerated plant species (Mengistu et al., 2005; Abebe et al., 2006; Yami et al., 2006; Berhane et al., 2007). Moreover, in semi-arid Kenya, Verdoodt et al. (2010) and Muchiru et al. (2009), in Iran, Hosseinzadeh et al. (2010), in semi-arid USA, Yeo (2005), in semi-arid Azerbaijan (Peper et al., 2010), and in southern Mongolian desert steppes (Wesche et al., 2010) demonstrated that woody and herbaceous species richness, diversity and aboveground biomass increased following the establishment of exclosures on communal grazing lands.

Plant species richness, diversity and aboveground biomass increased with exclosure age. This was supported: (1) by the decreasing trend of vegetation similarity between exclosures and grazing lands with the increase in exclosure age (Figure 2), and (2) by the increase in species turnover (or differences in species richness between exclosures and grazing lands, Table 3) with the increase in exclosure age. The strong increase in plant species richness during the first 5 years following exclosure establishment (Table 3) compared to the increase during the first 10 years following the establishment of exclosure on communal grazing lands is caused by the establishment of large numbers of herbaceous species resulting from the prohibition on harvesting grass and other herbaceous species. This result did support our expectation in which herbaceous species dominates in the younger exclosures and shrub and tree species in the older exclosures. A study in semi-arid Kenya (Oba et al., 2001) demonstrated that grazing exclosures may increase species richness to a certain level and the increase in species richness declines with age of exclosures indicating that long-term exclusion of grazing may not necessarily increase species richness in arid-zone grazing lands. The substantial increase in aboveground biomass and shrub and tree species with exclosure age could also explain the slower rate of the increase in plant species richness with exclosure age, as the increase in abundance of shrub and tree species and biomass reduce herbaceous and pioneer species richness. Similar results have been reported in other case studies in which plant species richness declines when biomass exceeds 500 g m^{-2} (Oba et al., 2001), and with the increase in abundance of shrub and tree species (e.g. Smit et al., 1999).

Yet, Angassa and Oba (2010) demonstrated that woody species richness, diversity and evenness were greater in younger exclosures than in the older exclosures in the southern Ethiopia rangelands. The difference could be attributed to the different environment, soils and type and management of the grazing lands. Osem et al. (2002, 2004) have also shown that exclosures may decrease plant diversity if the site is productive as productive sites promote higher plant growth and earlier canopy closure, which prevents further establishment of pioneer species. Anderson et al. (2007) demonstrated that excluding herbivores caused the strongest relative decreases in plant species richness where exclusion of herbivores increased available soil P. The increase in plant species richness and diversity in our exclosures may be due to the soil layer has been degraded and is therefore very unproductive (Mekuria et al., 2007, 2009). Our study illustrates that the level of degradation of communal grazing lands prior to the establishment of exclosures on communal grazing lands influences the effects of exclosures age in enhancing regenerated indigenous plant species richness and diversity.

The positive correlations in exclosures of % OM, % TN, % clay and available P with plant species richness, diversity and aboveground biomass show that soil fertility is important in enhancing the effectiveness of exclosures in restoring degraded indigenous vegetation in addition to rainfall and age of exclosures following exclosure establishment. This result is in consistent with our expectation in which we expected a higher plant species richness and diversity in relatively fertile sites as our sites are degraded. This was already suggested in an earlier study by Aerts et al. (2006), which indicated that the success of natural regeneration in exclosures will greatly depend on the quality of the physical environment and the proximity to forest remnants. Other studies conducted in other semi-arid ecosystems (e.g. Lane et al., 1998; Leger et al., 2007) have also shown that the positive influences of moisture and soil nutrient content on the number of woody species and aboveground net primary productivity. Similarly, the negative correlations of %sand and %silt with species richness, diversity and above-ground biomass illustrate the effect of poor soil fertility in reducing the effectiveness of exclosures to restore degraded native vegetations.

The total number of plant species recorded in the exclosures was comparable with the total number of species recorded in the church forests (Table 2), suggesting that in exclosures less than two decades was required to reach a level comparable to church forests. This was also evident from the relatively equivalent value of the Shannon-Wiener diversity index of the 15-year old exclosures (H= 1.6) compared with the church forests (H= 1.76). Quite a lot of species were unique to exclosures and church forests (that is, high beta-diversity) although the majority of these species are herbaceous and small shrub species. Compared with natural dry afromontane forests in Ethiopia, all exclosures had a lower number of woody species (e.g. Alelign et al., 2007). Also, the Shannon-Wiener diversity index of all exclosures (H ranging from 1.1 to 1.6) was much lower compared to dry afromontane forest (H ranging from 2.72 to 3.72) (Woldemariam et al., 2000; Alelign et al., 2007). This illustrates that even the 15-year-old exclosures are in a relatively early successional stage. Compared to native forests, church forests are very small forest fragments containing a species-poor tree and shrub communities. They are, however, important for their role in the landscape ecology of the area as species pools (Aerts et al., 2006). As their area increases, with time, exclosures may increasingly obtain a similar role to church forests and should be protected and managed accordingly.

Conclusions

The results of this study confirm that the establishment of exclosures on degraded communal grazing lands in the northern lowlands of Ethiopia is a viable option to restore native vegetation composition, richness, diversity and aboveground biomass. Our study showed that soil fertility and level of degradation of communal grazing lands prior to exclosure establishment moderates the effectiveness of exclosures in improving vegetation composition and diversity and enhancing aboveground biomass. Such information is necessary to optimize the selection of areas for the establishment of exclosures in the future and for policymakers to take into account the importance of level of degradation of communal grazing lands in their management decisions. Moreover, our study suggests that with time exclosures in the northern lowlands of Ethiopia may increasingly obtain an important role as source of seeds of indigenous plant species and species pool similar to church forests and should be protected and managed. We do believe that the regional and district level agricultural offices should work to expand exclosure land management in order to restore degraded communal grazing lands and maximize benefits from these degraded resources. Community participation in selecting areas for exclosure establishment should also continue to ensure the sustainability of exclosure land management. However, care should be taken in identifying the availability of land for livestock grazing while expanding exclosures because expansion of exclosures could increase grazing pressure on the remaining communal grazing lands and aggravate land degradation.

ACKNOWLEDGEMENTS

We are grateful to the German Academic Exchange Service (DAAD), British Ecological Society (BES) and Mekelle University for the financial and logistical support.

REFERENCES

Abebe MH, Oba, G, Angassa A, Weladji RB (2006). The role of area enclosures and fallow age in the restoration of plant diversity in northern Ethiopia. Afr. J. Ecol. 44:507-514.
Aerts R, Overtveld KV, Mitiku H, Hermy M, Deckers J, Muys B (2006). Species composition and diversity of small Afromontane forest fragments in northern Ethiopia. Plant Ecol. 187:127-142.
Alelign A, Teketay D, Yemshaw Y, Edwards S (2007). Diversity and status of regeneration of woody plants on the peninsula of Zegie, northwestern Ethiopia. Trop. Ecol. 48:37-49.
Anderson TM, Ritchie ME, Mcnaughton SJ (2007). Rainfall and soils modify plant community response to grazing in Serengeti national park. Ecology 85:1191-1201.
Angassa A, Oba G (2010). Effects of grazing pressure, age of enclosures and seasonality on bush cover dynamics and vegetation composition in southern Ethiopia. J. Arid Environ. 74:111-120.
Assefa D, Oba G, Weladji RB, Colman JE (2003). An assessment of restoration of biodiversity in degraded high mountain grazing lands in Northern Ethiopia. Land Degrad. Dev. 14:25-38.
Berhane E, Teketay D, Barklund P (2007). Enclosures to enhance woody species diversity in the drylands of Tigray. East Afr. J. Sci. 1:136-147.
Descheemaeker K, Nyssen J, Rossi J, Poesen J, Haile M, Moeyersons J, Deckers J (2006). Sediment deposition and pedogenesis in exclosures in the Tigray Highlands, Ethiopia. Geoderma 132:291-314.
Ethiopian Meteorological Service Agency (2007). Annual climate bulletin for the year 2007. Ministry of Water Resource, Addis Ababa, Ethiopia.
FAO (2001). Global Forest Resources Assessment 2000. FAO Forest Paper, No. 140. FAO, Rome.
Fukami T, Wardle DA (2005). Long-term ecological dynamics: reciprocal insights from natural and anthropogenic gradients. Proceedings of the Royal Society of London, Series B - Biological Sciences 272:2105.
Girma T (2001). Land degradation: a challenge to Ethiopia. Environ. Manag. 27:815-824.
Hoff C, Rambal S, Joffre R (2002). Simulating carbon and water flows growth in a Mediterranean evergreen *Quercus ilex* coppice using the FOREST-BGC model. For. Ecol. Manag. 164:121-136.
Hosseinzadeh G, Jalilvand H, Tamartash R (2010). Short term impacts of enclosure on vegetation cover, productivity and some physical and chemical soil properties. J. Appl. Sci. 10:2001-2009.
Jama B, Zeila A (2005). Agroforestry in the drylands of eastern Africa: a call to action. ICRAF Working Paper - no. 1. Nairobi: World Agroforestry Centre.
Kebrom T (2001). Natural regeneration of degraded hill slopes in Southern Wello, Ethiopia: A study based on permanent plot. Appl. Geogr. 21:275-300.
Lane DR, Coffin DP, Lauenroth WK (1998). Effects of soil texture and precipitation on aboveground net primary productivity and vegetation structure across the Central Grassland region of the United States. J. Veg. Sci. 9:239-250.
Leger EA, Howe KM, Gurevitch J, Eliza W, Jonathan H, Ashton IW, Manuel L (2007). The interaction between soil nutrients and leaf loss during early establishment in plant invasion. For. Sci. 53:701-709.
Lemenih M, Karltun E, Olsson M (2005). Soil organic matter dynamics after deforestation along a farm field chronosequence in southern highlands of Ethiopia. Agric. Ecosyst. Environ. 109:9-19.
Mekuria W, Veldkamp E, Corre MD, Mitiku H (2011). Restoration of ecosystem carbon stocks following exclosure establishment in communal grazing lands in Tigray, Ethiopia. Soil Sci. Soc. Am. J. 75(1):246-256.
Mekuria W, Veldkamp E, Mitiku H, Gebrehiwot K, Muys B, Nyssen J (2009). Effectiveness of exclosures to control soil erosion and local community perception on soil erosion in Tigray, Ethiopia. Afr. J. Agric. Res. 4:365-377.
Mekuria W, Veldkamp E, Mitiku H, Nyssen J, Muys B, Gebrehiwot K (2007). Effectiveness of exclosures to restore degraded soils as a result of overgrazing in Tigray, Ethiopia. J. Arid Environ. 69:270-284.
Mengistu M, Teketay D, Håkan H, Yemshaw Y (2005). The role of enclosures in the recovery of woody vegetation in degraded dryland hillsides of central and northern Ethiopia. J. Arid Environ. 60:259-281.
Muchiru AN, Western D, Reid RS (2009). The impact of abandoned pastoral settlements on plant and nutrient succession in an African savanna ecosystem. J. Arid Environ. 73:322-331.
Nedessa B, Ali J, Nyborg I (2005). Exploring ecological and socio-economic issues for the improvement of area enclosure management: a case study from Ethiopia. Drylands coordination group report no. 38. p. 63.
Oba G, Vetaas OR, Stenseth, NC (2001). Relationships between biomass and plant species richness in arid-zone grazing lands. J. Appl. Ecol. 38:836-845.
Osem Y, Perevolotsky A, Kigel J (2002). Grazing effect on diversity of annual plant communities in a semi-arid rangeland: interactions with small-scale spatial and temporal variation in primary productivity. J. Ecol. 90:936-946.
Osem Y, Perevolotsky A, Kigel J (2004). Site productivity and plant size explain the response of annual species to grazing exclusion in a Mediterranean semi-arid rangeland. J. Ecol. 92:297-309.
Peper J, Jabbarov AS, Manthey M (2010). Short-term effects of grazing abandonment on semi-arid rangelands in Azerbaijan. Ann. Agrar. Sci. 8:14-17.
Pickett STA (1989). Space-for-time substitution as an alternative to long-term studies. In Long-term studies in ecology: approaches and alternatives (ed. G. E. Likens), New York: Springer-Verlag. pp. 110-135.
Shannon CE (1948). A mathematical theory of communication. Bell Syst. Tech. J. 27:379-423.
Smit GN, Richter CGF, Aucamp AJ (1999). Bush encroachment: an approach to understanding and managing the problem. In: Tainton, N.M. (ed.), Veld Management in South Africa, University of Natal Press, Pietermaritzburg, South Africa. pp. 246-260.
Snowdon P, Raison J, Keith H, Ritson P, Grierson P, Admas M, Montagu K, Huiquan BB, Eamus D (2002). Protocol for sampling tree and stand biomass. National carbon accounting system technical report, Australian greenhouse office. No. 31.
Sorensen T (1948). A method of establishing groups of equal amplitude in plant sociology based on similarity of species content. Det Kong. Danske Videnskabernes Selskab, Biologiske Skrifter (Copenhagen), 5(4):1-34.
Taddese G, Saleem MA, Abyie A, Wagnew A (2002). Impact of grazing on plant species richness, plant biomass, plant attribute, and soil physical and hydrological properties of vertisol in East African highlands. Environ. Manag. 29:279-289.
Verdoodt A, Mureithi SM, Ranst EV (2010). Impacts of management and enclosure age on recovery of the herbaceous rangeland vegetation in semi-arid Kenya. J. Arid Environ. 74:1066-1073.
Wassie A, Sterck FJ, Teketay D, Bongers F (2009). Effects of livestock exclusion on tree regeneration in church forests of Ethiopia. For. Ecol. Manag. 257:765-772.
Wesche K, Ronnenberg K, Retzer V, Miehe G (2010). Effects of large herbivore exclusion on southern Mongolian desert steppes. Acta Oecologica 36:234-241.
Woldemariam T, Teketay D, Edwards S, Olsson M (2000). Woody plant

and avian species diversity in a dry Afromontane forest on the central plateau of Ethiopia: Biological indicators for conservation. Ethiop. J. Nat. Res. 2:255-293.

WRB (2006). World reference base for soil resources. 2nd edition. World Soil Resources Reports No. 103. FAO, Rome.

Yami M, Gebrehiwot K, Stein M, Mekuria W (2006). Impact of area enclosures on density, diversity, and population structure of woody species: the Case of May Ba'ati-Douga Tembien, Tigray, Ethiopia. Ethiop. J. Nat. Res. 8:99-121.

Yayneshet T, Eik LO, Moe SR (2009). The effects of exclosures in restoring degraded semi-arid vegetation in communal grazing lands in northern Ethiopia. J. Arid Environ. 73:542-549.

Yeo JJ (2005). Effects of grazing exclusion on rangeland vegetation and soils, east central Idaho. Western North Am. Nat. 65:91-102.

Zenebe G (2007). Household fuel consumption and resource use in rural-urban Ethiopia. PhD Thesis, Wageningen University. ISBN: 978-90-8504-745-2.

Zhao HL, Zhang TH, Zhao XY, Zhou RL (2011). Effects of sheep grazing and precipitation patterns on sandy grassland vegetation in Inner Mongolia, China. 2nd International Conference on Environmental Science and Development, IPCBEE, vol.4, IACSIT Press, Singapore.

Optimising a fall out dust monitoring sampling programme at a cement manufacturing plant in South Africa

Jacobus Engelbrecht[1] , Jacobus Joubert[1], Johannes Harmse[1] and Charles Hongoro[1,2]

[1]Department of Environmental Health, Faculty of Science, Tshwane University of Technology, Private Bag X680, PRETORIA 0001, South Africa.
[2]Health Systems Research Unit, South African Medical Research Council, South Africa.

The aim of this study at the specific cement manufacturing plant and open cast mine was to develop a positioning guideline for monitoring equipment and to optimise the fall out dust sampling programme. A baseline study was conducted on existing practices and legal requirements relating to fall out dust monitoring. The different methods and techniques for the identification of particulate emission sources and the calculation of emission rates were evaluated in order to identify the simplest and most cost effective options. Due to the complexity of sampling particulate emissions, emission factors and existing stack monitoring results were used to calculate the emissions. The quantified point and fugitive emissions rates were used to populate a dust dispersion model and the modeling results were compared with the existing monitoring program results. Due to the complexity of actual measurements or applying the principles of a mass balance in the total cement manufacturing process to calculate emission rates it was found that the selected use of emission factors is a simpler and less costly method. The study concluded that, after superimposing the dispersion model to the results from the existing sampling positions, existing sampling positions are not optimally located. A guideline for the optimisation of a fall out dust monitoring program is proposed.

Key words: Fall out dust monitoring, cement plant, optimising, air pollution sampling, fall out dust sampler locations.

INTRODUCTION

Fall out dust sampling and monitoring is becoming one of the preferred methods to determine dust pollution impact from industries/or mines on the receiving environment (Lodge, 1988). Particulates can be divided into the following categories, namely:

1. Condensable material: Material in the vapour phase at stack temperature which condenses to form particles as the plume disperses.

2. Inhalable particles: Particles with a diameter greater than 10 µm that are usually deposited in the upper respiratory system.

3. Respirable particles: Particles smaller than 10 µm but larger than 2.5 µm can generally get to as far as the larynx. PM10 (particulate matter) is particulate matter with a diameter of less than or equal to 10 µm. PM2.5 is particulate matter with a diameter of less than or equal to 2.5 µm (Loans, 2007).

Dust that precipitates and is deposited in the environment is defined as any particulate matter that has an aerodynamic diameter between 10 and 100 µm (Loans, 2007). Such particulate matter is also referred to as total suspended particulate (TSP).

Fall out dust monitoring is a useful and cost effective method of providing trend analysis of dust deposition over a period of time. It also provides an indication of the main areas of dust generation and can be conducted for both health and nuisance purposes. Particulate matter (dust) emissions resulting from the cement manufacturing process is usually as a result of point and fugitive emission sources (Davis, 2000). The nuisance dust close to cement processes is associated with haze and poor visibility. The dust from cement production contains small quantities of heavy metals (Vincent et al., 2006). Examples of such heavy metals include: lead, mercury and chromium. These elements in cement dust may pose a great threat to the health of plants, animals and residents in and around the factory (Gbadebe and Bankole, 2007). Pollution from cement dust can result in marked reduction in vitamin levels and also increase heavy metal concentrations in plants (Ade-Ademilua and Obalola, 2008). Fall out dust may negatively affect the soil in the area surrounding the cement manufacturing facility (Bilen, 2010). Cement dust can also cause damage to structures such as roof material by causing microbiological and chemical disintegration of such roof material (Tijani et al., 2005)

Standards for monitoring fall out dust monitoring stations exist. Examples of such standards include: American Society for Testing and Materials (ASTM) D 1739-98 and Australian Standards (AS) 3580.10.1 – 1991. Although the standards for dust deposition levels are set, the positioning of monitoring equipment in the receiving environment is not adequately prescribed. In many instances, industries/or mines position air quality monitor-ring equipment on the boundaries of its premises inter-preting ambient air quality impacts as dust deposition beyond its boundaries (Lodge, 1988). The loss of equipment beyond the fence due to vandalism or theft made this practice to be the best way out in the sampling of fall-out dust. Sampling sites to measure pollutant levels should be selected close to or representative of exposed populations, structures and plant (Vallero, 2008). Positioning of air quality monitoring equipment should be specific as per the results of an air quality modeling analyses that appropriately locates air quality monitors (Arizona Department of Environmental Quality, 2004). The National Framework for Air Quality Management in the Republic of South Africa (South Africa, 2007) requires that the number and location for fall out dust sampling points should be at representative locations around the dust source and will include air quality monitoring equipment located at human residences, sensitive business, industrial or agricultural locations within a maximum distance of two kilometers from the source boundary.

In order to design an air quality monitoring plan it is important to understand what is to be monitored and to also investigate redundancies, gaps and inadequacies in the existing monitoring programme (Knight, 2007). This provides the opportunity to improve or change to potential alternatives to get more value from the monitoring results.

Due to the fact that positioning of fall out dust sampling equipment at cement manufacturing plants are presently based on professional judgment and not on scientific research, a research study to develop a dust deposition guideline that will guide and assist the cement industry with the placement of monitoring equipment was conducted. The aim of the study at the specific cement manufacturing plant and open cast mine was to develop a positioning guideline for monitoring equipment and to optimise the fall out dust sampling programme. The objectives of the study were to: 1: Conduct a literature study to understand the background and existing practices of placement of fall out dust sampling points; 2: identify all the emission sources from the cement manufacturing process and related activities and record these in an emissions inventory; 3: quantify the emission rates from such sources; 4: generate a dust dispersion model from the results and use such a model to optimise the fall out dust monitoring programme.

This paper focuses on the development of a positioning guideline for monitoring equipment and optimising the fall out dust sampling programme at a cement manufacturing plant and open cast mine. More specifically, it presents a detailed review of literature on existing practices of placement of fall out dust sampling points; It then identifies all the emission sources from a selected cement manufacturing process and related activities recorded in an emissions inventory; quantify the emission rates from such sources; and generate a dust dispersion model from the results and used such a model to optimise the fall out dust monitoring programme.

MATERIALS AND METHODS

The study focused on the identification and quantification of all dust particulate emission sources, both point and fugitive related to the cement manufacturing operation at the Pretoria Portland Cement Plant De Hoek, located near the town of Piketberg, Western Cape, South Africa (Figure 1). The cement manufacturing plant included the following main processes and steps:

1. Mining- Open cast mine in which limestone is blasted and transported on haul roads
2. Crushing- Limestone is crushed and screened in primary and secondary crushers. The crushed limestone is transported and blended on material stockpiles.
3. Raw material handling- Raw materials received by road are stockpiled together with the limestone.
4. Raw material grinding- All raw materials are proportionally extracted and mixed. The mixed raw materials are ground and stored in silos.
5. Burning- The ground raw material is burned in thermal kilns to produce clinker.
6. Clinker grinding- Clinker is ground in finish mills to produce different types of cements.
7. Packaging and dispatching- Final products is transported, packed, stored and loaded to be dispatched. Figure 1 shows the

Figure 1. Main processes at the case study facility (location indicated using Google Earth, 2011).

locations of the various main processes at the cement manufacturing facility. The study excluded the sampling of dust particulate emission sources of external and neighbouring activities such as farming activities because it was not possible to quantify and qualify the emission rates from neighbouring activities. The main purpose of the monitoring programme is also to verify the impact of the cement manufacturing facility on its environment.

Quantification of particulate emissions

Emission rates from fugitive and point dust emission sources were calculated for use as input data to the dispersion model. These are described in turn.

Equipment used to monitor point sources

Point sources are easier to quantify due to existing legal monitoring requirements. In most cases, point sources require continuous or periodic in stack monitoring. Results from continuous measurement are reported on a standardised basis in mg/Nm^3 (normal conditions of 10% O_2, 273 Kelvin and 101.3 kPa). The results in mg/Nm^3 are converted into an emission rate of grams per second to be used as input to the dispersion model. The four point sources at the case study facility are shown in Table 1. These were from the cement milling and kiln processes. Each of the two finishing mills and two kilns has its own stacks. All four stacks are equipped with continuous in stack particulate emission monitoring equipment that can generate reports at two minute, hourly and 24 h averages.

In order to use results from the continuous in stack monitoring, it is important to ensure that the data is valid. In addition to a correct monitoring method, the accuracy of the monitoring equipment needs to be verified. The method used for the verification of the continuous measurement of dust particulate from a point source was ISO 9096: 2003. ISO 9096 describes a reference method for the measurement of particulate matter (dust) concentration in waste gases (ISO, 2003). The isokinetic sampling results are compared with that of the automated monitoring systems to make the necessary adjustments to ensure that it is calibrated to give more accurate results. To verify that in stack particulate matter monitoring equipment is properly calibrated the calibration certificates were obtained for all monitoring equipment at the time of sampling. These calibration certificates were available for the installation as well as annual calibration. The condition of the in stack monitoring equipment was also confirmed by inspection and reports from the maintenance system. The maintenance system covers all supplier minimum service requirements to ensure that equipment is operational. From the 24 h average particulate emission levels (mg/Nm^3) an average emission rate was calculated and used as input rate in the dispersion model. Annual emission data were used to calculate the average grams per second of dust emitted.

Estimation of fugitive sources

Due to the complexity of measuring fugitive particulate emissions in

Table 1. Identified point sources and continuous monitoring methods.

Reference Number	Section	Process	Monitoring equipment	Report
PRO2C	Burning (kilns)	Kiln 5	In stack continuous measurement instrument	2 min, 1 h and 24 h averages (mg/Nm^3)
		Kiln 6	In stack continuous measurement instrument	2 min, 1 h and 24 h averages (mg/Nm^3)
PRO2E	Finish milling (clinker grinding)	FM 5	In stack continuous measurement instrument	2 min, 1 h and 24 h averages (mg/Nm^3)
		FM 6	In stack continuous measurement instrument	2 min, 1 h and 24 h averages (mg/Nm^3)

in a process like cement manufacturing, the emission levels were estimated by applying emission factors such as US EPA AP42 (United States Environmental Protection Agency Air Pollution 42) and Australian National Pollutant Inventory (NPI). Most of the emission factors applied is from the US EPA AP42 and where more appropriate, the factors from the Australian NPI were used. The sources of emission factors used are listed in Table 3. Both US EPA AP42 and Australian NPI provide the community, industry and government with free information about emission factors from different sources. Making use of emission factors (US EPA, 1995) to calculate emission rates are the more practical and cost effective methods for example, EPA-454/R-95-015, due to the complexity and cost of actual measurement. An emission factor is expressed as the weight of pollutant emitted divided by the unit weight, volume, distance or duration of the activity. Emission factors are used to estimate a source emission rate and the following equations were used (USA EPA: 1995):

$$E = A \times EF \times [1 - (ER/100)]$$

Where: E = emissions, A = activity rate, EF = uncontrolled emission factor and ER = overall emission reduction efficiency (%).

The following were considered in the calculations: annual production tonnages, quantity of activities listed in Table 3, activity duration, material moisture content, material silt content, number of days raining >0.254 ml, mean wind speed, time wind speed exceeding 5.4 m/s, total hectares, certain prescribed constants or a predetermined emission factor as prescribed by the selected methods. After calculation of the emission rates, the reduction factors were applied for those areas where controls are in place.

Sampling of fall out dust

The sampling method applied is based on the internationally accepted ASTM D 1739-98 American Society for Testing and Materials: Standard Test method for Collection and Measurement of Dust fall method. This test method covers a procedure for collection of fall out dust and its measurement (ASTM, 2004). It also prescribes that buckets be placed two meters above ground level. This test method describes determination of both water-soluble and insoluble particulate matter.

The sampling period was 30 days and conducted during 2009 to 2011. Samples were collected, filtered and weighed and reported as $mg/m^2/day$ (total dust). A Global Positioning System (GPS) was used to confirm the location of the existing sampling points.

The location of the sampling points in relation to the cement manufacturing activities is illustrated in Figure 2. The sampling points are concentrated north-west of the facility and around the open pit mining area. Limitations of the existing sampling points are that they are focused around the open cast mining and manufacturing activities only. One specific sampling point (SS) is far north in a neighboring town. The background aerial photograph reveals the other activities surrounding the cement manufacturing facility such as agriculture.

Ten sampling points were used to cover all prevailing wind directions in opposite directions (Table 3). The sampling period was 30 days. Samples were collected, filtered and weighed and reported as $mg/m^2/day$. A GPS was used to confirm the location of the existing sampling points.

Modeling of air pollution monitoring and estimation results

Emission factor results were converted to actual concentrations in the various units which are required for the modeling process. With actual results from the fugitive dust emission levels reported, a modeling report was drafted using Screen View. It is a user friendly interface for U.S. EPA screening model, SCREEN3 and contains several features to make modeling as easy as possible (Lakes Environmental, 2008). The emission levels were reported as micro grams per cubic meter ($\mu g/m^3$). Results showed the emission levels at different distances from the source. The distances selected for this study were 5 to 5000 m to be able to confirm that the two kilometer radius is valid. The receptor height was selected as two meters as this is in line with the height of a typical fall out dust sampler. Other input data such as the height, length and width of emission generating source were also selected.

No direct or in directed interventions took place with individual human beings, therefore ethical approval was not necessary. Ethical approval for the project was however obtained from the University.

RESULTS AND DISCUSSION

Point source emission modelling

The dust emission levels for the period 2009 to 2011 and stack details obtained from the in stack monitoring results were used to calculate emission rates. These rates were used as input data for the dust emission modeling of the identified four point sources.

Point source particulate emission rates

Point source emission rates were calculated from the in stack monitoring results (Table 2). Limestone is received from the lime quarry and introduced into a burning process (rotary kilns Kiln 5 and 6). Clinker is produced which is fed into two cement mills, finished mill FM5 and

Table 2. Particulate emission rates from point sources.

Reference Number	Section	Process	Average emission level (mg/Nm^3)	Calculated emission rate (g/s)
PRO2C	Burning (kilns)	Kiln 5	168.17	7.38
		Kiln 6	73.73	4.34
PRO2E	Cement milling	FM 5	1	0.001
		FM 6	15.35	0.114

Table 3. Stack data input data into dispersion model.

Stack	Stack height (m)	Stack inside diameter (m)	Stack gas exit velocity (m/s^{-1})	Stack gas exit temperature (°K)
Kiln 5 (DK5)	61	2.19	11	393
Kiln 6 (DK6)	64.1	2.74	10.3	388
Finish Mill 5 (FM5)	30	0.7	6.8	364
Finish Mill 6 (FM6)	18	0.989	6.8	364

Figure 2. Location of existing fall out dust sampling points (Google Earth, 2011).

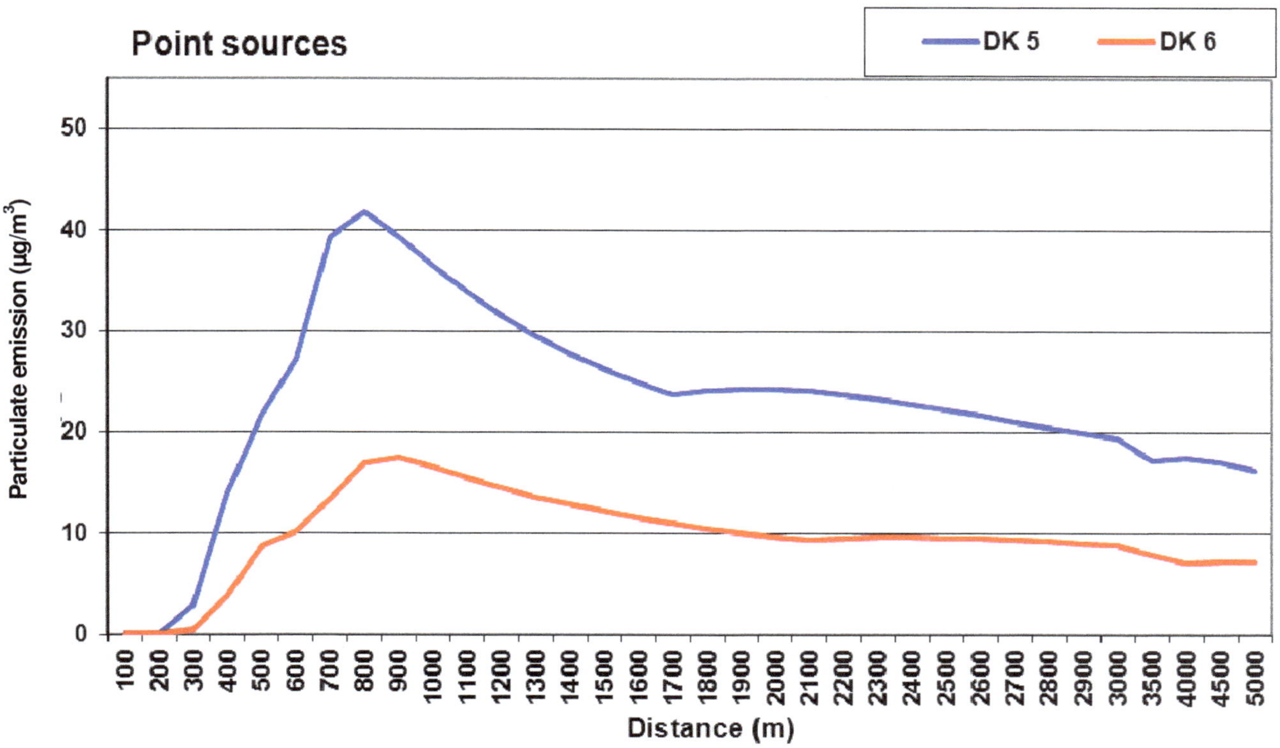

Figure 3. Kilns particulate emission modeling.

FM6. The variation in the emission levels of the different processes is mainly as a result of old versus newer technologies. Both kilns are equipped with electrostatic precipitators where as the best technology today is a bag filter which is commissioned at the cement mills. Low emission from FM5 is an example of effective bag filter technology.

According to Table 2, the variation in the emission levels of the different processes is mainly as a result of old versus newer technologies. Both kilns (FM5 and FM6) are equipped with electrostatic precipitators where as the best technology today is a bag filter. Low emission from FM5 (1 mg/Nm^3) is an example of the effectiveness of bag filter technology. Older technologies implemented at the burning kilns were electrostatic precipitators whereas bag filters that are commissioned at the cement mills (FM5 and FM6) have improved collection efficiency.

Point source emission modeling results

The emission rates and stack information were fed into the dispersion modeling software (Lakes Environmental, 2008). The results of the point source dispersion modeling were reported as micrograms per cubic meter (µg/m^3) at the set distances of between 600 to 1000 meter distances from the generating source. The particulate emission modeling from the four point sources is illustrated in Figures 3 and 4.

The highest concentrations of ambient dust levels from both kilns (DK5 and DK6) are between 600 and 1000 m from point source. This is mainly as a result of the concentration, height and velocity at the point of release. Table 3 provides a summary of stack data that was used in the dispersion model.

The highest concentrations of ambient dust levels from both mills (FM5 and FM6) are between 600 and 1000 m from the generating source (Figure 4). The results are much lower as compared to the kilns (Figure 3) due to the lower concentrations emitted by the milling process.

Point sources dust emission concentrations were all below the South African National Air Quality Standard of 300 µg/m^3 (24 h average) (SANS, 2005). Therefore, the maximum emission levels (peaks) were used as impact range. The impact ranges were mapped to compare with the existing fall out dust sampling locations. As illustrated in Figure 3, the maximum impact was from kiln 5 (DK5) at 900 m. Figure 5 illustrates the relationship between the existing fall out dust sampling points and the impact from the dust emissions from the four point sources.

From the modeling results shown in Figure 5, it is clear that the existing fall out dust sampling locations are correct (in relation to point sources) except for the one to the South close to the gate (sampling point GAT). The maximum impact was more to the South. The sampling location in town to the North (sampling point SS), is not required as it will not be representative of the impact from the cement manufacturing plant. The point source model on its own cannot be used to relocate sampling equipment as the results from the fugitive emission sources

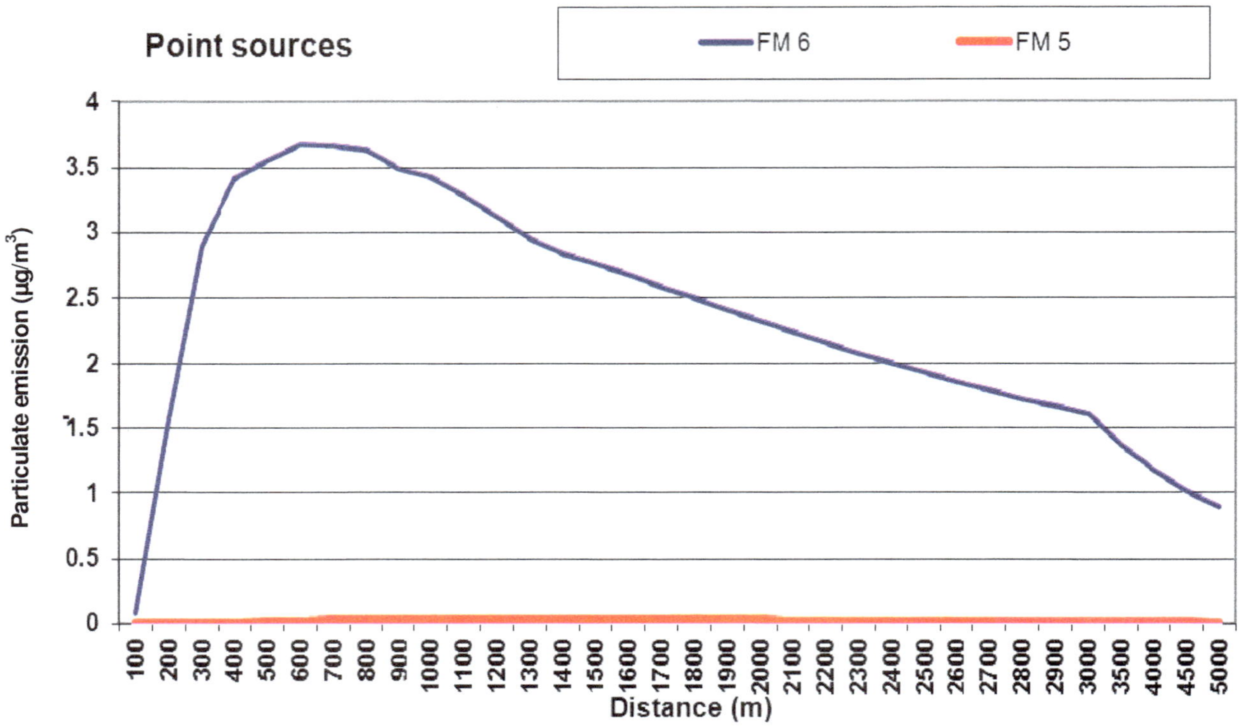

Figure 4. Cement mills particulate emission modeling.

Figure 5. Mapped point source dispersion modelling results (location indicated using Google Earth, 2011).

Table 4. Particulate emission rates from fugitive sources.

Source name	Activity	Source description	Emission rates after reduction factors (EF – ER %) g/s	Emission factor source
Mining	Drilling	Mark out blast holes and conduct drilling with a drill rig	0.0004917	US EPA
Mining	Blasting	Arranging and execute blasting	0.0019053	NPI
Mining	Crushing and screening	Discharge into crusher bin (feeding crusher)	0.0773824	NPI
		Crushing	0.2219685	NPI
Mining	Transportation	Transportation via belt conveyor (Transfer points) from primary crusher to surge pile	0.0773824	US EPA
Mining	Stockpiling	Discharge from belt conveyor to surge pile	0.1547648	US EPA
Mining	Crushing and screening	Secondary crushing	0.1331811	NPI
Mining	Stacking and storage of material	Transportation via belt conveyor (Transfer points)	0.3095297	US EPA
		Discharge from belt to screening stockpile	0.0116074	US EPA
		Stacker operations	0.0986626	US EPA
Production	Raw material handling	Sand offloading	0.0000143	US EPA
Production	Raw material handling	Reclaiming from stockpiles	0.0012485	US EPA
Production	Raw material handling	Transportation via belt conveyor (Transfer points) - Stockpile to proportioning	0.0397552	US EPA
Production	Raw material handling	Transportation via belt conveyor (Transfer points) Proportioning to Raw mill	0.0993880	US EPA
Production	Burning	Transportation via belt conveyor (Transfer points) - From stockpile to coal mill	0.0072091	US EPA
Cement Milling	Gypsum extraction and transportation	Transportation of Gypsum	0.0046857	US EPA
Packaging and Logistics	Material handling	FDG offloading at tippler	0.0000916	US EPA
		FDG transportation via belt conveyor	0.0002749	US EPA
Packaging and Logistics	Material handling	Coal offloading at tippler	0.0287043	US EPA
		Coal transportation (tippler to stockpile)	0.1435213	US EPA
		Coal discharge onto stockpile	0.1435213	US EPA
Packaging and Logistics	Material handling	Gypsum offloading from rail trucks	0.0010426	US EPA

can influence the model.

Fugitive source emission modelling

Fugitive source emission rates were calculated using emission factors as proposed by the US EPA 42 and the Australian NPI. Emission factor source is shown in Table 4.

Fugitive source emission modeling results

The emission rates were fed into the dispersion modeling software (Lakes Environmental, 2008). The results from the fugitive source dispersion modeling were reported as micrograms per cubic meter (μg/m^3) at different distances from the generating source. The selected distance range was 5 to 5 000 m, therefore all fugitive sources where the impact is five meters or less were excluded and regarded as insignificant.

From the modeling results shown in Figure 6, it is clear that the existing fall out dust sampling locations are inadequate especially towards the northern and west-south - western directions of the operation. It also clear that the sampler located towards the south-east (sampling point QD) is not adding value.

Conclusions and recommendations

The study concluded that to optimise a fall out dust monitoring programme, the following steps should be

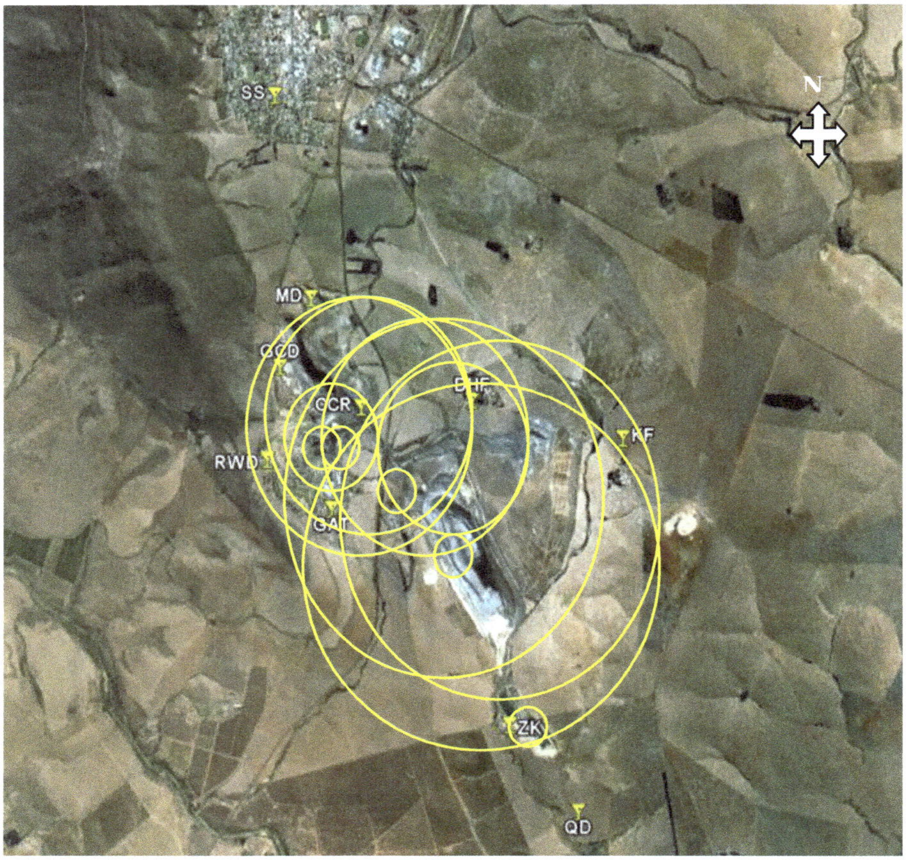

Figure 6. Mapped fugitive source dispersion modelling results (location indicated using Google Earth, 2011).

considered:

1. Identify all particulate emission generating sources.
2. Determine the particulate emission rates from fugitive sources by using emission factors.
3. Monitor emissions from point sources as a legal requirement. The results can be used to determine the emission rates of point sources.
4. Use the source characteristics and emission rates as input data for dispersion modeling.
5. Use the results of the dispersion modeling to determine the location of samplers.
6. Keep the programme relevant by considering activity or emission rate changes and at least annually review the monitoring programme accordingly.

Historically, locations of fall out dust samplers have been guided by experience, observed effects or physical property boundaries. This study, however, confirmed that it is possible to locate samplers by utilising all available information and emission factors without spending vast capital and time on actual sampling. In summary, the steps for the developing of a location model (steps 1 to 4) and optimising a fall out dust sampling programme (Step 5) are illustrated in Figure 7.

Optimising the fall out dust monitoring programme

Mapping the impact zones for both point and fugitive sources together with the existing sampling point locations revealed the adequacy of such sampling points in detecting the impact of fall out dust from the activities.

The mapping of the results from the dispersion model did not consider prevailing wind directions, but considered the location of the samplers. The prevailing wind directions were obtained from an onsite weather station, a Davis Vantage Pro2 type that is calibrated annually by a SANAS accredited facility.

The location of the existing fall out dust sampling points in yellow are adequate for the four point sources and do not require any change. With regard to fugitive source dust dispersion modeling results (Figure 6), it is also evident that the location of the sampling points is inadequate. According to the modeling results, the optimal location of sampling points is indicated in red. The sampling point in blue (Figures 8 and 9) is the ideal position for the relocation of sampling point CCR. GPS was used to confirm the coordinates of the changed locations.

The study concentrated on the cement manufacturing process and related activities and it did not consider the

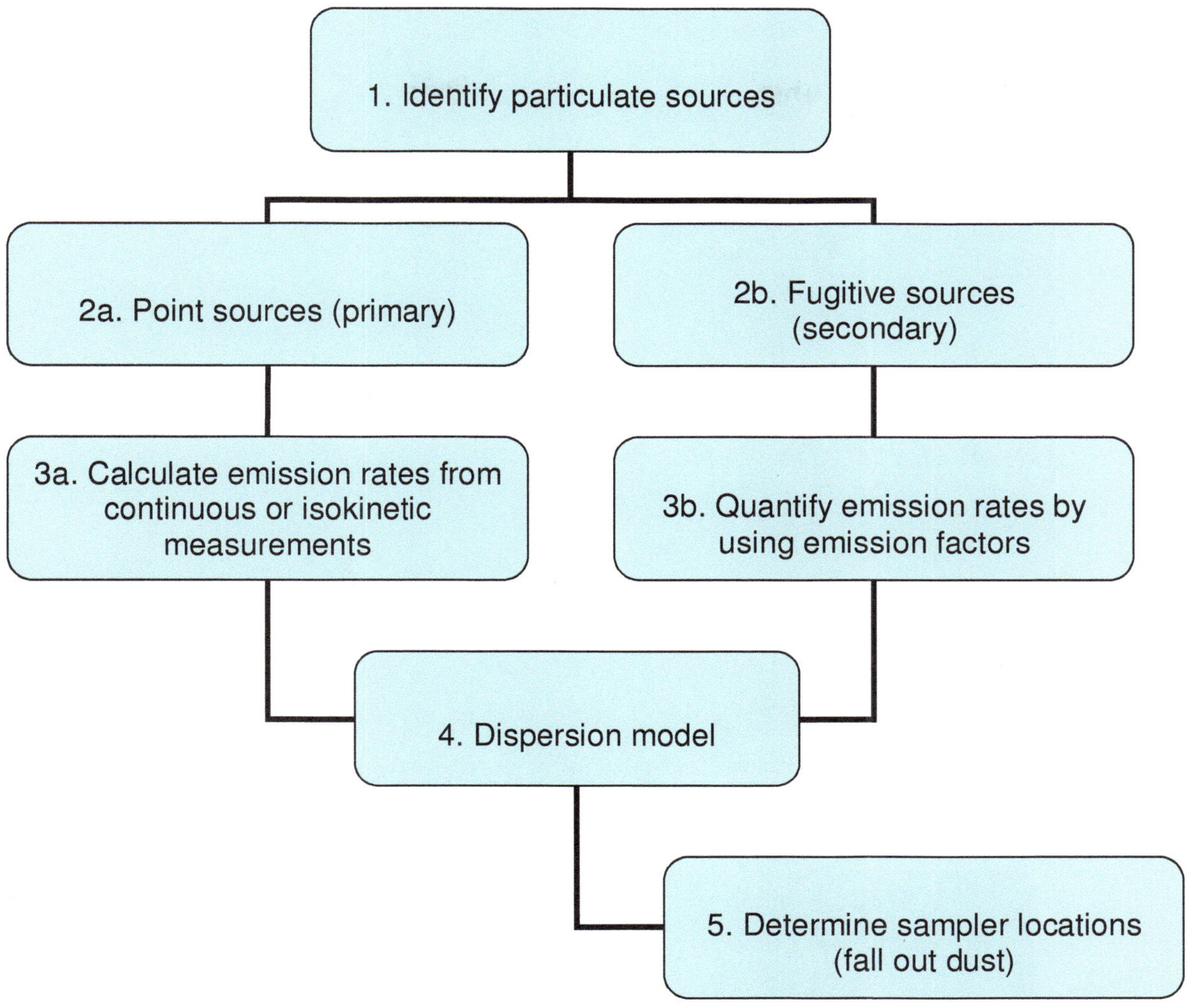

Figure 7. Steps for optimising fall out dust sampler locations.

impact of adjacent activities such as agricultural or other mining (apart from the cement plant) activities because these were not major contributors to dust emissions in that area. Nonetheless, results from this case study can be used as a guideline for the categorisation (zoning) of use for properties surrounding the cement manufacturing plant. In order to further optimise the monitoring program, the following research agenda is suggested:

1. Comparison of results from the existing fall out dust monitoring programme with those from the newly optimised programme to verify adequacy and appropriateness.
2. Analyses of samples from fall out dust samplers to determine composition and characteristics of the particulates to trace its origin.
3. Collection and analyses samples of raw materials and at particulate emission generating sources and comparison of the results with those of the fall out dust samples. The results can be used to confirm the origin of the particulates.
4. A comparative study between various dust emitting operations to reflect operational differences in South Africa.

The development of a guideline on how to optimise a fall out dust monitoring program for a cement manufacturing operation requires knowledge of all relevant activities and quantification of particulate emission rates that can be expected from such activities.

This information on sources and emission rates provide the input data for the dispersion model that will in turn reflect the impact areas of such particulate emissions. These models assist with the location of fall out dust samplers in order to adequately monitor and measure the impact from production activities. Due to the complexity of actual measurements or applying the principles of a mass balance in the total cement manufacturing process to calculate emission rates, it was found that the use of selected emission factors was a simpler and less costly method. Implementing the steps described above will

Figure 8. Location of optimised sampling points in relation to point sources (location indicated using Google Earth, 2011).

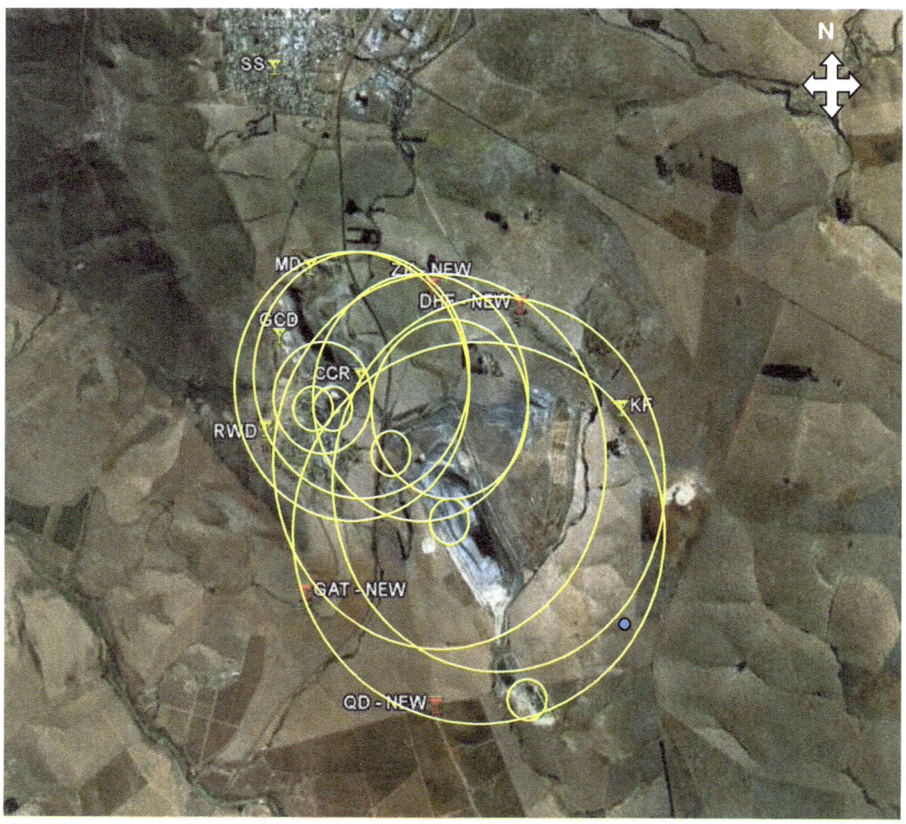

Figure 9. Location of optimised sampling points in relation to fugitive sources.

contribute to optimisation of the fall out dust monitoring program at a cement manufacturing plant and any such similar plant with significant benefits for environmental planning and management.

REFERENCES

Ade-Ademilua OE, Obalola DA (2008). The effect of Cement Dust Pollution on Celosia Argentea (Lagos Spinach) Plant. J. Environ. Sci. Technol. 6(22):1-24.

Arizona Department of Environmental Quality Air (ADEQ) (2004). Dispersion Modeling Guidelines for Arizona Air Quality Permits. Air quality division 1.2:5.

ASTM (2004). Standard Test Method for Collection and Measurement of Dustfall (Settleable Particulate Matter). American Society for Testing and Materials - ASTM D1739/ 98.

Bilen S (2010). Effect of cement dust pollution on microbial properties and enzyme activities in cultivated and no-till soils. Afr. J. Microbiol. Res. 4(22):2418-2425.

Davis WT (2000). Air Pollution Engineering Manual. United States of America.: John Wiley and Sons, Inc.

Gbadebe AM, Bankole OD (2007). Analysis of Potential Toxic Metals in Airborne Cement Dust Around Sagamu, Southwestern Nigeria. J. Appl. Sci. 7(1):35-40.

International Standards Organisation (2003). Stationary source emissions — Determination of concentration and mass flow rate of particulate material in gas-carrying ducts — Manual gravimetric method. International Organisation for Standardisation. ISO 9096.

Knight MS (2007). Historic air quality monitoring data review and proposed air quality monitoring plan – A report for the Asperance Port Authority [Online]. Available from: http://www.esperanceport.com.au/downloads/.../NickelShiploading_12Nov07.pdf [Accessed: 21/02/2009].

Lakes Environmental (2008). Screen View™ - Freeware [Online]. Available from: http//www.environmental- expert.com/stse_resulteach_product.aspx. [Accessed: 25/10/ 2010].

Loans C (2007). Fall-out dust levels around two enterprises in the Western Cape of South Africa from 2001-2005 [Online]. Available from:http://wiredspace.wits.ac.za/bitstream/handle/10539/4792/Research%20Report%20-%20MPH%20-%20Chris%20Loans%20-%200416886J.pdf?sequence=1 [Accessed: 18/10/2009]. 2-3.

Lodge JP (1988). Methods of air sampling and analysis. New York USA: CRC Press.

Sans (2005). South African National Standard. Ambient Air quality – Limits for common pollutants. SANS 1929:2005 Edition 1.1, Pretoria, South African Bureau of Standards.

South Africa (2007). The National Framework for Air Quality Management in the Republic of South Africa. pp.36-37.

Tijani AA, Ajobo O, Akinola AA (2005). Cement Production Externalities and Profitability of Crop Enterprises in Two Local Government Areas of Ogun State. Department of Agricultural Economics. 11(1):43-48.

US Epa (1995). Compilation of Air Pollutant Emission Factor. States Environmental Protection Agency, Office of Air Quality Planning and Standards. Volume 1: Stationary Point and Area Sources, fourth edition, AP-42. Section 11.6 Portland Cement Manufacturing [Online]. Available from: http://www.epa.gov/ttn/chief/ap42.html. [Accessed: 25/08/2010].

Vallero DA (2008). Fundamentals of Air Pollution. 4th ed. Academic press. p.469.

Vincent K, Passant N (2006). Assessment of heavy metal concentrations in the United Kingdom [Online]. Available from: http://www.uk-air.defra.gov.uk/reports/cat16/ 0604041205_heavy_metal_issue1_final.pdf. [Accessed: 06/02/2010].

3

Uptake kinetics of arsenic by lettuce cultivars under hydroponics

Mathieu Nsenga Kumwimba, Zeng Xibai, and Bai Lingyu

Institute of Environment and Sustainable Development in Agriculture, Chinese Academy of Agricultural Sciences / Key Laboratory of Agro-Environment, Ministry of Agriculture, Beijing 100081, P.R. China.

Arsenic (As) uptake ability based on kinetic parameters by two lettuce cultivars; Sijibaiye (SJBY) and Texuanyanlingsun (TXYLS) was investigated in nutrient solution containing eight levels of arsenic (As). Depletion of As from solution was monitored over a period of 24 h at regular time to estimate As uptake kinetics of the plants. K_m and V_{max} values were evaluated with the aid of the Lineweaver - Burk plot (that is the double reciprocal plot of initial velocity (1/v) versus substrate concentration (1/[s]). Both cultivars were capable to assimilate As(V), although with different kinetic parameters. The values obtained for K_m and V_{max} varied with plant cultivars. Using Lineweaver- Burk, values of K_m and V_{max} were found to be much lower in SJBY (26.721 mg L^{-1} and 0.7737 $mgAsg^{-1}$ shoot $d.wt.hr^{-1}$) than that of TXYLS (81.922 mg L^{-1} and 1.8488 mgAsg-1 shoot $d.wt.hr^{-1}$), indicating SJBY had the lowest uptake ability than TXYLS plants. Furthermore, TXYLS had the highest K_m value than SJBY, confirming that uptake in the plant cell plasma membrane of TXYLS had a higher affinity for As than SJBY plants. The time (24 h) affects the depletion of As in both cultivars. It can be concluded that among the cultivars, SJBY could be finally identified as the most suitable cultivar to be grown in polluted farmland. Therefore, it has been suggested that there is possibility to lower the As concentration in lettuce by selecting and breeding cultivars with less As concentration that can safely grown in contaminated soils with slight levels of As for safe consumption.

Key words: Arsenic (V), cultivars variation, kinetic parameters, Lineweaver Burk, uptake rate.

INTRODUCTION

Arsenic (As) is one of the most dangerous toxic farmland that can be found. Besides various adverse effects of As on humans and animals, arsenic could also hamper normal growth of plants (Cozzolino et al., 2010).

Lettuce plant (*Lactuca sativa L.*) is an important leafy vegetable in the world, especially for Chinese people, due to the large consumption. It is considered as an excellent nutritive source of minerals and vitamins as is consumed as fresh green salad (Mulabagal et al., 2010). Crop with such promising potentialities for local markets, would necessitate much research in order to improve agricultural products' safety and to reduce arsenic threat on human health.

In recent years, many researchers have paid much attention to the As uptake kinetic excised plant roots by different plant species, such as hyperaccumulation, rice (Mohamed and Andrew, 2008), *Brassica carinata* (Irtelli et al., 2008), Pteris *vittata* (Weifeng et al., 2011), maize (Abbas et al., 2008), adopting hydroponic and field experiments with intact roots, while there is lack of data on the kinetics of arsenic uptake by leafy vegetable. Plant species and even genotypes differ greatly in their ability to take up, transport and accumulate As within the plants. The genotypic variation in the different cultivars for

Table 1. Name, abbreviation, origin, and seed provider of 32 leafy vegetable cultivars used for hydroponic experiments.

Cultivar name	Abbreviation	Origin	Seed provider
Texuanyanling sun	TXYLS	China	BGSRDC
Sijibaiye	SJBY	China	BGSRDC
Huayesun jing yong woju	HYSJYWJ	China	BGSRDC
Yeyong woju	YYWJ	China	BGSRDC
Guayuanyiyesun	GYYYS	China	BGSRDC

IVF, CAAS: The Institute of vegetable and flowers, Chinese Academy of Agricultural Sciences; BGSRDC: Beijing Gardening Seeds of Research and Development Center;BGLATI: Beijing Golden Land Agricultural Technology Institute;BSVSPC: Beijing Special Vegetable Seed and Plant Co.; BSBC : Beijing seed and breed center.

arsenic concentration, as a strategy to reduce the movement of As from contaminated soils to edible parts of crop plants has been receiving close attention (Norton et al., 2012). Moreover, the fundamental requirement for breeding low As-accumulation cultivars is to know the genotypic differences and the mechanisms governing As accumulation in crops. It is therefore imperative to understand the variation in As accumulation of different lettuce cultivars and to evaluate the mechanisms involved in arsenic uptake kinetic and translocation in order to screen and find out excluder-cultivars which transport less As concentration to the shoot suitable for growing in the As-contaminated farmlands. Once identified, cultivars with high and less As accumulation will be tested for As uptake kinetic to confirm the result. The objective of this study was: (1) to screen and identify the cultivars with less As concentrations that can be grown in As contaminated farmland for food safety. (2) to compare the rates of As uptake ability between cultivars and to determine the kinetic parameters. (3) Because of the general lack of data on the uptake kinetics of arsenic in leafy vegetable cultivars, the main objective of this study was also to provide theoretical support and quantitative data that could offer an understanding of the differences in V_{max} and K_m among leafy vegetable.

MATERIALS AND METHODS

Plant material and preparation of seedlings

Seeds of five lettuce cultivars (Table 1) were purchased from Chinese Academy of Agricultural Sciences and cultivated at Institute of Environment and Sustainable Development in Agriculture CAAS, Beijing/China. In the experiment (I), seeds were surface sterilized with 30% H_2O_2 (w/w) solution for 15 min, rinsed five times with distilled water. The seeds were germinated for 10 days in the plate containing autoclaved vermiculite. During the seed germination period, half-strength modified Hoagland's nutrient solution was supplied once a week.

Hydroponic screening

On the 7th day, at third leaf stage, the healthy and uniform sized seedlings were carefully removed from vermiculite, and washed with distilled water and then transferred to hydroponic bottles each containing 1500 ml of full-strength of modified Hoagland and Arnon (Hoagland and Arnon, 1950) with pH adjusted to 6.3 by using NaOH or HCl for 10 days. Nutrient solution was renewed twice a week. On the 21st day, the As treatment consisted of the basic nutrient solution with addition of 6 mg As L^{-1}, As(V) (applied as $Na_3AsO4.12H_2O$) for 14 days. There were four replicates for each cultivar and each replicate comprises four plants. The As treated nutrient solution was renewed once a week. Plants were grown for 35 days, and then harvested. The shoots and roots of all harvested plants were separated carefully, washed with water, and then rinsed severally using deionizised water to remove adhering particles. These samples of roots and shoots were gently packed in envelopes. Plant parts were dried at 90°C for 1 h, then at 70°C in the oven until constant weight was obtained. Both dry and fresh weights were recorded. Dried plant materials were ground and digested and arsenic concentrations were determined using the atomic fluorescence spectrometry following the procedure described by Smith et al. (2009).

Kinetics of As(V) uptake

The experiment (II) was carried out under the same conditions as described above. Two lettuce cultivars which have both been shown to absorb the highest and lowest arsenic in their shoots (Exp I) were used as materials.

After 4 weeks of growth before the uptake kinetic experiment (II), the seedlings were transferred to the deionized water for 5 days, and then uniform seedlings from TXYLS and SIJBY were selected, washed from root system and transferred to plastic bottles containing 1200 ml of As uptake solution. Each bottle was wrapped with opaque plastic membrane. Eight different concentrations of arsenic (0, 0.5, 1, 2, 4, 6, 8, and 10 mg As L^{-1}) were used to study the uptake kinetics of arsenic. Each treatment was replicated three times. Each 1200-mL container had 4 plants. The pH of the nutrient solution was maintained at 6.3. At each time interval (0, 0.5, 1, 2, 4, 6, 8, 10, 12, 22, and 24 h), sample of 1 ml aliquot was withdrawn from each uptake solution to measure depletion in As, as a result of its uptake by plants, the depletion period was for 24 h, and 1.0 ml of deionized water was then added to each container at set time intervals (Wang et al., 2002). Water losses through transpiration were compensated by additions of deionized water at hourly intervals. After 24h of uptake experiment, the plants were harvested and quickly rinsed with distilled water. The plants were separated into roots and shoots, blotted dry with paper tissue and dried at 65°C for 72 h, and the dry weights were recorded. Arsenic concentrations were determined as previously described.

Substrate concentrations, with which the uptake rates were

Table 2. Mean As concentrations and As uptake of five cultivars (Hydroponic experiment, 2 weeks of As exposure).

Cultivar name	As concentration (mg kg^{-1} dwt)		Dry weight (g pot^{-1})		As uptake (μg pot^{-1} dwt)
	shoot	Root	Shoot	Root	Shoot
Texuanyanling sun	20.56a	255.91	2.70	0.21	55.51a
Sijibaiye	14.07c	442.52	2.14	0.10	30.25c
Huayesun jing yong woju	16.27b	478.31	2.77	0.11	45.07ab
Yeyong woju	17.67b	534.06	2.75	0.11	47.97ab
Guayuanyiyesun	16.22b	482.65	2.51	0.13	40.62bc

Values (mean n=4) with different letters in the same column are significantly different according to the duncan's test (P<0.05)

compared, was calculated as the arithmetic mean of the initial (t = 0) and final concentration for a time interval.

Calculations of As uptake rates and kinetic parameters

The amount of uptake rate (V [mg As g^{-1} shoot d.wt.hr^{-1}]) was calculated from the depletion of As in the uptake solution. As uptake by the plant per unit time per gram dry weight was calculated according to the following formula: (Shimono and Bunce, 2009; Ajaelu et al., 2011)

$$V = \frac{(Vi \times Ci) - (Vf \times Cf)}{t \times Bdwt} \quad (1)$$

Where, Vi, Ci, Vf, Cf and Bdwt are the volume (V) and quantities in mg L^{-1} of substrate in solution at the start and the end of uptake experiment, and t is the time in hours and Bdwt is the biomass dry weight of the plants, respectively. The reduction in the volume was very small, so this was not taken into account in the calculation.

The Michaelis-Menten Equation is:

$$V_0 = \frac{V\max.[S]}{Km + [S]} \quad (2)$$

Where, V_0 is the initial velocity of the reaction at substrate concentrations (S); [S]: is the substrate concentration; V_{max}: is the maximum uptake rate (mgAsg^{-1} shoot d.wt.hr^{-1}) achieved at which the enzyme is saturated with substrate, and K_m: is the Michaelis-Menten constant (Briggs and Haldane, 1925).

The parameters K_m and V_{max} characterize the ability of plant to absorb nutrients from their soil environment (Glass, 1989). The equation describes the relationship between the uptake rate of the nutrient and the nutrient concentration. Uptake rate reaches a constant or saturated rate (V_{max}) at high ambient concentration. Lineweaver and Burk (1934) pointed out that equation (2) becomes linear in form upon taking the reciprocal of both sides of the equation to calculate the key parameters in the Michaelis-Menten equation:

$$V_0 = \frac{V\max.[S]}{Km + [S]} \longrightarrow \left(\frac{1}{V}\right) = \left(\frac{K_m}{V_{\max}}\right)\cdot\left(\frac{1}{S}\right) + \left(\frac{1}{V_{\max}}\right)$$

That is (y= m x+ c), where (y=1/V and x= 1/[s])

In order to obtain the kinetic uptake parameters representing the maximum uptake rate (V_{max}) and the half saturation constant (K_m), by plotting 1/V against 1/[S] will give a straight line graph having a slope of K_m/V_{max} and a y- intercept on the ordinate at $1/V_{max}$. From this plot, the K_m and V_{max} values of arsenic uptake were determined by linear regression analysis, thus giving a convenient method for obtaining both V_{max} and K_m.

Statistical analysis

All data were subjected to analysis of variance (ANOVA) using SPSS 16.0. Differences at p<0.05 were considered significant. All curve fitting was completed using a computerized, linear regression on Lineweaver-Burk transformed data to estimate the V_{max} and K_m.

RESULTS

Genotypic variation in shoot As concentrations and accumulation

In this preliminary experiment (I) we designed to determine cultivars with less and high shoot As concentration. Significant genotypic variations were observed in the shoots and roots As concentration of 5 cultivars studied under As treatment (6 mg L^{-1}). Of the 5 cultivars, As concentration ranged from 14.07 to 20.56 mg kg^{-1} dry weight (1.5 variation). The main results were shown as follows: SJBY had the lowest shoot As concentration (14.065 mg kg^{-1} dry weight). While the highest shoot As concentration were detected in TXYLS (20.56 mg kg^{-1} dry weight) (Table 2). After taken up by roots, arsenic was accumulated and sequestered in the roots in much higher amounts than in the shoots (root>shoot), only about 4-6% of As was translocated to shoots of cultivars. The average As concentration in root was nineteen to twenty-six times higher than that in shoot (root>shoot) as an indicator for translocation ability from root to shoot. Therefore absolute majority of As accumulated by cultivar plants was restrained in root and only a very small portion was transferred into shoot.

Arsenic accumulation

The data presented in Table 2 shows the amounts of arsenic (As) accumulated in the shoot of 32 cultivars. The values are expressed in μg^{-1}dry weight each calculated from the formula: (dry weight × As concentration) and they show significant difference (p<0.05) ranging from 30.25 to 55.11 μg^{-1} pot dry weight. Arsenic accumulated

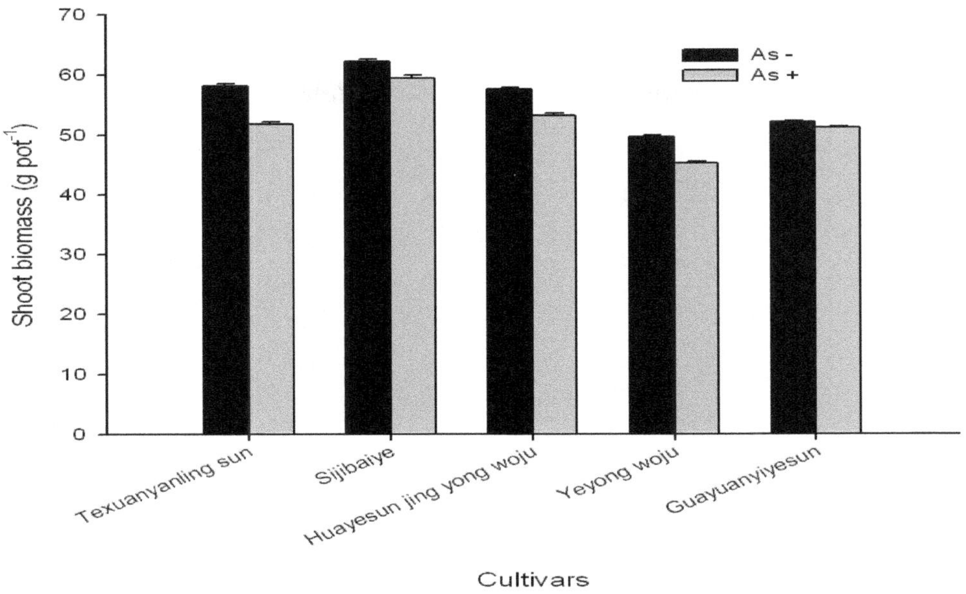

Figure 1. Shoot biomass (FW) of 5cultivars exposed to 6mg L^{-1} As for 14 days. Data are expressed as means±S.E. of four replicates. $p<0.05$.

Table 3. Tolerance index, % (TI) based on shoot biomass of 5 lettuce cultivars exposed to 6mg/L As for 14 days.

Cultivar	Shoot biomass
Texuanyanling sun	89.03c
Sijibaiye	95.58a
Huayesun jing yong woju	92.45b
Yeyong woju	91.26bc
Guayuanyiyesun	98.17a

Values (mean n=4 with different letters in the same column are significantly different according to the Duncan's test ($P<0.05$).

to the greatest extent in TXYLS had the highest As accumulation while SJBY had less arsenic accumulation.

Variation in shoot biomass trait under As stress

By growing 5 lettuce cultivars in the greenhouse, the responses of the selected cultivars differs. The shoot biomass (FW) of tested cultivars varied between 46.340-60.505, respectively under 6 mg L^{-1} (Figure 1). On the other hand GYYYS and SJBY showed greatly higher shoot biomass (Figure1). The maximal reduction in shoot biomass was observed in TXYLS (11%), followed by YYWJ (9%), HYSJYWJ (8%), SJBY (4.5%), and GYYYS (1.8%) (Table 3). The tolerance index, in terms of the biomass ratio of As treatment to control, was high in all cultivars except TXYLS. By contrast, 80% of tested cultivars showed a tolerance index higher than 90%. Therefore, the effects of As on leafy vegetable growth and development appear to be cultivar dependent.

Variation in root-to-shoot transport and nutrient solution-to-plant

Translocation factor (TF)

The TF can be used to evaluate the ability of the plant to translocate As from the root to shoot. The average TF of various cultivars differed among cultivars (Table 4). The root-shoot translocation factor ranged from 0.03 to 0.08. There were no significant ($P>0.05$) differences found in As TF of SJBY, HYSJYWJ, YYWJ and GYYYS. While TXYLS showed the highest TF similar to As accumulation and BF (Tables 2 and 4). The translocation factor (TF) was calculated by dividing the As concentration in the shoot by its concentration in the root.

.

Bioaccumulation factor (BCF)

The quotient of the As concentration in shoots to that in nutrient solution, can be used to evaluate the ability of plants to accumulate heavy metals. Table 4 shows the average arsenic bioaccumulation factor of 5 cultivars under As treatment (6mg L^{-1}), ranging from 3.43 to 2.34 in the cultivars. The lowest BCF was found in SJBY, similar to the TF, and the highest BCF was detected in TXYLS. There were no significant differences between HYSJYWJ, YYWJ and GYYYS.

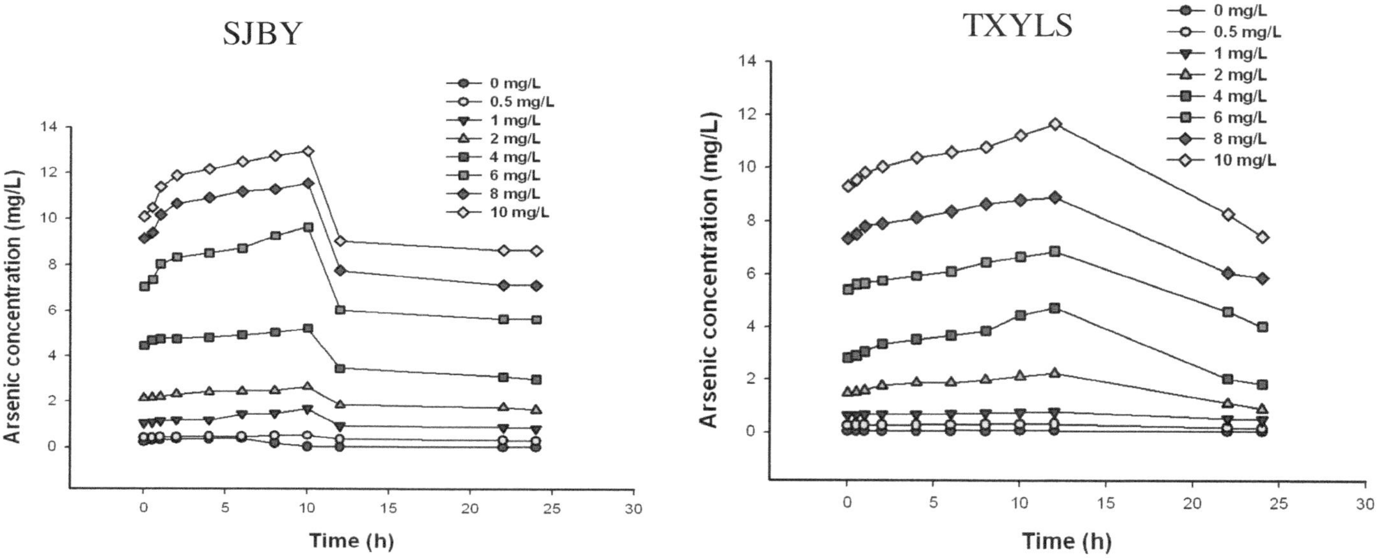

Figure 2. Changes of As concentration with time (h) in the cultivars with high and low-As Accumulation (TXYLS and SJBY) grown in nutrient solutions with different As(V) concentrations for 24 h.

Table 4. Translocation and bioaccumulation factor of As in 5 lettuce cultivars exposed to 6mgL^{-1} As.

Genotype	BCF	TF
Texuanyanling sun	3.43a	0.08a
Sijibaiye	2.34c	0.03b
Huayesun jing yong woju	2.71b	0.04b
Yeyong woju	2.95b	0.04b
Guayuanyiyesun	2.71b	0.03b

Different letters in a column indicate significant differences between the cultivars at the 0.05 level according to the Duncan's test.

Depletion of As concentration in nutrient solution

The uptake time is one of the most important factors affecting the uptake of heavy metals by plants. Generally, in the initial period of exposure metal is absorbed and accumulated at a high rate, and then the level stabilizes when equilibrium of metal uptake rates is attained. Short-term, time-dependent arsenic depletion from hydroponic solution was measured at a series of different time intervals (after 0 h, 0.5, 1, 2, 4, 6, 8, 10, 12, 22 and 24 h respectively) with various concentration of arsenic. Figure 2 show that the depletion of As in TXYLS gradually increased with increasing uptake time for at least 12 h and then starts decreasing rapidly with the passage of time until 24 h, while SJBY showed the same trend for the first 10 h then starts decreasing until 24 h. After 24 h of uptake, the depletion of As became slower for all cultivars and showed a significant difference in the arsenic accumulation. It can be explained that arsenic was absorbed by cultivars. Both the cultivars and treatments influenced concentrations of As in nutrient solution. At the end of the depletion (24 h), TXYLS and SJBY have up taken 23.09% and 28.40% of arsenic from nutrient solution (figure 2). Interaction between treatments and cultivars influenced depletion of As concentration.

Short-term Arsenic uptake kinetics

Short-term (24h) uptake rate of the cultivars over a range of arsenic concentrations (0.5, 1, 2, 4, 6,8 and 10 mg L^{-1}) was calculated as the ratio of the amount of absorption arsenic by cultivars in a unit time to the dry weight of cultivars and its unit is mg. g^{-1}. h^{-1}. The results showed that there was variation in uptake rate between two cultivars. Lower uptake rate was found at the rate of 0.5, 1 and 4 mg As L^{-1} in SJBY and at 2, 6, 8 and 10mg L^{-1} in TXYLS (Figure 3). The data for kinetic parameters were derived from a linear regression on Lineweaver-Burk plot transformations (Table 5), indicated that uptake characteristics were different between cultivars. TXYLS exhibited about 2-fold higher V_{max} value (1.8488 mg As g^{-1} shoot d.wt.hr^{-1}) than that (0.7737 mg As g^{-1} shoot d.wt.hr^{-1}) in SJBY, similar TXYLS possesses higher value of K_m (81.92 mg L^{-1}) as compared to that in SJBY (26.721 mg L^{-1}). The uptake ability is jointly decided by the V_{max} and K_m; hence the V_{max}/K_m represents the absorption ability of cultivars. The V_{max}/K_m values indicated that the arsenic uptake ability of cultivars followed TXYLS > SJBY. The slope of the linear component of TXYLS was higher (44.312) than SJBY (34.537) (Table 5). Therefore, we could conclude TXYLS 's cell wall had a relatively strong arsenic binding capacity, while SJBY had relatively lower.

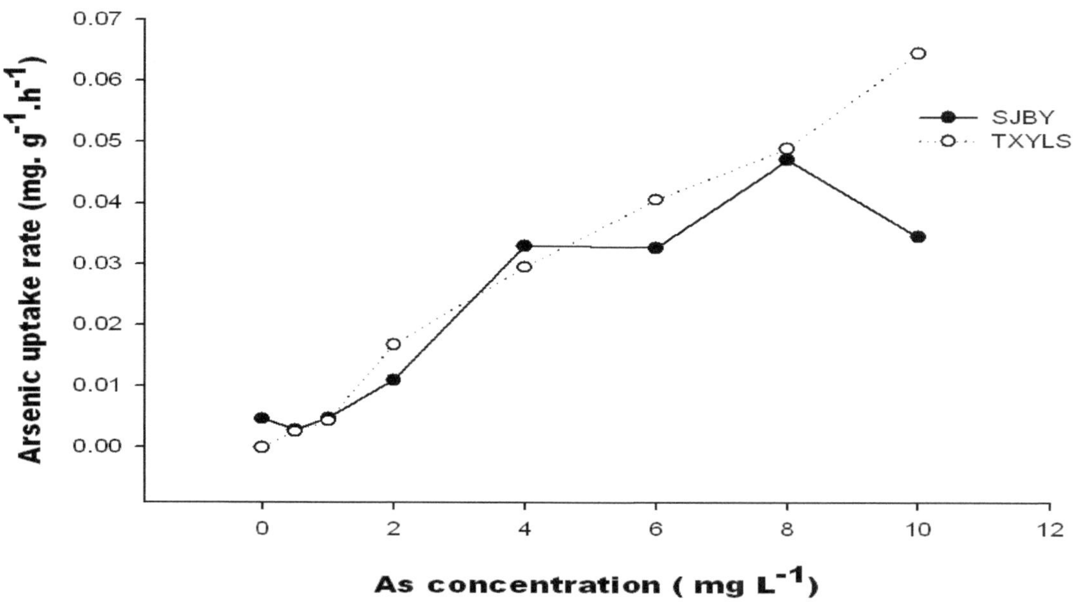

Figure 3. As uptake rate ($mg.g^{-1}.h^{-1}$) by plants of TXYLS and SJBY grown with 0, 0.5, 1, 2, 4, 6, 8 and 10 mg L^{-1} As for 24 h. (values are means of 3 replicates).

Table 5. Comparison in the low-As-accumulating cultivar 'SJBY' and the high-As-accumulating cultivar 'TXYLS' in kinetic parameters (K_m, V_{max} and K_m/V_{max}) for concentration-dependent uptake of As into intact plants immediately after 24 h of As.

Cultivar	V_{max} (mgAsg-1 shoot d.wt.hr-1)	K_m (mg/L-1)	K_m/V_{max}	R^2
TXYLS	1.8488	81.9227	44.312	1
SJBY	0.7737	26.7211	34.537	0.9999

Data are presented as mean (n=3).

DISCUSSION

In this study, the transport of As from nutrient solution to root and shoot of the cultivars were the important factors which lead to the cultivar differences (Tables 2 and 4). The plant is able to move the As from the root to the shoot, but the mechanism through which these leafy vegetable plants do this is unknown. The average As concentration in the roots was nineteen to twenty six times higher than that of shoot (root>shoot), this might be due to the inability of the plants to translocate the As beyond the roots (Smith et al., 2009). The result reveals that all the cultivars had TFs less than 1, (3 fold variation) in TF ability and (1.2 fold variation) in BCF, indicating that As was limitedly transported into the shoot. According to the study of Sinha et al. (2004), the plants act both as "accumulators" and "excluders". Therefore, these five cultivars can be considered as a As excluder, comprising avoidance of metal uptake and restriction of metal transport to the shoots (Sinha et al., 2004; Devos et al., 1991). This is consistent with the results reported by Huang et al. (2005) using lettuce and when compared with other plants such as *Pteris vittata* (Francesconi et al., 2002), the As concentrations were high in the shoot (aboveground) than in root, also the TFs were greater than 1.0. Wei et al. (2005) found that the exclusion of metals from shoot has been regarded as a metal tolerant strategy. Metals, once taken up by roots, can either be stored in the root or exported to the shoot. Resistance of plants to heavy metals can be achieved by an avoidance mechanism resulting in low metal concentration in the shoots as shown in this study.

The mean TFs for five cultivars were 0.043. Ming et al. (2012) found that TFs of rice were 0.20, 0.25, and 0.24 and Williams et al. (2007) reported that TF of As to the shoots of rice was 0.05–3.8, with mean 0.76. Su et al. (2010) demonstrated that rice had a higher efficiency of As translocation from root to shoot than wheat and barley. Our result revealed that lettuce cultivars had a less efficiency of TF than rice. These data suggest that As was transported at a much slower rate to the shoot of cultivars. Liu et al. (2004) reported that the large cultivars variation in rice was caused by As uptake and shoot As transport, which could also lead to genetic differences in

As uptake. In the present study cultivars may affect remarkably TF of As. The results indicated that As concentration in shoot was somewhat related to As uptake and the transport of As to shoot.

Our objective was to characterize the uptake kinetic aspects of arsenic in two cultivars which have both been shown to absorb the highest and lowest arsenic in the experiment I (Table 2) in order to verify the cultivar with less As accumulation through kinetic experiment II. Uptake kinetics characteristics can be considered as one of the important criteria for selecting a cultivar to use in As-contaminated areas because it gives the information on how fast arsenic is taken up by plant. There are a number of studies investigating uptake kinetics of arsenate, both in higher and lower plants (Weifeng et al., 2011; Mohamed and Andrew, 2008; Geng et al., 2006). In general, arsenic uptake kinetics has not been described for leafy vegetable. Researchers using the depletion method have used time periods ranging from 15 min to several days. Mcfarlane and Yanai (2006) selected two-hour interval because in earlier experiments they found that two hours were sufficient to cause a measurable change in solution concentrations. Wang et al. (2002) found between 20 and 40% of arsenic was depleted over the 8 h period. In this study, the result showed that at the end of the depletion experiment (24 h), between 23.09 and 28.40% of arsenic from nutrient were depleted. Interaction between treatments and cultivars influenced depletion of As concentration. The value of V_{max} represents the plant's ability to absorb a nutrient. The higher the V_{max} value, the more could be absorbed. The value of K_m corresponds to the plant's affinity for a certain element, and the lower the K_m value, the greater the affinity (Chao et al., 2008).

In this study, there was a clear difference between cultivars with regards to the uptake kinetic parameters over the short time period (24 h) investigated here. Interestingly, SJBY, which had low shoot As accumulation in experiment I (Table 2), had a lower As uptake kinetic than TXYLS which shows higher As accumulation. The value of V_{max} in SJBY was (0.7737 mg As g^{-1} shoot d.wt.hr^{-1}) and (1.8488 mg As g^{-1} shoot d.wt.hr^{-1}) in TXYLS, and the value of K_m was (81.92 mg/L^{-1}) in TXYLS and (26.721 mg/L^{-1}) in SJBY. These values varied with plant species (Xiaozhang et al., 2005). SJBY had the lowest V_{max}, explained the lowest uptake ability in SJBY and TXYLS plants had the highest V_{max}, which explains the higher uptake ability. Furthermore, TXYLS had the highest K_m value than SJBY, confirming that uptake in the plant cell plasma membrane of TXYLS have a higher affinity for As (Poynton et al., 2004). The V_{max}/K_m values (Table 5) indicated that the arsenic uptake ability of cultivars followed TXYLS > SJBY. The difference between these two Km values indicates that different cultivars of lettuce have different As affinity (Table 5). Yu and Gu, (2006) found that different cultivars had different kinetic parameters in the same substrate. The results show that TXYLS was more As efficient to accumulate arsenic than SJBY. The higher As efficiency of TXYLS can be explained by the higher As uptake rate of the shoot with the passage of time (Table 5).

Conclusion

This study clearly established the differences in K_m and $V_{ma}x$ between two cultivars. In addition, the present study shows that, during 24 h of exposure, arsenic was depleted from solution, suggesting uptake of arsenic by the plants. The present study provides insights into the underlying mechanisms of As uptake kinetic in two different cultivars of lettuce which have both been shown to absorb the highest and lowest arsenic in their shoots. It is concluded that SJBY is the suitable cultivar plant to be developed on arsenic contaminated areas, which has a lower ability to accumulate a lesser concentration of arsenic from the hydroponic solution. Therefore, results suggest the possibility to lower the As concentration in lettuce through local cultivar by selecting and breeding cultivars in order to produce cultivars for safe consumption from soils with slight and moderate levels of As.

ACKNOWLEDGEMENTS

We thank Dr Yi bing Ma, (China), Dr Liaqat Ali, scientist research officer(Pakistan) and Dr Paul Held Laboratory manager for their guidance in kinetic experiment and designing and performing Michaelis- Menten equation. This study was financially supported by the National Scientific and Technological Program of the "12th Five-Year" Plan of China (Grant No. 2011BAD04B01).

REFERENCES

Abbas MHH, Meharg AA (2008). Arsenate, arsenite and dimethyl arsinic acid (DMA) uptake and tolerance in maize (Zea mays L.). Plant Soil 304: 277-289.

Ajaelu CJ, Ibironke, Oluwafunke L, Adedeji, Victor, Olafisoye O (2011). Equilibrium and Kinetic Studies of the Biosorption of Heavy Metal (Cadmium) on Cassia siamea Bark. Am-Euros. J. Sci. Res. 6 (3): 123-130.

Briggs CE, Haldane JBS (1925). A note on the kinetics of enzyme action. Biochem. J. 19: 338-339.

Chao W, Xue Y, Pei FW, Chun MC (2008). Interactive Influence of N and P on their uptake by four different hydrophytes. Afr. J. Biotechnol. 7: 3480-3486

Cozzolino V, Pigna M, Dimeo V, Caporale AG, Violante A (2010). Effects of arbuscular mycorrhizal inoculation and phosphorus supply on the growth of Lactuca sativa L. and arsenic and phosphorus availability in an arsenic polluted soil under non-sterile conditions. Appl. Soil Ecol. 45:262-268

Devos CHR, Schat H, Dewaal MAM (1991).Increased resistance to copper-induced damage of root cell plasmalemma in copper tolerant Silene cucubalus. Physiol. Plant 82: 523-528.

Francesconi K, Visoottivieth P, Sridokchan W (2002). Arsenic species in an arsenic hyper accumulating fern, Pityrogramma calomelanos: a potential phytoremediation of arsenic-contaminated soils. Sci. Total Environ. 284: 27-35.

Geng CN, Zhu YG, Hu Y, Williams P, Meharg AA (2006). Arsenate causes differential acute toxicity to two Pdeprived genotypes of rice

seedlings (*Oryza sativa* L.). Plant Soil 279:297–306
Glass ADM (1989). Plant nutrition: an introduction to current concepts. Boston, MA, USA: Jones and Bartlett Publishers. pp. 234.
Hoagland DR, Arnon DI (1950). The water-culture method for growing plants without soil. Calif. Agric. Exp. Stn. Circ. 347:1-32.
Huang B, Kuo S, Bembenek R (2005). Availability to Lettuce of Arsenic and Lead from Trace Element Fertilizers in Soil. Water Air Soil Poll. 164(1-4), 223-239.
Irtelli B, Navari-Izzo F (2008). Uptake kinetics of different arsenic species by Brassica carinata. Plant Soil 303:105-113.
Lineweaver H, Burk D (1934). The determination of enzyme dissociation constants. J. Am. Chem. Soc. 56: 658-666.
Liu WJ, Zhu YG, Smith FA (2004). Do iron plaque and genotypes affect arsenate uptake and translocation by rice seedlings (Oryza sativa L.) grown in solution culture? J. Exp. Bot. 55: 1707- 1713.
Mcfarlane KJ, Yanai RD (2006). Measuring nitrogen and phosphorus uptake by intact roots of mature Acer saccharum Marsh., Pinus resinosa Ait., and Picea abies (l.) Karst. Plant Soil 279: 163-172.
Ming L, Baiqing T, Min Z (2012). An arsenic-contaminated field trial to assess the uptake and translocation of arsenic by genotypes of rice. Environ. Geochem. Health DOI 10.1007/s10653-012-9501-z
Mohamed HHA , Andrew AM (2006). Arsenate, arsenite and dimethyl arsinic acid (DMA) uptake and tolerance in maize (Zea mays L.). Plant Soil 304:277–289
Mulabagal V, Mathieu N, Ajay N, Muraleedharan G N, Yanjun Z, Aditya LG (2010). In Vitro evaluation of red and green lettuce (Lactuca Sativa) for functional food properties. Food Chem. 118(2):300-306.
Norton GJ, Pinson SRM, Alexander J, Mckay S, Hansen H, Duan GL, Islam RM (2012). Variation in grain arsenic assessed in a diverse panel of rice (Oryza sativa) grown in multiple sites. New Phytol. 193(3):650-664
Poynton CY, Huang JWW, Blaylock MJ, Kochian LV, Elless MP (2004). Mechanisms of arsenic hyperaccumulation in Pteris species: root As influx and translocation. Planta 219: 1080-1088.
Shimono H, Bunce JA (2009). Acclimation of Nitrogen Uptake Capacity of Rice to Elevated Atmospheric CO2 Concentration. Ann. Bot. 103:87-94.
Sinha RK, Herat S, Tandon PK (2004). phytoremediation: role of plants in contaminated site management. in Book of Environmental Bioremediation Technologies, Springer, Berlin, Germany, 2004. pp. 315–330
Smith E, Juhasz AL, Weber J (2009). Arsenic uptake and speciation in vegetables grown under greenhouse conditions. Environ. Geochem. Health 31(1):125-132
Su YH, Mcgrath SP, Zhao FJ (2010). Rice in more efficient in arsenite uptake and translocation than wheat and barley. Plant Soil 328: 27-34
Wang J, Zhao FJ, Meharg AA, Raab A, Feldmann J, McGrath SP (2002). Mechanisms of arsenic hyperaccumulation in Pteris vittata. Uptake kinetics, interactions with phosphate, and arsenic speciation. Plant Physiol. 130: 1552- 1561.
Wei SH, Zhou QX, Wang X (2005). Identification of weed plants excluding the absorption of heavy metals. Environ. Int. 31: 829-834.
Weifeng S, Xiaoqing C, Ming Y, Qi D (2011). Absorption and oxidation of arsenite by Pteris vittata roots and its kinetics. Procedia Eng. 18: 72 – 77
Williams PN, Villada A, Deacon C (2007). Greatly enhanced arsenic shoot assimilation in rice leads to elevated grain levels compared to wheat & barley. Environ. Sci. Technol. 41: 6854-6859.
Xiaozhang Y, Puhua Z, Xishi Z, Yunda L (2005). Cyanide removal by Chinese vegetation--quantification of the Michaelis-Menten kinetics. Environ. Sci. Pollut. R. 12(4):221-6
Yu XZ, Gu JD (2006). Differences for different cultivars of maize during cyanide removal. Ecotoxicol. Environ. Safety, (Article in press).

An insight into environmental laws in Canada

S. C. Jhansi and Santosh Kumar Mishra

[1]Gujarat Research Society's P.N. Shah Centre for Lifelong Learning, Dr Madhuri Shah Campus, Ramkrishna Mission Road, Corner of 16th & 17th Road, Khar (W), Mumbai, Maharashtra, 400052, India.
[2]Population Education Resource Centre (PERC), Department of Continuing and Adult Education and Extension Work, S. N. D. T. Women's University, Patkar Hall Buil, First Floor, Room. Number: 03, New Marine Lines, Mumbai - 400020, Maharashtra, India.

Concern for the environment has been increasing around the world since the early 1980s. In the year 1987, the Brundtland report heightened awareness of the need for significant legislative and other changes. People in Canada are gradually becoming more aware of the urgent need to protect the environment. Canadians are involved with many projects to protect fragile ecosystems and stop further environmental destruction. Some projects are individual efforts and some are carried out through Non-Governmental Organizations (NGOs) such as World Wildlife Federation and Greenpeace. Other projects are initiated by the Government of Canada. Enforcement of Canadian environmental law can involve three stages: voluntary abatement; mandatory rectification; and, as a last resort, prosecution and penalties. The federal agencies and most provincial ministries have designated abatement and enforcement personnel. Written enforcement policies set out criteria with respect to how and when each of the three enforcement stages is to be applied. Federal enforcement policies include the compliance and enforcement policy for the Canadian Environmental Protection Act, 1999 and the compliance and enforcement policy for the Habitat Protection and Pollution Prevention Provisions of the Fisheries Act. Many, but not all, provincial regulators have similar policies, which can be obtained by searching their websites or requesting a copy in writing. This paper has been prepared to present an overview of environmental laws in Canada. In this manuscript, an effort has been made to: (a) discuss Canadian environmental legislation at various levels such as federal, provincial and municipal levels; and (b) give an insight into the trends in Canadian environmental law, and emergency response and government investigations.

Key words: Federal, provincial, municipal environmental law, Canada.

INTRODUCTION

'Environment' is a term that includes physical place; resources of land, water and air. In order to prevent the harmful consequences and effects of the global warming and climate change, there is need to protect the environment. In recent years, international society has become more concerned about environmental pollution. Growing awareness of environmental problems has resulted in more governmental regulations.

Environmental law addresses the relationship between humans and the physical environment, and is an area of law with both domestic and international implications. It is a complex and interlocking body of treaties, conventions, statutes, regulations, and common law that operates to regulate the interaction of humanity and the natural environment, toward the purpose of reducing the impacts of human activity. Environmental law draws from and is influenced by environmental principles, such as ecology, conservation, stewardship, responsibility, and sustainability.

Pollution control laws generally are intended (often with varying degrees of emphasis) to protect and preserve both the natural environment and human health.

Resource conservation and management laws generally balance (again, often with varying degrees of emphasis) the benefits of preservation and economic exploitation of resources. From an economic perspective, environmental laws may be understood as concerned with the prevention and preservation of common resources from exhaustion. The limitations and expenses that such laws may impose on commerce, and the often unquantifiable (non-monetized) benefit of environmental protection, have generated and continue to generate significant controversy.

Concern for the environment has been increasing around the world since the early 1980's. In 1987, the Brundtland report heightened awareness of the need for significant legislation and other changes. People in Canada are gradually becoming more aware of the urgent need to protect the environment. Canadians are involved with many projects around the world to protect fragile ecosystems and stop further environmental destruction. Rephrase (suggestion): some projects are initiated by the Government of Canada while others are carried out through individual efforts and Non-Governmental Organizations (NGOs) such as World Wildlife Federation and Greenpeace.

This paper presents an overview of environmental laws in Canada. It discusses Canadian environmental legislation at various levels, such as federal, provincial and municipal levels. This paper also gives an insight into the trends in Canadian environmental law, and emergency response and government investigations.

ENVIRONMENTAL LEGISLATION IN CANADA

Environmental law in Canada applies to businesses across virtually all sectors of the economy and all regions of the country. The federal and ten provincial governments, as well as the three territorial governments, are active in the creation and evolution of environmental law to meet the changing environmental challenges of the day, such as: climate change, toxic substances management, waste reduction, urban renewal through brown fields redevelopment, and the facilitation of environmental assessments of infrastructure and renewable energy projects.

Canadian environmental law will continue to evolve to keep pace with the times. Water management challenges, adaption to climate change, and the interconnection of environmental regulation and global trade, promise to be the issues of the immediate future. A large and diversified country, with a significant industrial base, a wealth of natural resources in mining, forestry and fisheries, large expanse of agricultural lands, coastal frontage on three oceans, arctic and subarctic territories, Canada is sensitive to virtually the full range of environmental issues facing the planet. Canadian environmental legislation includes: Canada National Parks Act, Canada Water Act, Canada Wildlife Act, Canadian Environmental Assessment Act, Canadian Environmental Protection Act, Department of the Environment Act, Environment Week Act, Fisheries Act, International Boundary Waters Treaty Act, International River Improvements Act, Lac Seul Conservation Act, Lake of the Woods Control Board Act, Manganese-Based Fuel Additives Act, Migratory Birds Convention Act, National Wildlife Week Act, Resources and Technical Surveys Act, Species at Risk Act, Weather Modification Information Act, and Wild Animal and Plant Protection and Regulation of International and Interprovincial Trade Act.

The legislative powers of the federal parliament and provincial legislatures are expressly defined by the Constitution Act, 1867; Canada's founding legal document. However, the power to legislate with respect to the environment is not expressly included in the powers granted to the federal parliament or the provincial legislatures. The federal government's express authority over criminal law, fisheries, shipping and navigation, interconnected undertakings, peace, order and good government, has been used to justify a wide variety of federal environmental legislation. The provincial governments have generally relied on their authority over property and civil rights, matters purely of a local nature, and local works and undertakings in justifying the passage of environmental legislation.

MANAGEMENT OF ENVIRONMENTAL LEGISLATION AND IMPLEMENTING AGENCIES

Environment Canada (EC) is the department of the Government of Canada with responsibility for coordinating environmental policies and programs as well as preserving and enhancing the natural environment and renewable resources. Its ministerial headquarters is located in les Terrasses de la Chaudière, Gatineau, Quebec. Under the Canadian Environmental Protection Act (R.S., 1999, c. 33), EC became the lead federal department to ensure the cleanup of hazardous waste and oil spills for which the government is responsible, and to provide technical assistance to other jurisdictions and the private sector as required. The department is also responsible for international environmental issues (e.g. Canada-USA air issues).

Under the constitution of Canada, responsibility for environmental management in Canada is a shared responsibility between the federal government and provincial/territorial governments. For example, provincial governments have primary authority for resource management including permitting industrial waste discharges (e.g. to the air). The federal government is responsible for the management of toxic substances in the country (e.g. benzene). EC provides stewardship of the environmental choice program, which provides consumers with an ecolabelling for products manufactured within Canada or services that meet international label standards of Global Ecolabelling Network. EC continues to undergo a structural transformation to centralize authority and decision-making, and standardize policy implementation.

ENVIRONMENTAL LEGISLATION AT VARIOUS LEVELS

Federal

Canadian Environmental Protection Act (CEPA)

The Canadian Environmental Protection Act, 1999 (CEPA), is the federal government's primary environmental regulatory statute. It provides for broad federal regulatory authority over the management and control of toxic substances and a range of other issues from environmental emergencies to the cross-border movement of wastes and recyclable materials. CEPA is a consolidation of a number of pre-existing federal environmental statutes.

CEPA provisions concerning toxic substances provided for a wide range of controls for any substance which is classified as '*toxic*' and listed in Schedule 1 of the Act. The minister of the environment has the authority under CEPA to obligate any person to provide samples, information, and data regarding a particular substance. Special procedures are in place regarding substances which are new to Canada. Additionally, CEPA imposes a duty to report and take remedial action on those who are in control of or own a spilled toxic substance. The same duty is imposed on persons who contribute to the initial release of a toxic substance. Authority to issue orders in the case of environmental emergencies is also included in the CEPA.

CEPA is supported by a variety of enforcement powers. A person in breach of CEPA may be given '*monetary penalties*' (or in certain cases 'imprisonment'). Alternative dispositions through environmental protection and alternative measures agreements may be possible. Officers and directors may be prosecuted if they authorize, assent to or acquiesce in the commission of an offence by a corporation. They may also be prosecuted if they fail to take all reasonable measures to ensure corporate compliance.

National Pollutant Release Inventory (NPRI)

CEPA also establishes the NPRI. Facilities which release one or more of the substances found in the NPRI substances list in an amount greater than the reporting thresholds are required to submit a detailed inventory of their emissions to EC. The data collected through NPRI reporting is made accessible to the public.

Fisheries Act

Federal authority regulations with respect to water quality and pollution are primarily found in the Fisheries Act. While the act as a whole is generally concerned with the protection of commercial and recreational fisheries, provisions of the act prohibiting the deposition of deleterious substances into water frequented by fish, and the harmful alteration or destruction of fish habitat are enforcement which, in effect, establish virtual zero tolerance thresholds for unapproved water discharges, and disruptive water works.

Dangerous goods

Both federal and provincial law exists with respect to the transportation of dangerous goods. At the federal level, the Transportation of Dangerous Goods Act, 1992 (TDGA) establishes a comprehensive regulatory regime; all of the provinces have adopted an identical regime with respect to intra-provincial transportation. Eight classes of goods are regulated ranging from explosives to dangerous organisms.

The TDGA regulates on a wide range of issues outlined in the Transportation of Dangerous Goods Regulations. The Act contains a full range of enforcement powers including provisions for officer and director liability. Environmental assessment at the federal level is governed by the Canadian Environmental Assessment Act (CEAA). To trigger an assessment under CEAA, a federal authority must: either be the proponent of the project, provide financing to the proponent, give federal lands to be used for the project, or issue a license or approval required for the project to be carried-out.

The legislation discussed earlier includes several of the most prominent federal environmental statutes. However, it should be noted that there are several other federal statutes which contribute to the body of Canadian federal environmental law, such as: the Pest Control Products Act, the Navigable Waters Protection Act, the Marine Liability Act, and the Canada Shipping Act, 2001.

Provincial

The provincial legislatures are empowered by the Constitution Act, 1867 to legislate with respect to a very wide range of environmental issues. In general, each province has developed a complex web of regulations and statutes which address matters such as: waste management, waste disposal, air pollution, water pollution, fuel handling, contaminated site remediation, and environmental assessment.

The provinces each typically have a central environmental conservation and protection statute which establishes both general and specific prohibitions with respect to discharges of contaminants and the disposal of waste and provides for a means of approval by the administrative branch of government. Examples of these statutes include: Ontario's Environmental Protection Act, British Columbia's Environmental Management Act, Alberta's Environmental Protection and Enhancement Act, and Quebec's Environmental Quality Act.

Spills and clean-up

Spills of pollutants are generally given special treatment as a particular class of contaminant discharge. Under the provincial environmental protection statutes, persons who own or are in control of a pollutant at the time of a spill may be subject to a range of obligations regarding notification and the remediation of adverse impacts, regardless of whether they are at fault. Failure to respond adequately may result in an order to remediate and prosecution.

The provincial environmental protection statutes may also contain special provisions regarding the clean-up, use and control of contaminated sites. For instance, Ontario's Record of Site Condition Regulation sets out detailed requirements pertaining to the investigation and documentation of contaminated site conditions. Most importantly, in certain circumstances, there is a requirement for the filing of a Record of Site Condition (RSC). The regulations also specify who is qualified to certify a RSC, and what information must be included.

Waste management

Waste management is also generally regulated under provincial environmental protection statutes. Approval may be required for the operation, alteration or construction of a waste management system or disposal site. Applications for approval will need to disclose plans and specifications of the proposed site or system. In some situations, a hearing by an administrative tribunal may be required prior to approval. Regulations regarding waste management often include specific standards which must be maintained, reporting requirements and fees payable to the province. Provincial environmental protection statutes usually contain provisions which authorize the imposition of fines or other punishment for breaches of the act.

Environmental assessment statutes

The provinces have generally also enacted environmenttal assessment (EA) statutes or regulations. Examples include: Ontario's Environmental Assessment Act, British Columbia's Environmental Assessment Act, and Alberta's Environmental Assessment Regulations.

Typically, EA legislation applies to projects proposed by the province or provincial agencies, municipalities and private sectors where they are designated by the regulations as being subject to an EA. Ontario has also recently provided for more streamlined EA approvals for green energy projects through a wide range of amendments to environmental legislation as provided for under the Green Energy Act, 2009.

Generally speaking, proponents of projects are required to submit both proposed terms of reference and an EA for approval) in accordance with the approved terms of reference. The ministry administering the process will review the EA and coordinate with the proponent in an effort to have any outstanding issues resolved. Once reviewed, the public will be notified and given an opportunity to comment on the proposed project and EA. Final decisions regarding a project are often reviewable by an administrative tribunal.

Canada's three territories (the Northwest Territories, Yukon and Nunavut) do not have inherent jurisdiction entrenched in the Constitution Act, 1867 as do the provinces. The jurisdictions of the territories are wholly defined by the power granted to them by the federal government through enabling statutes.

Environmental laws in the territories generally are reflective of federal environmental law. Territorial specific environmental legislation is a relatively recent occurrence. The enabling statutes for the territories, such as the Yukon Act, contain provisions which expressly provide for the paramount of federal law in cases of conflict with territorial law. To the extent that territorial environmental law is similar to that of the provinces, it can generally be said that territorial regulation is less complex and extensive.

Municipal

Municipalities across Canada have used the law-making authority granted to them by the provinces to enact a variety of environmentally related by-laws. An example is the "City of Toronto's former Pesticide By-law" which banned the use of pesticide products for purely cosmetic purposes. The By-law was recently displaced by Ontario's Pesticides Act to the same effect. Most recently, the city of Toronto has enacted the "Environ-mental Reporting and Disclosure By-Law" which requires businesses to report: (1) their annual usage, production; and (2) release of 25 listed toxic chemicals where the quantity of the chemical exceeds a reporting threshold.

ENFORCEMENT

Environment Canada Enforcement Branch is responsible for ensuring compliance with several federal statutes. The Governor-in-Council appoints enforcement officers and pursuant to section 217(3) of the Canadian Environmental Protection Act, enforcement officers have all the powers of peace officers. There are two designations of enforcement officers: Environmental Enforcement and Wildlife Enforcement.

The former (Environmental Enforcement) administers the Canadian Environmental Protection Act and pollution provisions of the Fisheries Act and corresponding regulations. The latter (Wildlife Enforcement), on the other hand, enforces: Migratory Birds Convention Act, Canada Wildlife Act, Species at Risk Act, and The Wild Animal and Plant Protection and Regulation of International and Interprovincial Trade Act.

All officers wear dark green uniform with black ties and a badge (appear on the right). Environmental enforcement officers only carry baton whereas wildlife enforcement officers are also equipped with firearm. The minister may also:

1) appoint members of the Royal Canadian Mounted Police, fishery officers, parks officers, customs officers and conservation officers of provincial and territorial governments as enforcement officers; and
2) allow them to exercise the powers and privilege of Environment Canada officers.

On March 4, 2009, a bill to increase the enforcement capabilities of EC was introduced into the house of commons. The Environmental Enforcement Bill has the provision to: increase the fines for individuals and corporations for serious offenses, give enforcement officers new powers to investigate cases, and grant courts new sentencing authorities that ensure that penalties reflect the seriousness of the pollution and wildlife offences.

TRENDS IN CANADIAN ENVIRONMENTAL LAW

Environmental legislation in Canada is continuing to grow and evolve. Recently, there has been an effort to increase the use of administrative penalties, the imposition of fines for the commission of environmental offences on an absolute liability basis. This trend can be seen at both provincial and federal levels. Examples include Ontario's passage of the Environmental Enforcement Statute Law Amendment Act and federal Environmental Violations Administrative Monetary Penalties Act. Further government initiatives are expected to reduce the use and creation of toxic substances. For example, Ontario's Toxics Reduction Act, 2009 promotes: voluntary reductions by requiring reduction plans; voluntary reduction targets, and toxic substance use; and creation and reduction disclosure obligations to the public, the workforce and the government.

CANADIAN ENVIRONMENTAL ASSESSMENT ACT AND REGULATIONS

The Canadian Environmental Assessment Act (the Act) is the legal basis for the federal environmental assessment process. Regulations under the Act are used to determine when the Act applies and what type of environmental assessment is required, and to prescribe procedures for government departments and agencies to coordinate the environmental assessment process.

Regulations under the Canadian Environmental Assessment Act are mentioned as follows: Canada Port Authority Environmental Assessment Regulations (Comprehensive Study List Regulations; Crown Corporations Involved in the Provision of Commercial Loans Environmental Assessment Regulations; Establishing Timelines for Comprehensive Studies Regulations; Exclusion List Regulations, 2007; Federal Authorities Regulations; Inclusion List Regulations; Law List Regulations; Projects outside Canada Environmental Assessment Regulations; and Regulations Respecting the Coordination by Federal Authorities of Environmental) and Assessment Procedures and Requirements.

EMERGENCY RESPONSE AND GOVERNMENT INVESTIGATIONS

The Canadian environmental legislation provides that when an environmental mishap occurs, one must be prepared to respond in terms of controlling the escape of a contaminant or pollutant, obtaining the assistance of environmental specialists, and notifying government agencies.

One must also be prepared for the government investigation which will likely follow in an attempt to determine the degree of negligence involved in the occurrence. Depending on the outcome of such an investigation, regulatory charges of a criminal nature or civil claims for compensation may be brought against the corporation and/or individual employees. Brief description of the emergency responses as well as the government investigations involved under various settings is presented in the following.

Notification and clean-up

Any time there occurs a discharge of a contaminant into the natural environment "out of the normal course of events"; the Ontario Ministry of Environment must be notified 'forthwith'. If the discharge is "abnormal in quality or quantity in light of all the circumstances", thereby con-

constituting a 'spill' under the Ontario Environmental Protection Act, or involves a specific substance regulated by such federal legislation as the Transportation and Dangerous Goods Act and the Canadian Environmental Protection Act, notification of the local police, the local municipality, the federal Departments of Transport or Environment, Canadian Transport Emergency Centre (CANUTEC), and the Canadian Coastguard will also be required. The CANUTEC is operated by the Transportation of Dangerous Goods (TDG) Directorate of Transport Canada.

These statutes also require that the spill of a pollutant or release of a dangerous good or toxic substance, be cleaned up immediately. For example, the Ontario Environmental Protection Act states as follows:

"*The owner of a pollutant and the person having control of a pollutant that is spilled and that causes or is likely to cause an adverse effect shall forthwith do everything practicable to prevent, eliminate and ameliorate the adverse effect and to restore the natural environment*".

While the very occurrence of an environmental mishap may involve a regulatory offence, further offences will be committed if a spilled substance is not controlled, cleaned up and reported to the appropriate government authorities. In addition, the risk of personal injuries which can lead to civil claims will be increased if the company is slow to respond to the mishap. It is thereby imperative that the company have in place an emergency response plan which anticipates possible mishaps and the avenues by which spilled materials may escape the property. Other provisions are as follows:

1) Company personnel must be trained and equipped for responding to spills;
2) The names, addresses and telephone numbers of environmental specialists capable of assisting in the cleanup of the spill must be kept on file; and
3) Procedures for reporting environmental mishaps, both internally and to government agencies, must be established and communicated to all employees.

Getting the facts

While little time can be wasted before notifying government authorities of a significant occurrence, management must obtain, to the extent possible, all relevant facts relating to the event and the company's response thereto. If the event is in any way likely to give rise to regulatory charges or civil litigation, management should involve legal counsel as early as possible. In this way the making of internal reports describing the event and any internal investigation related thereto, can benefit from what is known as "solicitor-client privilege" and thereby, not be subject to government seizure. At a minimum, someone in the corporation should be designated to take charge of responding to an environmental mishap and handle all inquiries or requests for information from government agencies and other third parties. The sooner the company gets all the facts, the sooner it can make a well informed decision on how it wishes to respond to government agencies and the world at large.

Preparing for government investigations

It should come as no surprise that even run-of-the-mill environmental occurrences may become the subject of a criminal investigation. Environmental authorities at every level of government in Canada have been given a strong mandate to take tough action to enforce and be seen to enforce, environmental laws. In Ontario, the Ministry of Environment has two types of provincial officers who will generally have direct dealings with the private sector. They are known as: "Abatement officers" and "Investigation and Enforcement officers".

The former officers deal with relatively routine concerns such as responding to public complaints and carrying out periodic inspections of activities within their jurisdiction known to be environmentally sensitive. The Investigation and Enforcement officers are only engaged once the abatement officer decides that a responsible person or company is failing to take its environmental obligations seriously or that the circumstances of a particular event require criminal sanction. Once the investigation and enforcement officer is involved, it should be assumed that: (a) court proceedings will likely result, and (b) anything and everything stated to the officer can and will be used to secure a conviction against the company or individual employees. However, even statements made to an abatement officer at the initial stage of the government's involvement, may be used in a prosecution.

If at all possible, all dealings with government authorities should be channeled through a single person, in order to ensure that investigators do not go on a "fishing expedition" or speak to poorly informed employees. However, while every effort possible must be exercised to manage the situation and control the trans-mission of information within and beyond the company, one cannot hinder or obstruct a provincial officer in the lawful performance of his or her duties. This in itself constitutes an offence. Provincial officers acting under the Environmental Protection Act have very wide powers to: (a) inspect, (b) take copies of relevant documents, and (c) take samples and ask questions.

Once it becomes clear that a regulatory offence may have been committed, individual employees may choose not to provide information to the provincial officers, for fear that they may incriminate themselves. However, this is an individual decision and management cannot, in any way, be seen to coerce or intimidate an employee, because the employee may seek the enforcement of provincial environmental law. Unfortunately, a corporation does not have a right against self-incrimination.

As already indicated, management should involve legal counsel as early as possible in its investigation of the relevant events and in its response to any government investigation. Information regarding the company's environmental control or preventive maintenance systems must be identified and preserved in anticipation of formulating a due diligence defense. Such information may be given to the investigators early on, in order to avoid the commencement of a prosecution. However, this decision will inevitably require the benefit of legal counsel. It will also be important to monitor oral information and documentation provided by employees to the government investigators. Accordingly, employees should be trained in advance to notify management if and when they are contacted by government investigators. Copies of any written statements prepared by investigators must be obtained.

Given the fact that any information provided to the government authorities may be used against the company and individual employees, all employees must be advised of the seriousness of an investigation. The company itself cannot prevent its employees from cooperating with investigators. However, by stressing the fact that an individual can and will be prosecuted, it can be expected that the employees will think twice before dealing carelessly with investigators. Employees should be told that if they choose to cooperate, they are only obligated to provide information relevant to an environmental matter under investigation and they should only answer the questions. They should be discouraged from volunteering information; particularly, with respect to matters they do not have direct knowledge. They should never provide hearsay information. They are not obligated to sign a written statement and they should explicitly reserve the right or opportunity to revise or clarify information given to an investigating officer. If at any time during a government investigation, management wishes to seek legal advice, it should not hesitate to request an opportunity to do so before allowing the investigators access to the plant and its employees. Management should also decide to what extent it wishes to make available independent legal counsel to its employees.

Solicitor-client privilege

An important protection against the disclosure of company documents during a government investigation is the privilege respecting confidential communications between a solicitor and his or her client. This privilege operates at all times, subject to waiver by the client. In order to effectively assert this privilege, one must understand which documents are affected by it, have in place a system for segregating and controlling such documents and assert the privilege as early as possible.

Professional communications of a confidential nature between a solicitor and his or her client, which take place with the purpose of obtaining legal advice, are privileged. It is not necessary that the communications should be made either during or relating to an actual, or even an expected legal proceeding. It is sufficient if they pass as professional communications in a professional capacity, subject to one or two specific exceptions, such as communications between solicitor and client that had been made to facilitate the perpetration of a crime or fraud.

Information generated within the company should be directed to legal counsel and marked as being "Legal and confidential" and for the purposes of providing management with legal advice. Internal reports should be made only to legal counsel, with copies to a very select group within management. Such information must not be disclosed to third parties or in any way treated in a "non-confidential manner". Otherwise, the privilege may be lost.

A separate file should be created containing all such legal communications. In the event, the investigator should request a review of company files, any legal documentation should be sent immediately to legal counsel or withheld until the request has been discussed with legal counsel. If required, a legal procedure exists for dealing with any challenge to the company's position that the documentation is in fact privileged.

The protection of third party communications to a solicitor, made by an agent instructed by the client to communicate with the solicitor, is particularly important in environmental matters. Very often the specific source of a contaminant entering into the environment is unknown and environmental specialists may be retained to provide advice to the company and its legal counsel. Accordingly, where the company wishes to obtain such information or simply conduct a review or audit of its operations for the purpose of insuring compliance with environmental laws, it is recommended that (a) the consultants be retained by legal counsel and (b) final advice to the company be made through its solicitors.

Whether any written reports based on such a review or audit will in fact be protected by the privilege, will likely depend on the existence of circumstances which support the need to provide legal advice. One recent case indicates that as long as the dominant purpose of the specialists' or consultants' advice is to provide for the giving of legal advice, the privilege will be sustained (http://www.blakes.com/pdf/EnvLawOntCan.pdf, accessed on May 14, 2012). Naturally, the fact that a legal proceeding is imminent will strengthen the claim that the purpose of the communication is to provide for the giving of legal advice.

CONCLUSION

Under Canada's constitution, environmental regulation is

shared between the federal government and the 10 provinces/3 territories (referred to collectively as 'provinces'). Where there is a direct conflict between federal and provincial environmental statutes in relation to the same matter, federal law prevails, but such conflicts are rare, and overlapping requirements are common. Municipalities also play a growing role. Federal environmental laws are based on federal constitutional powers such as: international borders, international relations, trade and commerce, navigation and shipping, seacoasts and fisheries, and criminal law.

The federal government also has primary jurisdiction over federal works and undertakings, such as the land and activities of the federal government, its agencies and corporations, the armed forces, and a variety of federally regulated entities such as the railways, aviation, interprovincial transport, grain elevators, etc.

The principal federal statutes are: Canadian Environmental Protection Act, 1999 (CEPA), Fisheries Act, Canadian Environmental Assessment Act (CEAA), Species at Risk Act (SARA), Transportation of Dangerous Goods Act, 1992, Canada Shipping Act, Hazardous Products Act, and Pest Control Products Act.

Provincial environmental laws are based on provincial constitutional powers over municipalities, local works and undertakings, property and civil rights, provincially owned (public) lands and natural resources. In inland areas, most environmental laws that affect private activity are provincial. Each province has its own environmental statutes. For example, Ontario statutes include: Environmental Bill of Rights, Environmental Protection Act (EPA), Ontario Water Resources Act (OWRA), Clean Water Act, Environmental Assessment Act (EAA), Pesticides Act, Safe Drinking Water Act, Nutrient Management Act, Green Energy Act, and Toxics Reduction Act.

Acceptable activities, standards and thresholds vary between jurisdictions. As well, many sectoral laws (that is, those relating to a resource or industry) include substantial environmental requirements. In addition, Canada has a vibrant system of common law and administrative law, which often have a marked effect on environmental law.

REFERENCES

Canadian Environmental Law Research Foundation (1984). Preliminary Analysis of a Federal Environmental Bill of Rights. Toronto.

http://en.wikipedia.org/wiki/Environment_Canada, Accessed on December 23, 2011.

http://www.acc.com/legalresources/quickcounsel/elic_rjs.cfm, Accessed on December 22, 2011.

http://www.blakes.com/pdf/EnvLawOntCan.pdf, Accessed on May 13, 2012.

http://www.iclg.co.uk/index.php?area=4&country_results=1&kh_publications_id=187&chapters_id=4404, Accessed on December 24, 2011.

http://www.nrcan.gc.ca/minerals-metals/business-market/recycling/3313, Accessed on December 25, 2011.

http://www.saskschools.ca/~kidddlaw/environmental/lesson1/1a.html, Accessed on December 23, 2011.

http://www.tc2.ca/pdf/H3_Environmental.pdf, Accessed on December 23, 2011.

World Commission of Environment and Development (1987). Our Common Future (The Brundtland Report), Oxford University Press, London.

5

Positions of human dwellings affect few tropical diseases near Baghdad city, Iraq

Shihab A. SALMAN[1], Mohammed J. AL-OBAIDI[2] and Fadhaa O. SAMEER[2]

[1]The Central Environmental Laboratory, College of Science, University of Baghdad, Iraq.
[2]Biological Researches Unit for Tropical Diseases, College of Science, University of Baghdad, Iraq.

Some factors that possibly affect tropical disease distribution was investigated in about 500 randomize human dwellings. The studied factors include wild animals, domestic animals, wild plants, cultivated plants, nature of soil, nature of water, positions of human dwellings, nature of building material and position of animal shelters. The diseases that studied were malaria, Kala-azar (visceral leishmaniasis), cholera, Bilharzias (urinary schistosomiasis), diarrhea, typhoid fever, hydatid cysts, dysentery, jaundice and tuberculosis (TB). The study was conducted in the ruler area near Baghdad city in Taji, Abou-Greeb, Rhadwaniya, Rasheed, Youssifiya, Latifiya, Mahmmodiya, Madaien and Nahrwann districts. It appears that the studied factors have direct and indirect effects on distribution of the diseases.

Key words: Position, human dwellings, animal shelters, Baghdad.

INTRODUCTION

The top-geographical feature of Iraq is very different from the north to the south. The main fertile alluvial plain is the central region with elevation of about 30 to 360 m (above sea level). The northeast region is mountain, the west and southwest regions are deserts and the southern region is a marshy area (WHO, 2003).

Iraqi fauna is influenced by five geographical regions (Palearctic, Mediterranean, Central Asia, Eastern and Ethiopian) due to the geographical situation of this country. Therefore, the fauna and flora are very different from the north to the south of the country (Al-Barwari and Saeed, 2007). Vegetation of Iraq is also different from the north to the south; however, the central and south regions are well-known with date palm trees. Cultivated plants are mainly wheat, berry, corn and vegetables (Tengberg, 2011; Shabani et al., 2012).

Poultry farm have increased in the last decade. The estimation indicates that the number of cattle increased as well. Man activities are mainly limited to agriculture. Because of problems with sanctions from 1990, the number of rodents has highly increased (Cortes et al., 2012).

Human dwellings are made from cement, mud and brick. Those for animals are mainly made of mud. Since many centuries ago, people in Iraq benefited from their environment because of the sun energy in winter and the wind direction in summer, thus the traditional direction of villages in Iraq is toward the southeast (Wisborg et al., 2008). In addition, the mud and plants branches were widely used in the construction of both human dwellings and animal shelters. Now, villagers choose their dwellings with regards to different factors such as exposure to sun, wind direction, roads, canals, trees and other factors (Salman and Mohammed, 2010).

Due to the problem of sanctions, high number of citizens are exposed to different illnesses and risks including the endemic diseases, especially in the rural area due to deficit in food, sanitation and the decries of infrastructures capability of Iraq (Hamid and Everett, 2007). Malaria for example reappears in different localities in the north and in the southeast of Iraq due to the deficit of preliminary control programs as mosquito control

Table 1. Relationship between directions of dwellings and total diseases in study area.

Direction	House building		Total diseases	
	No.	Percentage	No.	Percentage
North	54	10.8	48	13.8
North-east	26	5.2	35	10
East	145	28.9	16	4.5
South-east	4	0.8	23	6.5
South	54	10.8	28	8
South-west	95	19	52	14.9
West	118	23.7	84	24
North-west	4	0.8	64	18.3
Total	500	100	350	100

The table shows the total percentage of the reported cases of illness in the dwelling in the north-west, south-west and north directions, respectively.

Table 2. Relationship between positions of animal shelters and total diseases in the study area.

Direction	Animal shelters		Total diseases	
	No.	Percentage	No.	Percentage
North	173	34.6	35	10
North-East	8	1.6	14	4
East	133	26.6	68	19.4
South-East	30	6	79	22.5
South	76	15.2	36	10.3
South-West	17	3.4	14	4
West	55	11	47	13.5
North-west	8	1.6	57	16.3
Total	500	100	350	100

The table shows the highest percentage of diseases in the positions of the south-east, east and north-west directions, respectively.

control in the marshy area which is considered as the main endemic area with malaria.

Number of visceral leishmaniasis patients also increased, diarrhea and enteric malady was remarkably high due to the bad water quality (Zolnikov, 2013).

Climate of Iraq is so hot and dry in summer and cold and rainy in winter. The variation in temperature between day and night is about 20°C in summer (Barreto et al., 2011). Thus, the aim of this study was to determine any possible relationships between ecological factors, endemic diseases and the position of villagers dwelling in rural area near Baghdad.

MATERIALS AND METHODS

The study area

About 500 human dwelling were randomly selected from the following area around Baghdad:

1. Taji (about 20 km in north of Baghdad).
2. Abou Greeb (about 25 km north-west of Baghdad).
3. AL Rhadwaniya (about 20 km westward of Baghdad)
4. AL Youssifiya (about 20 km southward of Baghdad).
5. AL Latifiya (about 35 km south ward of Baghdad).
6. AL Mahmmodiya (about 25 km southward of Baghdad).
7. AL Rasheed (about 25 km southward of Baghdad).
8. AL Madaien (about 25 km south-east of Baghdad).
9. AL Nahrwann (about 29 km eastward of Bagdad)

A standard forms was completed for each dwelling. The form included:

a. Cultivated and wild plants.
b. Domestic and wild animals.
c. Nature of soil.
d. Source of water.
e. Types of diseases in study area.
f. Tower of houses at East or West.
g. Positions of animal shelters to the houses.

Statistic analysis

Chi-square analysis was performed to limit the relationship between study parameters and prevalence of diseases. The software statistical program used in this study was SPSS version 11.

RESULTS

The examination of about 500 human dwelling was studied in order to find possible relationship between different ecological factors and the prevalence of endemic diseases in the study localities in Iraq.

The houses building toward to the east and west (28.9 and 23.7%, respectively) were more than toward the north and south (both 10.8%). The highest rate of houses was building toward the east (24%) while the lowest rate of these was building toward the south-east and north-west (both 0.8%). Generally, the main examinations of dwellings were as follows:

North, 10.8%; north-east, 5.2%; east, 28.9%; south-east, 0.8%; south, 10.8%; south-west, 19%; west, 23.7%; north-west, 0.8% (Table 1).

The villager's dwellings in the study area are divided into three main categories as follows:

1. Cement made dwellings.
2. Brick made dwellings.
3. Mud made dwellings (Table 2).

Villagers live in the study area in grouped or separated dwellings. Now, these villages are rear while 85% of dwellings were built in respect to the directions of the highway and network of roads and irrigation canals. The study showed relationship between direction of house building and the diseases. The houses with building tower to the west recorded the highest rate of diseases (24%).

Table 3. Classification of dwellings in the study area.

Type	Number	Orientation	Open-land	Orchard	Vegetation
Cement	374	ALL	238	337	Cultivated and wild
Brick	81	ALL	137	93	Cultivated and wild
Mud	45	ALL	125	70	Cultivated and wild
Total	500	-	500	500	-

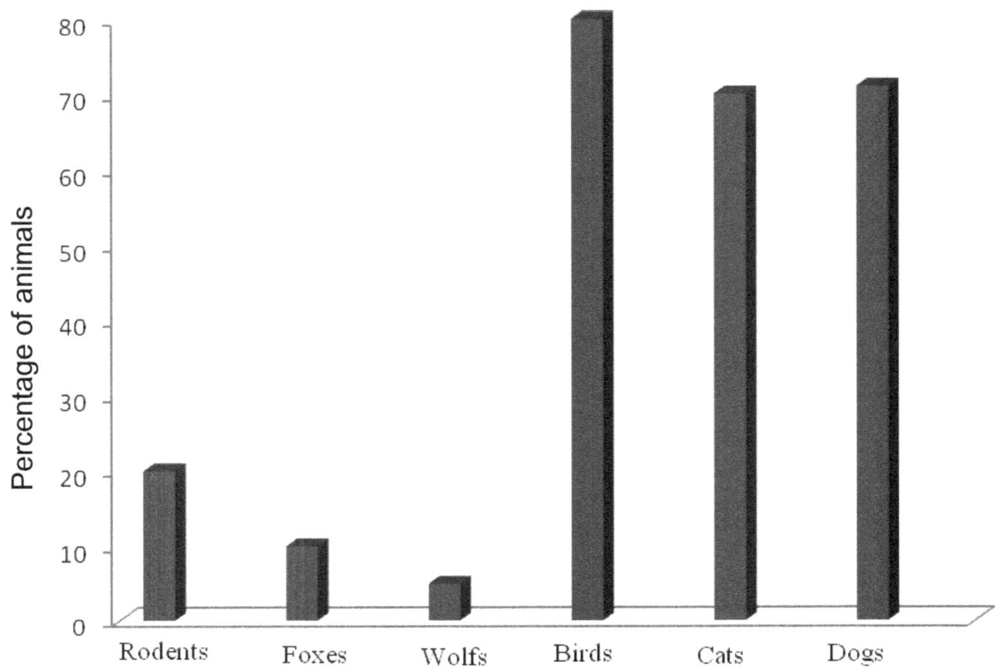

Figure 1. Common animals in study area.

The study proves that the place of animal shelters to the house was in correlation with the diseases. The highest rate of shelters was building with tower to the north which was 34.6%. The highest rate of diseases (22.5%) was recorded in the shelters with building tower to south-east. Generally, the main examinations of animal shelters were as follows:

North, 34.6%; north-east, 1.6%; east, 26.6%; south-east, 6%; south, 15.2%; south-west, 3.4%; west 11%; north-west, 1.6% (Table 3).

All the shelters of animals were mainly mud made. The ruler of the area near Baghdad have a big variety of animals as rodents, cats, dogs, foxes, wolves and birds which were the highest density in the study stations (Figure 1).

The total number of disease recorded in the study area was 350 cases. High number of children suffered from diarrhea (120 cases). Cases of visceral leishmaniasis (VL) recorded in the study area were six cases. Only three cases of malaria were report. Typhoid fever (50 cases) and Brucellosis (7 cases) were common. The reported cases of Echinococcosis were few. No case of Schistosomiasis was reports (Figure 2).

The common diseases reported during the study were significantly different according to the value of Chi-square (541.88) under probability level ($p > 0.05$).

The highest rate of the diseases was in Al-Mahmmodiya station (85 cases) while the station of Al-Taji had lowest rate of diseases (20 cases) (Figure 3). More than 90% of the study area soil was good type but more than 75% of the study area drinking water was the bad type. The rate of diseases was related to the nature of soil and source of drinking water (Figure 4).

DISCUSSION

The traditional Iraqi village in the rural area has building

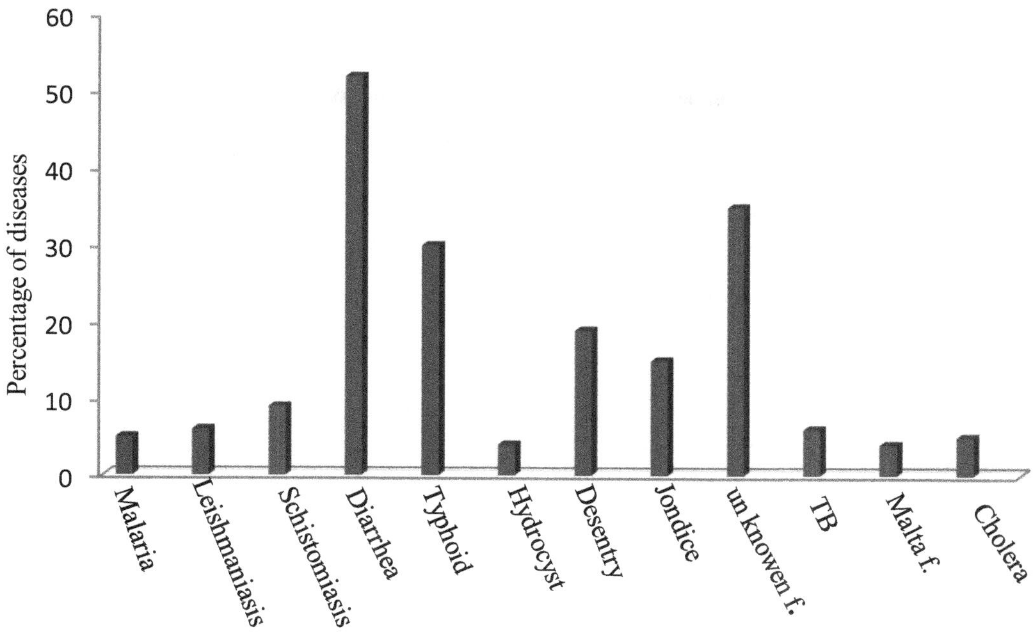

Figure 2. Common diseases in the study area.

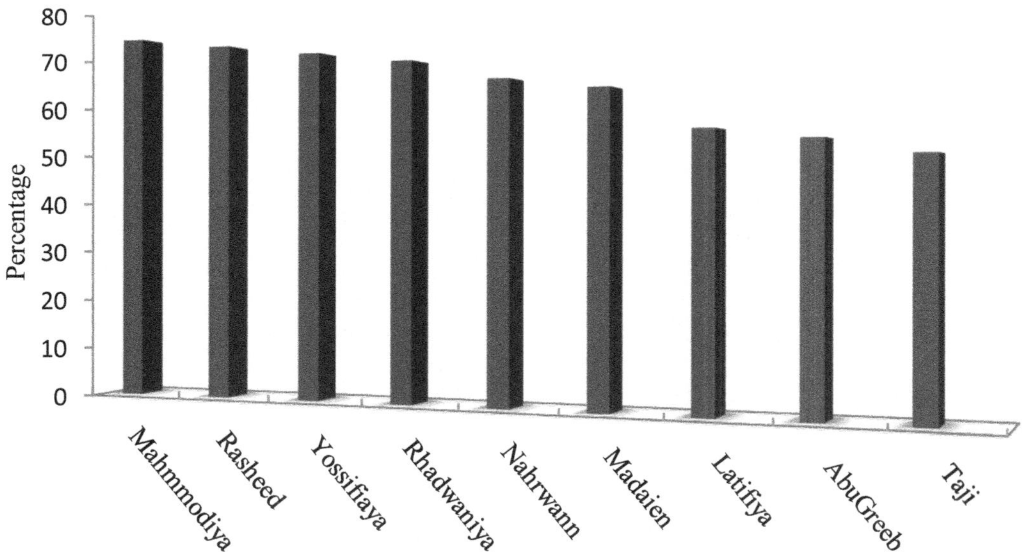

Figure 3. Percentage of diseases in each station of the study area.

in respect to many traditional considerations including the direction of the wind. The main wind direction in Iraq is usually northwest, thus a very high number of villagers consider this effect in the construction of their dwellings. Many villagers were building their dwellings for a religion reason, in this case, the direction of dwellings were toward the saint place of Muslims AL-QAABA. Others choose the dwellings in the orchards of date palms or near the orchard. The relationship between house building and diseases is due to the sterling of sun on the house built toward the east, the sunrays make kill some of the germs and insect larvae.

The animal shelters were a few meters from the human dwellings; this is due probably to the security. The animal's shelters play a very important role as good places for feeding and breeding for insects such as sand flies and flies (Bray and Hamilton, 2013). Dick of irrigation canals, date palms trees and farms are very suitable

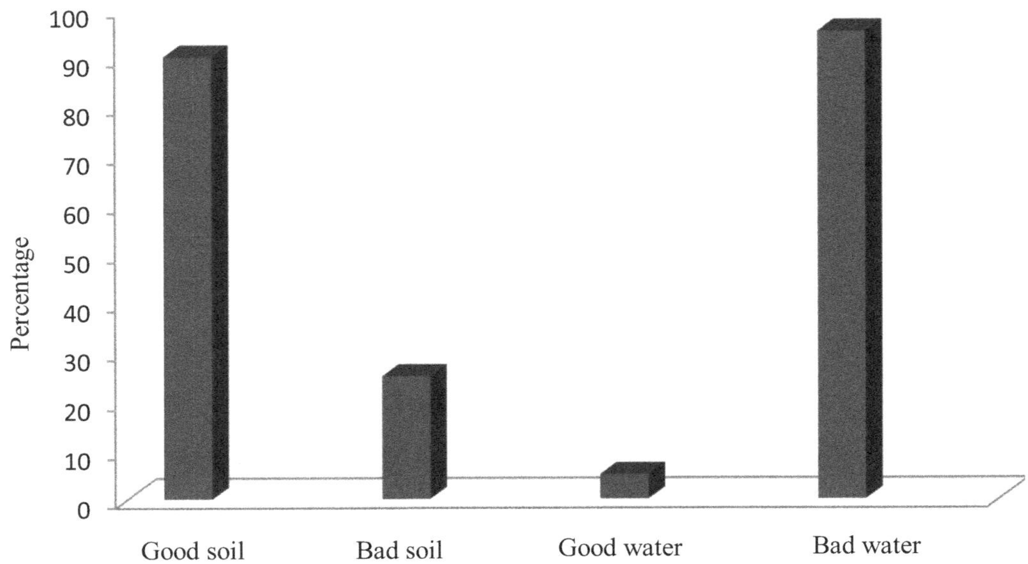

Figure 4. Sources of water and nature of soil in the study area.

places for rodents (Tavares and da Fonseca, 2013; Daniel, 2012). There is an association between socio-economic factors and dwelling without regarding the climate or something that affect the incidence of the reported cases.

The constructions of dwellings were changed. Thus, the cement replaced mud which may affect the ecological aspect of vector diseases. Style and fusion of dwellings were change during the last two decades as the use of new mode of design without considering the environment and the daily exigency of villagers and their domestic animals. The position of animal shelters was a determinant factor because the construction of mud consists of a suitable place for both feeding and resting for the sandflies which is the vector of visceral leishmaniasis.

It seems that the effect of microclimate was not on the considerations of villagers to build their dwellings as the very dense of date palms trees, vegetable farms and irrigation canals which are a determinant factors in the prevalence of diseases. Some dwellings had been constructed in respect to the wind direction. No significant mean was noted for the distance from dwelling and drainage canals except rodent barrows which are suspected as the intermediate host of both cutaneous and visceral leishmaniasis disease in Iraq.

The study area is a suitable place for the breeding and feeding of rodents and according to villagers, their number was highly increased in the last decade. Use of manure as a source of energy is a common habit in the study area (Tauseef et al., 2013). The sanction was a very important factor in the degradation of ecological factors, which directly affect the prevalence of many endemic diseases leading to the degradation of both sanitation services and environmental conditions (Al-Hilfi et al., 2013). It seems clearly that this kind of study is very important to find alternative for more healthy dwellings.

Recommendation

An epidemiological study needs a high number of samples before definitive conclusion and we hope to achieve that in the future. The study may be oriented toward this vital problem and we suggest that the cooperation between different official authorities, universities and international organization as WHO and UNICEF can be of help to investigate this big problem in Iraq in order to reduce the high number of reported cases of the main endemic disease in the rural area in the country.

REFERENCES

Al-Barwari SE, Saeed I (2007) On the helminthic fauna of some Iraqi reptiles'.Turkey Parazitol Derg; 31(4):330-336.

Al-Hilfi TK, Lafta R, Burnham G (2013) Health services in Iraq. Lancet 16; 381(9870):939-948.

Barreto ML, Teixeira MG, Bastos FI, Ximenes RA, Barata RB, Rodrigues LC (2011) Successes and failures in the control of infectious diseases in Brazil: social and environmental context, policies, interventions, and research needs. Lancet 28:377(9780):1877-1889.

Bray DP, Hamilton JG (2013) Insecticide-impregnated netting as a potential tool for long-lasting control of the leishmaniasis vector Lutzomyia longipalpis in animal shelters. Paras. Vectors 4:6(1):133.

Cortes S, Vaz Y, Neves R, Maia C, Cardoso L, Campino L (2012) Risk factors for canine leishmaniasis in an endemic Mediterranean region. Vet. Parasitol. 189(2-4):189-196.

Daniel T C (2012) Contributions of cultural services to the ecosystem services agenda. Proc. Natl. Acad. Sci. USA 109:8812-8819.

Hamid HI, Everett A (2007) Developing Iraq's mental health policy. Psychiatr Serv. 58(10):1355-7.
Salman SA, Mohammed HA (2010) Field Study of Ecological Factors Influencing Visceral Leishmaniasis Foci. Ibn Al-Haitham, J. for Pure and Appl. Sci.Vol.23, No (1).
Shabani F, Kumar L, Taylor S (2012) Climate change impacts on the future distribution of date palms: a modeling exercise using climex. PLoS 7(10):e48021.
Tauseef SM, Premalatha M, Abbasi T, Abbasi SA (2013) Methane capture from livestock manure. J Environ. Manage 15;117:187-207.
Tavares AL, Da Fonseca IP. (2013) Natural infection of synathropic rodent species Mus musculus and Rattus norvegicus by Leishmania infantum in Sesimbra and Sintra - Portugal. Parasit Vectors 8;6:88.
Tengberg M (2011) Beginnings and early history of date palm garden cultivation in the Middle East. J. Arid Environ. 5:1-9.
WHO (2003) Communicable diseases toolkit Iraq crisis .Health survey sample forms.
Wisborg T, Murad MK, Edvardsen O, Brinchmann BS (2008) Life or death the social impact of paramedics and first responders in landmine-infested villages in northern Iraq. Rural Remote Health 8(1):816.
Zolnikov TR (2013). The maladies of water and war; addressing poor water quality in Iraq. J. Public Health. 103(6):980-7.

Effect of rainfall on cropping pattern in mid Himalayan region

R. K. Aggarwal

Department of Environmental Science, Dr Y S Parmar University of Horticulture and Forestry, Nauni (Solan) 173 230 India.

The analysis of effect of rainfall during the last 20 years is needed to evaluate cropping pattern in the rain-fed region. In this study, trends in annual, seasonal and monthly rainfall of district of Himachal Pradesh in India over the past 20 years were examined. The annual rainfall varies from 863.3 to 1470.0 mm. During the summer season, the highest total rainfall was 1089.7 mm during 2000, whereas lowest rainfall was 287.9 mm. The average rainfall for 20 years was found to be highest in the month of June whereas the rainy days were highest in the month of August. Fifteen years yield data of four crops in the area was analyzed. The trends in wheat and rice are upward whereas barley trends were downward and maize trends were constant.

Key words: Rainfall, cropping pattern, rain-fed region.

INTRODUCTION

India ranks first among the rain-fed agricultural countries of the world in terms of both extent (86 M ha) and value of produce. Rain-fed areas in India are highly diverse, ranging from resource-rich areas with good agricultural potential to resource-constrained areas with much more constrained potential. It is in the rain-fed regions where cultivation of nutritious (coarse) cereals (91%), pulses (91%), oilseeds (80%) and cotton (65%) predominates. Rosegrant et al. (2002) employing the IMPACT model have estimated that even by 2025, one-third of India's cereal production shall be contributed by rain-fed areas. Rain-fed agriculture supports 40% of India's population. Earlier, rain-fed farming systems were dependent upon locally available inputs and grew traditional drought-resistant crops. But over-time cropping systems have changed (Kanwar, 1999) and farmers have started cultivating high-value crops requiring intensive use of costly inputs.

Besides several other factors related to the agriculture sector as a whole, adverse meteorological conditions resulting in long dry spells and droughts, unseasonal rains and extended moisture-stress periods with no mechanisms of storing and conserving the surplus of rain to mete out during the deficit periods were the major causes of non-remunerative yields. It is only recently that the Government of India has constituted a National Rain-fed Area Authority (2006) to address these issues and develop and implement a comprehensive single-window program for the development of rain-fed areas in the country. There have been some studies by the various authors to analyse rainfall pattern in India. (Subash et al. 2012) revealed that there was a slight increase in 0.5% of rainfall in Bihar during 1999-2008 as compared to 1989-1998. (Kumar et al, 2010) studied the trend analysis of rainfall data of all India for 135 years (171-2005) found that there was no significant trend for annual, seasonal and monthly rainfall on an all-India basis. (Choudhury et al., 2012) in their study in mid Altitude Meghalaya, North-East India analyzed long time (1983-2010) weather variables to detect trend changes using non-parametric

Table 1. 15 years wise yield of different crops in Solan district.

Year/yield (kg/hectare)	Crop			
	Wheat	Barley	Maize	Rice
1996-97	1266.905	1010.442	1842.626	1826.779
1997-98	1760.977	1681.799	2123.829	1921.200
1998-99	1562.293	1240.735	2674.233	1688.163
1999-00	1679.132	1070.720	2162.918	1851.040
2000-01	368.645	457.938	2031.878	2114.312
2001-02	1789.423	1136.197	2356.827	1862.672
2002-03	1435.621	831.443	1624.775	1592.283
2003-04	1525.479	719.256	2071.319	1468.820
2004-05	1922.838	1143.716	1763.141	2193.449
2005-06	1577.559	600.571	1671.576	1754.544
2006-07	1877.701	1267.392	2166.276	2366.944
2007-08	1254.410	859.394	2364.519	2327.317
2008-09	1442.840	595.247	2408.814	2479.996
2009-10	1683.555	984.237	1896.514	2199.139
2010-11	1988.198	970.531	2440.690	2467.049

Table 2. Monthly average rainfall data (mm) for the period of January 1991 to December 2010.

Season	1991	1992	1993	1994	1995	1996	1997	1998	1999	2000
Winter	190.2	214.1	184.0	168.9	178.5	210.6	154.2	102.2	215.3	186.2
Spring	204.0	101.6	162.0	114.2	103.5	119.6	248.2	235.1	211.9	184.2
Summer	534.9	622.7	480.8	578.2	910.9	614.6	585.7	492.3	743.8	1089.7
Autumn	127.4	197.3	192.7	118.6	239.8	235.0	148.0	463.7	149.4	9.9
Total	1056.5	1135.7	1019.5	979.9	1432.7	1179.8	1136.1	1293.3	1320.4	1470
	2001	**2002**	**2003**	**2004**	**2005**	**2006**	**2007**	**2008**	**2009**	**2010**
Winter	43.2	128.6	178.2	127.8	221.6	88.4	238.0	83.8	50.6	178.7
Spring	179.6	235.4	134.0	99.8	73.3	210.1	199.8	134.3	99.9	51.9
Summer	686.6	513.0	613.2	518.6	453.8	599.5	539.7	625.0	287.9	824.8
Autumn	1.0	245.6	135.8	188.6	157.6	71.4	112.6	367.8	424.9	410.1
Total	910.4	1122.6	1061.2	934.8	906.3	969.4	1090.1	1210.9	863.3	1465.5

Mann Kendall test. Results revealed that total annual rainfall trend increased non-significantly at the rate of 3.72 mm/year.

Climate is the most significant environmental variable affecting the production of crops. The distribution of crops in Himachal Pradesh is influenced mainly by climate than any other factor.

In Himachal Pradesh, major rivers like Beas, Satluj and Ravi are mostly snow fed. Hence, regional climate change will influence not only the current production systems but also the river hydrology. Any adverse change in meteorological parameters, be it annual or seasonal (winter, spring, summer and autumn), will influence production systems of the region. The rains are the main source of water in Himachal Pradesh. The kinds of crops that can be successfully produced in any area are selected on the basis of the amount of water needed by the crop. It is evident that in addition to rainfall, the temperature and character of soil may add in determining which crops are to be grown.

METHODOLOGY

The variations in precipitation were analyzed season wise, that is, summer (June, July and August), autumn (September, October and November), winter (December, January and February) and spring (March, April and May). The rainfall data for 1991-2010 for Nauni (Solan) has been taken for analysis. In addition, the trends in 15 years of yield data of the four major crops in the area has been analyzed and presented in Table 1. The monthly average rainfall data (mm) for the period Jan 1991 to December 2010 are presented in Table 2.

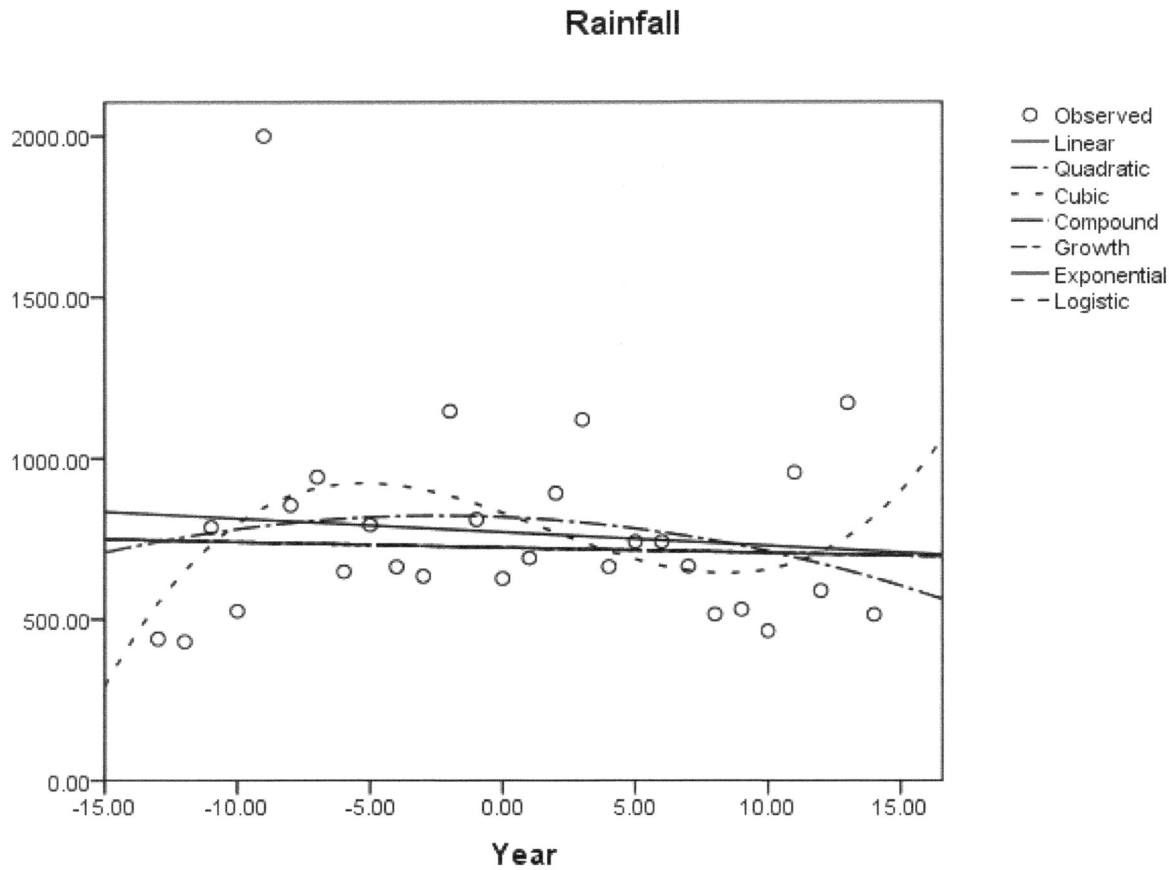

Figure 1. Annual rainfall variation.

RESULTS AND DISCUSSION

Annual, seasonal and monthly trends

Rainfall trends are presented in this section. The annual rainfall variation shows a downward trend during this period (Figure 1). The spring season shows a large downward trend, the autumn season shows an increasing trend, the summer season shows a decreasing trend and the winter season shows a significant downward trend (Figure 2). The average rainfall during last 20 years along with total rainy days is presented in Table 3. Month wise rainfall trends are presented in Table 4. The month wise rainfall trends are shown in Figure 3.

During the last two decades, the annual highest rainfall was 1470.0 mm in the year 1999 (highest annual rainy days, 108 in 2007) whereas the lowest annual rainfall was 863.3 mm in the year 2008 (lowest annual rainy days, 54 in 2001). The table shows hardly any relationship between rainfall and rainy days.

Season wise analysis of recorded data shows that during winter the highest total rainfall was 238.0 mm during 2007 (highest total rainy days, 17 in 1993) whereas lowest rainfall was 43.2 mm in 2001 (lowest total rainy days, 5 in 2001 and 2009). During spring the highest total rainfall was 248.2 mm during 1997 (highest total rainy days 30 in 1997) whereas lowest rainfall was 51.9 mm in 2010 (lowest total rainy days, 5 in 2010). During summer, the highest total rainfall was 1089.7 mm during 2000 (highest total rainy days 55 in 2010) whereas lowest rainfall was 287.9 mm in 2009 (lowest total rainy days 22 in 2002). During autumn, the highest total rainfall was 463.7 mm during 1998 (highest total rainy days 21 in 1997) whereas lowest rainfall was 1.0 mm in 2001(lowest total rainy days 0 in 2001).

Effect on cropping pattern

Figure 1 shows the deceasing trend of annual rainfall in the region, whereas Figure 4 shows the increasing trend of yield in wheat and rice. The increasing trend in rice may be due to the fact that rainfall in the month of May to July has increasing trend (Figure 3) which is the sowing period of rice in the region. However, the winter season has downward trend in rainfall (Figure 3) which is the sowing period for wheat, whereas wheat yield shows increasing tend (Figure 4) in the region. This may due to fact that the farmers have adopted some other sources of irrigation.

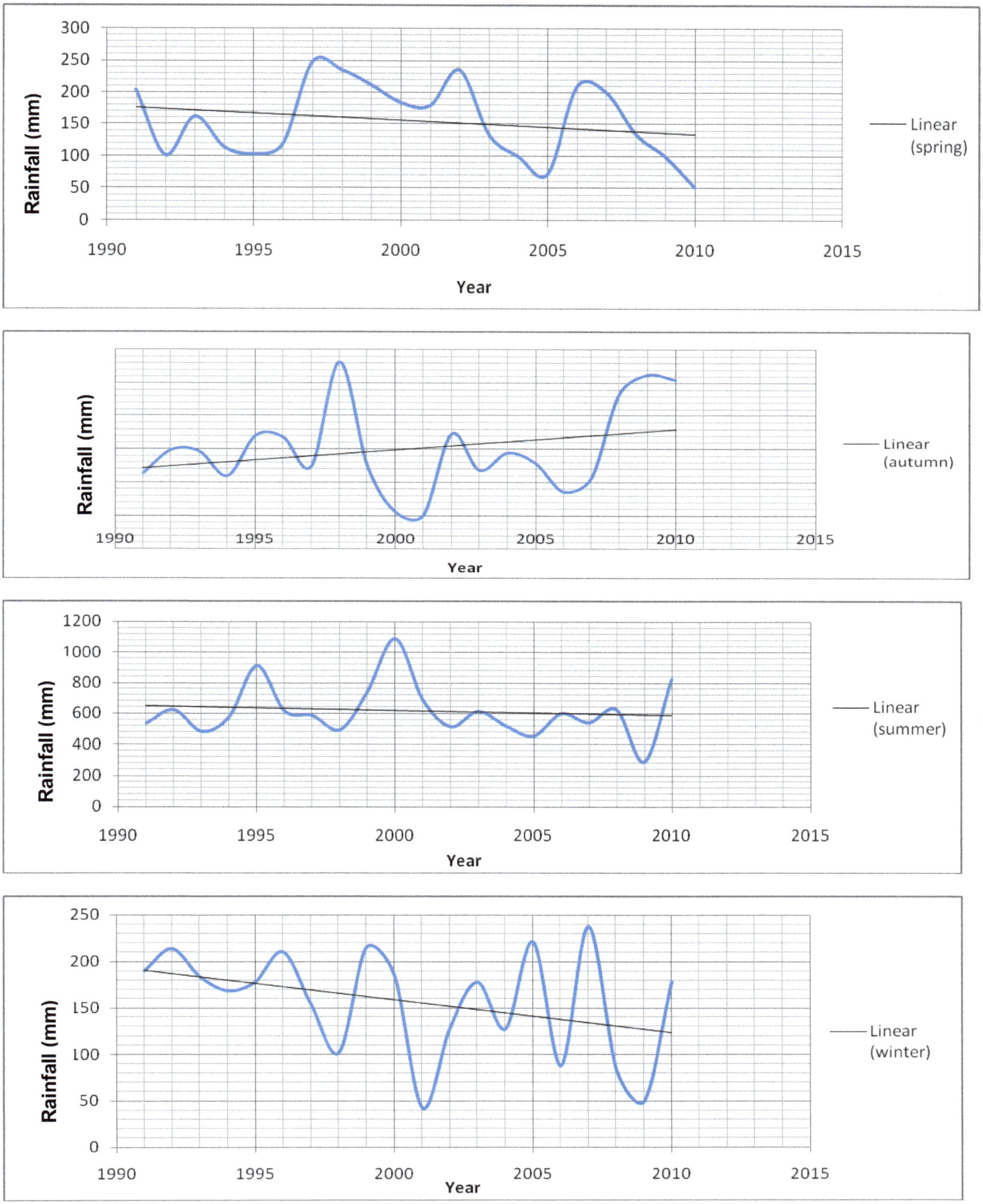

Figure 2. Seasonal rainfall variation.

Table 3. Average rainfall during the last 20 years along with total rainy days.

Month	Rainfall (mm)	Rainy days
January	76.13	41.3
February	61.0	77.9
March	38.67	145.9
April	55.45	222.5
May	142.4	290.9
June	242.2	335.2
July	231.1	363.6
August	159.2	435.5
September	32.4	301.8
October	8.3	185.6
November	19.4	106.8
December	61.6	52.2

Table 4. Month wise rainfall trends.

Month	Trend
January	Sharp downward
February	Straight
March	Downward
April	Sharp downward
May	Slightly upward
June	Straight
July	Upward
August	Sharp deceasing
September	Sharp upward
October	Slightly downward
November	Slightly downward
December	Downward

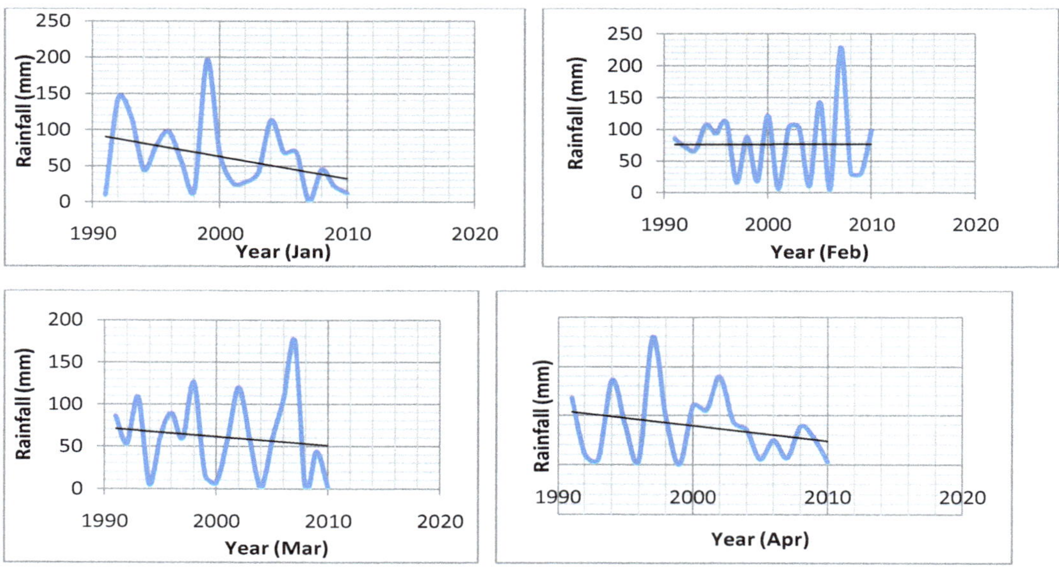

Figure 3. Month wise variation of rainfall.

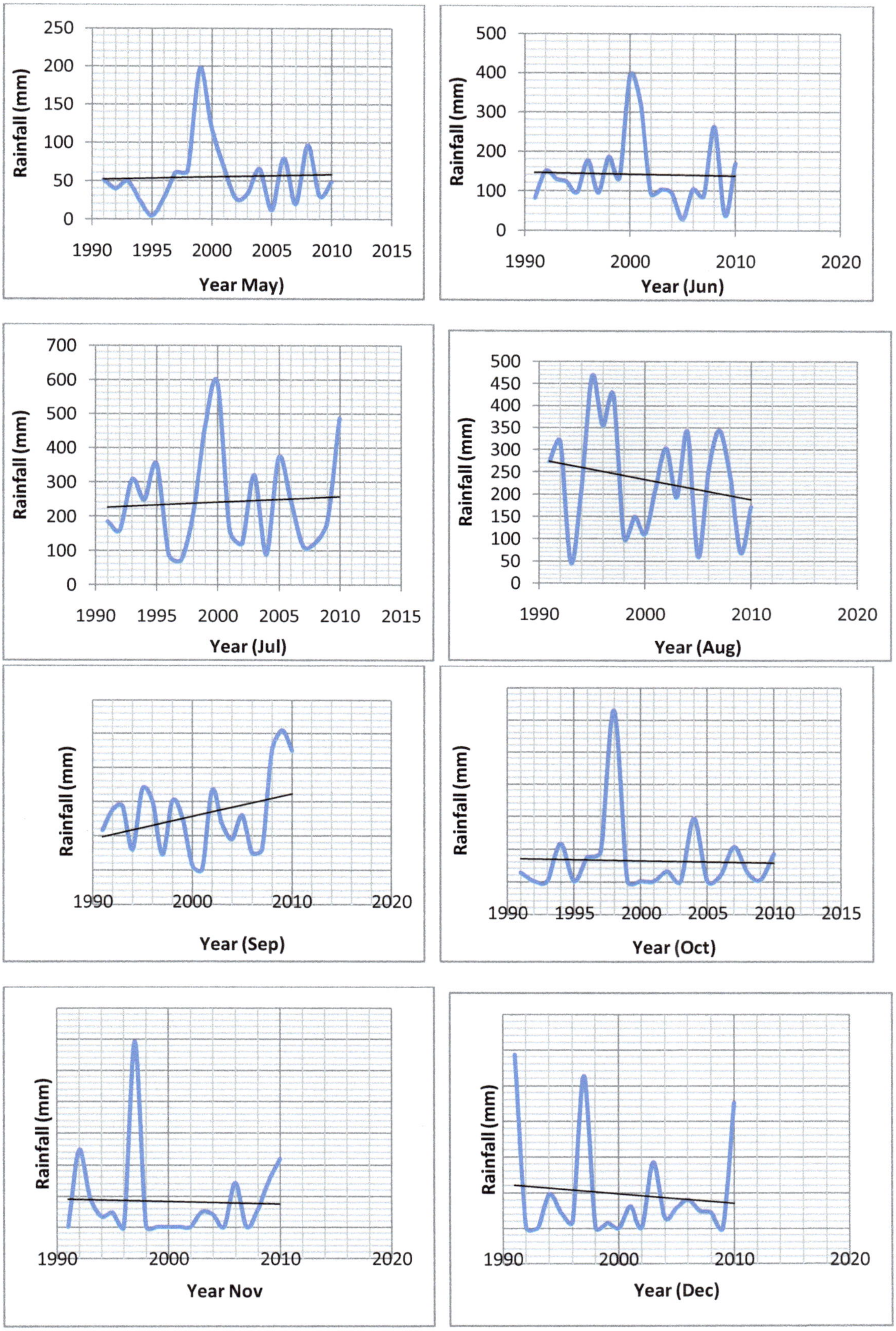

Figure 3. Contd.

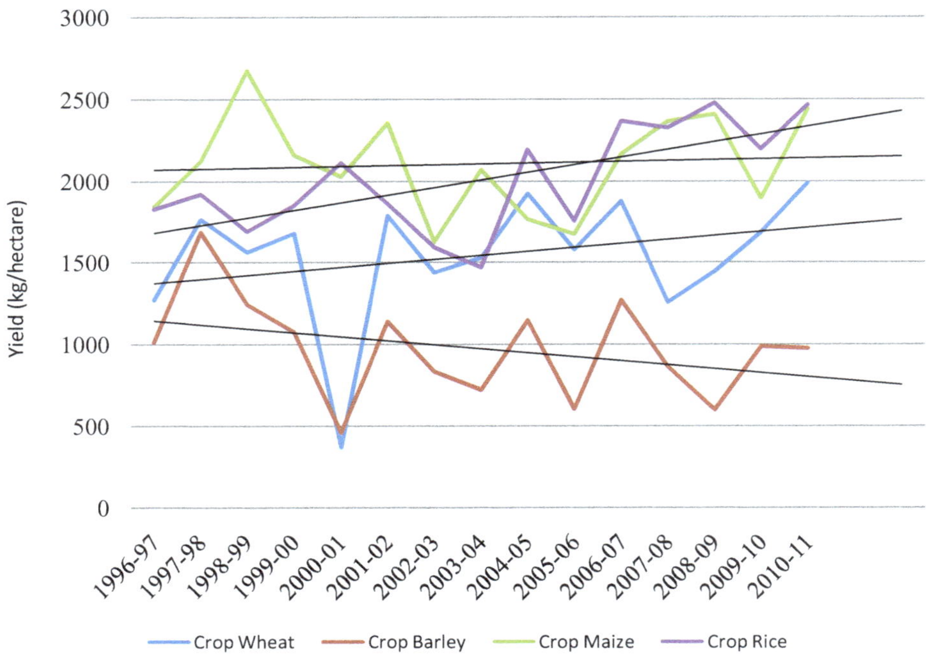

Figure 4. Year wise yield of different crops in Solan district.

Rainfall does not have any effect on the yield of maize which has constant trend (Figure 4). The barley shows deceasing trend in yield. The statistically analyzed rainfall trend is shown in Figure 1. This shows that annually, there is no particular trend in the region during the last 20 years.

Conclusion

The rainfall is a major requirement for crop production not only in this region but also in India. With the change in climate, the rainfall pattern has also changed. In spite of change in rainfall pattern, the crop yield does not fall sharply. This implies that the crop yield is not affected by rainfall significantly. The study also reveals that there is no trend in rainfall in the region.

REFERENCES

Kanwar, J.S. (1999). Need for a future outlook and mandate for dryland agriculture in India. In. Fifty Years of Dryland agricultural research in India, eds.

Rosegrant M, Ximing C, Cline S, Nakagawa N (2002). The role of rain-fed agriculture in the future of global food production. EPTD Discussion Paper 90. Washington D.C., USA: International Food Policy Institute. (www.ifpri.org/divs/eptd/dp/papers/eptdp90.pdf). pp. 252.

Kumar V, Jain S K and SinghY (2010). Analysis of long-term rainfall trends in India. Journal of Hydrol.Sc. 55:484-496

Subas N, Singh S S and Priya Neha (2012). Rainfall variability and its impact on changes of cropping system in Bihar. India journal of soil conservation. 40(1)33:40.

Choudhury B U, Anup Das, Ngachan S V, Slong A, Bordoloi L J Andchowdhury P (2012). Trend Analysis of Long Term Weather Variables in Mid Altitude Meghalaya, North-East India. Journal of Agricultural Physics. 12(1):12-22.

Multi-temporal assessment of forest cover, stocking parameters and above-ground tree biomass dynamics in Miombo Woodlands of Tanzania

Almas Kashindye[1], Elifuraha Mtalo[2], Mathew M. Mpanda[4], Evaristo Liwa[3] and Richard Giliba[1]

[1]Forestry Training Institute, Olmotonyi, P.O Box 943, Arusha, Tanzania.
[2]College of Science, Informatics and Built Environment, University of Bagamoyo, Dar es Salaam, Tanzania.
[3]School of Geospatial Sciences and Technology, Ardhi University, Dar es Salaam, Tanzania.
[4]World Agroforestry Centre, Tanzania Country Program, Dar es Salaam, Tanzania.

Miombo woodlands form the widespread ecosystem in Tanzania. The ecosystem of these woodlands serves as a carbon sink and source containing majority of the above-ground terrestrial organic carbon. The study assessed forest cover, stocking parameters and above-ground tree biomass dynamics in the Miombo woodlands of Bereku and Duru Haitemba in Tanzania. The data were obtained from conventional forest inventory and remote sensing and GIS techniques. Results show that forest cover has increased for the two consecutive assessment period of 1988 to 2000 and 2000 to 2009, mostly by higher canopy cover of 6.82 and 0.79%, respectively. Stocking parameters: stand density (N) stand volume (V) and stand basal area (G) were found to be 1909.5 ± 9.4 stem/ha, 12.3 ± 0.6 m^{2}ha^{-1} and 71.0 ± 6.8 m^{3}ha^{-1}, respectively. Vegetation indices (NDVI, ARVI and ND54) were then combined with forest inventory data for computation of average above ground biomass which was found to increase from 64 ± 6.53, 67.8 ± 5.42 to 79.218 ± 2.75 t/ha for 1988, 2000 and 2009, respectively. The results suggest that, the Miombo woodland resources have been consistently improving over the years of assessment. This could be attributed to reduction in negative anthropogenic factors that are known to be the major cause of resources degradation in Miombo woodlands. However, this improvement may be partial as plot level information revealed higher exploitation of the dbh class of 4 to 5 cm. Increase of management effectiveness through involvement of local people under participatory forest management, strengthened bylaws and provision of other incentives might have contributed greatly to the improvement of the forest resources.

Key words: Forest resources management, forest inventory, remote sensing, aboveground biomass stock.

INTRODUCTION

Forests and woodlands have been highly affected by land use changes across sub-Saharan Africa. This has caused change in vegetation cover especially conversion from forests and woodlands to other land uses like cropland and grazing land (Bukombe, 2007; Madulu, 2004). In the extreme cases, land use and land cover changes have been cited as one of the factors contributing to global climate change as well (Gong and Howarth, 1990). Miombo woodlands which are found in the Eastern and Southern Africa are one of vegetation types that have suffered due to land use changes. They cover approximately 320 million hectares. They are tropical seasonal woodlands and dry forest formation of African forests, most widespread fire-adapted and closed-canopy (Shackleton and Clarke, 2007). They occur in altitude from near sea level to 1600 m.a.s.l. with annual rainfall ranging from 500 to 1200 mm (Jeffers and Boaler 1966; Nshubemuki and Mbwambo, 2007).

In Tanzania, Miombo woodlands constitutes the largest single vegetation type, covering an area of about 30.93 million ha, forming nearly 90% of the total forest land or 13% of the entirely land (URT, 1998).

The physiognomy of Miombo woodlands in Tanzania consists of three layers: the tree canopy, under-wood layer of smaller trees and ground layer mainly shrubs and herbs (Nshubemuki and Mbwambo, 2007). With over 80% of population living in rural areas exploiting forest resources, Tanzania is witnessing rapid decline of forests and woodlands estimated at 97,260 hectares per year (URT, 2002; NBS, 2002). Clarke et al. (1996) reported that Miombo ecosystem supplies plenty of products and services that are essential to the wellbeing of rural communities. Miombo's woody products are normally extracted for fuelwood, poles, timber, carving, medicine, withies, food, ropes, live fences, ritual items and charcoal (Luoga et al., 2002; Abdallah and Monela, 2007).

In late 1980s, up to mid-1990s Tanzania engaged in introducing participatory forest management, which was a total shift from long standing paradigm of state control of forest resources (Kajembe et al., 2008). Soon, the promotion and adoption of the decentralisation model which was termed as Participatory Forest Management (PFM) practiced either as joint forest management or community-based forest management were accelerated country-wide and supported by National Forest Policy of 1998 (URT, 1998). However, the decentralization in a broader sense were in terms of forestland tenure changes and access to forest resources, witnessing some forests changing from general lands to village land forests reserves (VLFR) and others changing from private ownership to central government under the joint forest management (Mpanda et al., 2011).

It should be noted that in Tanzania there are two land acts (the Land act 1999 and the Village land act 1999). In the Land Act, the legal framework for the two land categories is defined, namely General Land and Reserved Land. Reserved Land, is that land set aside for specific purposes, under different legislation, it includes forest reserves, national parks and game reserves, etc, where as the General Land is a residual category, which includes all land that is not Reserved Land or Village Land, the Village land on the other hand is that land which is vested in the Village Assembly, and that the Village Council administers the land through the authority of the Village Assembly (Sundet, 2005)

In order to assess impact of these changes of forestland tenure and management arrangement that was spearheaded two decades ago, and taking into consideration new emerging ideas on REDD+, this study attempted to evaluate the status of forest resources in Bereku Forest Reserve (FR) and Duru-Haitemba Village Forest Reserves (VLFR) in the northern Tanzania. Forest biomass estimation have been estimated based on field inventory data which are normally labour intensive, time consuming, difficult to implement on vast areas and sometime they become destructive in cases where it is necessary to fell sample trees (Deo, 2008; Lu et al., 2005).

Remote sensing techniques have primarily been used in forest mapping, a real extent estimation and determination of deforestation rate (Mayaux and Lambin, 1995). The difficulty associated with traditional techniques of estimating forest biomass as expressed above has opened a door for the use of remote sensing techniques uses in biomass estimation. Remote sensing has been employed in many studies related to biomass estimations. For example, Solicha (2007) used remotely-sensed images in estimating aboveground biomass by employing three different ways based on classification of vegetation cover and generation of a vegetation type map. Indirect estimation of biomass through some form of quantitative relationship and partition the spatial variability of vegetation cover into relatively uniform patches.

A study conducted by Gibbs et al. (2007) showed that above ground biomass can be evaluated using remote sensing techniques supplemented by ground based data (inventory data). This approach has an advantage of providing wall to wall observation of carbon stock proxies. However, it should be noted that remote sensing does not measure biomass directly, but rather it measures other forest characteristics (spectral reflectance of trees), therefore it is necessary to supplement field measurement data with remote sensing data in estimating above ground biomass to develop statistical relationships between ground based measurements and satellite-observed vegetation indices (Gibbs et al., 2007).

Specifically, the study had three objectives including; i) understanding the forest vegetation cover changes at a landscape level, ii) determining stocking parameters at plot level depicting state of the forest resources, and iii) estimating temporal changes of total above ground biomass. The study employed remote sensing and GIS techniques (Abd El-Kawya et al., 2011; Kashaigili and Majaliwa, 2010) and thus complemented the standard forest inventory (Luoga et al., 2002). Combination of the two methods can provide more useful insights of the ecosystem at landscape scale (Mpanda et al., 2011), especially in assessing above ground biomass dynamics (Baccini et al., 2008).

MATERIALS AND METHODS

Study site

The study was conducted in Bereku Forest Reserve (6,111 ha) and Duru-Haitemba Village Forest Reserve (9,000 ha) located between

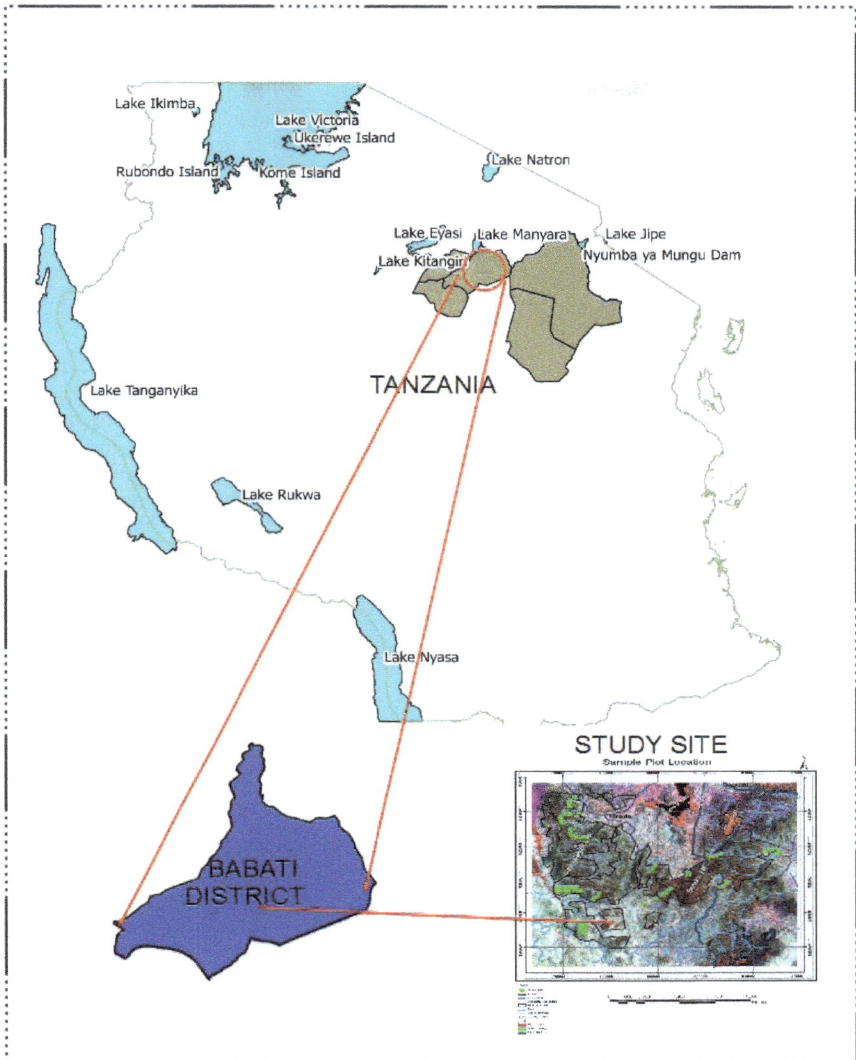

Figure 1. Location of sample plots in Bereku and Duru- Haitemba forests.

4°19' to 4°23' S and 35°43' to 35°51' E, respectively (Figure 1). The two forests cover the Bereku ridge and Gedabosh Mountain in Babati district, Manyara region, Tanzania (Lovett and Pócs, 1993). Population for the entire district stands at 303,013, with growth rate of 3.8% per annum and a density of 54 people per km^2 (NBS, 2002). The forests lie between altitudes of 1,000 and 2,500 m above sea level. Rainfall in the area is characterised by bi-modal and irregular rains ranging from 300 to 1,200 mm per year (Lovett and Pócs, 1993; Kajembe et al., 2003), and temperature ranges from 18 to 28°C (Lovett and Pócs, 1993). The soil is mostly loamy with good drainage and vegetation is mostly dry Miombo (Lovett and Pócs, 1993).

The two reserves are covered by woodland dominated by *Brachystegia microphylla*. The woodland consists of a continuous light canopy of 8 to 10 m tall grassland. The trees include: *Acacia* spp., *Albizia versicolor, Brachystegia microphylla, Brachystegia spiciformis, Cussonia arborea, Markhamia obtusifolia, Parinari curatellifolia, Strychnos* sp., *Syzygium guineense* subsp. *guineense, Vitex* sp. Bamboo is reported from the upper valleys of the Kikore and Madege rivers. Seasonally waterlogged grassland are associated with small tract of riverine forest which is mainly covered by *Bridelia micrantha* and *Syzygium cordatum* along the many undulating valleys.

Methods

The forests in Bereku Forest Reserve and Duru-Haitemba Village Forest Reserve were stratified into relatively small homogenous stands through visual interpretation of the geo-referenced satellite imageries at a scale of 1:50000 with the help of standard sheets of topographical maps of scale 1:50000 and Land Cover and Land Use map of Singida of scale 1:250,000 which shows the distribution of land cover and land use in Singida that includes some parts of Babati district.

In each stratum, sample plots were systematically laid out along transects at an interval of 250 m with an allowable error of 10%. Transect orientation was based on spatial distribution on road network passing through the reserves. The first sample plot was randomly determined; to minimize the edge effect, the plot centre for the first plots was half the prescribed plot interval from the road edge. Each sample plot was set out on the ground and its corresponding coordinate were recorded using a hand held GPS (Garmin GPS map etrex 60 CSx) receiver of a precision of ±3.

A total of 80 concentric sample plots of 0.07 ha or (700 m^2) was used, with 2, 5, 10 and 15 m radii respectively, measurement in the first concentric were taken on all trees with diameter at breast height (Dbh) ranging from 1 to 5 cm, in the second concentric it was 6 to 10 cm, in the third 11 to 20 cm and in the last concentric it was ≥ 20 cm (Figure 1), with the maximum radius of the plot of 15 m which is comparable with the spatial resolution of Landsat TM and ETM+. In each sample plot, all trees were identified by local people and a botanist. Additionally, two trees nearest (fattest and the small sized tree which were easily to be determined and time saving) to the plot centre were identified, dbh and height measured for establishing equations were used to estimate height of the rest of trees. Three satellite imageries were used including Landsat TM of 1988 and 2009 and Landsat ETM+ for 2000.

Standard forest inventory analyses were applied as used by Mpanda et al. (2011), where number of trees per plot was summarized and extrapolated into per hectare values (N). Basal areas for individual trees (g) were calculated according to DBH, summarized and transformed into per hectare values for the plot (G). A height-dbh equation developed by Malimbwi (2003) was used to estimate height of unmeasured trees as shown in Equation 1:

$$Ht = Exp\ (0.58048 + 0.602965 \times Ln(Dbh)) \quad (1)$$

with R^2= 83, SE = 1.32,

Where: Ht is total tree height in meters; Dbh = is tree diameter at breast height; Ln is the natural logarithm; SE is the standard error of estimate; the volume of individual tree was calculated using the following equation:

$$v = g \times Ht \times f \quad (2)$$

where v = tree volume, g = tree basal area, Ht = estimated tree height and f = form factor with value 0.5.

A form factor of 0.5 has traditionally been used for Miombo forests in Tanzania (Chamshama et al., 2004; Luoga et al., 2002). The volume of individual trees was summarized within plots and transformed into per hectare values (V). Biomass was calculated using the adopted equation from (3), which is given as:

$$y = 0.0625Dbh^{2.553} \quad (3)$$

With R^2 = 97, SE = 1.79, where, y = biomass ($kgha^{-1}$).

Landsat TM (1988 and 2009) and Landsat ETM+ (2000) of path/row 168/63 and 169/63 were geometrically corrected so as to transform from WGS84 UTM coordinate system to Tanzania UTM coordinate system (Clarke 1880 ellipsoid and Arc 1960 Datum) in UTM zone 36 South. The first order polynomial transformation for the geometric correction and resampling all images to pixel size 30 x 30 m using the nearest neighbour method was carried out. The images were also radiometrically corrected to remove any bias due to sensor parameters. The digital numbers (DN) were converted into radiance value and then to the surface reflectance in order to calculate the spectral vegetation indices.

Supervised classification was used to process the imageries using information collected at plot level using hand held GPS. Sample plot data (ground truth data) were used to train the computer to recognize patterns in the data by selecting pixels that represents pattern or recognizable land cover features. Normally in supervised classification, the analyst has a prior knowledge of some land cover types, then identifies and locates these features through a combination of field work, map analysis and personal experience. The resulting signature file created was used in the classification process, where by each pixel is categorized into the land cover class it mostly belongs to. The land cover classes developed were based on the predetermined classification scheme which adheres to the principle that the classification scheme for land use and land cover should be mutually exclusive, exhaustive and hierarchical (Jensen, 2007). Ten land cover class/category were developed namely as waterbodies, open grassland, grassland with scattered crop lands, woodlands with scattered cropland, closed woodland, open woodland, bushed grassland, Marshaland, cropland with scattered trees and bushland with scattered cropland.

Land cover maps for the years 1988, 2000 and 2009 were developed, and later used in change detection using image differencing (Singh, 1989; Coppin and Bauer, 1996; Lillesand et al., 2004; Jensen, 2007). In this study, image differencing was used as a method of choice for change detection where by the thematic classified images were differenced. This algorithm involves subtracting one epoch of the classified imagery from a second epoch of classified imagery that has been precisely registered to the first. It normally results into positive and negative values representing areas of change and zero values representing no change.

Regression analysis of Biomass and Spectral Indices was performed involving various vegetation indices to tested the correlation of field estimated biomass with spectral data for all three scenes. The regression equation developed was based on Vegetation Indices (VI) which relates to optical measures of vegetation canopy 'greenness' resulting from the composite property of total leaf chlorophyll, leaf area, canopy cover, and structure. Three VIs were tested i.e. Atmospherically Resistance Vegetation Index (ARVI), Normalized Difference Vegetation Indices (NDVI) and Normalized Different Vegetation index that use band 5 and 4 instead of band 1 and 2 (ND54) were tested to relationship between biomass and spectral data.

Vegetation Indices (VIs) were computed for each scene after the application of geometric and radiometric correction of the images. Forest inventory data were then integrated with satellite data (VIs). These values were then linked with field measured biomass data using empirical analysis employing the ordinary least square (OLS) regression to study the relationship between per plot above ground biomass and Vegetation indices, using the following modified equation;

$$AGB = \beta_0 + \beta_1 X_1 + \beta_2 X_2 + \beta_3 X_3 + \varepsilon i \quad (4)$$

Where, AGB = Above-ground biomass, β_0 = Intercept, β_1........ β_4 = slopes of respective explanatory variables, ε_i = error term, X_1 = NDVI, X_2 = ND54 (Normalized Different Vegetation Index that use band 5 and 4 instead of band 1 and 2) and X_3 = ARVI

The resulting regression equation from the analysis was used to predict Total Above-ground Tree Biomass (TAGB) for the entire study area as shown in Table 1.

RESULTS

Forest cover changes

Results of forest cover changes using post-classification change detection techniques over a period of 21 years from 1988 to 2009 at Bereku Forest Reserve and Duru-Haitemba Village Forest Reserve (Tables 1 and 2; Figure 2a and b) shows positive increment for the Closed woodland. The second highest cover class Open woodland has been unstable as some of its cover im proved

Table 1. Above-ground biomass estimation models.

Variable	Scene	Model	R^2 (%)
	1988	$-10.28X_1 + 20.33X_2 + 22.58X_3 + 30.81$	59.48
Spectral indices	2000	$-0.581X_1 + 20.484X_2 + 28.842X_3 + 68.73$	35.52
	2009	$4.752X_1 - 25.2X_2 - 12X_3 + 59.5$	51.9

Table 2. Areas of land cover changed forest cover changes from 1988 to 2009.

Class Name	1988	2000	2009	1988-2000	2000 -2009	Change (%) (1988 - 2000)	Change (%) (2000- 2009)
	Area of respective land cover classes in hectares						
Closed woodland	4690.93	6555.15	6970.77	2886.12	415.62	6.82	0.79
Open woodland	16208.59	11626.8	11721.9	2739.28	95.1	-6.47	0.18
Bushed grassland	4997.31	4720.59	6441.84	35.97	1721.25	-0.08	3.29
Unclassified	0	0	225.18	0	225.18	0.00	0.43
Total	25896.83	22902.5	25359.7	-	-	-	-

to a closed woodland while others deteriorated to a bushed grassland.

Forest stocking parameters

Results for the stocking parameters are shown in Tables 2 and 3. A total of 77 species were recorded of which the dominant species were *Braschytegia* species, *Jubernadia globiflora* and *Combretum molle*. The distribution of stand biomass and volume by diameter class indicates a J-shaped curve which implies that larger sized tree has more biomass as compared to small trees, where as that of stand stocking density indicated and inverted J-shaped curve, which implies the distribution trend is more dominated by small trees; which is a normal tendency for naturally growing forests. Furthermore, this study observed irregular regeneration and over-exploitation of certain preferred diameter classes (4 to 5 cm).

Above ground biomass estimation

Forest parameters delivered from inventory data were integrated with vegetation indices (satellite data). A window of 3 by 3 was used to extract pixels corresponding to plots using the shape file of the plots, and then the mean value at the centre of the pixel where the coordinates of the sample were located was determined. These values werethen linked with field measured biomass data using empirical analysis employing the ordinary least square (OLS) regression to study the relationship between per plot above ground biomass and vegetation indices.

The OLS regression is an empirical approach used in modeling the relationship between two observed variables, X and Y (Cohen et al., 2003). The OLS regression is in form of:

$$Y=\beta_0+\beta_1X+\varepsilon_i \qquad (5)$$

Where, Y = variable to be predicted (dependent variable); X= the variable to be predicted from explanatory or independent variable; β_0 = intercept; β_1 = slope of explanatory variables; ε_i = error term.

Cohen et al. (2003) suggested that data for the analysis in OLS regression should be supplied from paired observation of the two variables (dependent and explanatory variable). Commonly, one variable is difficult or costly to measure (forest or vegetation parameters from field sampling) and the other is relatively easy or inexpensive to observe (vegetation indices).

The regression analysis for both TM and ETM+ tested revealed a high correlation of spectral indices (ARVI, NDVI and ND54) with above-ground biomass (fAGB). The model for 1988 scene (Equations 5 and 6) (Table 1) was selected for use in computing TAGB for the entire study area since it has a relatively higher coefficient of determination (r^2):

$$-10.28X_1+20.33X_2+22.58X_3+30.81\,(r^2R^2=0.59) \qquad (6)$$

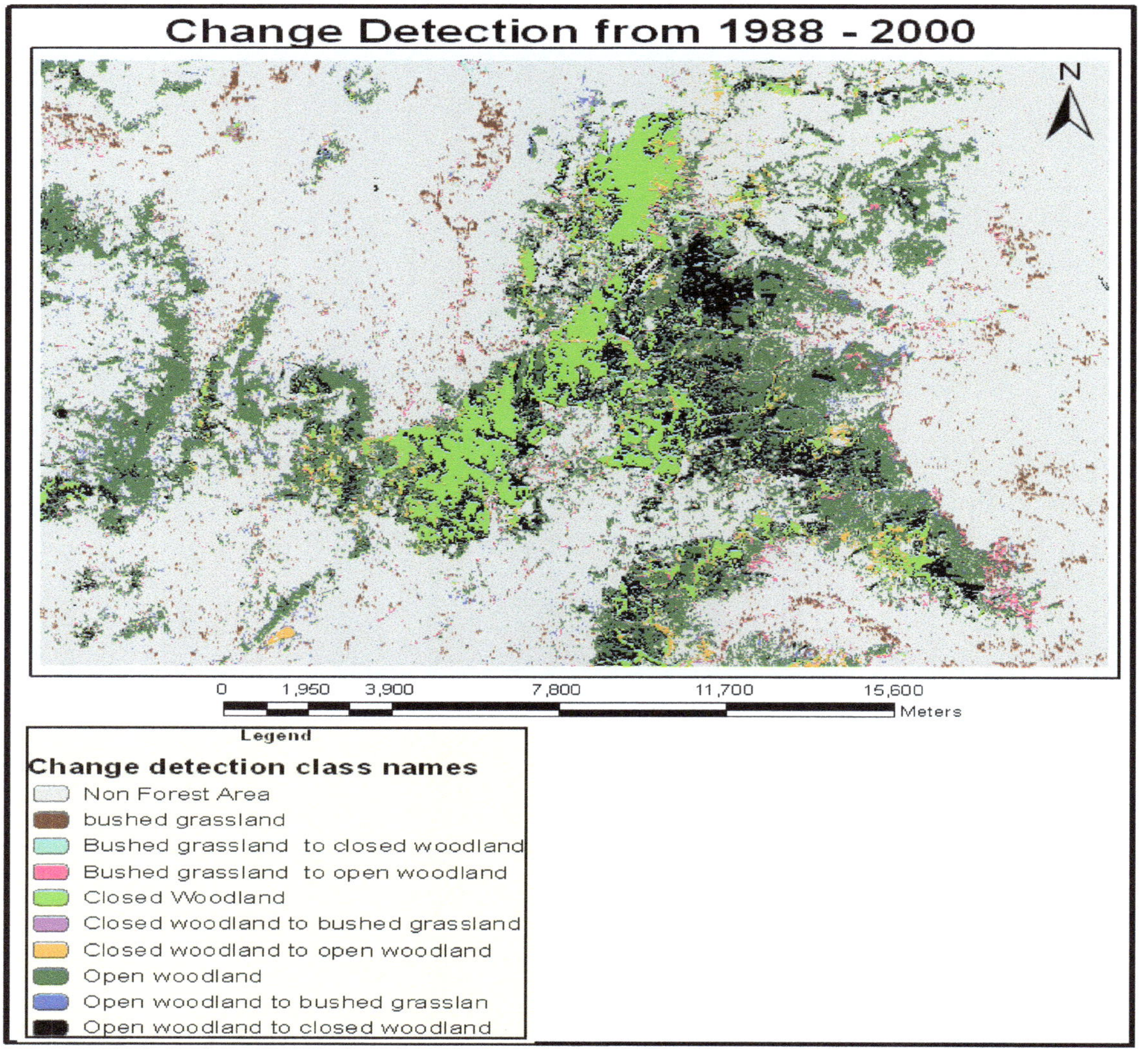

Figure 2a. Forest cover change detection map for the period of 1988 to 2000.

Where; x_1 =ARVI, x_2 =NDVI, and x_3 = ND54.

The biomass values obtained from stocking parameters and vegetation indices were used to prepare biomass maps using ArcGIS (Figure 3a, b, c and d), and further computing biomass of the entire area. The average biomass of all trees ≥ 1cm of the entire area has increased from 64 ± 6.53 t/ha in 1988) to 67.8 ±5.42 t/ha in 2000 and 79.218±2.75 t/ha in 2009.

DISCUSSION

Forest cover changes

Miombo woodlands in the Eastern and Central Africa have been observed to have dynamic systems whose multiple states are driven by disturbances, environmental variability and feedbacks (Forst, 1996). This study unveiled a similar situation where various cover types have been changing from 1988 to 2009 in both Duru Haitemba VFR and Bereku FR.

This study unveils changes in forest cover types and biomass from 1988 to 2009 in both Duru Haitemba VFR and Bereku FR, expressing the dynamic nature of the Miombo ecosystem. Similar observation was made by Forst (1996) resulting from multiple drivers including disturbances, environmental variability and feedbacks.

This study unveils changes in forest cover types and biomass from 1988 to 2009 in both Duru Haitemba VFR and Bereku FR, expressing the dynamic nature of the Miombo ecosystem. Similar observation was made by Frost (1996) resulting from multiple drivers including disturbances, environmental variability and feedbacks. Of the four cover classes, closed woodland which was the highest has been shown to increase, first at higher mag-

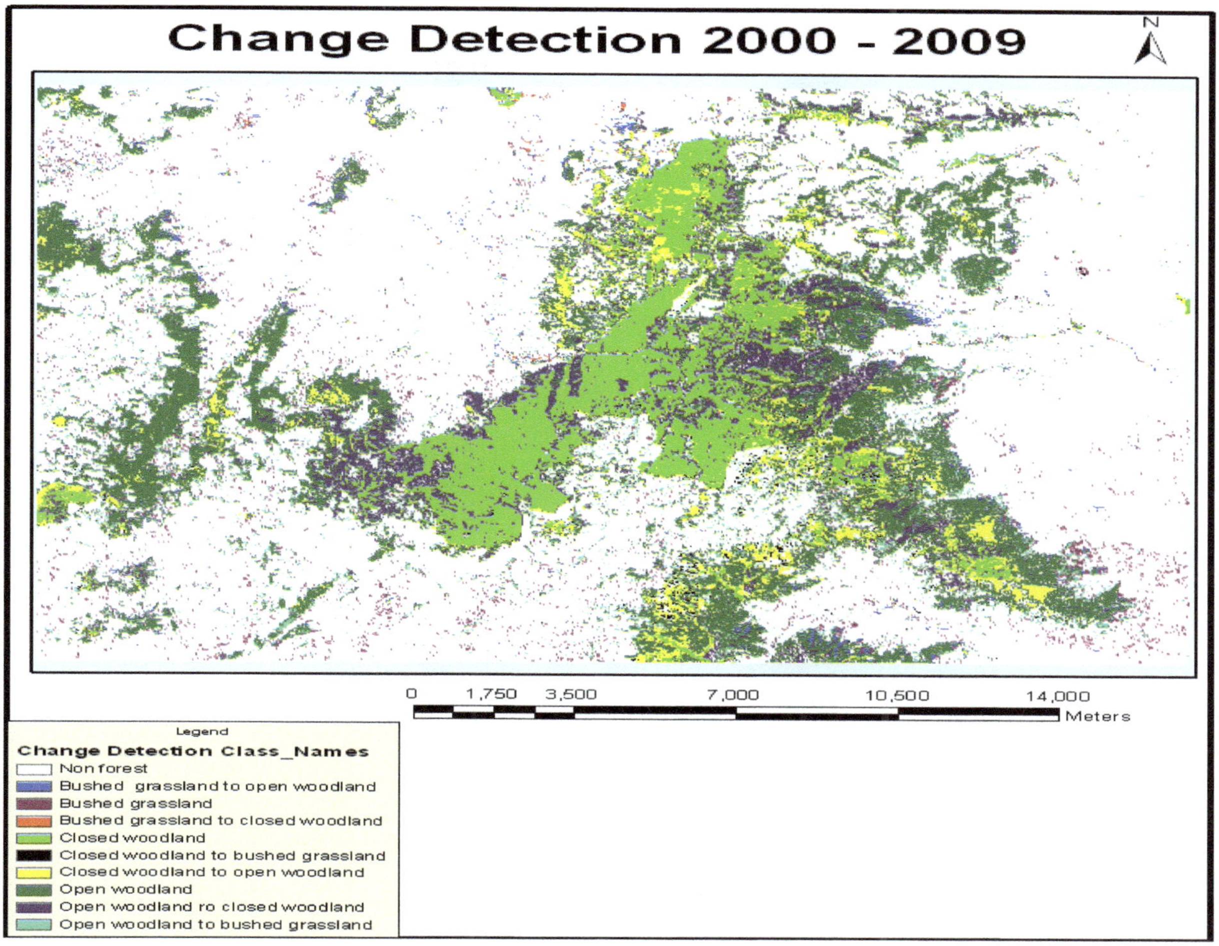

Figure 2b. Forest cover change detection map for the period of 2000 to 2009.

Table 3. Stocking parameter as per dbh classes in Bereku and Duru-Haitemba forests in 2010.

Stocking parameter (n=80)	Diameter class (in cm)						Mean*
	1 -4.9	5-9	10-19	20-29	30-39	≥40	
	1 -4.9	5-10	10-20	20-30	30-40	≥40.1	
N (no. ha^{-1})	1497.3	41126.1	50455.6	38360.3	13832.8	7487.6	1909.5 (9.4)
G (m^2ha^{-1})	0.28	29.3	178.1	301.1	212.6	262.7	12.3 (0.6)
V (m^3ha^{-1})	0.68	133.3	1300.4	2813.1	2297.1	3375.2	71.0 (6.8)
fAGB ($tonha^{-1}$)	0.40	80.1	790.8	1718.4	1406.0	2067.9	75.7 (2.4)

*Mean with standard error (in parenthesis) of the stocking parameters of all trees ≥ 1cm. Where: N = stocking per hectare, G = basal area per hectare, V = volume per hectare and fAGB = aboveground biomass in tonnes per hectare.

magnitude (between 1988 and 2000) and later at a low extent (between 2000 and 2009). This positive change especially that occurred between 1988 and 2000 can be attributed to improvement in management regime, which minimized human disturbances. In early 1990s, Tanzania witnessed a paradigm shift in forestry resource management where decentralized model was introduced in the name of participatory forest management (PFM) (Zahabu, 2008). This has empowered local communities living adjacent to forests and woodlands to either enter into Joint Forest Management (Bereku FR) or establish their own Community-Based Forest (Duru- Haitemba) (Kajembe et al., 2008; Mpanda et al., 2011). A number of studies have reported halting of deforestation and forest degradation in most areas of the country as result of PFM implementation.

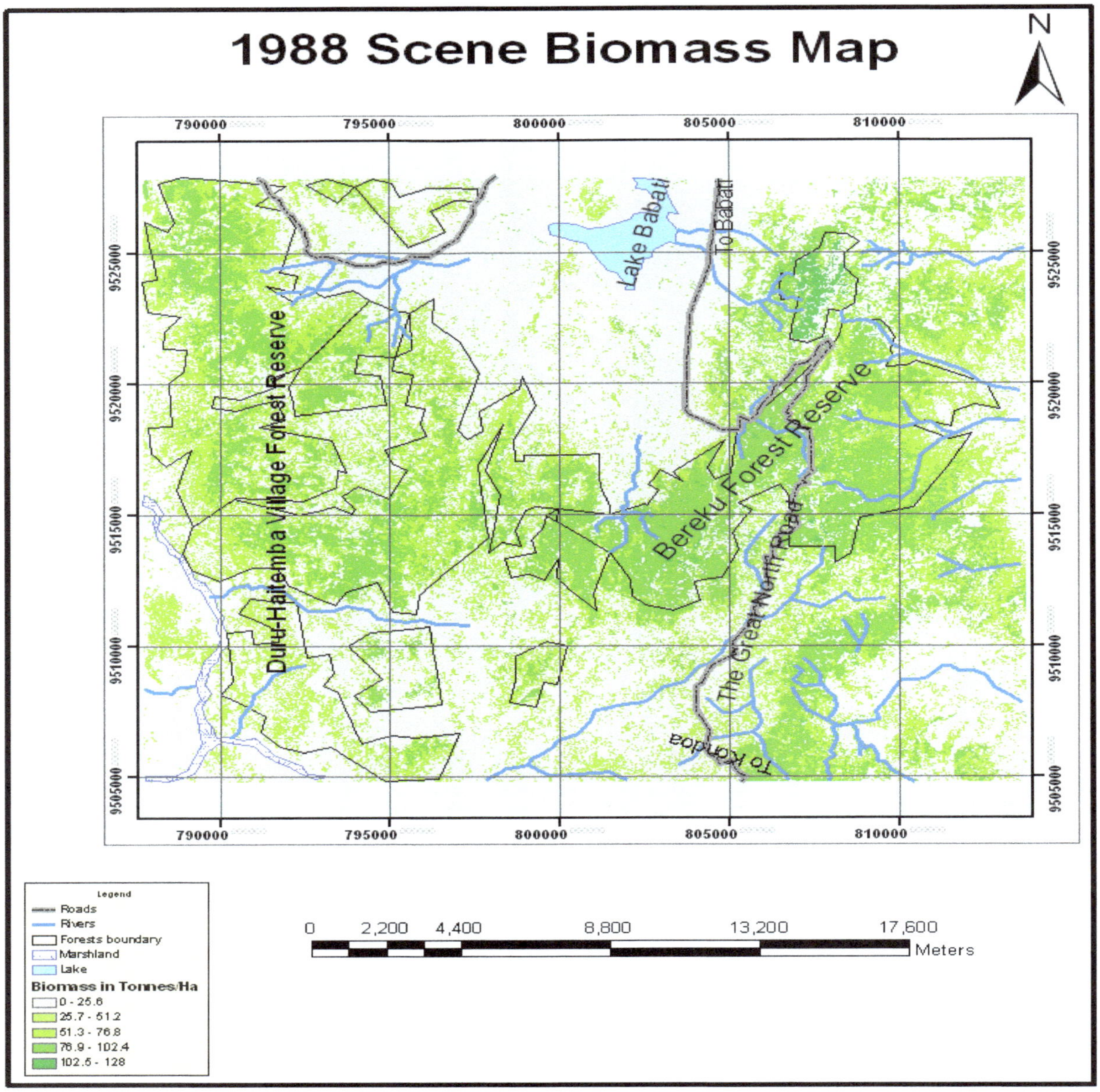

Figure 3a. Above-ground biomass distribution map for the 1988 scene for both path 168, row 63 and 169 row 63.

Support from bilateral and multi-lateral agrees with the government of Tanzania to support Participatory Forest Management ensured availability of finances and human resources to better manage the forests, mostly from Norway, Denmark, the World Bank, Germany and Finland (URT, 2003). It is important to note that one village named Ayasanda bordered the two forests (Bereku and Duru Haitemba) in the study area and hence benefited from both JFM and CBFM.

Results from the study further indicates that bushed grassland which is among the lower cover classes has increased in the period of 2000 to 2009, this is opposed to slight decrease in period of 1988 to 2000. This recent increase which translates to deforestation is probably attributed to local people's harvesting pressure on these forests. As both JFM and CBFM do not advocate total exclusion but rather rational use of the forest resources, it is probably this factor with minimum supervision that led to the decline of forest cover. A balance between utilization and protection need to be stricken otherwise deterioration of forest resource can result. Kajembe et al. (2008) and Vihemäki (2009) have argued that a complex situation exists between decentralization and local livelihoods, where in some cases instead of decentraliza-

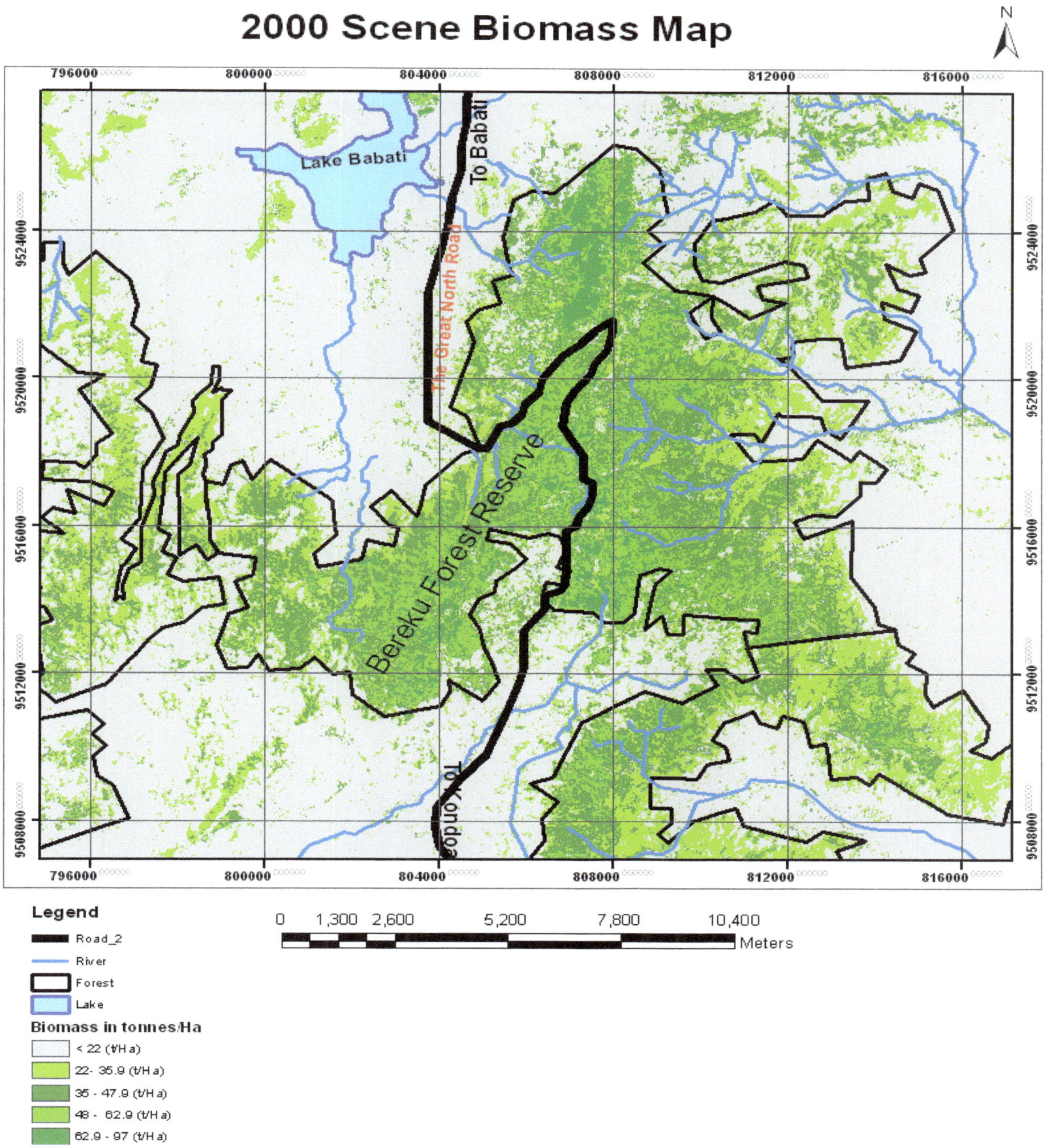

Figure 3b. Above-ground biomass distribution map for the 2000 scene path 168 row 63.

lization the situation becomes further exclusion which in turn deterioration of livelihood and forest resources can be witnessed.

Forest stocking

Average stocking parameters in Bereku FR and Duru Haitemba VFR found in this study are within the range reported by other researchers in various areas of Miombo woodlands (Malimbwi, 2003; Mugarura, 2007; Isango, 2007; Mwase et al., 2007; Zolho, 2005). Number of stems per hectare of tree species at diameter class 1 to 4.9 cm was lower than expected. This is probably due to excessive exploitation of that particular class. It was expected

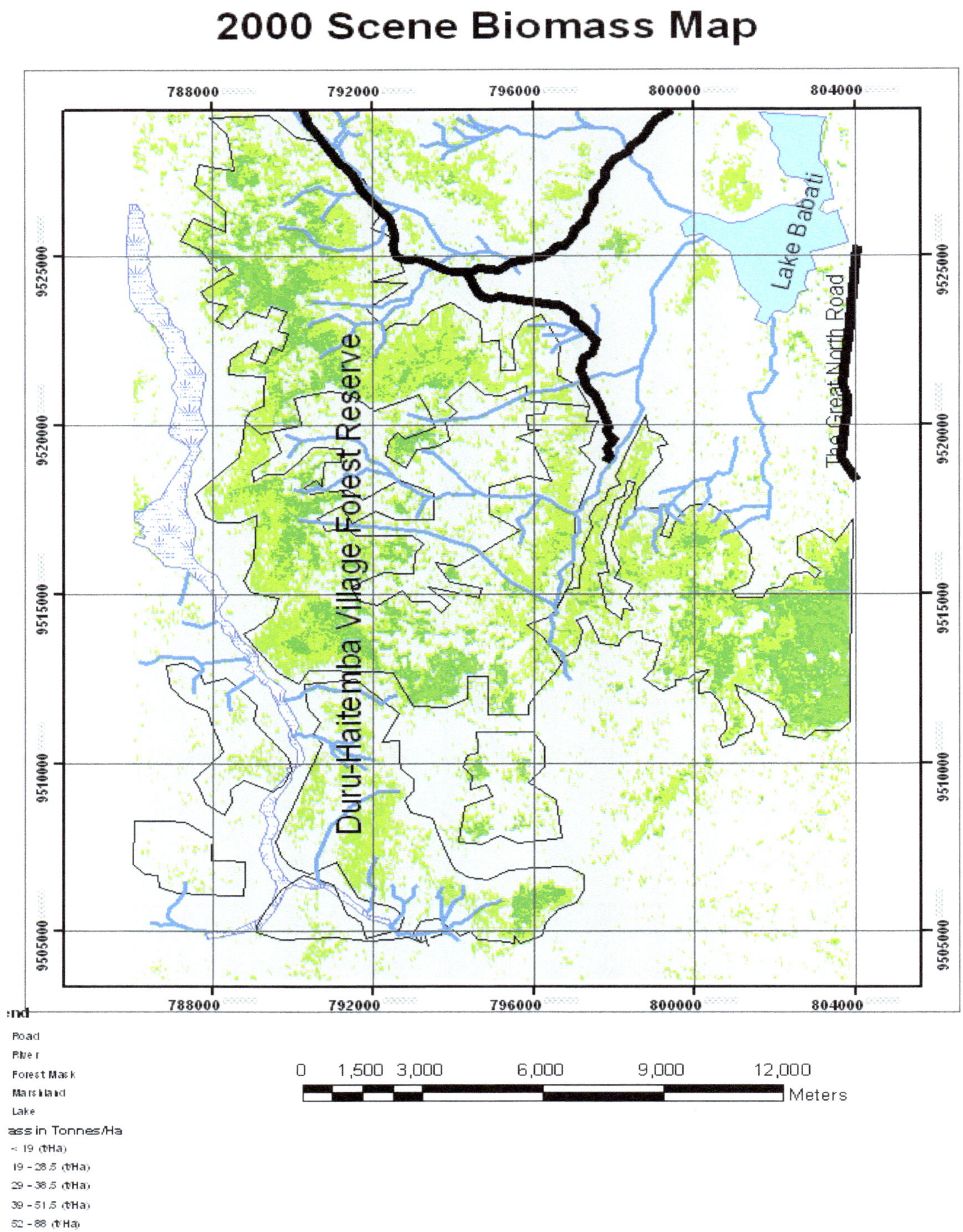

Figure 3c. Above-ground forest biomass distribution map for the 2000 scene path 169 row 63.

that the diameter classes at this dbh class would have similar trend with their nearby classes, as Chidumayo (1997) noted that there is positive correlation between age and wood annual increment in Miombo woodlands at least for the first ten years. Signs of excessive removals were observed for this diameter class, and their corresponding uses were identified in the surrounding villages. Withies are highly used in construc-tion of shelters and household compound fences in the study area, prompting extraction of this diameter class. Furthermore, they are used for baking bricks and as poles for construction of houses.

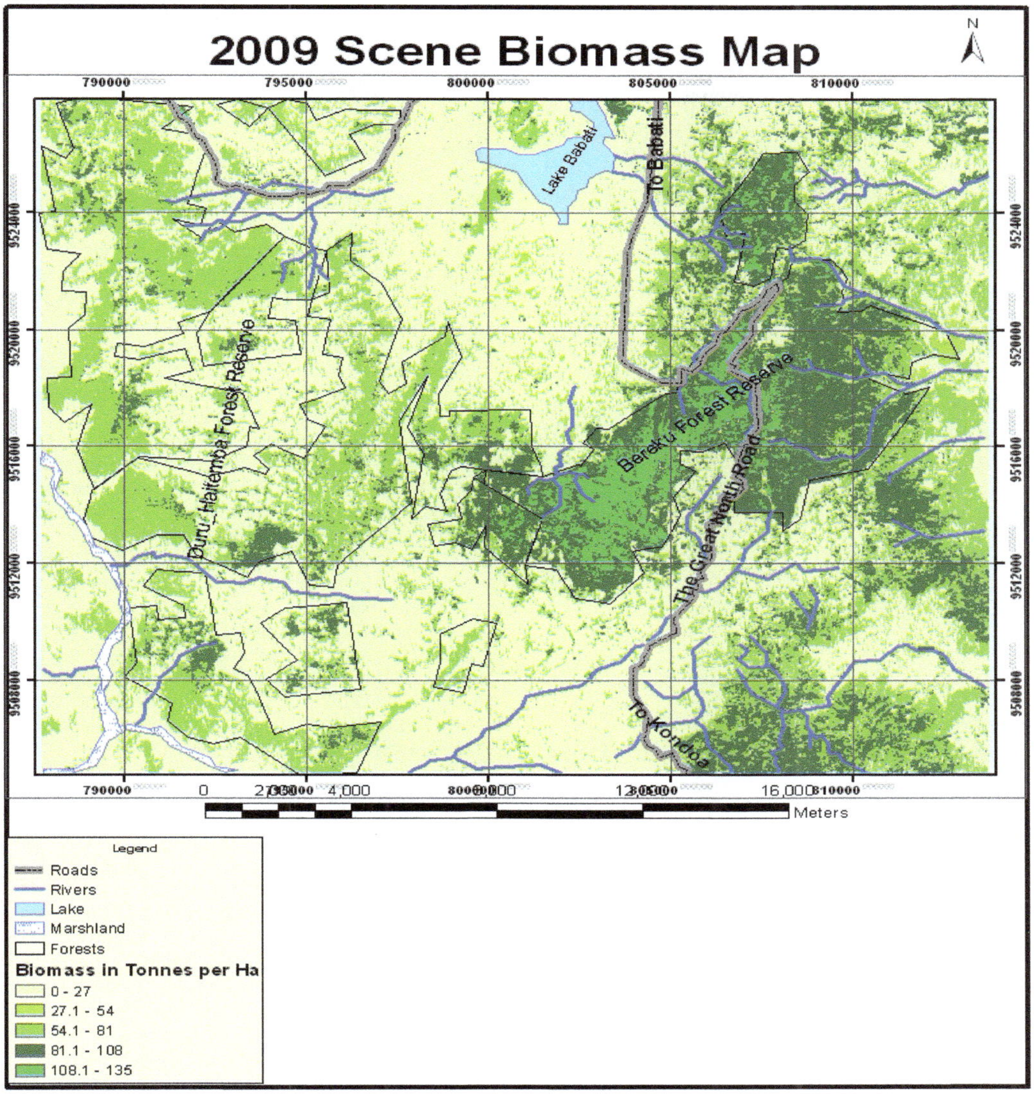

Figure 3d. Above-ground biomass distribution map for the 2009 scene for both path 168, row 63 and 169 row 63.

Ecologically, Miombo woodlands co-exist with fire but seedling shoots undergo a series of death and re-sprouting until they reach a threshold height and diameter to escape fire (Frost, 1996; Zolho, 2005). This may be one of the factors undermining the seedling layer at diameter class of 1 to 4.9 cm in terms of both survival and early growth as is still under the recruitment stage before escaping to sapling stage. Furthermore, as noted by Akinnifesi et al. (2008) early growth of Miombo vegetation differs with species. Frost (1996) reported that the mean biomass in Miombo woodlands tends to increase with increasing mean annual rainfall of a site. However, for this study area there were no significant spatial climatic differences which could have resulted into biomass differences. What is therefore observed in the study site could be a result of direct forest management and application of local forest governance.

Above ground biomass

Total biomass of the study area has been found to increase consistently from 1988, 2000 to 2009. This trend suggests that the forests (Bereku and Duru-Haitemba) are still undergoing growth and have not reached climax stage. As noted earlier, at one point during firm state control without Participatory Forest Management, these forests were overexploited. For instance, Duru-Haitemba was previously managed under traditional customs and taboos, with sometimes punishment less that damage sustained. As a result, the harvesting of the forest products intensified in the late 1970s and early 1980s and thus forests were highly degraded. The state intervened and gazetted the forests under state ownership between 1987 and 1994, and signified some improvement, and furthermore it was later in 1994 that Village Forest Reserve was made to exercise full decentralization (Zahabu, 2008).

Another factor that could have also contributed to this observation is perhaps the forest ecosystem succession, whereby the degraded parts in the forest recovered upon decrease of anthropogenic factors that were undermining the forest cover. Miombo woodlands have high coppicing and re-sprouting ability which make it cope with fire and other detrimental regimes. Luoga et al. (2002) reported that 83% of the 30 harvested species in Kitulangalo forest reserve (Morogoro, east-central Tanzania) were re-sprouting; with the average stump height for all preferred species at 40.6 cm. This means with cessation of excessive human pressure, the Miombo woodlands rapidly regenerates to recover its healthy state.

Conclusion

There was consistently positive change in forest resources from results of forest cover, stocking parameters and above ground biomass. This general trend signifies the improvement in favourable conditions which had positive effect on the state of the forests. Some dynamics in forest condition has also been observed, indicating that there is still a room for improvement of the overall forest resources by restricting unwise use. Increase of management effectiveness through involvement of local people under both joint forest management and community based forest management, strengthened bylaws and provision of other incentives might have contributed greatly to the improvement of the forest resources.

The methodological approach presented in this study offers a simple and fast way to estimate the temporal and spatial dynamics of aboveground biomass and assessment of forest cover changes especially when the ancillary data are unavailable. These findings are therefore an important baseline data source for future scientific research and knowledge in the area of geosciences and forestry biometrics in Tanzania.

ACKNOWLEDGEMENTS

Fund for this study was provided by Ministry of Natural Resources and Tourism, Tanzania, through a project titled Tanzania Forest Conservation and Management Project (TFCMP). Tanzania Forestry Training Institute provided useful logistic support, and special thanks are due to the then Principal, Isaiah Mnang'wone. Mathew Mpanda received independent support from the World Agroforestry Centre. The authors also wish to thank the editor and anonymous reviewers for the contribution to finalization of this work.

REFERENCES

Abd El-Kawya OR, Rød JK, Ismail HA, Suliman AS (2011). Land use and land cover change detection in the western Nile delta of Egypt using remote sensing data. Appl. Geogr. 31:483-494.

Abdallah JM, Monela GG (2007). Overview of Miombo Woodlands in Tanzania. Working Papers of the Finnish Forest Research Institute 50:9-23

Akinnifesi FK, MhangoJ, Sileshi G, Chilanga T(2008). Early growth and survival of three Miombo woodland indigenous fruit tree species under fertilizer, manure and dry-season irrigation in southern Malawi. For. Ecol. Manage. 255: 546-557.

Asner GP (1998). Biophysical and Biochemical Sources of Variability in Canopy Reflectance. Remote Sens. Environ. 64(3): 234-253.

Aziz NM, Hashim M (2007). Mapping of Urban Above-ground Biomass with High Resolution Remote Sensing Data. http://www.isprs.org/istanbul. Accessed on 24th August 2009.

Baccini A, Laporte N, Goetz SJ, Sun M, Dong H (2008). A first map of tropical Africa's above-ground biomass derived from satellite imagery. Environ. Res. Lett. 3:045011 (9 pp).

Bukombe J (2007). "Land Use Impact on Woody Species Composition and Diversity in Miombo Woodland - The Case of Minyughe Woodland in Singida District - Tanzania". In: Proceedings of the Sixth TAWIRI Scientific Conference. Consequences of Global Environmental Changes to Natural Ecosystems. (Julius D. Keyyu and Victor Kakengi, eds.). pp. 95-107.

Chamshama SAO, Mugasha AG, Zahabu E (2004). Stand Biomass and Volume Estimation for Miombo Woodlands at Kitulangalo, Morogoro, Tanzania. S. Afr. For. J. 200:59-64.

Chidumayo E (1997). Miombo Ecology and Management: An Introduction. Practical Action Publishing. 22883 Quicksilver Drive, Sterling, VA 20166-2012. p. 166.

Clarke J, Cavendish W, Coote C (1996). Rural Households and Miombo Woodlands: Use, Value and Management. In The Miombo in Transition: Woodlands and Welfare in Africa (ed B. Campbell), Center for International Forestry Research, Bogor, Indonesia. pp. 101-135.

Cohen WB, MaierspegerTK, Gower ST, Turner DP (2003). An Improved Strategy for Regression of Biophysical variables and Landsat ETM+ Data. Remote Sens. Environ. 84:561-571.

Coppin PR, Bauer ME (1996). "Digital change detection in forest ecosystems with remote sensing imagery". Remote Sens. Rev. 13:207-234.

Danson FM, Bowyer P (2004). Estimating Live Fuel Moisture Content from Remotely Sensed Reflectance. Remote Sens. Environ. 92(3):309-321.

Deo RK (2008). Modelling and Mapping of Above-ground Biomass and Carbon Sequestration in the Cool Temperate Forest of North East China. MSc Thesis. International Institute for Geo-information Science and Earth Observation Enschede, N.L.

Foody GM, Boyd DS, Cutler MEJ (2003). Predictive Relatioship of Tropical Forest Biomass From Landsat TM Data and Their Transferability Between regions. Remote Sens. Environ. 85:463- 474.

Frost P (1996). The ecology of miombo woodlands. In The Miombo in Transition: Woodlands and Welfare in Africa (ed B. Campbell), Center for International Forestry Research, Bogor, Indonesia. pp. 11-57.

Gemmell F, McDonald AJ (2000). View Zenith Angle Effects on the Forest Information Content of Three Spectral Indices. Remote Sens. Environ. 72:139-158.

Gibbs HK, Brown S, Niles JO, Foley JA (2007). Monitoring and estimating tropical forest carbon stocks: making REDD a reality. Available at http://stacks.iop.org/ERL/2/045023. Accessed on 6th January 2009.

Gitelson AA, Kaufman YJ, Stark R, Rundquist D (2002). Novel Algorithms for Remote Estimation of Vegetation Fraction. Remote Sens. Environ. 80:76-87.

Gong P, Howarth PJ (1990). A graphical approach for the evaluation of land-cover classification procedures. Int. J. Remote Sens. 11(5):899-905.

Heiskanen J (2005). Estimating Aboveground Tree Biomass and LAI in Mountain Birch Forest Using ASTER Satellite Data. Int. J. Remote Sens. 27(6):1135-1158.

Isango JA (2007). Management of Indigenous Tree Species for Ecosystem Restoration and Wood Production in Semi-Arid Miombo Woodlands in Eastern Africa. Proceedings of the First MITMIOMBO Project Workshop held in Morogoro, Tanzania, 6th- 12th February 2007. Working Papers of the Finnish Forest Research Institute 50.

Jensen JR (2007). Remote Sensing of the Environment: An Earth Resource Perspective. 2nd Edition. Upper Saddle River, NJ: Pearson Prentice Hall, USA. ISBN: 0131889508. 9780131889507.

Kajembe GC, Nduwamungu J, Luoga EJ (2003). The impact of community-based forest management and joint forest management on forest resource base and local peoples' livelihoods: Case studies from Tanzania: CASS/PLAAS occasional paper series. URL http://www.plaas.org.za/pubs/op/cbnrm/CBNRM08.pdf/ Accessed on 9th November 2009.

Kajembe GC, Ngaga YM, Chamshama SAO (2008). Participatory forest management in Tanzania: Opportunities, challenges and lessons Learned. URL http://www.CERES_formagmt/site.html. Accessed on 9/12/2008.

Kashaigili JJ, Majaliwa AM (2010). Integrated assessment of land use and cover changes in the Malagarasi river catchment in Tanzania. Phys. Chem. Earth 35:730- 741

Kaufman YJ, Tanré D (1992). Atmospherically Resistant Vegetation Index (ARVI) for EOS-MODIS. IEEE Trans. Geosci. Remote Sens. 30:261-270.

Lillesand TM, Kiefer RW, Chipman JW (2004). Remote Sensing and Image Interpretation. 5th Edition. John Wiley & Sons, Inc. New York. USA.

Lovett JC, Pócs T (1993). Assessment of the Condition of the Catchment Forest Reserves. A Botanical Appraisal. The Catchment Forestry Project. Forestry and Beekeeping Division.

Lu D, Mausel P, Brandzio E, Moran E (2004). Relationships Between Forest Stand Parameters and Landsat TM Spectral Responses in the Brazilian Amazon Basin. For. Ecol. Manage. 198:149-167.

Luoga EJ, Kajembe GC, Shemweta DTK, Zahabu E, Mwaipopo CS, Kweka DL (2002). Assessment of Tree Stocking and Diversity for Joint Forest Management (JFM) in Nkweshoo Village Forest Management Area, Kilimanjaro, Tanzania. Forests, Trees Livelihood 15:259-273

Luoga EJ, Witkowski ETF, Balkwill K (2002). Harvested and standing wood stocks in protected and communal Miombo woodland of eastern Tanzania. For. Ecol. Manage. 164:15-30.

Madulu NF (2004). Assessment of Linkage Between Population Dynamics and Environmental changes in Tanzania. AJEAM-RAGEE vol 9:88-102.

Malimbwi RE (2003). Inventory reports of Ayasanda, Bubu, Duru, Endagwe, Endanachan, Gidas, Hoshan and Riroda Village Forest Reserves in Babati, Manyara, Tanzania. Manyara, Tanzania: Land Management Programme, Babati District Council. Unpublished Consultancy Report.

Mayaux P, Lambin EF (1995). Estimation of tropical forest area from coarse spatial resolution data: A two-step correction function for proportional errors due to spatial aggregation. Remote Sens. Environ. 53(1):1-15.

Mpanda MM, Luoga EJ, Kajembe GC, Tron EH (2011). Impact of forestland tenure changes on forest cover, stocking and tree diversity in Amani Nature Reserve, Tanzania. Forests, Trees Livelihoods. 20:215-230.

Mugarura S (2007). Valuing Forest Ecosystem Goods. A case study of Duru-Haitemba Forest Reserves in Tanzania. MSc Thesis. International Institute of Geo-information Science and Earth Observation. Enschede, N.L.

Mwase WF, Bjørnstad A, Bokosi JM, Kwapata MB, Stedje B (2007).The role of land tenure in conservation of tree and shrub species diversity in miombo woodlands of southern Malawi. New For. 33:297- 307

NBS (2002). National Census 2002 Statistical Abstract. Dar es Salaam, Tanzania.

Nshubemuki L, Mbwambo L (2007). Trees to Promote in the Management of Miombo Woodlands in Tanzania: Species, Sizes and Qualities. Proceedings of the First MITMIOMBO Project Workshop held in Morogoro, Tanzania, 6th- 12th February 2007 Working Papers of the Finnish Forest Research Institute 50.

Okuda T, Suzuki M, Numata S, Yoshida Y, Nishimura S, Adachi N, Miyama K, Manokaran N, Hashim M (2004). Estimation of Aboveground Biomass in Logged and Primary Lowland Rainforest Using 3-D Photogrammetric Analysis. For. Ecol. Manage. 203:63-75.

Qi J, Chehbouni A, Heute AR, Kerr H, Sorroshim S (1994). A Modified Soil Adjusted Vegetation Index. Remote Sens. Environ. 48:119-126.

Samalca IK (2007). Estimation of Forest Biomass and its Error. A case in Kalimantan, Indonesia. MSc Thesis. International Institute for Geo-information Science and Earth Observation. Enschede. N.L.

Shackleton CM, Clarke JM (2007). Research and Management of Miombo Woodland for Products in Support of Local Livelihoods. For the World Bank. Version 2. Genesis Analytics. Johannesburg, SA.

Singh A (1989.). Digital Change Detection Techniques Using Remotely Sensed Data; Int. J. Remote Sens. 10(6):989-1003.

Solicha N (2007). Estimating Aboveground Biomass and Carbon Stock of Forest Cover Using Multitemporal Satellite Images in Lore Lindu National Park-Central Sulawesi. MSc. Thesis. Graduate School, Bogor Agricultural University.

URT (1998). National Forest Policy. Ministry of Natural Resources and Tourism - Forestry and Beekeeping Division. Dar es Salaam, Tanzania. Government Press.

URT (2002). Forestry in Figures, Ministry of Natural Resources and Tourism-Forestry and Beekeeping Division. Dar es Salaam, Tanzania. Government press.

URT (2003). Participatory Forest Management: A report on Lesson learnt. Ministry of Natural Resources and Tourism. Forestry and Beekeeping Division. Dar es Salaam. pp. 80.

Vihemäki H (2009). Participation or further exclusion? Contestations over Forest Conservation and Control in the East Usambara Mountains, Tanzania. Ph.D. Thesis in Development Studies. Interkont Books 18, Institute of Development Studies, University of Helsinki.

Wahyuningrum N (2006). Foliage Biomass Estimation in Tropical Logged Over Forest. East Kalimantan, Indonesia. MSc Thesis. International Institute for Geo-information Science and Earth Observation. Enschede.

Zahabu E (2008). Sinks and Sources: A Strategy to involve forest communities in Tanzania in global climate policy. PhD dissertation.University of Twente. 249 pp.

Zolho R (2005). Effect of Fire Requency on the Regeneration of Miombo Woodland in Nhambita, Mozambique. A Dissertation Presented for the Degree of Master of Science University of Edinburgh.

Ecosystem recovery analysis of mine tailings at Mhangura copper mine through normalised difference vegetation index

Gregory M. Dowo and Shakkie Kativu

Tropical Resource Ecology Programme, Department of Biological Sciences, University of Zimbabwe, P. O. Box MP167, Mount Pleasant, Harare, Zimbabwe.

Remote sensing techniques are increasingly being employed in monitoring environmental change. Vegetation indices such as normalised difference vegetation index (NDVI), are useful in estimating primary production, an important component of ecosystem function. The success of rehabilitation on mine tailings may be indirectly assessed through NDVI changes. Mine tailings at Mhangura copper mine, Zimbabwe, have not been monitored since their rehabilitation following the mine closure in 2000. The present study indirectly assesses primary productivity over the tailings dams as a way of establishing the extent of ecosystem recovery. NDVI images were obtained from the United States geological survey (USGS), global visualisation viewer (GloVis) website. These were processed using the Integrated Land and Water Information System (ILWIS) Version 3.3 GIS software. Two-way ANOVA showed significant differences in NDVI between the sites ($p < 0.05$). Actual NDVI trends over the years could not be established. NDVI values over the dams were also significantly different from those in adjacent natural Acacia woodland. These observations were supported by principal components and hierarchical cluster analyses. The NDVI analysis thus showed a good establishment of grass on the dams with poor tree cover when compared with the Acacia woodland. This suggests a positive movement towards recovery. However, factors preventing tree establishment need to be investigated.

Keywords: Ecosystem development, normalised difference vegetation index (NDVI), remote sensing, tailings rehabilitation.

INTRODUCTION

The increasing scale of human induced environmental disturbance has necessitated the use of remote sensing techniques in monitoring environmental change (Kerr and Ostrovsky, 2003). Vegetation indices based on differing spectral reflectance in the red and near infrared bands have recently been widely used in assessing vegetation productivity from canopy to global scale (Gamon et al., 1995). Normalised difference vegetation index (NDVI) has proved to be a useful tool in estimating net primary productivity (NPP), a major component of ecosystem function (Kerr and Ostrovsky, 2003). This is mainly due to its linear relationship with absorbed photosynthetic active radiation (APAR) which then provides a "theoretical connection" between NDVI and net primary production (NPP) (Paruelo et al., 1997). NDVI remains the most commonly employed and studied vegetation index (Kerr and Ostrovsky, 2003), and can be represented by the following equation:

$$NDVI = (RNIR - RRED)/(RNIR + RRED)$$

Where, RNIR and RRED designate reflectance in the near infrared and red bands, respectively (Gamon et al., 2005).

Mine tailings rehabilitation usually starts when a mine is still operational. Much of the monitoring is done during this operational period as mines are required to provide evidence of stability before making a case for closure (Haagner et al., 2008). In Zimbabwe, once a mine is closed, miners are exonerated from any legal liability concerning their mining activities. Thus, there is usually no follow up on long term ecosystem recovery once vegetation cover establishes on the tailings dams. The Zimbabwe Environmental Management Act [Act 13 of 2002 Chapter 20:27] explicitly calls for the restoration of biota and ecological processes on mine tailings dams. Hence, rehabilitation of mine tailings becomes a major step towards land restoration. The Society for Ecological Restoration International (SER) defines ecosystem restoration as "the process of assisting the recovery of an ecosystem that has been degraded, damaged, or destroyed with respect to its health, integrity and sustainability" (SER, 2004). This, therefore, entails the need to establish a self sustaining ecosystem (Aronson et al., 2003; SER, 2004).

The primary objective of the present study was to track the environmental restoration course of the rehabilitated Mhangura Copper Mine through an assessment of primary productivity on mine tailings dams using NDVI as a proxy. Primary productivity acts as a surrogate for ecosystem development at a given period (Raich et al., 1997). When expressed as a ratio of community respiration, primary production can serve as an excellent indicator of ecosystem maturity (Odum, 1969). The use of remote sensing techniques for mapping and planning mine waste disposal sites is gathering impetus as evidenced by their application across southern Africa (Limpitlaw, 2006) and Australia (Lau et al., 2008; Ong et al., 2002). Remote sensing techniques are also increasingly applied in re-vegetation monitoring in China (Wei et al., 2009). Thus, changes in primary productivity on the slimes dumps was assessed by the use of NDVI as an indicator of productivity in order to establish trends in ecosystem development. It was hypothesised that any change in primary production (NDVI), especially an increase towards reference site values over time, indicates an increase in ecosystem functionality.

MATERIALS AND METHODS

Study area

Mhangura Copper Mine lies approximately 186 km northwest of the city of Harare. It is located in the Mashonaland West Province of Zimbabwe, at approximately 16° 53' S and 30° 09' E (Figure 1). The mine area is surrounded by farmland that is mainly used for maize crop cultivation and, to a lesser extent, livestock grazing. Mhangura lies in Natural (Agro-climatic) Region II (Vincent and Thomas, 1960) that is characterised by high rainfall (750 to 1000 mm year^{-1}), and intensive farming. This includes a dry, cool season from May to September when very little rain falls. The wet season is usually characterized by heavy rainfall from November to March. Geologically, Mhangura lies in the largest, old stratabound copper occurrences within the Proterozoic metasediments of the Lomagundi basin. The mine's original size was estimated to contain some 60 metric tons of copper deposit, at 1.2% Cu, with 15 to 20 ppm silver. There are minor occurrences of gold, platinum and palladium.

Stratification

The study focused on rehabilitated slimes dams (Dams 4, 5 and 6) within the Miriam Shaft Complex. Though they were constructed separately, Dams 4 and 5 coalesce, such that they appear to be a single landform. Hence, for simplicity, in the present study, Dams 4 and 5 were referred to as Dam 5. This particular slimes dam is older, having been constructed in the early 1970's, whereas Dam 6 was constructed in the early 80's. The substrate on the dams is homogenous for all sites, although a cladding of organic matter in the form of domestic waste was also used. Sites were initially stratified according to dam number, which is related to dam age, thus resulting in a pseudo-chronosequence. If succession was taking place on the dams, it was assumed that different vegetation types would be observed based on the different ages of the dumps. Aspects provided a second criterion for stratification. Sampling sites were located on the North, South, East and West sides of the dumps (Figure 2). Such stratification bore significance on the naming of the study sites. There was delineation between sampling sites and units. The whole side of a dam was considered as a sampling unit in which several sites were delineated, e.g. D5N-01, D5N-02 and D5N-03 were different sampling sites within sampling unit D5N (northern slope of Dam 5). Stratified random sampling techniques were employed for the delineation with the aid of Google Earth® software. In each case, sampling sites were named in accordance with the slimes dam number first, followed by the aspect and the transect number. Thus, D5N-02 would be the second transect/site surveyed on the northern side of Dam 5.

Reference sites

An acacia woodland lying to the north of and adjacent to Dam 8 was used as a reference site, representing a natural functional ecosystem (Figure 2). Reference sites are supposed to occur in similar biotic zones, in close proximity to the rehabilitated site, and exposed to similar natural disturbances (SER, 2004). Though the site was flat, and did not have similar topography as the slimes dams, it served in providing the biogeochemical potential for rehabilitation. It was also important to compare the rehabilitated sites with a site representing a dysfunctional ecosystem. For this purpose, the top of Dam 7 (also flat) was used (Figure 2). The dam

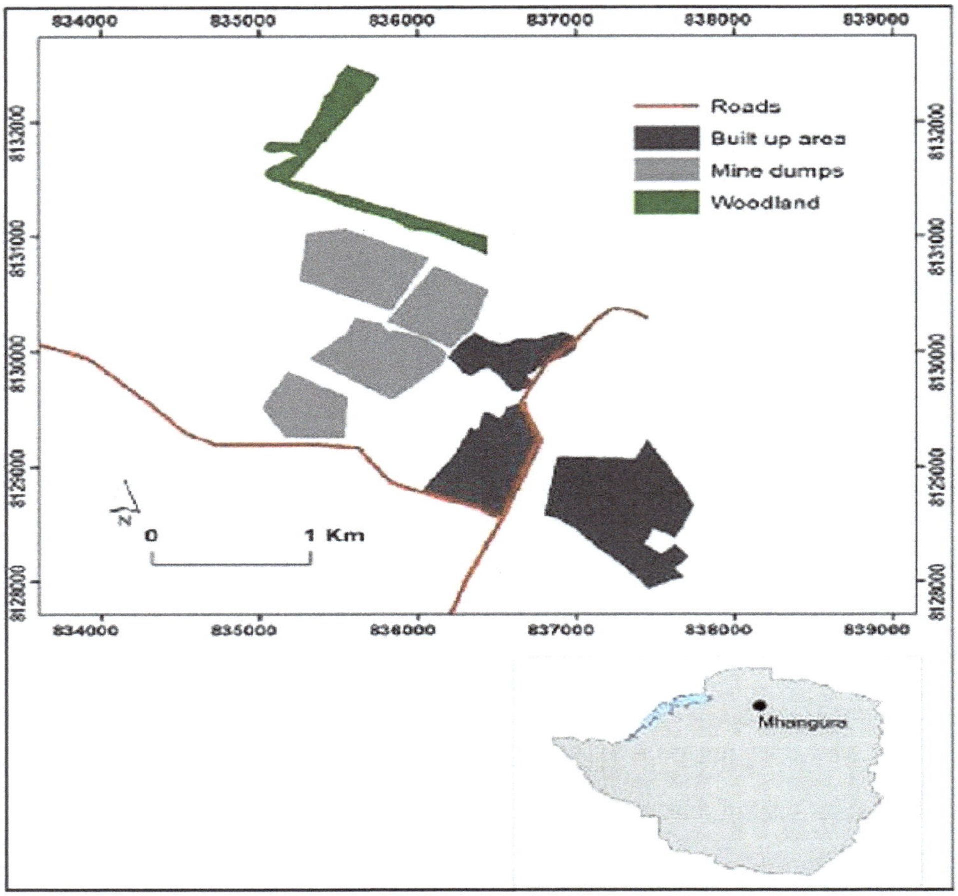

Figure 1. Map of Mhangura Mine showing the location of mine dumps (slimes dams).

was constructed in the late 1980's as an infill between Dams 3 and 5. The mine closed before the dam was fully rehabilitated.

Randomisation

Sampling points were randomly selected on each side of a dump and in the reference sites with the aid of Google Earth® satellite imagery software. The points demarcated positions of sampling sites. Coordinates of each point were recorded and then located in the field using a hand held Global Positioning System (GPS).

Data acquisition and image processing

NDVI data were collected with the assistance of staff of the Department of Geography and Environmental Sciences, University of Zimbabwe. NDVI images were downloaded from the United States Geological Survey (USGS), Global Visualisation Viewer (GloVis) website (www.glovis.usgs.gov). Images from the following dates were available from the USGS archives, and were thus used to show changes over the years: 06 May 1990, 22 September 1994, 30 September 2000 and 05 May 2010. In Zimbabwe, the months of September and May occur within the dry season. This avoids any differences in NDVI which is attributed to differences in season. The images were processed using the Integrated Land and Water Information System (ILWIS) Version 3.3 GIS software which employed map-value function to extract NDVI values from sample locations.

Classification of NDVI values

The classification of NDVI values was carried out according Kriegler et al. (1969) and Weier and Herring (2011). Calculations of NDVI for a given pixel always result in a number that ranges from minus one (-1) to plus one (+1); however, no green leaves gives a value close to zero. A zero means no vegetation and close to +1 (0.8 - 0.9) indicates the highest possible density of green leaves. Very low values of NDVI (0.1 and below) correspond to barren areas of rock, sand or snow. Moderate values represent shrub and grassland (0.2 to 0.3), while high values indicate temperate and tropical rainforests (0.6 to 0.8).

Statistical analysis

Available NDVI data were not sufficient for a full time series analysis. There was still need, however, to assess the existence of any significant changes over the years, and to establish whether any differences existed in terms of NDVI between slimes dams and reference sites. A two-way ANOVA was, therefore, carried out on the data set using the MS Excel Data Analysis package (Windows 2007). The observations were the different sites and the variables

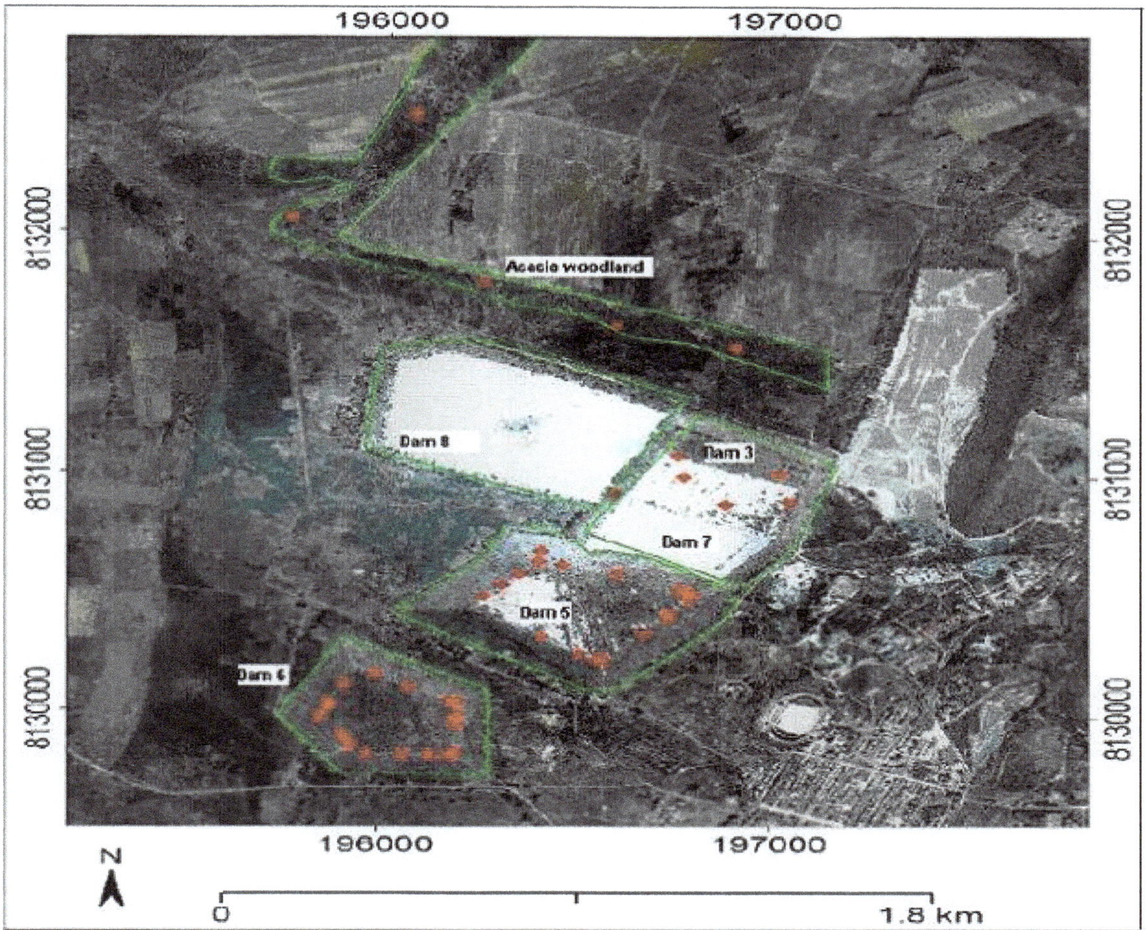

Figure 2. A Google Earth® satellite image showing sampling sites (red dots) on the slimes dams and in the Acacia woodland reference site.

were the NDVI values. To establish whether there were any relationships among the sites in terms of NDVI changes over the years (1990, 1994, 2000 and 2010), a principal components analysis (PCA) was carried out on the data set using program package PAST (Hammer et al., 2001). To further explore the relationships between the sites, a Hierarchical Cluster Analysis (HCA) was carried out on the data set using program package STATISTICA (StatSoft, 1995).

RESULTS

NDVI imagery

Vegetation index images are useful in tracking progress of dump rehabilitation (Limpitlaw, 2006). Thus, NDVI images were used for monitoring vegetation changes on slimes dams at Mhangura Copper Mine. The green shade on the images represents green vegetation. The white-bluish shade represents the bare tailings. Black areas represent burnt vegetation. Figure 3 shows vegetation development on the slimes dams over the years from 1990 to 2010. Dams 6 and 5 showed a considerable amount of greenness by 2010.

NDVI values

Dam 5

There are 4 sets of data corresponding to 16 location points, that is, a total of 64 data points. Among them only one value is more than 0.3 (D5W02 - 1990); there are six other values (2 in 1990; 0 in 1994; 1 in 2000; and 3 in 2010) ranging between (0.2 and 0.3) representing shrub and grassland.

Dam 6

There are 4 sets of data corresponding to 18 location points, that is, a total of 72 data points. Among them only one value is more than 0.3 in 2010 (D6E04) while there

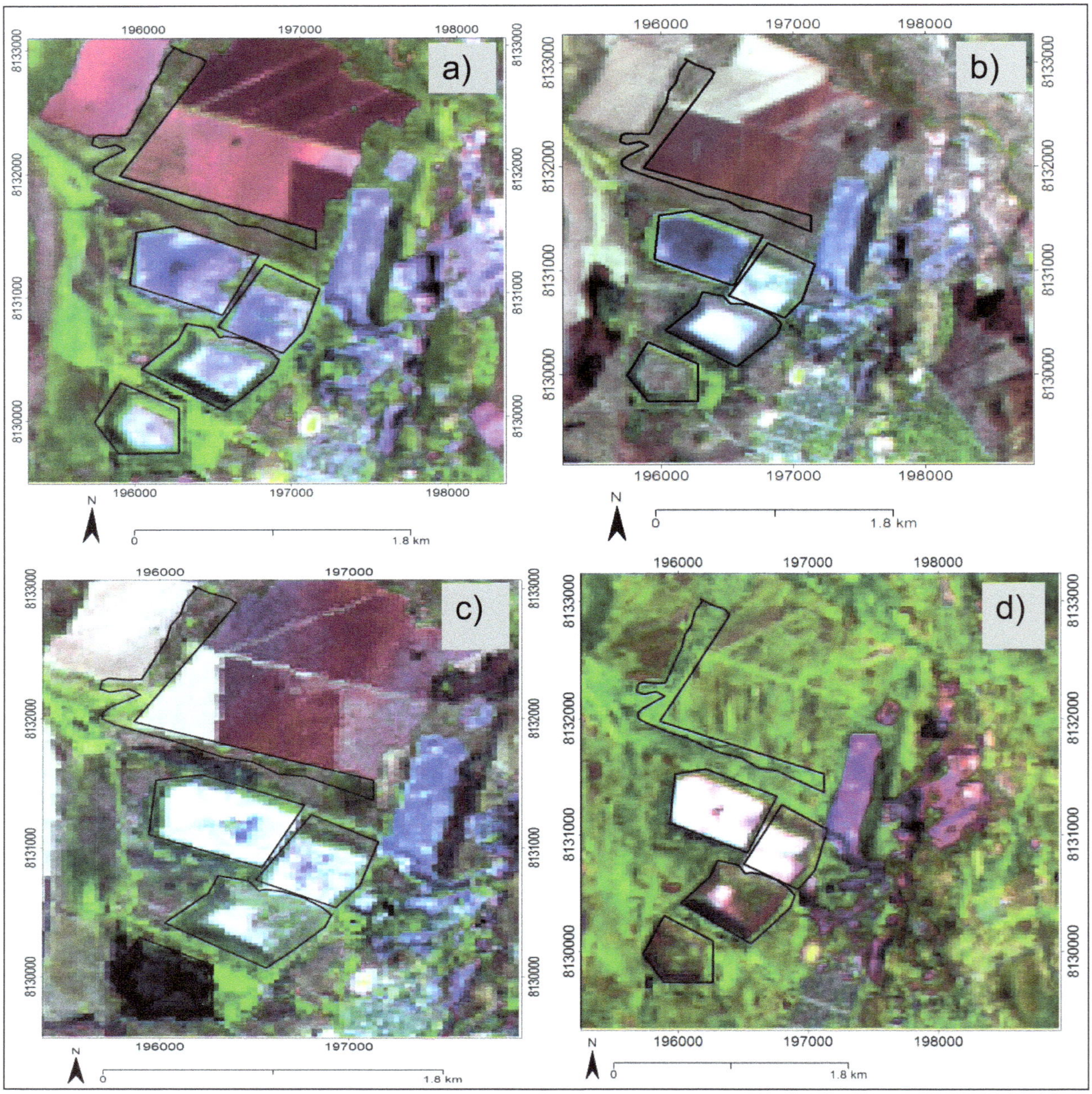

Figure 3. Colour composite Landsat satellite images covering the Mhangura area for the years 1990, 1994, 2000 and 2010 (a, b, c and d, respectively). Also refer to Figure 2.

were 7 values greater than 0.3 in 1990 (D6W02, D6W03, D6W04, D6NW01, D6NW02, D6N01 and D6N02); there are 15 other values (2 in 1990; 0 in 1994; 2 in 2000; and 11 in 2010) ranging between (0.2 and 0.3) representing shrub and grassland.

Dam 7

There are 4 sets of data corresponding to 4 location points, that is, a total of 16 data points. Among them only one value is more than 0.3 (D701 - 2010); there are 2 other values (D701 - 1994; D704 - 2010) ranging between (0.2-0.3) representing shrub and grassland.

Acacia woodland

There are 4 sets of data corresponding to 6 location points, that is, a total of 24 data points. Among them, there are 6 values greater than 0.5 (AW01 - 06 for 2010) representing dense woodland. There are 4 values between 0.2 and 0.3 representing shrub and grassland, 2

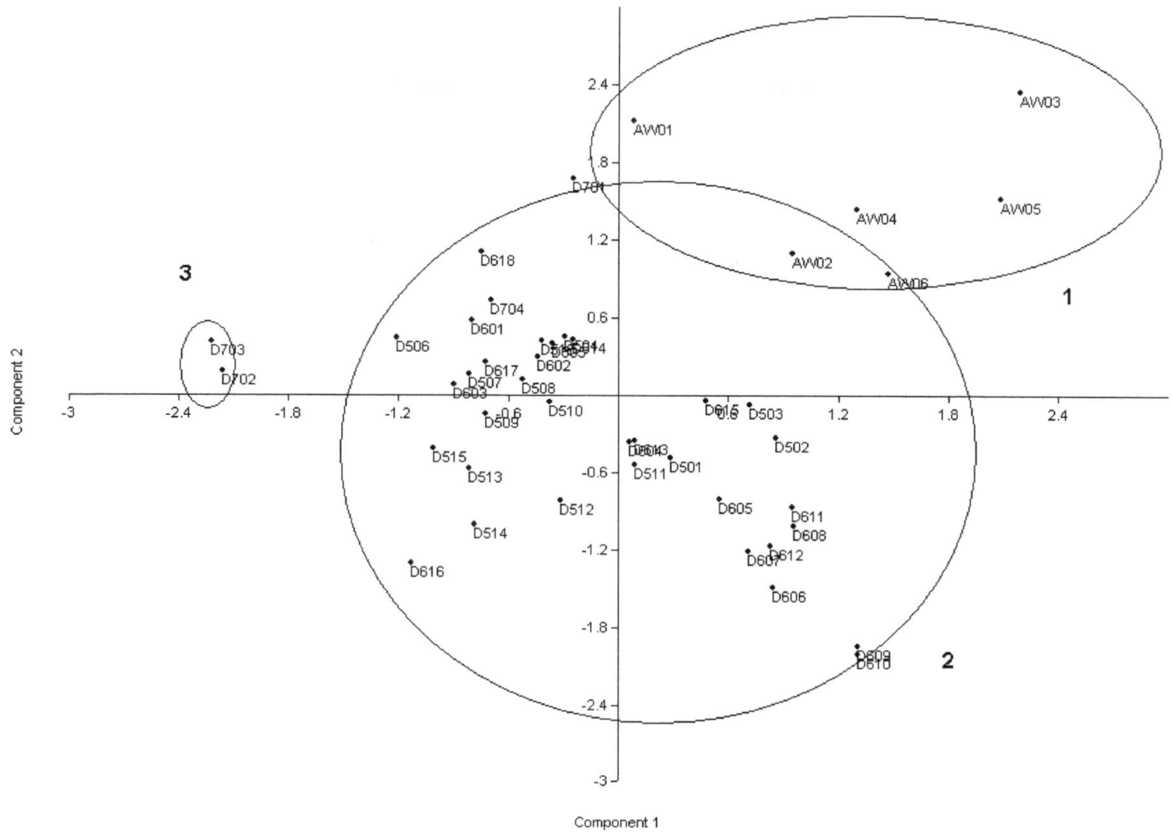

Figure 4. Principal components analysis (PCA) of the NDVI data. The numbers indicate clusters of related sites.

of them occurring in 1990 and the others in 2000. All values for 1994 were less than 0.1 with 1 being negative, representing a bare area.

Statistical analysis

A two-way ANOVA on the data set revealed significant differences between years and sites ($p < 0.05$). The trend implied by these differences, however, could not be established. To further investigate whether there were any differences among the sites in terms of NDVI changes over years 1990, 1994, 2000 and 2010, a principal components analysis (PCA) was carried out on the data set. With an eigenvalue of 0.065093, Component 1 explained 45.271% of the variance, and with an eigenvalue of 0.0402642, Component 2 explained 28.003% of the variance. Together, these two components accounted for 73.274% of the variance.

The PCA (Figure 4) produced three clusters of related sites. The Acacia woodland sites grouped on their own, separate from the slimes dams' sites. Two sites from Dam 7, that is, D703 and D704, also produced a separate cluster. This shows that although the dam sites (Dams 5 and 6) had vegetation cover, their NDVI levels were different from those of the Acacia woodland. Sampled sites of Dam 7 that formed a third cluster (Cluster 3) had no vegetation cover. To further illustrate the differences between the sites, a Hierarchical Cluster Analysis (HCA) was carried out on the data set (Figure 5).

In contrast to the PCA, the HCA produced four clusters. These were: Acacia woodland, Dam 5, Dam 6 and Dam 7. The HCA also had the Acacia woodland clustered furthest from the other sites. It was followed by Dam 7. Although Dams 5 and 6 were different from each other, they lay closer to each other and away from the other sites. Site D616, though exhibiting some difference from the other Dam 6 sites, it was considered to be a sub-cluster of Cluster 1.

DISCUSSION

A visual interpretation of the images in Figure 3 shows that by 2010, the green cover on top of Dam 5 and 6 was extensive (plate d) except for a bare patch on the left corner of Dam 5 (see Figure 2 for dam labels). Other plates show considerable bareness especially 1990 and 1994. The low values recorded in 1994 are attributed to

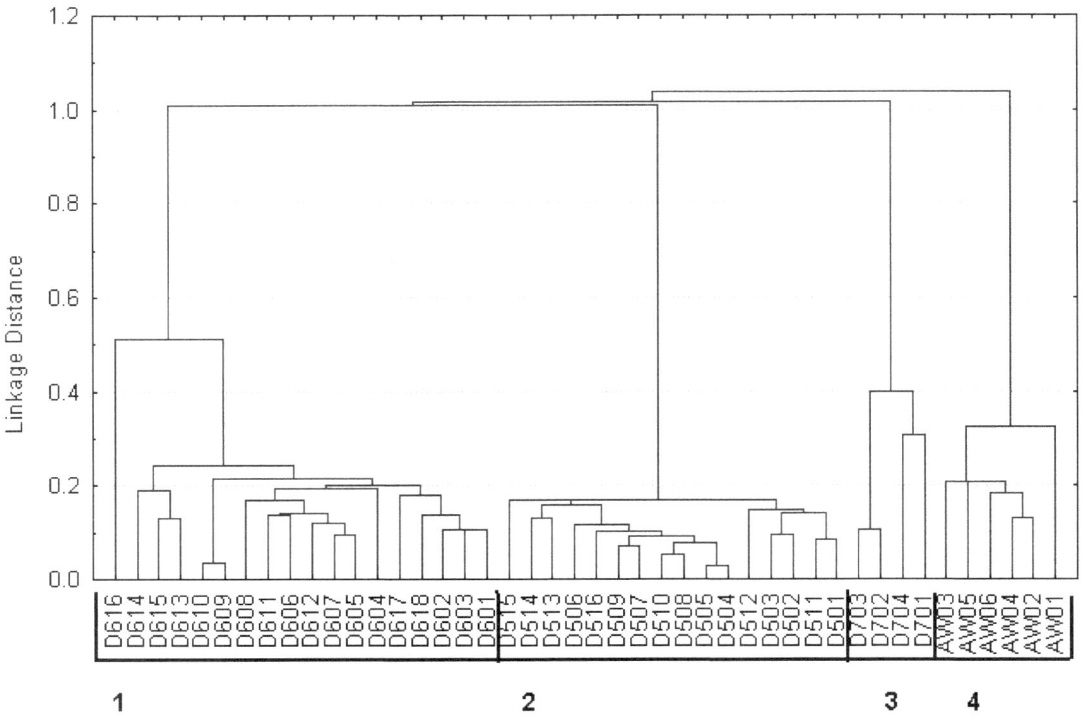

Figure 5. A Hierarchical cluster analysis of the NDVI data. Numbers indicate clusters of related sites.

the 1992 drought that resulted in the death of some plant species as reported by Northard and Figg (1992). A time series analysis which could elucidate the actual trend in NDVI between 1990 and 2010 was, however, not carried out. NDVI images obtainable from Landsat archives (www.glovis.usgs.gov) were only available for a few selected years and months. Other years had images that were not interpretable due to obscurity caused by cloud cover.

The statistical analyses, however, indicate that in terms of NDVI, the slimes dams are still different from the natural Acacia woodland. This is attributed to the differences in tree cover that produced higher NDVI values for the Acacia woodland. There was no significant difference between Dams 5 and 6, although by 2010, Dam 6 recorded more sites with higher NDVI values. Both dams contain sparse tree vegetation and are mostly covered with grass. Dam 7 is devoid of any vegetation as shown by the negative NDVI values and its apparent difference from the rest of the sites. However, it should be noted that a small data set was available for Dam 7 and Acacia woodlands reference sites affecting statistical robustness. Nevertheless, Raich et al. (1997) suggested that primary production of an ecosystem could act as a surrogate for ecosystem development trends over time. Thus, it is evident that grass cover has sufficiently developed over the slimes dams. The same cannot be said for trees and shrubs which are sparsely distributed and very few.

Conclusions

NDVI trends through 1990 to 2010 could not be established. However, it could be established that the levels of NDVI between the tailings dams and Acacia woodland are significantly different. By 2010, there was a considerable grass cover on the slimes dams with a poor tree cover. There is a need to investigate further, the factors preventing the establishment of trees on the tailings dams. Human-induced disturbances such as fire and wood harvesting are possible factors. The establishment of grass on the tailings dams points to ecosystem recovery to a significant extent.

ACKNOWLEDGEMENTS

Much thanks to Dr. A. Murwira and Mr G. Mutowo, Department of Geography, University of Zimbabwe for help with the NDVI image acquisition and analysis. Heartfelt gratitude also goes to the anonymous reviewers for their insightful comments and suggestions.

REFERENCES

Aronson J, Floret C, Le Floc'h E, Ovalle C, Pontanier R (1993). Restoration and rehabilitation of degraded ecosystems in arid and semi-arid lands. I. A view from the South. Restor. Ecol. 1:8 - 17.

Environmental Management Act [Chapter 20:27] Act 13 of 2002. URL http://www.law.co.zw/downloads/statutes/20/Environmental%20Management%20Act.pdf [accessed on 21 May 2010].

Gamon JA, Christopher B, Field ML, Goulden, KL, Griffin AE, Hartley GJ, Josep P, Riccardo V (1995). Relationships between NDVI, canopy structure, and photosynthesis in three Californian vegetation types. Ecol. Appl. 5:28-41.

Haagner ASH, Tongway DJ, Kellner K (2008). Enhancing conventional rehabilitation monitoring by adding landscape function characteristics. In: Fourie A, Tibbett M, Weiersbye I, Dye P (Eds.). Mine Closure 2008: Proceedings of the 3rd International Conference on Mine Closure, 14 - 17 October 2008, Johannesburg, South Africa. Australian Centre for Geomechanics, Perth.

Hammer O, Harper DAT, Ryan PD (2001). PAST: Paleontological Statistics Software package for Education and Data Analysis. Palaeontol. Electron. 4:9.

Kerr JT, Ostrovsky M (2003). From space to species: ecological applications for remote sensing. Trends Ecol. Evol. 18: 299 - 305.

Kriegler FJ, Malila WA, Nalepka RF, Richardson, W (1969). Preprocessing transformations and their effect on multispectral recognition. Proceedings of the Sixth International Symposium on Remote Sensing of Environment. University of Michigan, Ann Arbor, Michigan. pp. 97-131.

Lau IC, Hewson, RD, Ong, CCH, Tongway, DJ (2008). Remote minesite rehabilitation monitoring using airborne hyperspectral imaging and landscape function analysis (LFA). Int. Arch. Photogramm. Remote. Sens. Spat. Inf. Sci. 37:325-330.

Limpitlaw D (2006). Use of remotely sensed imagery and methods for mapping and planning mine waste facilities. SAIMM Colloquium: Mine waste disposal and achievement of mine closure - what does it take? SA Military History Museum, Johannesburg.

Northard WP, Figg I (1992). Slimes dam stabilisation and rehabilitation as practised at Mhangura Mine. Chamb. Min. J. 34: 11 - 15.

Odum EP (1969). The strategy of ecosystem development. Sci. 164: 262- 270.

Ong C, Tongway D, Caccetta M, Hindley N (2002). Deriving Ecosystem Function Analysis Indices from Airborne Hyperspectral Data. CSIRO Exploration and Mining Technical, Canberra.

Paruelo JM, Epstein HE, Lauenroth WK, Burke IC (1997). ANPP estimates from NDVI for the Central Grassland Region of the United States. Ecol. 78:953-958.

Raich JW, Russell AE, Vitousek PM (1997). Primary productivity and ecosystem development along an elevational gradient in Mauna Loa, Hawai'i. Ecol. 78:707-721.

Society for Ecological Restoration International Science and Policy Working Group (SER) (2004). The SER Primer on Restoration Ecology. Society for Ecological Restoration International. URL http://www.ser.org/pdf/primer3.pdf [accessed on 1 May 2010]

StatSoft Inc. (1995). STATISTICA for Windows [Computer Program Manual]. Tulsa, Oklahoma.

Wei S, Zhong-Ping S, Dao-Liang S, Raaj R, Xiang Z, Xiang-Yun G (2009). Vegetation recovery monitoring over the waste dump area in Haizhou Opencast Coal Mine area during 1975-2000 using NDVI and VF Index. J. Indian Soc. Remote. Sens. 37: 631-645.

Weier J, Herring, D (2011) "Measuring Vegetation (NDVI & EVI)" (http://earthobservatory.nasa.gov/Features/MeasuringVegetation/ [accessed on 23rd Sept, 2011]).

9

Market conduct and performance of wild and semi-wild food plants traded in Bunyoro-Kitara Kingdom, Uganda

Jacob Godfrey AGEA[1] , Clement Akais OKIA[1], Bernard Bonton OBAA[1], James Munga KIMONDO[2], Prossy ISUBIKALU[1], Dino Andrew WOISO[3], Joseph OBUA[4] and Zewge TEKLEHAIMANOT[5]

[1]Department of Extension and Innovation Studies, College of Agriculture and Environmental Sciences, Makerere University, P.O. Box 7062 Kampala, Uganda.
[2]Kenya Forestry Research Institute, P.O Box 20412-00200, Nairobi, Kenya.
[3]Department of Biological Science, Sokoine University of Agriculture, P.O. Box 3038, Morogoro, Tanzania.
[4]Inter-University Council of East Africa, P.O Box 7110, Kampala Uganda.
[5]School of Environment, Natural Resources and Geography, Bangor University, Bangor-Gwynedd, LL57 2UW, United Kingdom.

This study assessed the market conduct and performance of wild and semi-wild food plants (WSWFPs) traded in Bunyoro-Kitara Kingdom, Uganda. A rapid market survey (RMS) was conducted in 17 local markets in Kibanda County. Market prices and weekly volumes of traded WSWFPs were compared with some of the selected conventional food plants traded in the same locality. Weekly volumes of traded WSWFPs based on the usual units of the measurement (including bundles and heaps) within the markets were estimated per species sold. The profit margin was computed per traded species. Transport expenses were excluded in the cost computation because only 4% of the traders incurred transport expenses in form of hired bicycles. Out of 62 WSWFPs belonging to 31 botanical families documented as edible in the Bunyoro-Kitara Kingdom, about 47% belonging to 12 botanical families were traded in formal and informal markets. Market information system was largely rudimentary and undeveloped, and traders rely mainly on information from fellow traders as well as their customers to make market decisions. Traded products were primarily delivered to markets on foot and using bicycles. Currently, there are no definite or formal mechanisms of setting prices of traded WSWFPs; most traders relied on the daily market demand, time and risks involved in gathering process, information of the price of substitute food and prices from other areas, knowledge of the past seasons' prices, and on the costs incurred from the suppliers. With exception of few species such as *Physalis peruviana* and *Basella alba*, weekly volumes of traded WSWFPs were low as compared to most conventional food crops. On the other hand, prices of most traded WSWFPs were generally similar to those of alternative conventional food plants marketed in the area. Some WSWFPs like *Hibiscus acetosella*, *B. alba* and *Hyptis spicigera* (seeds), had higher market prices per unit measurement as compared to the related conventional food plants. Average weekly profits yielded from the trade of various WSWFPs were moderate and ranged from UGX 764.5 to 6754.2 (USD 0.38 to 3.36). The highest return came from species such as *H. spicigera, Hibiscus sabdariffa*, *Aframomum angustifolium, Borassus aethiopum*, *B. alba*, *Solanum nigrum*, *Aframomum alboviolaceum* and *Canarium schweinfurthii.*

Key words: Market conduct, market performance, wild edibles, wild foods, Uganda.

INTRODUCTION

Wild and semi-wild food plants (WSWFPs) have significant cultural, biological and economic value at local, regional and national levels (Agea, 2010; Agea et al., 2011). People who utilize WSWFPs to meet these needs often rely on organized exchange systems to obtain them (De Beer and McDermott, 1989; Moreno-Black, 1991; 1994; Moreno-Black and Price, 1993; Ngarmsak, 1987; Scoones et al., 1992). An important aspect of recent ethnobiological studies has been the focus on how resources are defined, appropriated and distributed. Most subsistence-based communities are linked to larger economic and political systems through markets. Consequently, markets are a valuable arena for gathering information on people-resource relationships (Agea et al., 2008; Kimondo et al., 2010). Recent marketplace studies have shown the importance of this domain for monitoring changing selection pressures on specific resources, that is, selection by people because of culturally defined qualities of the items (Bye and Linares, 1983, 1990; Jacquat, 1990; Johnson and Johnson, 1976; Schlage, 1969; Scoones et al., 1992; Wester and Chuensanguansat, 1994; Whitaker and Cutler, 1966). Indeed the importance and complexity of food marketing systems is normally associated with economic and socio-political development. In efficient and balanced marketing systems, consumers get the highest food value at the lowest possible price, farmers or gatherers (in case of WSWFPs) obtain the highest possible returns from food sales, and the market agents earn the greatest profit possible (Kohls and Uhl, 2002).

The markets for wild edible products have been compared with cultivated products in some parts of Africa. For instance, Okafor (1980) found out that the prices for wild products are comparatively higher, especially during the seasonal shortfall Johnson and Johnson (1976) reported that the periods when other cultivated products are unavailable. marketing of *Irvingia gabonenesis*- a wild fruit is particularly important in this period in southern Nigeria where traders even import this wild fruits from Cameroon. Many more wild products have potential economic values, which are yet unexploited (Arnold et al., 1985; Rasoanaivo, 1990). Although these wild and semi-wild food resources make a significant contribution to rural livelihoods and to the national economies of many countries, the magnitude of the income derived from them and particularly wild food plants is not well known, due to a lack of systematic and rigorous data collection at country level (FAO, 2000). Besides this, a recent analysis of markets in sub-Saharan and southern Africa revealed that markets for indigenous and wild edible plants are largely informal, small and volatile (Russell and Franzel, 2004) which often escape the attention of policy makers. As such, these resources are often ignored in official nutritional investment projects.

In Uganda, there are amazingly few accounts of documented trade in WSWFPs, despite growing importance of the WSWFPs in the country. Worst still, there is a dearth of information on the market conducts and performance of wild or semi-wild food resources in the country. In this paper, we reported the findings of an assessment of the market conduct and performance of WSWFPs traded in Bunyoro-Kitara Kingdom, Uganda. The assessment was guided by the following research questions: Which WSWFPs are traded in Bunyoro-Kitara Kingdom? What are the sources of market information? How are the traded WSWFPs transported to the market locations? How are their market prices determined? What quantities of WSWFPs are traded weekly in the market during the harvest seasons? What are the average unit prices of the traded WSWFPs and how is the market demand of traded WSWFPs like? Lastly, what are profit margins for traded WSWFPs during the harvest seasons?

MATERIALS AND METHODS

A rapid market survey (RMS) was conducted in 17 local markets in Kiryandongo and Mutunda sub-counties of Kibanda County (Figure 1). Kibanda is one of the counties of Bunyoro-Kitara Kingdom, which is located in the western region of Uganda. It lies between 0°36' and 2°20' N, and 30°30' and 32°23' E (UDIH, 2005). The surveyed markets included Kiryadongo, Kattulikire, Tecwa, Chopelwor, Bweyale, Kalwala, Pumuzika, Kiryampungura, Karungi, all in Kiryandongo sub-county, and Diima, Karuma, Kawiti, Laboke, Mutunda, Nanda, Okwece and Teyango in Mutunda sub-county. Five mobile hawkers and eleven home-based/roadside markets were also surveyed.

RMS is a procedure for analysing commodity markets using a combination of techniques such as structured and semi-structured questionnaires with key informant informants, direct observations, and other participatory rural appraisal tools (Simmons et al., 1994). The method is very useful in trying to identify and understand the current market trends, opportunities and constraints (Simmons et al., 1994). The framework is best suited to research studies in which either little research has been done before, poorly conducted or needs an update (Holtzman et al., 1993). Its strength includes its practicability on identifying and sharpening research problems and ranking them according to importance. It can also detect emerging issues, themes and opportunities facing marketing systems. So far, the framework has been widely used to analyse market systems. Examples are the analysis of fuelwood and charcoal markets in Asia (Padoch, 1988; FAO, 1993).

It is obvious that probability sampling allows a random selection of elements, each with non-zero chance of being selected for the sample and hence produces good representation of the population. However, due to the descriptive nature of this study, purposive sampling was used to select traders selling WSWFPs for informal interviews. Semi-structured questionnaire was administered face-to-face to sixty six (66) traders that were encountered selling WSWFPs in the formal markets: five (5) mobile hawkers that were met selling mainly wild fruits, and eleven (11) home-based/roadside traders selling WSWFPs. Because of their small numbers, all traders that were found selling WSWFPs were interviewed.

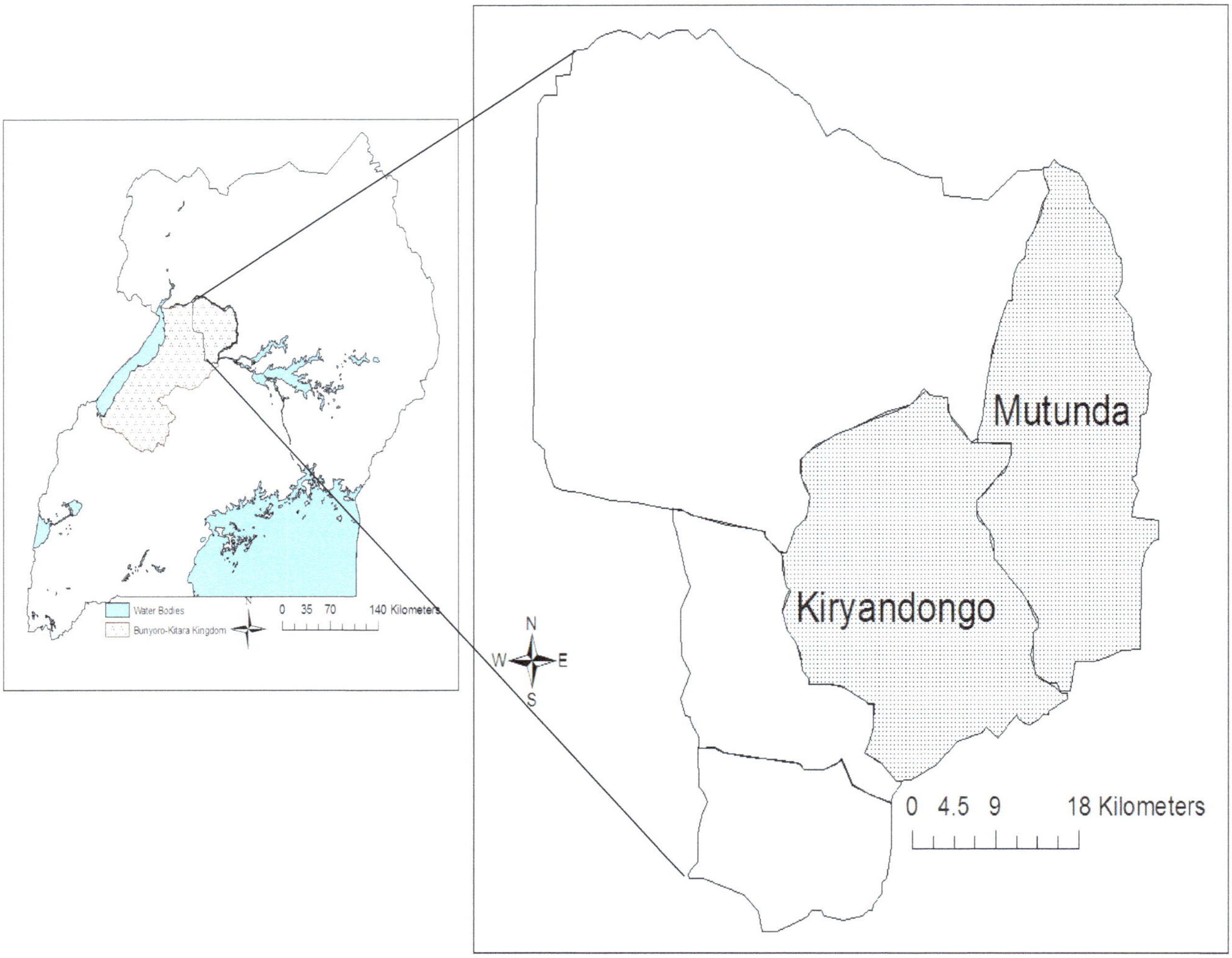

Figure 1. Location of Bunyoro-Kitara Kingdom and the study sites.

Data analysis

In order to describe the market potential for WSWFPs; data from RMS which included market characteristics, market conduct and performance were analysed using simple descriptive statistics in Excel spreadsheet and MINITAB statistical package. Data were coded to obtain a limited set of attributes for a variable (Babbie and Mouton, 2001), cleaned (checked) for mistake and entered into the computer. As a coding process, lists of responses were made for variables, groups identified and numbers assigned to these groups. However, some data were not coded but used descriptively. Market prices and weekly volumes of traded WSWFPs were compared with some of the selected conventional food plants (*Abelmoschus esculentus*, *Brassica oleracea var capitat*, *Mangifera indica* and *Sesamum indicum*) traded in the same locality. Weekly volumes of traded WSWFPs based on the usual units of the measurement (including bundles and heaps) within the markets were estimated per species sold. The profit margin, which is the dollar value difference in the selling price and total cost (Holland, 1998), was computed per traded species. Transport expenses were excluded in the cost computation because only 4% of the traders incurred transport expenses in the form of hired bicycles.

RESULTS

Socio-demographic characteristics of the traders

Socio-demographic characteristics of the interviewed traders (vendors) selling WSWFPs are presented in Table 1. The majority (82%) of the traders were women. Traders' ages ranged from 13 to 75 years, although the majority (59%) were above 36 years old. About 15% were less than 18 years old and the rest were aged between 18 and 36 years. About 74% of traders were married, 15% were not yet married (single) and the rest were either widow/widower or divorced/separated. The majority (51%) of the traders had attained primary level of education and only 9% had no formal education. The rest were either secondary school leavers or had attained tertiary level of education.

Although, all respondents interviewed were traders, only 51% reported trading as their main occupation and

Table 1. Socio-demographic characteristics of the traders selling WSWFPs.

Variable	Response (%)
Sex	
Male	18.3
Female	81.7
Age	
<18	14.6
18-36	26.8
>36	58.6
Marital status	
Single	15.7
Married	73.9
Divorced/separated	2.4
Widow/widower	8.0
Education level	
No formal education	9.4
Primary	51.4
Secondary	31.7
Tertiary	7.5
Major occupation	
Market vendor/trader	55.1
Subsistence farming	42.6
Others (Housewifery, students, brick making)	2.3
Family size	
<3 people	15.3
3-6 people	18.5
>6 people	66.2
Annual cash income (UGX)*	
<200,000 (≈USD 100)	7.4
200,000-400,000 (≈USD 100-200)	31.0
>400,000 (≈USD 200)	61.6
Main sources of cash incomes	
On-farm	70.8
Off-farm	29.2

*USD1 = 2010 Uganda shilling (UGX).

43% said their chief occupation was subsistence farming. The rest were mainly occupied in housewifery, in school as students, and in small scale brick making activities. Most (66%) traders had more than six persons per household and the average family size was seven persons. The majority (62%) of the traders had annual cash income greater than UGX 400,000 (USD 200). Thirty-one percent had annual cash income ranging from UGX 200,000 to 400,000 (≈USD 100 to 200) and the rest earned less than 200,000 (≈USD 100) per year. The majority (71%) of the traders earned their cash income mainly from on-farm activities and the rest (29%) from off-farm activities.

Which WSWFPS traded in the market?

Twenty nine (29) WSWFPs belonging to 12 botanical families were recorded to be traded during the rapid market survey (RMS) (Table 2). The most frequently recorded species were *Amaranthus dubius* (42.7%), *Hibiscus sabdariffa* (39.0%), *Cleome gynandra* (32.9%), *Solanum lycopersicum* (31.7%), *Solanum nigrum* (29.3%), *Tamarindus indica* (26.8%), *Amaranthus spinosus* (24.4%), *Basella alba* (24.4%), *Amaranthus graecizans* (22%), *Canarium schweinfurthii* (22%) and *Physalis peruviana* (22.0%). In terms of botanical family, members of Amaranthaceace, Solanaceae and Malvaceae were the most represented in the markets (Table 2).

What are the sources of market information?

Most (62%) traders of WSWFPs rely on communication from other traders either from within the same market location or from different markets as the main sources of market information (Figure 2). Other traders (33%) depended more on interaction with their customers (buyers) as the source of market information such as what they prefer, and how much they pay for similar products elsewhere. Reliance on media and service providers are very limited. Only 3% of traders ever received market information through media sources (local FM radios). Similarly, only 2% received market information from service providers operating within the area.

How are the traded WSWFPs transported to the market locations?

Majority (74%) of the traders transported WSWFPs they sell to the market locations on foot (walking) carrying them on their heads. Others transported their products to the markets using family bicycles (10%), hired bicycles (4%) and family wheelbarrows (2%) (Figure 3). Ten percent of traders did not transport their products to the formal market locations as customers come to buy from their homestead stalls. Only 4% of the traders incurred transport costs in the form of hired bicycles.

How are the market prices of WSWFPs determined?

Results presented in Figure 4 indicate that most (73.2 ±

Table 2. WSWFPS traded in the market within the study locality.

WSWFPs	Local names	Botanical family	Percentage of traders selling the plant
Amaranthus dubius Mart. *ex* Thell.	Doodo	Amaranthaceace	42.7
Hibiscus sabdariffa L.	Bamya, Ekikenke	Malvaceae	39.0
Cleome gynandra L.	Eyobyo	Brassicaceae	32.9
Solanum lycopersicum L.	Bunyanya bunyoro	Solanaceae	31.7
Solanum nigrum L.	Enswiga	Solanaceae	29.3
Tamarindus indica L.	Mukoge	Fabaceae	26.8
Amaranthus spinosus L.	Doodo y'amahwa	Amaranthaceace	24.4
Basella alba L.	Enderema	Basellaceae	24.4
Amaranthus graecizans L.	Nyabutongo, Ocoboro	Amaranthaceace	22.0
Canarium schweinfurthii Engl.	Empafu	Burseraceae	22.0
Physalis peruviana L.	Ntuutu	Solanaceae	22.0
Aframomum angustifolium (Sonnerat) K.Schum.	Amatehe, Kongo amor	Zingiberaceae	20.7
Crotalaria ochroleuca G.Don	Kumuro, Alaju	Fabaceae	20.7
Solanum anguivi Lam.	Obuhuruhuru, Katukuma	Solanaceae	20.7
Phaseolus lunatus L.	Amaijalero, Okuku	Fabaceae	19.5
Borassus aethiopum Mart.	Ekituugu, Tugo	Arecaceae	18.3
Hibiscus acetosella Welw. *ex* Hiern	Makawang kulo, Gwanya	Malvaceae	18.3
Corchorus tridens L.	Eteke	Malvaceae	17.1
Corchorus trilocularis L.	Otigo lum	Malvaceae	17.1
Cleome hirta (Klotzsch) Oliv.	Akayobyo akasajja	Brassicaceae	15.9
Vernonia amygdalina Del.	Kibirizi	Asteraceae	15.9
Aframomum alboviolaceum (Ridley) K.Schum	Amasaasi, Ocao	Zingiberaceae	14.6
Amaranthus lividus L.	Bwora, Mboog'ennene	Amaranthaceace	14.6
Mondia whitei (Hook.f.) Skeels	Omurondwa	Apocynaceae	14.6
Vitex doniana Sweet	Muhomozi, Owelo	Verbenaceae	13.4
Bidens pilosa L.	Obukurra	Asteraceae	12.2
Solanum macrocarpon L.	Bugorra	Solanaceae	12.2
Amaranthus hybridus (L.) Thell. subsp. *Cruentus*	Omujuiga	Amaranthaceace	9.8
Hyptis spicigera Lam.	Amola, Lamola	Lamiaceae	9.8

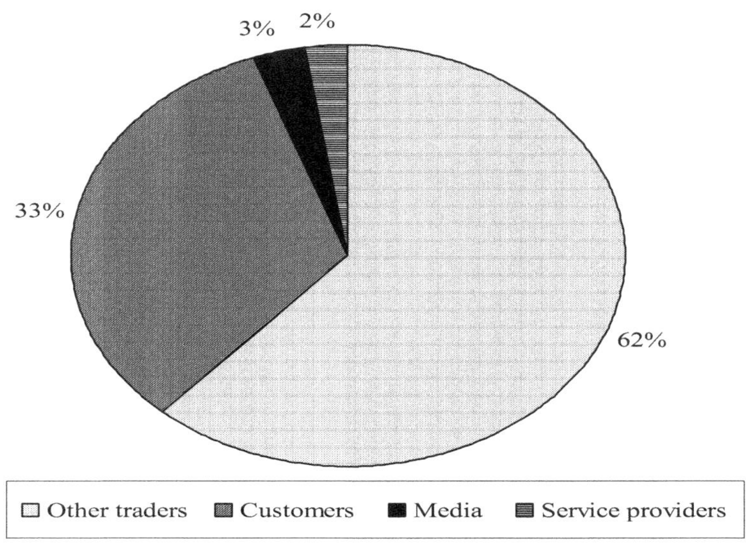

Figure 2. Sources of market information for traders selling WSWFPs.

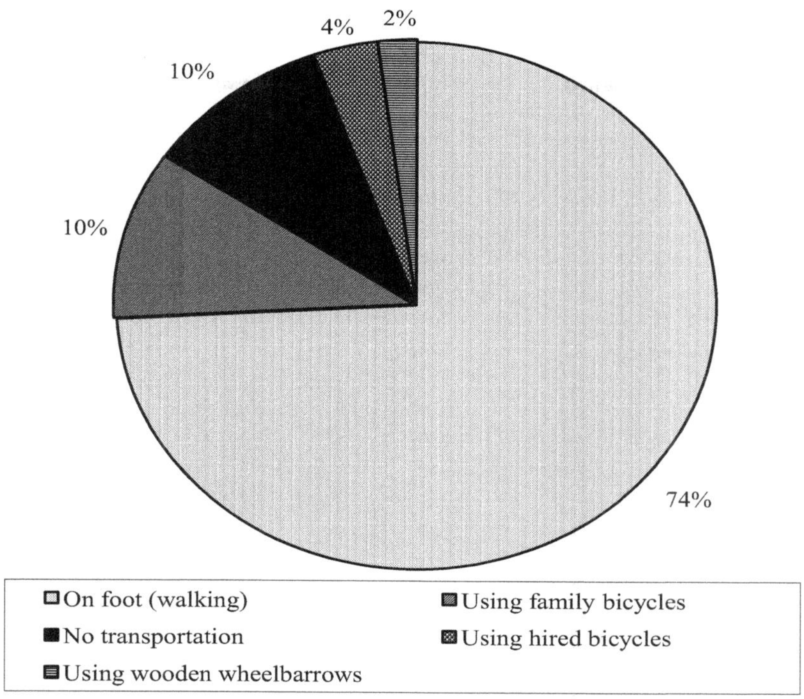

Figure 3. Modes of transport for traded WSWFPs from collection sources to the market locations.

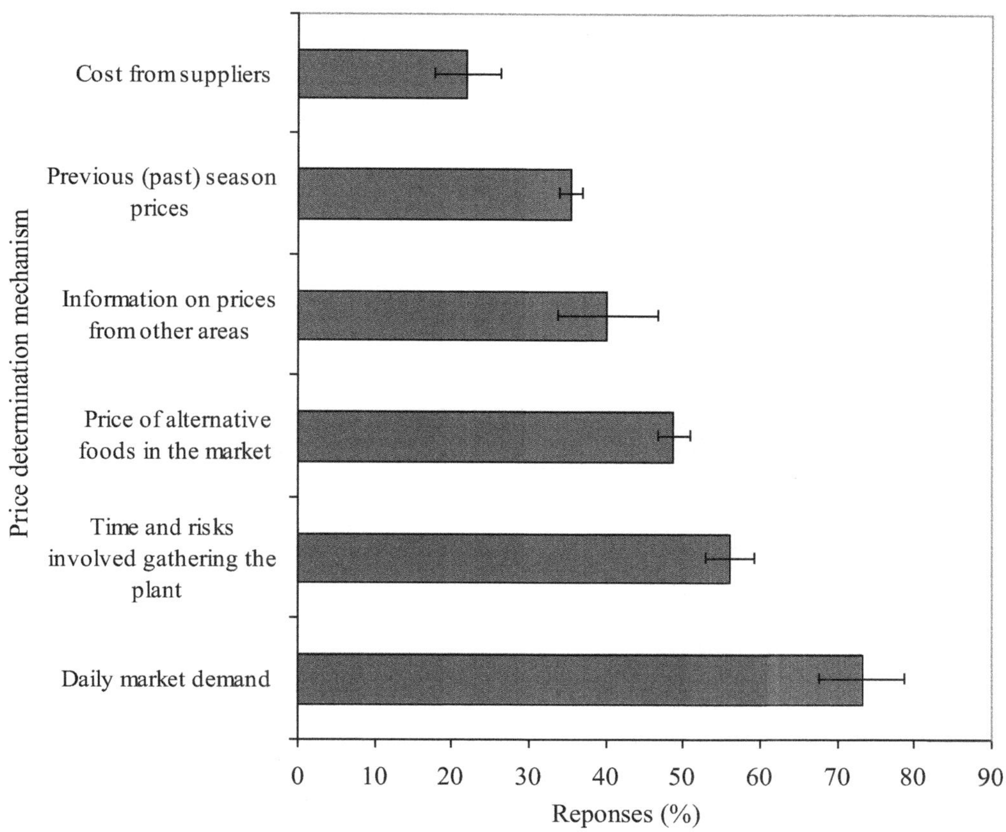

Figure 4. Avenues through which traders determine market prices of WSWFPs sold.

5.5%) traders assign prices to WSWFPs they sell based on the daily market demand of each product as well as on the time and risks involved in gathering of the plant (56.1 ± 3.2%). Other traders rely on price information of alternative (substitute) food plants being sold in the market (48.8 ± 2%) as well as price information from other areas (40.2 ± 6.4%). In addition, 35.4 ± 1.4% of the traders depend on price information of the previous or past seasons; while some (22 ± 4.4%) traders particularly those who are supplied by the gatherers simply rely on costs incurred from suppliers to determine the prices upon which they can sell their products.

What quantities (estimates) of WSWFPs are traded weekly in the market during the harvest seasons?

Among WSWFPs that are often sold in terms of bundles, *H. sabdariffa*, *S. nigrum*, *A. dubius*, *A. graecizans* and *C. gynandra* were sold in the highest quantities per trader weekly during their harvest seasons (Table 3). Whereas among those that are measurable and sold in terms of heaps, *P. peruviana*, *Solanum anguivi*, *B. alba*, *T. indica* and *S. lycopersicum* (Bunyanya bunyoro) were sold in highest quantities per trader weekly during their harvest seasons. Estimated quantities of *Mondia whitei* roots sold per trader weekly was high while the quantities of *C. schweinfurthii* fruits and *B. aethiopum* fruits marketed weekly were low. Similarly, the amount of fresh seeds of *Phaseolus lunatus* and seeds of *Hyptis spicigera* sold weekly were low. The rest WSWFPs were generally sold in very small quantities per trader per week (Table 3).

What are the average unit selling prices of the traded WSWFPs and how is the market demand of traded WSWFPs like?

The market price of WSWFPs sold varied from species to species as well as on the unit of measurements. Among the leafy WSWFPs, the most valuable species were *H. acetosella* with an average price of UGX 493.3 ± 88.4 bundle^{-1}, followed by *B. alba* selling at UGX 440 ± 80.3 heap^{-1} (Table 4). Majority of other leafy WSWFPs including *Crotalaria ochroleuca*, *H. sabdariffa*, *C. gynandra*, *Cleome hirta*, *A. dubius*, and *Amaranthus hybridus* subsp. *Cruentus*, *S. nigrum* and *Amaranthus lividus* had the average market price ranging from UGX 370.6 ± 84.9 to 308.3 ± 79.3 bundle^{-1}. Among leafy WSWFPs sold per bundle, *Bidens pilosa* had the lowest average market price of UGX 160 ± 51.6 bundle^{-1}.

H. spicigera (dry seeds) and *P. lunatus* (fresh seeds) locally called Amaijalero commanded the highest average market price of UGX 1187.5 ± 241.6 and 675 ± 85.6 kg^{-1}, respectively. Fruits of *Aframomum alboviolaceum* and *Aframomum angustifolium* were traded at an average market price of UGX 641.7 ± 202.1 and 558.8 ± 127.8 heap^{-1} respectively. *Borassus aethiopum* fruits also commanded a high average market price of UGX 620.0 ± 178.1fruit^{-1}. Fruits of *C. schweinfurthii* and *Vitex doniana* were traded at average market price of UGX 433.3 ± 45.4 and 363.6 ± 50.5 plastic mug^{-1} respectively. While those of *P. peruviana* were sold at a relatively low market price (UGX 305.6 ± 72.5).

What are profit margins for traded WSWFPs during the harvest seasons?

Average weekly profit yielded from the trade of various WSWFPs per trader ranged from UGX 764.5 to 6754.2 to (USD 0.38 to 3.36) (Table 5). Highest profit comes from selling *H. spicigera* seeds (UGX 6754.2 trader^{-1}), *H. sabdariffa* (5944 trader^{-1}), *A. angustifolium* (5914.2 trader^{-1}), *B. aethiopum* (5749.2 trader^{-1}), *B. alba* (5410.1 trader^{-1}), *S. nigrum* (5343.2 trader^{-1}), *A. alboviolaceum* (5191.9 trader^{-1}) and *C. schweinfurthii* (4922.4 trader^{-1}). Average weekly profit margins for most plants ranged from UGX 4922.4 to 2314.7. However, species such as *A. spinosus* (UGX 1873.4 trader^{-1}), *Corchorus tridens* (UGX 1749.4 trader^{-1}), *Corchorus trilocularis* (UGX 1608.9 trader^{-1}) and *Bidens pilosa* (UGX 764.5 trader^{-1}) had the lowest weekly profit margins.

Weekly average cost incurred per trader during marketing process varied depending on the species. *B. aethiopum*, *H. sabdariffa*, *C. schweinfurthii*, *A. dubius*, *A. graecizans*, *H. Acetosella* and *P. lunatus* had the highest average costs of over UGX 1000 (USD 0.5) trader^{-1} (Table 5). Costs incurred included mainly market dues as well as packing materials (polythene bags). Transport expenses were excluded in the cost computation because only 4% of the traders incurred transport expenses in the form of hired bicycles.

DISCUSSION

WSWFPs traded in the locality

Out of 62 WSWFPs belonging to 32 botanical families reported to be eaten in Bunyoro-Kitara Kingdom (Agea 2010), about 29 (47%) belonging to 12 botanical families were traded in the formal and informal markets in this area, which is a fairly high number considering that many species are consumed either at the household levels or at their collection sites only. About seven to ten WSWFPs species were always available in the markets, showing their high diversity in the markets. As compared to other places elsewhere, the number of WSWFPs traded in the present study is quite high. For instance in Sikkim Himalaya, Sundriyal and Sundriyal (2004) reported that out of 190 wild edible species that are consumed as food in the region, only about 24% are brought to the local markets. In nearly all instances, the findings of this study

Table 3. Estimated quantities of WSWFPs traded weekly in the market during the harvest seasons in Bunyoro-Kitara Kingdom.

WSWFPs sold	Local names	Parts sold	Estimated quantity/week in the harvest season				
			No. of traders selling	Units	Total	Range	Average/ trader
Aframomum alboviolaceum (Ridley) K.Schum.[¥]	Amasaasi, Ocao	Fruits	12	Heaps	105	5 - 15	8.8
Aframomum angustifolium (Sonnerat) K.Schum.[¥]	Amatehe, Kongo amor	Fruits	17	Heaps	197	5 - 20	11.6
Amaranthus dubius Mart. *ex* Thell.*	Doodo	Leaves and shoots	35	Bundles	571	8 - 30	16.3
Amaranthus graecizans L.*	Nyabutongo, Ocoboro	Leaves and shoots	18	Bundles	290	8 - 25	16.2
Amaranthus hybridus subsp. *Cruentus* (L.) Thell.*	Omujuiga	Leaves and shoots	8	Bundles	72	5 - 13	9.0
Amaranthus lividus L.*	Bwora, Mboog'ennene	Leaves and shoots	12	Bundles	122	5 - 15	10.2
Amaranthus spinosus L.*	Doodo y'amahwa	Leaves and shoots	20	Bundles	218	5 - 15	10.9
Basella alba L.[##]	Enderema	Leaves and shoots	20	Heaps	288	5 - 25	14.4
Bidens pilosa L.[#]	Obukurra	Leaves and shoots	10	Bundles	55	3 - 08	5.5
Borassus aethiopum Mart.[†]	Ekituugu, Tugo	Fruits	15	Fruits	172	5 - 15	11.5
Canarium schweinfurthii Engl.[††]	Empafu	Fruits	18	Plastic mugs	255	8 - 20	14.2
Cleome gynandra L.*	Eyobyo	Leaves and shoots	27	Bundles	376	8 - 20	13.9
Cleome hirta (Klotzsch) Oliv.*	Akayobyo akasajja	Leaves and shoots	13	Bundles	107	5 - 15	8.2
Corchorus tridens L.**	Eteke	Leaves and shoots	14	Bundles	150	5 - 15	10.7
Corchorus trilocularis L.**	Otigo lum	Leaves and shoots	14	Bundles	119	5 - 13	8.5
Crotalaria ochroleuca G.Don*	Kumuro, Alaju	Leaves and shoots	17	Bundles	191	5 - 15	11.2
Hibiscus acetosella Welw. *ex* Hiern*	Makawang kulo, Gwanya	Leaves and shoots	15	Bundles	177	8 - 15	11.8
Hibiscus sabdariffa L.*	Bamya, Ekikenke	Leaves and shoots	32	Bundles	635	10 - 35	19.8
Hyptis spicigera Lam.	Amola, Lamola	Seeds	8	Kg	51	3 - 10	6.4
Mondia whitei (Hook.f.) Skeels[Ω]	Omurondwa	Roots	12	Pieces	283	15 - 35	23.6
Phaseolus lunatus L.	Amaijalero, Okuku	Fresh seeds	16	Kg	124	3 - 20	7.8
Phaseolus lunatus L.[ΩΩ]	Amaijalero, Okuku	Fresh mature seeded pods	10	Heaps	88	6 - 16	8.8
Physalis peruviana L.[♀]	Ntuutu	Fruits	18	Heaps	287	10 - 25	15.9
Solanum anguivi Lam.[♀♀]	Obuhuruhuru, Katukuma	Fruits	26	Heaps	395	10 - 20	15.2
Solanum lycopersicum L.[®]	Bunyanya bunyoro	Ripe fruits	17	Heaps	217	8 - 20	12.8
Solanum macrocarpon L.[®]	Bugorra	Unripe fruits	10	Heaps	98	5 - 15	9.8
Solanum macrocarpon L.*	Bugorra	Young leaves	8	Bundles	83	7 - 18	10.4
Solanum nigrum L.*	Enswiga	Leaves	24	Bundles	459	10 - 30	19.1
Tamarindus indica L.[®®]	Mukoge	Fruits	22	Heaps	293	5 - 20	13.3
Vernonia amygdalina Del.[#]	Kibirizi	Leaves and shoots	13	Bundles	160	8 - 20	12.3
Vitex doniana Sweet[††]	Muhomozi, Owelo	Fruits	11	Plastic mugs	109	3 - 15	9.9

Table 3. Contd.

Conventionally cultivated food plants							
Abelmoschus esculentus (L.) Moench.[♀♀]	Okra	Fruits	27	Heaps	270	5-20	10
Brassica oleracea var capitata. L. (Alef.)[ϴ]	Cabbages	Head	42	Heads	689	10 - 30	16.4
Mangifera indica L.[¥¥]	Mangos	Fruits	50	Heaps	1100	15 - 35	22.7
Passiflora edulis Sims.[¥]	Passion fruits	Fruits	35	Heaps	534	8-25	15.5
Phaseolus vulgaris L.	Common beans	Fresh seeds	45	Kg	653	9 - 18	14.5
Sesamum indicum Linn.	Simsim	Seeds	39	Kg	531	5-22	13.6
Vigna unguicullata (L.)Walp*	Cowpeas	Leaves	30	Bundles	537	15 - 25	17.9

[¥]1 Heap = 8-15 fruits; *1 bundle = 300-450 g; **1 bundle = 200-300 g; [#]1 bundle = 300-500 g; [##]1 heap = 250-400 g; [†]1 fruit = 1.4-1.6 kg; [††]1 plastic mug (cup) = 0.5 kg; [Ω]1 piece of *M. whytei* root ≈ 0.01-0.02 m in diameter and 0.3-0.4 m in length; [ΩΩ]heap = 500-800 g; [♀]1 heap = 150-300 g; [♀♀]1 heap = 200-400 g; [®]1 heap = 200-400 g; [®®]1 heap = 300-400 g; [¥¥]1 heap = 300-600 g; [ϴ]1 head = 400-1200 g.

indicated that WSWFPs were not the primary product sold by traders that deal in WSWFPs. Many traders sold WSWFPs in combination with conventional farmed products such as tomatoes, okra, oranges, mangoes, banana of different types, cabbages and beans. Similar observation was also reported by Karaan et al. (2005) who noted that traders of wild fruits such as *Vitex mombassae, Vitex doniana* and *Strychnos cocculoides* in Tanzania, Zimbabwe and Zambia were engaged primarily in selling farmed products such as tomatoes, pumpkins, lemons and even fish. Therefore, any attempt to promote trade in WSWFPs should perhaps, be integrated into those strategies aimed at promoting trade of conventional crops.

Transportation of WSWFPs to market location

In marketing, products have to undergo a series of transfers or exchanges from one hand to another before finally reaching the consumer. This movement is always facilitated by transport. Transportation thus plays an important role in market conduct and performance as it helps in assembling and dispersing the products to the market locations, thereby linking the producers/gatherers with the buyers who might be located at different places (Jain, 2009). Availability of cheap transport is therefore, a key factor that influences the market conduct and performance of most traded goods (Kotler and Armstrong, 1999). In the present study, walking was by far the most predominant mode of transport for traded WSWFPs to the market locations. Generally, women and girls carry the harvested produce on top of their heads by walking while men especially the mobile hawkers largely carry them in polythene or locally made baskets bags by handles. Although, walking plays a considerable role in marketing of WSWFPs in this area, this mode of transport is often restricted by weight carried or distance to the market.

A few people also use bicycles to deliver the stocks to the markets. Bicycles are seldom used for a number of reasons. The main reason being that most men who own bicycles cannot entrust them with the wives who are the main traders of WSWFPs. Besides, the 2005-2006 Uganda National and Housing Survey (Uganda Bureau of Statistics, 2006) showed that only 33.4% of the households in western Uganda possessed bicycles. Therefore, most women depend on the cheapest form of transport to the markets, which is walking. None of the traders in this locality reported using the public transport means such as taxis to deliver their stocks to the market. Similar findings were reported by Kadzere et al. (1998) who noted that delivery of wild fruits to the market in remote areas of southern Africa was predominantly by foot (walking) besides occasional use of ox-carts, bicycles and buses. Because of the meagre returns from the sale of wild food plants, gatherers would not want to rely on costly modes of transport to deliver their stocks to the market places.

Price setting and market information for WSWFPs

Pricing is one of the main elements of the marketing mix and is an important strategic issue in market conduct and performance because it affects products positioning and sales (Tomek and Robinson, 1990; Panigyrakis, 1997). There are generally three basic types of pricing strategies:

Table 4. Selling price and market demand of commonly sold WSWFPs in Bunyoro-Kitara Kingdom.

WSWFPs sold	Local names	Parts sold	Market price (Ugandan Shillings)[@]	Average price (±StDev)	SEM	Market demand
Aframomum alboviolaceum (Ridley) K.Schum.[¥]	Amasaasi, Ocao	Fruits	500 - 1000 $heap^{-1}$	641.7 (202.1)	58.3	High
Aframomum angustifolium (Sonnerat) K.Schum.[¥]	Amatehe, Kongo amor	Fruits	400 - 800 $heap^{-1}$	558.8 (127.8)	31	High
Amaranthus dubius Mart. *ex* Thell.*	Doodo	Leaves and shoots	200 - 500 $bundle^{-1}$	345.7 (103.9)	17.6	Very high
Amaranthus graecizans L.*	Nyabutongo, Ocoboro	Leaves and shoots	100 - 300 $bundle^{-1}$	211.1 (75.8)	17.9	Low
Amaranthus hybridus subsp. *Cruentus* (L.) Thell.*	Omujuiga	Leaves and shoots	300 - 400 $bundle^{-1}$	337.5 (51.8)	18.3	Very high
Amaranthus lividus L.*	Bwora, Mboog'ennene	Leaves and shoots	200 - 400 $bundle^{-1}$	308.3 (79.3)	22.9	low
Amaranthus spinosus L.*	Doodo y'amahwa	Leaves and shoots	100 - 300 $bundle^{-1}$	215.0 (74.5)	16.7	Low
Basella alba L.[##]	Enderema	Leaves and shoots	300 - 600 $heap^{-1}$	440.0 (80.3)	19.7	High
Bidens pilosa L.[#]	Obukurra	Leaves and shoots	100 - 200 $bundle^{-1}$	160.0 (51.6)	16.3	Very low
Borassus aethiopum Mart.[†]	Ekituugu, Tugo	Fruits	500 - 1000 $fruit^{-1}$	620.0 (178.1)	46	High
Canarium schweinfurthii Engl.[††]	Empafu	Fruits	400 - 500 plastic mug^{-1}	433.3 (45.4)	10.7	High
Cleome gynandra L.*	Eyobyo	Leaves and shoots	300 - 500 $bundle^{-1}$	363.0 (79.2)	15.2	High
Cleome hirta (Klotzsch) Oliv.*	Akayobyo akasajja	Leaves and shoots	200 - 500 $bundle^{-1}$	346.2 (112.7)	31.2	Low
Corchorus tridens L.**	Eteke	Leaves and shoots	100 - 300 $bundle^{-1}$	200.0 (78.4)	21	Low
Corchorus trilocularis L.**	Otigo lum	Leaves and shoots	100 - 300 $bundle^{-1}$	228.6 (82.5)	22.1	Low
Crotalaria ochroleuca G.Don*	Kumuro, Alaju	Leaves and shoots	300 - 500 $bundle^{-1}$	370.6 (84.9)	20.6	Low
Hibiscus acetosella Welw. *ex* Hiern*	Makawang kulo, Gwanya	Leaves and shoots	400 - 700 $bundle^{-1}$	493.3 (88.4)	22.8	Very high
Hibiscus sabdariffa L.*	Bamya, Ekikenke	Leaves and shoots	300 - 500 $bundle^{-1}$	368.7 (73.8)	13	High
Hyptis spicigera Lam.	Amola, Lamola	Seeds	1000 - 1600 kg^{-1}	1187.5 (241.6)	85.4	High
Mondia whitei (Hook.f.) Skeels[Ω]	Omurondwa	Roots	100 - 200 $piece^{-1}$	158.3 (46.9)	13.5	High
Phaseolus lunatus L.	Amaijalero, Okuku	Fresh seeds	600 - 800 kg^{-1}	675 (85.6)	21.4	Low
Phaseolus lunatus L.[ΩΩ]	Amaijalero, Okuku	Fresh mature seeded pods	300 - 500 $heap^{-1}$	370 (82.3)	26	Low
Physalis peruviana L.[♀]	Ntuutu	Fruits	200 - 400 $heap^{-1}$	305.6 (72.5)	17.1	High
Solanum anguivi Lam.[♀♀]	Obuhuruhuru, Katukuma	Fruits	200 - 300 $heap^{-1}$	265.38 (48.5)	9.51	High
Solanum lycopersicum L.[®]	Bunyanya bunyoro	Ripe fruits	100 - 400 $heap^{-1}$	247.1 (80.0)	19.4	Low
Solanum macrocarpon L.[®]	Bugorra	Unripe fruits	200 - 500 $heap^{-1}$	380.0 (91.9)	29.1	Low
Solanum macrocarpon L.*	Bugorra	Leaves and shoots	200 - 300 $bundle^{-1}$	262.5 (51.8)	18.3	Low
Solanum nigrum L.*	Enswiga	Leaves	200 - 400 $bundle^{-1}$	316.7 (70.2)	14.3	Very high
Tamarindus indica L.[®®]	Mukoge	Fruits	100 - 300 $heap^{-1}$	231.8 (71.6)	15.3	Low
Vernonia amygdalina Del.[#]	Kibirizi	Leaves and shoots	200 - 300 $bundle^{-1}$	261.5 (50.3)	14	Very low

Table 4. Contd

Vitex doniana Sweet[††]	Muhomozi, Owelo	Fruits	300 - 400 plastic mug^{-1}	363.6 (50.5)	15.2	High
Conventionally cultivated food plants						
Abelmoschus esculentus (L.) Moench.[♀♀]	Okra	Fruits	200 - 600 heap^{-1}	355.5 (98.4)	22.3	High
Brassica oleracea var capitata. L. (Alef.)[ϴ]	Cabbages	Head	300 - 700 head^{-1}	489.9 (67.8)	15.1	High
Mangifera indica L.[¥¥]	Mangos	Fruits	100 - 500 heap^{-1}	365.0 (70.1)	28.5	Very low
Passiflora edulis Sims.[¥]	Passion fruits	Fruits	400 - 1000 heap^{-1}	600.9 (157.3)	42.6	High
Phaseolus vulgaris L.	Common beans	Fresh seeds	700 - 1200 kg^{-1}	845.4 (100.5)	33.4	High
Sesamum indicum Linn.	Simsim	Seeds	1200 - 2200 kg^{-1}	1670.6 (125.8)	38.8	Very high
Vigna unguicullata (L.)Walp*	Cowpeas	Leaves	200 - 500 bundle^{-1}	315.3 (49.4)	17.7	Low

[¥]1 Heap = 8-15 fruits; *1 bundle = 300-450 g; **1 bundle = 200-300 g; [#]1 bundle = 300-500 g; [##]1 heap = 250-400 g; [†]1 fruit = 1.4-1.6 kg; [††]1 plastic mug (cup) = 0.5 kg; [Ω]1 piece of *M. whytei* root ≈ 0.01-0.02 m in diameter and 0.3-0.4 m in length; [ΩΩ]heap = 500-800 g; [♀]1 heap = 150-300 g; [♀♀]1 heap = 200-400 g; [®]1 heap = 200-400 g; [®®]1 heap = 300-400 g; [¥¥]1 heap = 300-600 g; [ϴ]1 head = 400-1200 g; [@]USD1 = 2010 Uganda shilling (UGX).

pricing based on the total cost incurred, pricing based on demand and pricing based on competition (Panigyrakis, 1997). However, information garnered from this study shows that there are no definite or formal mechanisms of setting prices of WSWFPs traded by the vendors. Most traders relied on daily market demands to determine the price upon which they sell their products. Others considered time and risks involved in gathering process, price information of the substitute food plant, price information on WSWFPs from other areas and markets, knowledge of the past season prices or the costs incurred from their suppliers. However, in all these cases, price fixing follows a sort of action-reaction sequence, generally beginning with the interested buyer asking for the price, followed by the naming of a price by the trader. Based on this, the bargaining process would begin until a final price, which the buyer is ready to pay, is reached. So even after a trader has set a fair market price, negotiation with buyers would sometimes lower the price further.

These findings are similar to the situation in most southern African countries, where available information (Ham et al., 2008) indicates that there is no defined mechanism for setting market prices of wild edible plants especially the indigenous fruits. Similarly, in west and central Africa, there is no consistent mechanism (Tchoundjeu et al., 2008) upon which traders selling wild food plants determine the market price of their stocks; some traders, for instance, consider the customers affordability to buy the product. In that case, the prices are set depending on the sellers' assessment of the buying power of individual customers. A mere look at the customer would sometimes help sellers to gauge how much to charge them.

However, efforts to trade in WSWFPs can only thrive better in an environment where market information is freely available. As discussed by Shepherd (1997), market information can help traders, farmers and gatherers to decide whether they should sell their products immediately or whether storage is necessary or not; where and whom to sell; whether to add value to their products or not; and to know if there is more demand for one product or another. Such information can be used to check whether the prices the traders get are reasonable. Shortage of market information is disadvantageous for traders, farmers, and or gatherers in negotiating with buyers and therefore, weakens their bargaining power (Poole, 2001). Therefore, the selling price of traded wild food plants could be improved based on available and reliable price information. Unfortunately, in the present study, market information system for WSWFPs was largely rudimentary and undeveloped. Traders and gatherers rely mainly on information from fellow traders as well as their customers to make market decisions. There is very little market information passed on to traders and gatherers by public and private media as well as the service providers.

Estimated volumes, selling price and profit margins of traded WSWFPs

Generally, findings from the present study indicated that volumes of WSWFPs weekly traded in the markets are still low, as compared to some conventionally cultivated food crops. The low weekly volumes of WSWFPs such as

Table 5. Average weekly profit margins (Ugandan shillings) for the traded WSWFPs during the harvest seasons in Bunyoro-Kitara Kingdom.

WSWFPs sold	Local names	Parts sold	Average quantity/trader	Average price/unit	Average cost incurred/ trader*	Average income/trader	Average profit/trader[@]
Aframomum alboviolaceum (Ridley) K.Schum.	Amasaasi, Ocao	Fruits	8.8 heaps	641.7	455.1	5647.0	5191.9
Aframomum angustifolium (Sonnerat) K.Schum.	Amatehe, Kongo amor	Fruits	11.6 heaps	558.8	567.9	6482.1	5914.2
Amaranthus dubius Mart. *ex* Thell.	Doodo	Leaves and shoots	16.3 bundles	345.7	1152.4	5634.9	4482.5
Amaranthus graecizans L.	Nyabutongo, Ocoboro	Leaves and shoots	16.2 bundles	211.1	1097.8	3419.8	2322.0
Amaranthus hybridus subsp. *Cruentus* (L.) Thell.	Omujuiga	Leaves and shoots	9.0 bundles	337.5	520.3	3037.5	2517.2
Amaranthus lividus L.	Bwora, Mboog'ennene	Leaves and shoots	10.2 bundles	308.3	458.6	3144.7	2686.1
Amaranthus spinosus L.	Doodo y'amahwa	Leaves and shoots	10.9 bundles	215.0	470.1	2343.5	1873.4
Basella alba L.	Enderema	Leaves and shoots	14.4 heaps	440.0	925.9	6336.0	5410.1
Bidens pilosa L.	Obukurra	Leaves and shoots	5.5 bundles	160.0	115.5	880.0	764.5
Borassus aethiopum Mart.	Ekituugu, Tugo	Fruits	11.5 fruits	620.0	1380.8	7130.0	5749.2
Canarium schweinfurthii Engl.	Empafu	Fruits	14.2 plastic mugs	433.3	1230.5	6152.9	4922.4
Cleome gynandra L.	Eyobyo	Leaves and shoots	13.9 bundles	363.0	751.7	5045.7	4294.0
Cleome hirta (Klotzsch) Oliv.	Akayobyo akasajja	Leaves and shoots	8.2 bundles	346.2	422.3	2838.8	2416.5
Corchorus tridens L.	Eteke	Leaves and shoots	10.7 bundles	200.0	390.6	2140.0	1749.4
Corchorus trilocularis L.	Otigo lum	Leaves and shoots	8.5 bundles	228.6	334.2	1943.1	1608.9
Crotalaria ochroleuca G.Don	Kumuro, Alaju	Leaves and shoots	11.2 bundles	370.6	669.6	4150.7	3481.1
Hibiscus acetosella Welw. *ex* Hiern	Makawang kulo, Gwanya	Leaves and shoots	11.8 bundles	493.3	1078.5	5820.9	4742.4
Hibiscus sabdariffa L.	Bamya, Ekikenke	Leaves and shoots	19.8 bundles	368.7	1356.3	7300.3	5944.0
Hyptis spicigera Lam.	Amola, Lamola	Seeds	6.4 kg	1187.5	845.8	7600.0	6754.2
Mondia whitei (Hook.f.) Skeels	Omurondwa	Roots	23.6 pieces	158.3	296.7	3735.9	3439.2
Phaseolus lunatus L.	Amaijalero, Okuku	Fresh seeds	7.8 kg	675	1010.4	5265.0	4254.6
Phaseolus lunatus L.	Amaijalero, Okuku	Fresh mature seeded pods	8.8 heaps	370	600.7	3256.0	2655.3
Physalis peruviana L.	Ntuutu	Fruits	15.9 heaps	305.6	790.5	4859.0	4068.5
Solanum anguivi Lam.	Obuhuruhuru, Katukuma	Fruits	15.2 heaps	265.4	875.6	4033.8	3158.2
Solanum lycopersicum L.	Bunyanya bunyoro	Ripe fruits	12.8 heaps	247.1	785.5	3162.9	2377.4
Solanum macrocarpon L.	Bugorra	Unripe fruits	9.8 heaps	380.0	650.0	3724.0	3074.0
Solanum macrocarpon L.	Bugorra	Young leaves	10.4 bundles	262.5	415.3	2730.0	2314.7

Table 5. Contd

Solanum nigrum L.	Enswiga	Leaves	19.1 bundles	316.7	705.8	6049.0	5343.2
Tamarindus indica L.	Mukoge	Fruits	13.3 heaps	231.8	425.5	3082.9	2657.4
Vernonia amygdalina Del.	Kibirizi	Leaves and shoots	12.3 bundles	261.5	510.9	3216.5	2705.6
Vitex doniana Sweet	Muhomozi, Owelo	Fruits	9.9 plastic mugs	363.6	488.2	3599.6	3111.4

* Costs incurred included mainly market dues and packing materials (polythene bags). Transport expenses were excluded in the cost computation because only 4% of the traders incurred transport expenses in the form of hired bicycles. @USD1 = 2010 Uganda shilling (UGX).

B. pilosa and *V. doniana* could perhaps, be attributed to low demand, low purchasing power and general negative public perceptions of WSWFPs or lack of trade promotion in wild food plants (Sharma et al., 1992). However, there is some light at the end of the tunnel. The volumes of some traded species such as *H. sabdariffa*, *S. nigrum*, *A. dubius*, *C. gynandra*, *P. peruviana*, *B. alba*, *T. indica* and *M. whitei* traded weekly in the markets are reasonably high and are potential sources of cash incomes to the gatherers and traders. In Sikkim Himalaya, Sundriyal and Sundriyal (2004) also found that volumes of wild edible plants traded in three weekly markets of Gangtok, Singtam and Namchi were generally low with few exceptions of species such as *Spondias axillaris*, *Dendrocalamus hamiltonii*, *Urtica dioica, Diplazium esculentum*, *Eleagnus latifolia* and *Machilus edulis* that had very high weekly and annual volumes marketed. Similarly, Shanley et al. (2002) reports low volumes of traded wild edible plants in Capim region of Brazil with exception of *Caryocar villosum*, *Platonia insignis* and *Endopleura uchi* fruits, which were found to be traded in sufficient quantities, and whose high volumes were attributed to their high demand for flavouring ice cream, yogurt, jams and juice by the natives. Therefore, one can deduce that WSWFPs sold in large quantities are those that command high demands, otherwise most traders prefer selling small quantities of different WSWFPs, as they are well aware of their low market demand.

Concerning market prices, the results of the present study revealed that the selling prices of the traded WSWFPs varied according to species marketed and unit of measurements. As compared to conventional food plants, the prices of most traded WSWFPs were generally alike. Some WSWFPs such as *H. acetosella*, *B. alba*, *H. spicigera* (seeds), *P. lunatus* (fresh seeds), *A. alboviolaceum* and *A. angustifolium*, *B. aethiopum* (fruits) had higher selling prices per unit measurement as compared to the related conventional cultivated food plants traded in same locality. Their high market prices could perhaps, be attributed to their perceived nutritional and medicinal properties. Market demands of other traded WSWFPs, were noted to be rising either due to the same reason as above and other factors such as the repeated occurrence of drought in the study area that often results in crop failure hence heavy reliance on wild resources by poor households. Elsewhere, there is a report that some traded wild edible food plants command higher market prices. For instance, in Sikkim Himalaya region of India, most traded wild food plants (*Tupistra nutans*, *Dendrocalamus hamiltonii*, *Diploknema butyracea* and *Eleagnus latifolia*) had very high market prices per unit measurement, accredited to mainly excessive labour costs involved in the marketing process, high demand as well as well as higher income of the people in the region (Sundriyal and Sundriyal, 2004). Thus, increasing demand and prices of WSWFPs especially in the context of the present study could perhaps, be considered an opportunity to promote sustainable utilization and management of edible wild food plants.

Marketing margin is an indicator of the profitability of WSWFPs marketing. The finding from this study indicated that the average weekly profit yielded from the trade of various WSWFPs were low to moderate and ranged from UGX 764.5 to 6754.2 (USD 0.38 to 3.36). High return came from few species such *H. spicigera, H. sabdariffa, A. angustifolium, B. aethiopum, B. alba*, *S. nigrum*, *A. alboviolaceum* and *C. schweinfurthii.* According to most traders, profit from WSWFPs would be much higher, if it were not because they have to sell these plants at a low price. Besides, the rather low marketing margins could have probably stemmed from the focus on low-income local markets, where consumer spending powers were limited. This implies that any attempt to add value or increase the marketing margin must be associated with targeting higher income consumers who are able to compensate for this.

Very little costs ranging from UGX 115.5 to 1380.8 (USD 0.06 to 0.69) were incurred by the trader in the marketing process. These costs included mainly market dues and packing materials (polythene bags). Transport expenses were excluded in cost computation because only 4% of traders incurred it. Similar to Dixon et al. (1989), it was not possible to compute time and labour costs involved in gathering process. In view of these costs, it is plausible to say that the profit margin would have been very low if the time and labour employed in gathering process were taken into account. However, given the very poor economic status of most traders of WSWFPs and the fact that majority of them survive on less than 1 USD a day, even a small amount of earnings, which may only be payment for the labour and time involved in the gathering processing, has a significant value, as it helps to fulfil some subsistence requirements.

Conclusions

Out of the 62 WSWFPs belonging to 31 botanical families documented as edible in the Bunyoro-Kitara Kingdom (Agea, 2010), about 47% belonging to 12 botanical families were traded in formal and informal markets within the study area. Market information system for WSWFPs was largely rudimentary and undeveloped, and traders rely mainly on information from fellow traders as well as their customers to make market decisions. Traded WSWFPs were primarily delivered to markets on foot and using bicycles. Currently, there are no definite or formal mechanisms of setting prices of traded WSWFPs; most traders relied on the daily market demand, time and risks involved in gathering process, information of the price of substitute food and prices from other areas, knowledge of the past seasons' prices, and on the costs incurred from the suppliers. With exception of few species such as *P. peruviana* and *B. alba*, weekly volumes of traded WSWFPs were low as compared to most conventional food crops. On the other hand, prices of most traded WSWFPs were generally similar to those of alternative conventional food plants marketed in the area. Some WSWFPs like *H. acetosella*, *B. alba* and *H. spicigera* (seeds), had higher market prices per unit measurement as compared to the related conventional food plants. Average weekly profits yielded from the trade of various WSWFPs were moderate and ranged from UGX 764.5 to 6754.2 (USD 0.38 to 3.36). The highest return came from species such as *H. spicigera, H. sabdariffa*, *A. angustifolium, B. aethiopum, B. alba*, *S. nigrum*, *A. alboviolaceum* and *C. schweinfurthii.*

Recommendations

In light of the findings and the conclusions presented above, there should be a deliberate attempt to encourage both intra and inter-networking among the traders and gatherers dealing on WSWFPs. The importance of networking cannot be overemphasized. It is through this process that the traders and gatherer of WSWFPs can find where demand for certain products is high. They can also discover the going prices for various products. They can learn the latest profitable marketing methods and production practices. Traders should also be encouraged to price their products based not only on cost and demand or size of the market, but also on the economic and emotional value to customers. This means that, in addition to some other factors that associated into price determination, the traders should attempt to find out the economic and emotional value of his/her product to the customers. A trader, who wishes to market his/her product successfully, should also be able to have good customer relations. Therefore, the attributes and motivations that underlie the market demand and the current appeal of WSWFPs documented in this study should be explored further. Those attributes and motivations could be useful as selling gimmicks by traders to attract more buyers. At the moment, traders merely employ the art of display (arranging products in an interesting and persuasive manner) and persuasion (calling customers with familiar names, such as mother, father, uncle, sister, brother and auntie). It is equally important to identify and understand the psychology of consumers when they purchase and consume products of this nature. These would differ in different market segments. Therefore, consumer studies must be undertaken to gather the relevant information followed by testing and tasting studies.

REFERENCES

Agea JG (2010). Use and potential of wild and semi-wild food plants in alleviating household poverty and food insecurity: A case study of Bunyoro-Kitara Kingdom, Uganda. PhD Thesis. School of Environment, Natural Resources & Geography, Bangor University, Bangor, U K. 360 pp (including 15 preliminary pages, i - xiv).

Agea JG, Katongole B, Waiswa D, Nabanoga GN (2008). Market Survey of Mondia whytei (Mulondo) Roots in Kampala City, Uganda. Afr. J. Trad. CAM 5 (4):399-408.

Agea JG, Okia CA, Obua J, Hall J, Teklehaimanot Z (2011). Wild and semi-wild food plants in Bunyoro-Kitara Kingdom, Uganda: cultural significances, local perceptions and social implications of their consumption. Int. J. Med. Aromat. Plants 1(2):137-152.

Arnold TH, Wells MJ, Wehmeyer AS (1985). Khoisan food plants: taxa with potential for future economic exploitation. In: GE Wickens, JR Gooddin, and DV Field (eds.). Plants for arid lands. Proceedings of the Kew International Conference on Economic plants for arid lands 23-27 July 1984, George Allen and Unwin, London. pp. 69-86.

Babbie E, Mouton J (2001). The practice of social research. South African Edition. Oxford University Press, Cape Town. p. 674

Bye R, Linares E (1983). The role of plants found in the Mexican markets and their importance in ethnobotanical studies. J. Ethnobiol. 3:1-13.

Bye R, Linares E (1990). Mexican market plants of 16th century. 1. Plants recorded *in* Historia Natural De Nueva Espana. J. Ethnobiol. 10:151-168.

De Beer, JH, McDermott MJ (1989). The Economic Value of Non-timber Forest Products in Southeast Asia With Emphasis on Indonesia, Malaysia, and Thailand. Committee for the IUCN and WWF, Amsterdam.

Dixon JA, James DE, Sherman PB (1989). The Economics of dry land management. Earth Scan Publication Ltd. London. pp. 109-140.

FAO (2000) Global forest resource assessment. FAO Forestry Paper,

No.140.
FAO-RWEDP (1993). Charcoal production and marketing in Gujarat. Cited by Simmons R, A. Gordon and J. Cropley (1994) Characterising markets: basic research tools and methods and their application to NTTPs. Paper prepared for the workshop on non-timber tree product marketing research, December 12-14, Annapolis, Maryland.
Ham C, Akinnifesi FK, Franzel S, Du PS Jordaan D, Hansmann C, Ajayi OC, De Kock C (2008). Opportunities for commercilazation and enterprise development of indigenous fruits in southern Africa. In: Akinnifesi FK, Leakey RB, Ajayi OC, Silenshi G, Tchoundjeu Z, Matakala P, Kwesiga FR (eds.) indigenous fruit trees in the tropics: domestication, utilization and commercialization. Reading (CABI).
Holland R (1998). Selling price, gross margin and mark-up determination. Agricultural Development Centre (ADC) information no. 12. Agriculture Extension Service, The University of Tennessee Institute of Agriculture, USA.
Holtzman J, Martin J, Abbot R (1993). Operational guidelines for rapid appraisal of agricultural marketing systems. The agricultural marketing improvement strategies project. AID, R&D/EID Project 936-5447. Abt Associates Inc.
Jacquat C (1990). Plants from the markets of Thailand. Editions Duang Kamol, Bangkok. p. 251.
Jain A (2009). Principles of marketing. V.K publications. ISBN-13: 9788188597987. p. 474
Johnson E, Johnson TJ (1976). Economic plants in a rural Nigerian Market. Econ. Bot. 30:375-381.
Kadzere I, Watkins CB, Merwin IA, Akinnifesi FK, Saka JDK, Mhango J (2006). Harvesting and postharvest handling practices and characteristics of *Uapaca kirkiana* (Muell. Arg.) fruits: a survey of roadside markets in Malawi. Agroforestry Syst. 68:133-142.
Karaan M, Ham C, Akinnifesi F, Moombe K, Jordaan D, Franzel S, Aithal A (2005). Baseline marketing surveys and supply chain studies for indigenous fruit markets in Tanzania, Zimbabwe and Zambia. World Agroforestry Centre and CPWild Research Alliance.
Kimondo JM, Agea J.G, Okia C.A, Abohassan R.A.A, Mulatya J, Teklehaimanot Z (2010). Vitex payos (Lour.) Merr fruit trees in the drylands areas of Eastern Kenya: use, marketing and management. Bot. Res. J. 3 (1-4):14-21.
Kohls RL, Uhl JN (2002). Marketing of Agricultural Products. Ninth Edition. Prentice Hall, chapters 11 p. 544.
Kotler P, Armstrong G (1999). Principles of marketing, 8th Edition, Prentice-Hall International, Inc. New Jersey, USA. (various pagings)
Moreno-Black G (1991). Traditional foods in northeastern Thailand: Environmental cultural and economic factors. Paper presented at the American Anthropological Association Meetings, Chicago.
Moreno-Black G (1994). Gathered food and culture change: Traditions in transition in northeastern Thailand. J. Home Consum. Hortic. 1:93-107.
Moreno-Black G, Price LL (1993). The marketing of gathered food as an economic strategy in north-eastern Thailand. Hum. Organ. 52:398-404.
Ngarmsak T (1987). Status and nutritional importance of unconventional food crops in the Thai diets. Report submitted to Regional Office For Asia and The Pacific (RAPA), FAa. Bangkok.
Okafor JC (1980). Edible indigenous woody plants in the rural economy of the Nigerian forest zone. For. Ecol. Manage. 3:45-55.
Padoch C (1988). The economic importance of marketing of forest and fallow products in the Iquitos region. Adv. Econ. Bot. 5: 74-89.
Panigyrakis G (1997). Pricing policy, In: Padberg D, Ritson C and Albisu L (eds.) Agro-food marketing. CAB International Publishers (ISBN: 0-85199-144-0), UK. pp. 295-318.
Poole N (2001). Market information in theory and practice: Institutional perspectives' in technical consultation on the integration of statistical and agricultural market information services, Proceedings of a CTA workshop in Wageningen, the Netherlands, http://www.agricta.org/pubs/wd8026/wd8026.pdf, accessed on 8th December 2009.
Rasoanaivo P (1990). Rain forests of Madagascar: Sources of industrial and medicinal plants. Ambio 19 (8):421-423.
Russell D, Franzel S (2004). Trees of prosperity: agroforestry, markets and the African smallholder. Agroforestry Syst. 61:345-355.
Schlage C (1969). Observations at a local market in Usambara. in Investigations into Health and Nutrition in East Africa. H. Kraut and H.-D. Cremer (editors). Weltforum Verlag, Munich. pp. 71-85.
Scoones I, Melnyk M, Pretty J (eds.) (1992). Hidden harvest: wild foods and agricultural systems. A literature review and annotated bibliography. IIED, London.
Shanley P, Luz L, Swingland RI (2002). The faint promise of a distant market: a survey of Belém's trade in non-timber forest products. Biodivers. Conserv. 11:615-636.
Shepherd WA (1997). Market information services: theory and practice. FAO, Rome.
Simmons R, Gordon A, Cropley J (1994). Characterising markets: basic research tools, methods, and their application to NTTPs. Paper Prepared for the workshop on non-timber tree product marketing research, December 12-14, Annapolis, Maryland.
Sundriyal M, Sundriyal RC (2004). Wild edible plants of the Sikkim Himalaya: marketing, value addition and implications for management. Econ. Bot. 58(2): 300-315.
Tchoundjeu Z, Atangana A, Asaah E, Tsobeng A, Facheux C, Mbosso C, Degrande A, Sado T, Kanmegne J, Mbile P, Tabuna H, Angbeh P, Useni M (2008). Domestication, utilization and marketing of indigenous fruit trees in west and central Africa. *In*: Akiniifesi FK, Leakey RB, Ajayi OC, Silenshi G, Tchoundjeu Z, Matakala P, Kwesiga FR (eds.) Indigenous fruit trees in the tropics: domestication, utilization and commercialization. Reading (CABI).
Tomek GW, Robinson LK (1990). Agricultural product prices (3rd ed.). Cornell University Press, London. pp. 360.
Uganda Bureau of Statistics (2006). The 2005/2006 Uganda national and housing survey. Kampala, Uganda.
Wester L, Chuensanguansat D (1994). Adoption and abandonment of Southeast Asian food plants. J. Home Consum. Hortic. 1: 83-92.
Whitaker TW, Cutler CH (1966). Food plants in a Mexican market. Econ. Bot. 20:6-16.

10

An economic assessment of the Ramsar site of Massa (Morocco) with travel cost and contingent valuation methods

Mohammed El-Bekkay[1], Abedellatif Moukrim[2] and Faiçal Benchakroun[3]

[1]Department of Biological Sciences, Faculty of Sciences, AGADIR, Morocco.
[2]Department of Biological Sciences, Faculty of Sciences, AGADIR, Morocco.
[3]High Commissary of Waters and Forests and Fighting Against Desertification, RABAT, Morocco.

As the only protected area in the whole south west of Morocco, the Souss Massa National Park (SMNP) is characterized by a remarkable biodiversity, with more than 300 plants species, 250 bird species, 20 mammal species and also by the high diversity of its ecosystems, such as the *Aragnia spinosa* forest, steppes, dunes and coastal wetlands. This park can play a leading role in the region by enhancing its ecotourism potential and therefore contributing to its economic development. We intended to demonstrate in this study, by estimating the recreational value, one popular area, the RAMSAR site of the Estuary of Massa River (EOM). It is located 50 km from the urban center of Agadir City and it attracts about 30,000 visitors a year. To determine its recreational value, the contingent valuation method and travel cost method were applied. The results from 480 surveys, conducted during 2010, showed that the consumer surplus per person per visit is estimated at DH 490,196 ($US 65.36) and the willingness to pay per visitor is about DH 46,523 ($US 6.20). According to this economic valuation, any future local development must take into account, the opportunity that EOM offers as a recreational site, with high ecotourism potential.

Key words: Estuary of Massa River, recreational value, contingent valuation, travel cost method, willingness to pay, consumer surplus.

INTRODUCTION

Since the Earth Summit in Rio in 1992, the attention paid to the biodiversity conservation grows in importance in the strategies and socioeconomic development programs undertaken by the international community. Protected areas (PAs) are one of the most used models for the protection and preservation of biodiversity, and a real research field for the development of approaches and plans for conservation and sustainable use of natural resources.

National parks have been established in many countries to preserve ecosystems and provide for the recreation, wilderness and leisure demands of the population (Herath, 2004). Indeed, National parks or generally protected areas normally hold a high value as a recreational resource or destination (Navarro et al., 2010). Like many others countries, Morocco has been committed from the nineties, to protect and preserve the environment and biodiversity, through the elaboration and implementation of national strategies, which has led to establishing a national network of PAs that actually include nine national parks, with a total area of 600,000ha (Ribi,2007).

Despite different ecosystem services provided to society, the natural resources are often underestimated in terms of their contribution to the local and regional economic and social development (Ruzzier, 2010). Protected areas are considered by some stakeholders as an obstacle to local development despite their roles in the conservation of biodiversity and the potential opportunities they present for recreation and ecotourism development.

Economic valuation of national parks is an important source of information both for park managers and for society in general (Alvarez and Larking, 2008). Several economic methodologies have been developed in the last decades to approximate the value of non-markets environmental goods and services such as recreational value. Economic valuation techniques can be used to measure the benefits associated with environmental conservation projects and nature tourism activities (Mathieu et al., 2000).

Mehmet and Türker (2006) claim that the Travel Cost and Contingent Valuation Methods (TCM and CVM) are the most widespread and used methods, to estimate the recreational value of forests and protected areas around the world. A series of studies on this topic was conducted at the continental level but few results have been published. Recently in Ghana, Twerefou and Adjei Ababio (2012) used the TCM to estimate the value of recreation at Kakum National Park.

There is very little knowledge on the recreational value in protected areas in Morocco. The aim of this research was to assess the recreational value of the Ramsar site of Estuary of Massa River (EOM) and to show policy makers the economic importance of protected areas.

MATERIALS AND METHODS

To achieve the study objectives of the economic assessment of recreation in the Ramsar site of EOM, wed use the following two methods:

The contingent valuation method (CVM)

The contingent valuation method was applied for the first time by Quotes and Robert in 1963 for the valuation of natural assets. It is used by economists because of its relative simplicity (Cumming et al., 1986). It is based on the reality of an investigation during which it is shown directly to individuals surveyed what they would pay to preserve or restore an environmental good. In other words, this is to simulate a fictitious market in which the individual can directly reveal his preferences and willingness to pay for goods or services from nature.

Bateman and Turner (1993) noted that the CVM requires that individuals express their preferences according to some environmental resources or change in resource status, by answering questions on hypothetical choices. For example, for individual i:

$$WTPi = F(Qi, Yi, Ti, Si) \quad (1)$$

Where: Qi, the quantity or quality of environmental good; Yi, the income of the individual I; Ti, the preferences of the individual i; Si, other socio-economic variables of the individual i.

The travel cost method (TCM)

This method has been applied to estimate the demand and consumer surplus for wildlife and nature conservation at recreation sites. The demand for a park is estimated by determining the change in visits as the cost per visit is changed (Siti, 2009). TCM studies have consistently shown that as the price of access (cost of travel) increases, the visit rate to site falls (Garrod and Willis, 1999). The TCM simple model is usually estimated as a trip generating function such as the following:

$$V=f(C, X) \quad (2)$$

Where: V, Number of visits to a recreation site; C, costs per visit; X, other socio-economic variables, which significantly explain V.

There are two types of data used in the TCM: Zonal and Individual TCM. For the first one, the visitors are grouped into different categories or zones based on certain similar characteristics such as geographical origin. This is the oldest form of the travel cost method (Timah, 2011). The second was the precise number of site visits made by each visitor over a specific period. The ITCM uses survey data from individual visitors in the statistical analysis, rather than data from each zone. However, the ZTCM has been under serious criticism recently for its vagueness as a non-market valuation tool. For this reason, most researchers and economists have now turned to the ITCM as a better option (Bell and Leeworthy, 1990). In this study, the ITCM was applied to evaluate the recreational value of the RAMSAR Site of the estuary of Massa River. The Equation 2 can be rewritten as:

$$Vij=f(Cij, Xi) \quad (3)$$

Where, Vij: Number of visits made per year the individual i to recreation site j; Cij: visit cost by the individual i to recreation site j; Xi: all other socio-economic variables determining individual visits.

The demand curve produced by the ITCM relates individual's annual visits to the costs of these visits. Integrating under this curve gives us the consumer surplus per individual (ICS). Multiplying the ICS by the number of individuals visiting the site annually helps to estimate the consumer surplus for the recreational site:

$$\text{Total consumer surplus} = Nj.\int f(Cij,Xi).dCij \quad (4)$$

Where, Nj is the number of individual visits to recreation site j per year, Cij and Xi are defined as in Equation 3 (Bateman, 1993).

Another way to calculate Individual consumer surplus is to use the following formula for the semi-log function (Willis and Garrod, 1991):

$$CS = -1/\beta ij \quad (5)$$

Where, CS is the consumer surplus per person trip and βij is the coefficient of travel cost Cij.

The total annual consumer surplus obtained from the recreation site can be calculated by multiplying the ICS with the number of visits made in a year.

The case study

This study shows the RAMSAR site of the estuary of Massa River,

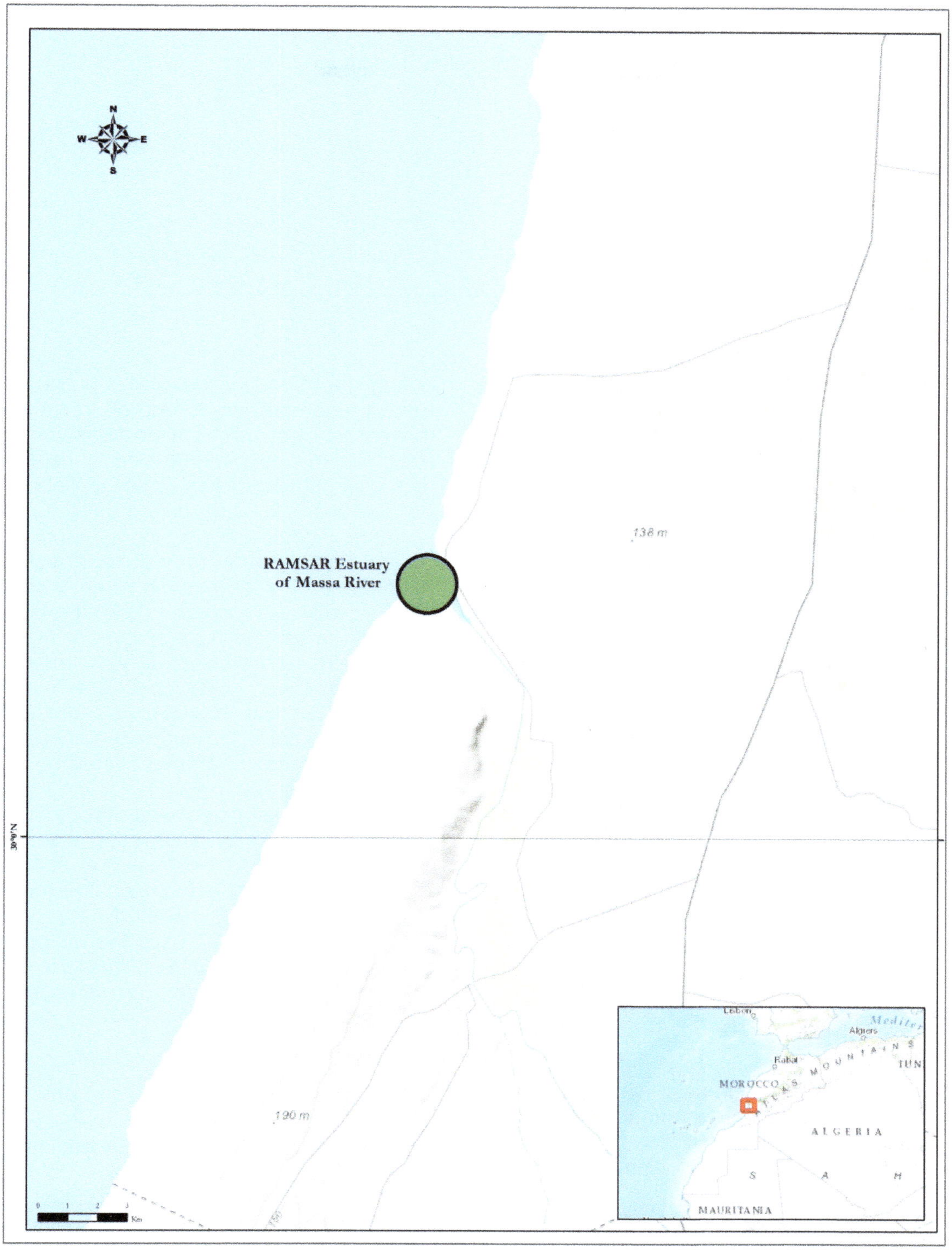

Figure 1. Localization map of the studied site.

located on the Atlantic coast of Morocco on an area of 1200 ha(Figure 1). This site forms the central part of the Souss Massa National Park (SMNP) and one of the richest areas in the park, as far as natural resources are concerned. The site seasonally hosts over thirty species of waterfowl, some of which are among the rarest as spoonbills, flamingos, cranes, glossy ibis and marbled teal. It also holds some of the most endangered birds in the world: the bald ibis and the Audouin's Gull (El Bekkay, 2009).

Table 1. General data related to recreation at the estuary of Massa River.

Variable	Units	Minimum	Maximum	Mean	SD
Age	Year	20	76	43.823	8.252
Annual visit number	Number	1	4	1.433	0.664
Distance from site	km	10	60	49.006	5.394
Time at recreation site	Hour	1	6	2.119	0.865
Number of children	Number	0	5	1.040	1.131
Total travel cost	Number	15	900	355.110	99.724
Monthly income	MDH	0	36000	16086.458	5538.511
WTP	Number	0	120	41.708	20.116

Managed by the High Commissary for Water and Forests and the Fight against Desertification, the estuary of Massa River is visited, all year round, by tourists in search of recreation in natural site. Open freely to the public, this estuary offers to visitors an original and outstanding landscape that combines different components, not often observed in nature: river, dunes, vegetation and sea, which draw about 30,000 visitors annually. In order to estimate the recreation value of the estuary of Massa River, a total of 480 surveys were carried out at the site, during a face to face interview. The monitoring started in Januaryand finished in December 2010. Surveys dating were conducted monthly to consider the different categories of visitors.

In this study, we used a survey form with 32 questions and conducted 480 surveys which helped to build a database large enough to estimate the recreational benefits of the RAMSAR site of the estuary of Massa River.

The first part of the survey form focuses on the identification of visitors and their social and economic data. The second part is on the behavior of visitors, their activities on the site, the number of the visits and their recommendations to improve the quality of the recreation at the Ramsar site of the estuary of Massa River. Finally, the third part aims to identify the level of awareness of interviewed people about the environment protection. The WTP was asked through an open question: What is the maximum amount you are willing to pay for improvements of the recreation site quality? An index card showing the proposed entrance fee, from 0 to 120Dh, was shown to the respondents; this card is the visual support of this valuation.

RESULTS

The contingent valuation method based was on a survey in which we seek, with appropriate information questions, the preferences of individuals and the amount each would be willing to pay for the preservation or restoration of an environmental good.

The descriptive results of the study sample of the Ramsar site of the Massa River are data of great value which can help in understanding visitor behavior. 7% of visitors were under 30 years old, 71% between 30 and 50 years and 20% were over 50 years. An effort should be made to develop an environmental education program to increase the rate of young visitors. Also, the distribution of the sample per months showed the presence of certain trends in terms of the variable visitor's age: those aged less than 30 years discover the Ramsar site of EOM during the period between March and July. For visitors with an age between 30 and 60 years old, it seems that the frequency of visits is almost the same during the year. People with age over 60 years, visit the EOM mainly during the period of November to February, which could be explained by the period of the presence of the wintering migratory birds.

Fifty four percent of visitors are married and 220 interviewed persons have children, with a dominance of couples with two children (42%). It should be noted that 78% of the sample have a level of technical training or higher. When asked whether they knew the Ramsar site of Massa River, 47% of those surveyed gave an affirmative answer. However, 53% of visitors had never heard of the park, showing that a special work is needed in communication and advertising to improve the earnings of the site.

The survey results showed that 77% of visitors do not exceed two hours in discovering the site. Visitors were also interviewed on the quality of services and desired improvements: 71% of them would liked the ecomuseum to be open, 60% wished the site includes a cafeteria/restaurant, 45% of the visitors suggested the installation of viewpoints in the site fauna and 43% expressed their desire to see the access road maintained.

Calculation of the willingness to pay

The WTP in our study is the dependent variable and the independent variables remain the socio-economic characteristics of the individual and specific characteristics of the studied site. Table 1 summarizes the key variables from our sample.

To develop the model on the WTP, this study was based on the same sample studied previously. The bid curve provides insight into the role of socio-economic characteristics of those interviewed on their willingness to pay for environmental goods and services.

The Recourse to use the regression, according to the Tobit model, is justified by the presence of several null offers: 69 people have expressed their willingness to pay

Table 2. The contingent valuation results.

Dependent variable	Model variable	Coefficient	t
	CONSTANT	-19.497	-3.809
	INCOME	0.003	24.572
WTP	NB_CHILD	-1.171	-2.125
	SATISFACTION	23.246	4.876
	TIME_VISIT	-6.991	-3.683
F	R	R^2	Adjusted R^2
186.871	0.796	0.634	0.631

Table 3. The results of regression analysis.

Model summary		**R**	**R^2**	**Adjusted R^2**		**SD**
Semi-Log		0.6532	0.4266	0.4218		0.3419
Analysis of variance						
Model type	Model	Sum of squares	df	Mean square	F	Sig.
Semi-Log	Regression	41.314	4	10.328	88.357	0
Coefficients	Residual	55.525	475	0.117		
Dependent variable	**Model variables**	**Coefficients**	**SD**	**t**		**Sig.**
Ln(AVN)	Constant	1.1663	0.0951	12.26394		0
	TTC	-0.00176	0.00014	-12.9425		0
	AGE	-0.00539	0.0024	-2.24073		0.0255
	NBCHILD	0.0621	0.01868	3.32468		0.001
	EDU	-0.17802	0.04539	-3.92242		0.0001

0 dirhams to improve the site characteristics of the EOM, 44 results correspond to offers of protest. The results for the regression are presented in the Table 2: The bid function is therefore written as follows:

$$WTP = -19.497 + 0.003INC - 1.171NBCHILD \quad (-3.809)^* \quad (24.572)^* \quad (-2.125)^*$$
$$+ 23.246SAT - 6.991\ VISITTIME \quad (4.876)^* \quad (-3.683)^* \qquad (6)$$

where, WTP is the willingness to pay, INC, the income of the respondent, NBCHILD, the number of children of the respondent, SAT, the satisfaction of the respondent, VISITTIME, the time spent on site by the respondent. Numbers in parentheses correspond to the values of t-Statistics; a cross indicates that the variable is significant at 1%. According to the value of R^2, f and t-Statistics, the willingness to pay was found to have a good correlation with some socio-economic variables of the visitors. As expected at the beginning of this study, it appears clearly from the bid function that the visitors with higher income and satisfied with their visit were willing to pay more for the conservation and improvement of the site quality. Also, the number of children of the visitors is a variable negatively related to the WTP. By setting variables in the Equation 6 to their average, we obtained the average WTP which is in the range of 46.523 Moroccan dirhams ($US 6.20).

Calculation of the recreational value

One of the purposes of this research is to determine the use value of recreation site at the estuary of Massa River via the use of the estimated consumer surplus means. The semi-log function was used by specifying the number of visits per year which is the dependent variable. Age, income, total expenses of the trip, the distance traveled to visit the EOM, etc., are the independent variables. Table 3 relates the results of the regression of the variation in the number of annual visits with the explanatory variables.

According to the Fisher F-values and t-Student and R^2, the semi-log function is significant at a level of 99%. The function of the application may be written as follows:

$$Ln(AVN) = 1.1663 - 0.00176TTC - 0.00539AGE + 0.0621NBCHILD - 0.17802EDU \qquad (7)$$

It appears from this demand curve, that the sign of the

coefficient of the variable "total travel cost" (TTC) is negative and therefore it is inversely proportional to the exponential increase of number of annual visits, which is consistent with the results of ITCM found in the literature.

Calculation of consumer surplus

Correcting the model of the demand curve in order to maintain TTC as the main explanatory variable, the Equation 7 becomes:

$$\text{Ln(AVN)} = 0.95778 - 0.00204\text{TTC}$$

It is necessary to calculate at first the consumer surplus per person per visit, and then the total consumer surplus will be estimated by using the formula referred to above (5):

$$\text{CS} = -1 / (-0.00204) = 490.196 \text{ DH } (65.36 \text{ US\$})$$

The total annual consumer surplus obtained by site visitors from the estuary of Massa River can be calculated by multiplying the individual consumer surplus by the number of visitors to the site of the EOM (30,000 visitors a year). Then, the value of individual consumer surplus per visit is about 490.196 DH (65.36 USD), and the value of total consumer surplus is estimated at 14.7 million DH (1.96 million dollars).

DISCUSSION

One of the most visited sites in the Souss Massa National Park, the estuary of Massa River offers considerable bio-ecological value, to develop recreation and ecotourism. The contingent valuation method and travel cost method are among the most often used techniques that attempt to assign monetary values to goods and services provided by environmental resources. As part of this research, 480 interviews were conducted to assess the monetary value of recreation site, using the two methods mentioned above.

According to our results, the ITCM has enabled first, the calculation of the consumer surplus per person per visit at about 490.196 DH (US$ 65.36), and second the estimation of the value of recreation to be about 14.7 million dirhams (1.96 million US$). In parallel, the willingness to pay for the improvement of services, offered by the site of the EOM was calculated by using the CVM, the WTP was estimated by the bid curve to 46.523 dirhams (US$ 6.20).

Comparable studies have been undertaken to approach the recreational value of protected areas, especially national parks in different regions of the world. For example, in Egypt, using the ITCM, the CS related to the visit of the coral reefs in Ras Mohamed National Park was estimated at US$2,280, and the WTP, according to tobit regression, was US$ 6.57 (Rady, 2006). In India, the individual consumer surplus for visiting the Periyar National Park was evaluated to be US$ 36.18 (Bulov and Lundgren, 2007). Twerefou in 2012 indicated that the consumer surplus for visiting Kakum National Park in Ghana was about US$ 46.40. In Australia, the average WTP for visiting Mont Buffalo national park was assessed at US$ 13.15 (Herath, 2004). The WTP to moderate the environmental impact of inland development in marine parks in Malaysia was estimated to be US$ 7.89 (Siti, 2009).

Taking into account these values and in the absence of similar studies in the other Moroccan protected areas, we have tried to compare our results with those found in some Mediterranean and Arab countries where comparable studies was conducted. The first example is related to recreation in Kayabaşi Forest (KF) in Turkey (Mehmet and Türker, 2006), the second one shows the recreational benefits of Dibeen National Park (DNP) in Jordan (Amer and Said, 2006). Using the TCM, the average value of the recreation was respectively estimated at US$ 22.11 and US$ 100. On the other hand, the willingness to pay, to conserve and to improve the services in the two sites has been estimated in KF to be US$ 1.01 and in DNP to be US$ 7.9. According to these cases, it appears clearly that the recreational value of the estuary of Massa River is at the same level of recreation as those recorded in other sites in the developing countries. The relative high values of WTP and CS of Massa Estuary can be explained by the attractiveness of the site and also by the lack of other possibilities to discover the wildness in the Agadir region.

The results show that the value of recreation-based on TCM is higher than the one generated by the CVM. Several authors have already compared the results obtained by the contingent valuation method with those of the travel cost method. They concluded that the economic value of the contingent valuation method is lower than the one obtained by the method of travel cost (Carson et al., 1996).

Conclusion

The monetary values of the recreation in the Estuary of Massa River, estimated by the TCM and CVM, reflects the exceptional potential of the site for economic development in the region, which must be taken into account, by both managers as local communities of the EOM. These values, as well as the other results in this research, should be taken into consideration, at every decision-making regarding any intended use of the site. In fact, this kind of study should be extended to other similar sites, dedicated to preservation and development of natural resources in Morocco, in order to demonstrate to all stakeholders, through economic criteria, the benefit which could be drawn from the valuation of natural resources.

REFERENCES

Alvarez S, Larkin L (2008). Valuation Recreational Benefits of a National Park in Andean Columbia. Southern Agricultural Economics Association Annual Meetings. Dallas, February 2-6, 2008.

Amer SJ, Said AD (2006). Estimation the Recreational Benefits of Dibeen National Park in Jordan Using Contingent Valuation and Travel Cost Methods. Pak. J. Biol. Sci. 9(12):2198-2206.

Bateman IJ (1993). Valuation of the Environment, Methods and Techniques: Revealed Preferences Methods, Sustainable Environmental Economics and Management, Bellhaven Press, London. pp. 194-265.

Bateman IJ, Turner RK (1993). Valuation of the Environment, Methods and Techniques: the Contingent Valuation, Sustainable Environmental Economics and Management, Bellhaven Press, London. pp. 121-191.

Bell FK, Leeworthy VR (1990). Recreational demand by tourists for Salt Water Beach Days. J. Environ. Econ. Manage. 18:189-205.

Bulov S, Lundgren T (2007). An Economic Valuation of Periyar National Park a Travel Cost Approach, Lulia University of Technology, India. pp. 55.

Carson RT, Flores NE, Martin KM, Wright JL (1996). Contingent Valuation and Revealed Preferences Methodologies: Comparing the Estimates for Quasi-Public Goods. Land Econ. 72:80-99.

Cumming RG, Brookshire DS, Schultze WD (1986). Valuing public goods: the contingent valuation method. Totowa (N.J.), Rowman and Allanheld Publishers. pp. 110.

El Bekkay M (2009). Valorization and Management Plan of Souss Massa National Park. Waters and Forests Department. pp. 48.

Garrod G, Willis KG (1999). Economic Valuation of the Environment: Methods and Case Studies. UK: Edward Elgar Publishing. pp. 384.

Herath G (2004). Estimating the Economic Value of Mount Buffalo National Park with the Travel Cost and Contingent Valuation Models. Tourism Econ. 10(1):63-78.

Mathieu L, Langford I, Kenyon W (2000). Valuing Marine Parks In A Developing Country: A Case Study Of The Seychelles. CSERGE Working Paper GEC 2000-27:31.

Mehmet P, Türker MF (2006). Estimation of Recreational Use Value of Forest Resources by Using Individual Travel Cost and Contingent Valuation Methods (Kayabaşı Forest Recreation Site Sample). J. Appl. Sci. 6(1):1-5.

Navarro GM, Paca ED, Rimas CS (2010). The Economic Value Of Mt Pulag: An Eco-Tourism Perspective The Economic Value Of Mt. Pulag: An Eco-Tourism Perspective 11th National Convention on Statistics (NCS), EDSA Shangri-La Hotel, October 4-5, 2010. pp. 24.

Rady Talaat T (2006). Recreational Value Of Coral Reefs An Application To Coral Reefs In Ras Mohammed National Park, the Forth Medpan Workshop on Marine Protected Areas, June 2006, Italy. pp. 72-87.

Ribi M (2007). International Workshop on sustainable tourism in Moroccan Protected Areas, Waters and Forests Department. pp. 56.

Ruzzier M (2010). Guidelines For The Economic Evaluation Of Natural Assets Of The Protected Areas, Institute of the Republic of Slovenia for Nature Conservation. pp. 66.

Siti Aznor A (2009). Visitors Willingness to Pay for an Entrance Fee: a Case Study of Marine Parks In Malaysia, University of Leeds UK. pp. 289

Timah P (2011). Non Market Valuation of Beach Recreation Using the Travel Cost Method in the Context of Developing World: An Application To Visitors of the Naoe Beach in Kribi. Cameroon. Swedish University of Agricultural Sciences. pp. 88.

Twerefou DK, Adjei Ababio DK (2012). An Economic Valuation of the Kakum National Park: an Individual Travel Cost Approach. Afr. J. Environ. Sci. Technol. 6(4):199-207

Willis KG, Garrod GD (1991). An Individual Travel Cost Method of Evaluating Forest Recreation. J. Agric. Econ. 42(1): 33-42

11

Mass of *Prunus africana* stem barks on Tchabal mbabo and Tchabal Gang Daba Mountain Forests, Cameroon

Jean Lagarde BETTI[1] and Joseph AMBARA[2]

[1]Department of Botany, Faculty of Sciences, University of Douala, BP 24 157 Cameroon.
[2]Ministry of Environment and Nature Protection, Cameroon.

Prunus africana **is a species of the Rosaceae family, known under its common name as pygeum or African cherry. The bark is the major source of an extract used to treat benign prostatic hyperplasia, an increasingly common health problem in older men in the western world. A study which aimed to produce a formula for establishing the mass of the bark of *Prunus africana* specimens was carried out in May 2011 on Tchabal Mbabo and Tchabal Gang Daba mountain forests, in the Adamaoua region of Cameroon. The diameter at breast height (DBH), the height of the tree and the thickness of the stem bark of each pygeum tree were recorded in order to establish the cubic volume of each specimen. This data was used to calculate the relationship between the diameter and the volume of the bark. Samples of bark were collected in order to establish the relationship between the volume of the bark and its mass (measure mass per cubic metre). A total of 105 pygeum trees were sampled, including 50 trees in Gang Daba considered as an un-exploited production site and 55 trees in Mbabo referred to an exploited production site. The best equation which links the volume (Vb) of fresh barks to the diameter (D) of each pygeum tree is Vb=a/(1+b*exp(-cD)) with a = 1.79588278896E-001, b = 5.29124992540E+002, c = 1.45488065368E-001. The average thickness of the bark is 13.01 ± 4.8 mm. This value is comprised between that of unexploited (16.99 ± 3.7 mm) and exploited (9.40 ± 2.07 mm) pygeum trees. Considering that for all trees above 30 cm DBH, only two quarters of the bark are taken from the main stem up to the first branch, the average sustainable mass of pygeum tree in Adamaoua region will be about 69.3 kg of fresh bark per tree. This value is comprised between that of unexploited (80 kg) and exploited (60 kg) pygeum trees in Ganga Daba and Mbabo respectively.**

Key words: Threatened species, *Prunus africana*, pygeum, Tchabal mountain forests, Bitterlich's Relascope, cubic tariff, CITES.

INTRODUCTION

Prunus africana (Hook.f.) Kalkman (formerly *Pygeum africanum* Hook.f.) is a species of the Rosaceae family, known under its trade/pilot name as pygeum or African cherry. It is a mountain tree species of the tropical Africa including Côte d'Ivoire, Bioko, Sao Tome, Ethiopia, Kenya, Uganda, South Africa, Madagascar, Congo, the Democratic Republic of Congo, Mozambique, Tanzania, Burundi and Cameroon. *P . africana* grows well in the sub-mountain and mountain forests at an altitude of 800 – 3000 m. In Cameroon, the plant is largely found in five regions including Adamaoua, North west, Littoral, South west, and West. *P. africana* is an evergreen canopy tree to 30 m tall with thick, fissured bark and straight bole that can reach a diameter of 1.5 m. It is light demanding and responds well to cultivation (Hall et al., 2000; Vivien and Faure, 1985; Fraser et al., 1996; Tchouto, 1996).

The bark is black to brown, corrugated or fissured and scaly, fissuring in a characteristic rectangular pattern. The fruits of *P. africana* are drupaceous, fleshy and red-purple in colours and are said to be eaten by a variety of birds and mammals (Cunningham and Mbenkum, 1993).

The bark is the major source of an extract used to treat benign prostatic hyperplasia, an increasingly common health problem in older men in the western world. Prostate-related diseases increase in prevalence as men age. And as the average age of the world's population increases, the incidences of prostate diseases will increase as well, triggering a corresponding rise in demand for therapies. According to the World Cancer Research Fund International, prostate cancer is the second most common cancer in men worldwide. Around 910,000 cases of prostate cancer were recorded in 2008, accounting for approximately 14% of all new cancer cases in men (World Agroforestry Centre, 2012). Bark extracts contain fatty acids, sterols and pentacyclic terpenoids (Cunningham and Mbenkum, 1993). The drugs processed from the bark extracts are sold under the brand-name of "Tadenan" in France by Laboratoire Debat, "Pygenil" in Italy by Indena Spa, and "Proscar" in UK by Merck Sharp and Dohme Ltd (ICRAF cit. Ndam, 1996). The United Nations Food and Agriculture Organization (FAO) reported as far back as 1996 that the demand for the species' bark, which is used to produce treatments for prostate gland disorders, could lead to its over-exploitation (FAO, 1996). In 1997, the global need is about 4 000 tons of dried barks per year for a value of 220 millions of USD. Two hundred kilogram of dried bark yield 5 kilogram of extract (Cunningham et al., 1997). The trade in dried pygeum bark and bark extract is in the order of 3 000 to 5 000 tonnes a year (Page, 2003) and the main sources are in Cameroon, Madagascar, Equatorial Guinea, Kenya, Uganda, and Tanzania. Cameroon was one of the major sources of Pygeum supply from the early 1970s, when the French company laboratoire Debat established a factory at Mutengene, on the lower slopes of Mount Cameroon. Operating as Plantecam Medicam, the company prepared bark extracts in tablets form from bark harvested in the wild. Between 1972 and 1985, Plantecam had a monopoly of the harvesting licenses for Pygeum and as a result was able to control exploitation. After 1985, when Plantecam's monopoly was revoked and many additional harvesting licenses were provided to Cameroonian entrepreneurs, the level of regulation of bark harvesting declined (Ndibi and Kay, 1997). Harvesting permits for a total of 2 558.37 tons of pygeum dried bark was granted to 33 trade companies by the Cameroon forest administration in 2005 (2 000 tons) and 2006 (1 260 tons). A total of 2 558.37 tons of pygeum bark (78.5% of the granted pygeum) exported from the Douala port was recorded by the national database on trade forest products. The most important quantity of the barks was exported in 2005 (1 498.5 tons) and the remaining (1 059.87 tons) was exported in 2006 (Betti, 2008).

P. africana is classified by the World Alliance for Nature (IUCN) as vulnerable species, which led to its listing in the Appendix II of the Convention on International Trade in Endangered Species of Fauna and Flora (CITES) in 1994, becoming effective in 1995 (Sunderland and Tako, 1999). Once a species is listed in Appendix II of CITES, its exportation is regulated in terms of quota. In Cameroon the minimum exploitable diameter (MED) for *Prunus* trees is 30 cm and till 2006, the national CITES quota for *P. africana* dried barks was 2000 tons/year. The data are not available to give precise figures, however it is clear from the increasing level of exploitation that the pygeum population is declining over time in terms of tree size, density of stand and overall tree population, resulting in declining habitat quality (Betti, 2008). *P. africana* can be considered at least as an endangered plant species in Cameroon according to population reduction as outlined in the IUCN check list for Non-Detriment Findings (IUCN, 2001). This explains the ban pronounced on October 2007 by the European Commission on Cameroon's pygeum. The pygeum ban impacts both the economic operators and the local people for whom *P. africana* represents an important non timber forest product. Cameroon was proposed for ban as there are concerns that some provisions regarding the sustainable harvesting of *P. africana* barks are not being fully met.

A key requirement of CITES is the non-detriment findings (NDF) made by the Scientific Authority of the range State prior to export, certifying that export is not detrimental to the survival of the species in its natural habitat. This requires information on the location, stocking, growth and condition of the species and on its ecology, regeneration and subsequent protection. Such information is often lacking, incomplete or imprecise making a proper evaluation of the sustainable levels of utilisation, establishment of quota and conditions attached to be difficult.

Recognizing the shortcomings in scientific information related to the sustainable harvesting of *P. africana*, the government of Cameroon submitted to the International Tropical Timber Organization (ITTO) for funding the project entitled "Non-detriment findings (NDF) for *P. africana* (Hook.f.) Kalman in Cameroon". The first important outputs of the project were the production of the partial NDF reports on *P. africana* for the mount Oku and mount Cameroon, respectively in the North west and South west regions of Cameroon. An annual quota of 350 tons of dried barks were estimated for the two regions (Akoa et al., 2010, 2011), which lifted the partial ban on the exportation of *P. africana* barks from Cameroon.

This paper aims to estimate the mass of the fresh bark of *P. africana* on Tchabal Mbabo and Tchabal Gang Daba mountain forests, Adamaoua region of Cameroon, as a contribution for making non-detriment findings on *P. africana* for that region.

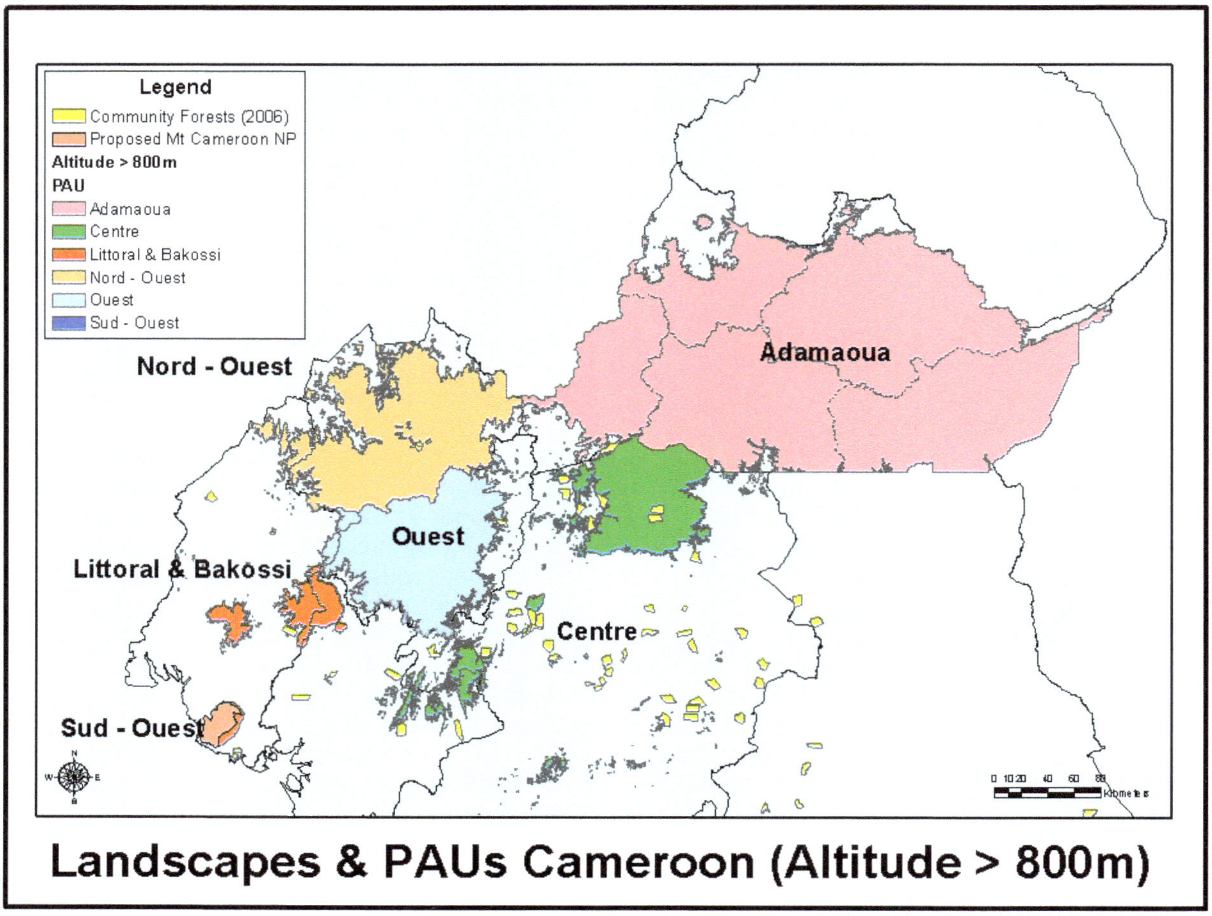

Figure 1. Landscapes and *Prunus* Allocation Units (PAUs) in Cameroon.

MATERIALS AND METHODS

Study site

Cameroon is located in Central Africa, in the bottom of the Guinean gulf. Cameroon is divided into ten regions including: centre, east, far north, north, north west, south, south west, west, and Adamaoua. Among the ten regions, six, namely Adamaoua (Ngaoundéré being the capital), the Centre (Yaoundé), Littoral (Douala), North West (Bamenda), Southwest (Buea), and West (Bafoussam), are regions where populations of *P. africana* occur. But the main reserves of *P. africana* are found in the Adamaoua, North West and South West regions. The major landscapes of Cameroon containing *P. africana* have been agreed, defined and consolidated into *Prunus* Allocation Units (PAUs) that cover six mountain areas (Figure 1).

The Adamaoua region is vast of 63 701 km² and comprises five divisions including: Djerem, Mbéré, Vina, Faro and Déo, and Mayo – Banyo. Faro and Déo and Mayo – Banyo are the two divisions which contain *Prunus* forests, in Tchabal Gang Daba and Tchabal Mbabo mountain forests respectively to be précised. These areas are composed of a succession of mountains or "Horé" in local language.

Tchabal Mbabo mountains are up to 2 240 m, situated between 7°10' to 7°40' latitude Nord, and 11°7' to 12°20' longitude East. The main high levels include: Horé Lassel, Horé Mayo Kélélé, Horé Yangaré, Horé Ngouri, Horé Garbaya, Fongoy, Nanaré. Tchabal Gang Daba mountain is up to 1 960 m, situated between 7°40' to 7°55' latitude Nord and 12°38' to 12°56' longitude East. Gang Daba is located at 10 km from Tignère city, the capital of the Faro and Déo division.

The climate is a transition - subtropical type, characterized by two equal seasons: A dried season from November to March and a rainy season from April to October. The average temperature is 23°C. The maximum average temperature is 30°C in March and the minimum is 15°C in December – January. The wind is dried and wet in rainy season and dried and hot in dried season. The average annual rainfall varies between 1 000 and 2 000 m. August and September being the most rainy months. Two types of soils are found in the region. Red and yellow iron soils resulting from the decomposition of metamorphic rocks on the slopes, and black alluvial soils in the bottom and in forest galleries found along rivers. These soils are regularly degraded due to the fluvial erosion, the bush fires, and the grazing. The Tchabal Mbabo and Tchabal Gang Daba mountains have a high diversity of plant species.

In Tchabal Mbabo there are four forest types: Forest galleries, herbaceous savannahs, dried mountain forests, and woody savannahs. Forest galleries are found in valleys and along rivers. They contain *P. africana* and other plant species. Herbaceous savannahs are composed of meadows dominated by *Hyparhenia* sp and *Andropogon* sp. Dried mountain forests are found in Dodéo, Fongoy, Nanaré, and Yangaré. They contain *P. africana* and characteristic savannah species such as *Khaya senegalensis, Daniella oliveri, Isoberlima doka, Cedrela odorata, Combretum* sp, *Burkea africana, Lophira laceolata, Prosopis* sp, *Syzygium guineense, Terminalia laviflora* and *T. mcroptera.* Woody savannahs are found in Dodéo, Fongoy and Mbabo. They are also composed of the same species found in dried mountain forests. The harvesting of *P. africana* started in Tchabal Mbabo by the year 2000 by two national trade companies including AFRIMED and ERIMON (Akagou and Betti, 2007).

In Tchabal Gang Daba, there are no high trees in the summits. There are some forest galleries which contain *P. africana*. The Samlekti valley gets forests which are dominated by *Isoberlima doka and I. tomentosa.* Tchabal Gang Daba has never been subject to *P. africana* exploitation.

Method

Selection of trees

This study was conducted in may 2011 in Tchabal Mbabo and Tchabal Gang Daba mountain forests, Adamaoua region, Cameroon. The study was conducted in the frame of the project "Non-detriment findings for *P. africana* (Hook.f.) Kalman in Cameroon" implemented by the Cameroon's National Forestry Development Agency, with the support of the International Tropical Timber Organization (ITTO), the Convention on International Trade in Endangered Species of Fauna and Flora (CITES), and the European Commission (EC). The study started one month later after the team in charge of assessing the abundance of *P. africana* trees in Mbabo and Gang Daba has finalised their work. The sheets of the team in charge of counting were used to select the best trees that can allow to better appreciate the mass of the stem bark of *P. africana* trees. Practically, trees were randomly selected according to their diameter, their accessibility, their healthy and their conformity (only straight stems were selected). Once a given stem was identified in a counting sheet, we noted its geographical data (latitude and longitude) and we went with the global position system (GPS) in the field to take appropriated measures of that stem. A total of 105 *P. africana* trees were sampled including: 50 trees in Gang Daba and 55 trees in Mbabo.

To estimate the mass of the stem barks of *P. africana*, we used an indirect method and proceeded in three steps: establishment of the relation linking the diameter at breast height (DBH) of each tree with the thickness of its stem bark, establishment of the relation linking the volume of the stem bark and its mass (weigh), and determination of the mass of the fresh bark for exploitable trees. Exploitable trees being trees with DBH ≥ 30 cm. A DBH is high at 1.5 m.

Relation diameter at DBH – thickness of volume of the stem bark: volume based tariff

The volume based tariff is a mathematical formula which gives the unit volume of a given tree according to different variables. These variables can be the diameter, the circumference, the height,...... The tariff is more valid for the area where the samples were collected (College of Forest Engineers of Quebec, 1996).

The diameter and the height of each *P. africana* tree section were measured using the "Bitterlich's Relascope with large bands" or "SPIEGEL RELASKOP". For each tree selected, we chose the best position which allows to see clearly the trunk. In this paper, the trunk refers to the distance between the DBH and the first big branch of the tree. We measured the horizontal distance and the distance according to the slope between our position and the tree. We measured the logging high which is the breast height and the useful height which is the height of the trunk. We measured the diameter and the percentage of the slope at logging height (DBH) and the diameter and the percentage of the slope at the first big branch level. After these measures, we determined the intermediary levels or measures for the percentage of the slope (P_1, P_2, P_3) using the equi-distance formula which is: $Eq = (P_u - P_a)/4$, with $P_1 = P_a + Eq$, $P_2 = P_1 + Eq$, $P_3 = P_2 + Eq$, and $P_u = P_3 + Eq$. We measured the slopes and diameters at intermediary levels. For short trees, we estimated directly the volume by measuring the DBH, the height of the tree, and the diameter at the first big branch level. We did not measured the intermediary levels, which explains the absence of some data as shown in Table 1. Their volumes were determined directly through the Smalian's formula (see below).

For each tree, we measured the thickness of the stem bark using the "Tarrière of Pressler". The unit volume (on or under the bark) was calculated using the formula of Relascope of Bitterlich of large bands (National Office for Forest Development, 1992) which is:

$$V = (\pi dh^2/8.106)*((P1-Pa)*(D1^2+Da^2) + (P2-P1)*(D2^2+D1^2) + (P3-P2)*(D3^2+D2^2) + (Pu-P3)*(Du^2+D3^2)).$$

Where, V is the unit volume (on or under the bark); $d_h = d_s \times \cos\alpha$ = corrected horizontal distance; d_s is the distance following the slope; P_a, P_u, P_1, P_2, P_3 are the percentage of the slope at the logging level (or useful level), at the first big branch level, at point 1, 2, and 3 respectively; D_a, D_u, D_1, D_2 and D_3: diameter in Relascope unit (RU) obtained at the logging level, at the first big branch level, at point 1, 2, and 3 respectively.

We also used the Smalian's formula to estimate the volume:

$$V = \pi/8\ (D_a^2 + D_u^2) \times h$$

Where, D_a is the diameter at the logging level of the tree = DBH; D_u is the diameter at the first big branch level; h is the height of the trunk = distance between the logging level and the first big branch.

The volume of the bark was deduced from the following equation:

$$V_b = V - V_o,$$

Where: V_b is the volume of the stem bark; V is the volume of the tree over the bark = with the bark; V_o is the volume of the tree under the bark = without the bark.

Assuming that $D_o = D - 2e$ with e = the thickness of the bark, D_o = diameter of the tree under the bark and D = diameter of the tree over the bark, the volume of the bark can be determined through the following equation:

$$V_b = (\pi h/2)*(e \times (D_a + D_u) - 2e^2).$$

Relation linking the volume of the stem bark and its mass = cubic mass

The cubic mass of an entity is the ratio of the mass of that entity and the volume occupied by that mass. Samples of the fresh bark were collected in all exploitable trees of the sample used to establish the volume based tariff (see above). For each sample we noted the length (cm), the width (cm), and the thickness (mm). The thickness of the bark in this section was measured using the calliper rule. The three measures including the length, the width, and the thickness allowed us to obtain the volume of the bark sample. We weighted the sample and found the equivalent mass. Equivalencies were made between the average volume of the samples and their corresponding fresh mass/weight. From those equivalencies, we deduced the cubic/volumic mass of *P. africana* barks in Tchabal Mbabo and Tchabal Gang Daba mountain forests. The cubic mass is $Cm = m/V$ with m = mass in kilogramme (kg) and V = volume of the stem bark in cubic meter (m^3).

Table 1. Data recorded and the corresponding volume per tree in Tchabal Mbabo and Tchabal Gang Daba.

N°_tree	DBH (cm)	P_a	P_u	d_h	h (m)	LB	SB	D_u (cm)	e (mm)	V (m^3)	Site
1	39.49	-39	10	7	3.43	2	1	31.50	11	0.041	Gang Daba
2	59.84	-28	37	10	6.50	2	3	55.00	20	0.226	Gang Daba
3	36.13				4.00			31.20	10	0.041	Gang Daba
4	59.68	-24	52	8	6.08	3	0	48.00	22	0.217	Gang Daba
5	37.05	-42	14	9	5.04	1	2	27.00	10	0.049	Gang Daba
6	43.42	10	127	9.3	10.88	1	3	32.55	15	0.187	Gang Daba
7	31.20	27	62	10	3.50	1	1	25.00	11	0.033	Gang Daba
8	59.21	-28	57	8	6.80	2	3	44.00	20	0.212	Gang Daba
9	65.89	-19	28	11	5.17	2	3.5	63.25	23	0.233	Gang Daba
10	33.42	-29	42	10.4	7.38	0	4	20.80	12	0.072	Gang Daba
11	58.16	-17	38	10.8	5.94	2	1	48.60	19	0.183	Gang Daba
12	30.43	-86	-16	8.23	5.76	1	1	20.58	11	0.049	Gang Daba
13	62.87	-16	57	11.3	8.25	2	1	50.85	16	0.229	Gang Daba
14	37.88	-13	80	6	5.58	2	2	30.00	11	0.063	Gang Daba
15	41.40	-11	36	10.3	4.84	1	3	36.05	9	0.052	Gang Daba
16	64.62	-16	39	11	6.05	2	1	49.50	20	0.209	Gang Daba
17	40.64	17	64	9.6	4.51	1	2	28.80	10	0.048	Gang Daba
18	41.85	35	137	8	8.16	1	3	28.00	15	0.129	Gang Daba
19	51.57	-36	68	8	8.32	2	3	44.00	14	0.170	Gang Daba
20	67.32	-24	48	10.5	7.56	2	1.5	49.88	18	0.243	Gang Daba
21	66.53				5.80			63.19	20	0.229	Gang Daba
22	43.29	-23	57	11.5	9.20	1	2	34.50	14	0.152	Gang Daba
23	82.76	-24	39	7.4	4.66	5	2	81.40	20	0.235	Gang Daba
24	65.73	-35	25	9.8	5.88	3	1	63.70	20	0.232	Gang Daba
25	46.28	11	82	10.7	7.60	1	3	37.45	16	0.154	Gang Daba
26	45.36	3	69	10	6.60	1	3	35.00	17	0.136	Gang Daba
27	37.94	-69	4	7.55	5.51	2	1	33.98	14	0.084	Gang Daba
28	59.72	-37	23	11	6.60	2	2	55.00	19	0.218	Gang Daba
29	50.45	-26	51	10.5	8.09	2	1	47.25	15	0.180	Gang Daba
30	61.12	42	108	10.5	6.93	2	2	52.50	19	0.227	Gang Daba
31	41.85	-27	28	11.5	6.33	1	3	40.25	15	0.118	Gang Daba
32	60.04	1	63	12	7.44	1	3	42.00	19	0.218	Gang Daba
33	49.66	30	101	8	5.68	1	4	32.00	14	0.099	Gang Daba
34	59.21	-16	49	9.7	6.31	2	2	48.50	21	0.215	Gang Daba
35	119.69	53	162	7	7.63	8	1	115.50	20	0.554	Gang Daba
36	30.08	-22	26	8.5	4.08	1	3	29.75	12	0.044	Gang Daba
37	48.19	-8	69	10	7.70	1	2	30.00	14	0.128	Gang Daba
38	38.20	-17	56	9	6.57	1	2	27.00	14	0.090	Gang Daba
39	58.63	-36	36	10	7.20	2	1	45.00	19	0.215	Gang Daba
40	37.88	8	73	8.5	5.53	1	2	25.50	14	0.074	Gang Daba
41	66.37	-84	36	7.13	8.56	2	4	42.78	16	0.228	Gang Daba
42	59.94	-31	17	13	6.24	1	3	45.50	22	0.218	Gang Daba
43	43.07	-20	37	10	5.70	1	4	40.00	13	0.094	Gang Daba
44	50.71				9.60			49.65	14	0.206	Gang Daba
45	55.26	16	83	11.6	7.77	1	3	40.60	19	0.214	Gang Daba
46	56.98	-17	61	9.3	7.25	2	2	46.50	19	0.216	Gang Daba
47	30.34	-31	16	9.5	4.47	1	1	23.75	12	0.044	Gang Daba
48	44.09	0	74	9	6.66	1	3.5	33.75	16	0.125	Gang Daba
49	41.95	-62	12	10.42	7.71	1	3	36.47	13	0.119	Gang Daba
50	44.98	7	76	10.8	7.45	1	2	32.40	15	0.131	Gang Daba
51	46.60	41	84	10.6	4.56	1	3	37.10	8	0.047	Mbabo

Table 1. Contd.

52	58.76	-26	18	12.5	5.50	1	3	43.75	7	0.061	Mbabo
53	36.80	37	94	11.7	6.67	1	1	29.25	7	0.047	Mbabo
54	52.30	3	36	11.6	3.83	1	3	40.60	12	0.065	Mbabo
55	50.80				3.40			46.98	10	0.051	Mbabo
56	49.43	3	41	12	4.56	1	1	30.00	8	0.045	Mbabo
57	78.05	31	120	12	10.68	1	3	42.00	8	0.159	Mbabo
58	93.14	12	78	14	9.24	2	2	70.00	7	0.164	Mbabo
59	48.83	-17	38	8	4.40	2	1	36.00	9	0.052	Mbabo
60	79.68	-77	11	11.51	10.13	2	0	46.04	8	0.158	Mbabo
61	44.15	-16	52	9.5	6.46	1	4	38.00	7	0.057	Mbabo
62	94.54	-26	78	9	9.36	4	1	76.50	6	0.150	Mbabo
63	101.73	-23	56	13.5	10.67	3	1	87.75	6	0.189	Mbabo
64	47.59	8	38	11.6	3.48	1	1	29.00	11	0.045	Mbabo
65	81.08	-33	48	11	8.91	2	3	60.50	7	0.137	Mbabo
66	47.65	3	57	10	5.40	1	4	40.00	9	0.066	Mbabo
67	82.25	5	73	11.4	7.75	2	1	51.30	9	0.144	Mbabo
68	35.24	-23	38	9.7	5.92	1	1	24.25	9	0.048	Mbabo
69	103.00	21	86	12.5	8.13	2	4	75.00	7	0.158	Mbabo
70	90.84	9	87	13.7	10.69	2	1	61.65	7	0.178	Mbabo
71	31.32				3.60			27.47	10	0.032	Mbabo
72	96.32	-17	84	11	11.11	3	1	71.50	6	0.174	Mbabo
73	39.38	-24	41	10.5	6.83	1	1	26.25	10	0.068	Mbabo
74	51.25				2.29			50.29	12	0.043	Mbabo
75	62.39	-44	37	9.5	7.70	2	2	47.50	7	0.092	Mbabo
76	67.61	-13	42	12.5	6.88	2	1	56.25	8	0.106	Mbabo
77	86.10	-34	48	13.4	10.99	2	3	73.70	7	0.191	Mbabo
78	69.87	6	84	13	10.14	2	1	58.50	9	0.181	Mbabo
79	38.00				2.36			36.48	11	0.029	Mbabo
80	83.30	14	107	10.6	9.86	3	1	68.90	6	0.140	Mbabo
81	74.71	-13	64	11	8.47	2	3	60.50	11	0.195	Mbabo
82	91.17	3	81	15	11.70	2	1	67.50	6	0.174	Mbabo
83	89.00	22	81	13	7.67	2	2	65.00	7	0.129	Mbabo
84	60.58	27	143	11	12.76	1	4	44.00	9	0.185	Mbabo
85	37.56	-23	18	10.5	4.31	1	1	26.25	10	0.042	Mbabo
86	70.22	-6	87	9.6	8.93	3	0	57.60	9	0.159	Mbabo
87	75.82	27	86	13.7	8.08	2	0	54.80	12	0.195	Mbabo
88	81.56	7	109	10.6	10.81	2	4	63.60	6	0.147	Mbabo
89	74.07	-11	52	15	9.45	1	4	60.00	10	0.196	Mbabo
90	92.50	-3	53	10.4	5.82	3	2	72.80	7	0.105	Mbabo
91	72.74	-30	81	8.5	9.44	2	2	42.50	7	0.118	Mbabo
92	30.56				2.53			26.10	10	0.022	Mbabo
93	93.11	17	68	13.6	6.94	2	3	74.80	9	0.163	Mbabo
94	74.04	-4	69	11.4	8.32	2	3	62.70	11	0.193	Mbabo
95	43.13	-72	14	6.89	5.93	2	3	37.90	8	0.059	Mbabo
96	90.85	-42	51	12.6	11.72	2	3	69.30	7	0.205	Mbabo
97	79.39	18	82	9.7	6.21	2	3.5	55.78	9	0.117	Mbabo
98	80.28	11	71	13	7.80	2	2	65.00	7	0.123	Mbabo
99	48.38				2.70			46.47	12	0.047	Mbabo
100	70.99	8	101	10.6	9.86	1	4	42.40	7	0.121	Mbabo
101	88.02	-34	23	14.7	8.38	2	1	66.15	7	0.141	Mbabo
102	46.25	-10	34	8	3.52	2	3	44.00	10	0.049	Mbabo
103	70.16	-13	51	13	8.32	1	3	45.50	13	0.192	Mbabo

Table 1. Contd.

104	76.91	26	84	12.6	7.31	2	0	50.40	11	0.158	Mbabo
105	83.94	8	104	8.7	8.35	3	2	60.90	8	0.150	Mbabo

DBH: Diameter of the tree at breast height; P_a: percentage of the slope at the logging level or at breast high; P_u: percentage of the slope at the first big branch level; d_h: horizontal distance; h: high; LB: number of large bands in relascope unit SB: small band: number of small bands; D_u: diameter of the tree at the first big branch level; e: thickness of the bark; V: volume of the bark.

Table 2. Comparison of the thickness of the *Prunus* bark measured using the Tarrière de Pressler between Gang Daba and Mbabo in the Adamaoua region.

Parameter	Mean	sd	n F value = 162.63 Pr < 2.2e-16 ***
Gang Daba	15.84	3.73	50
Mbabo	8.56	1.89	55
All sites	12.03	4.6	105

The F value and Pr are from the comparison between Gang Daba and Mbabo. Signifiance. codes: 0 '***' 0.001 '**' 0.01 '*' 0.05 '.' 0.1 ' ' 1.

Mass of the fresh bark for exploitable trees

In Cameroon, the minimum exploitable diameter (MED) for *P. africana* is 30 cm; this means that the stem bark can only be harvested from trees with diameter at DBH equal or more than 30 cm. We selected exploitable trees from the sample of trees used to yield the cubic tariff. Having the volume of the bark for each tree, we estimated the mass using the formula of the cubic mass (m = V x Cm). The average mass of all the exploitable trees was considered as the mass of an exploitable tree of *P. africana* for Tchabal Mbabo and Tchabal Gang Daba mountain forests.

In this paper, indirect method was preferred to direct method. A direct method would require diameter tape and bark thickness measurements at the critical heights (BH, first branch, etc), with some degree of verification through destructive sampling. This can yield précised data compared to indirect method. We preferred indirect method because it is non-destructive, and this method can easily be applied in the field by foresters during forest management inventories.

Data analysis

The volume based tariff or the relation between the diameter and the volume of the stem bark was performed using linear regression with the CurveExpert 1.4 package. The best equation is that for which the correlation coefficient (R) is near to 1.

Data analysis was performed using the R2.10.0 statistical package. Different sites, volumes, and materials used for measuring the thickness of the stem bark, were compared using the One way ANOVA. For what concerns the comparison of the two materials used for measuring the thickness of the *P. africana* barks, or the comparison of the observed (measured) and theoretical (obtained from the cubic tariff equation) volumes of the bark, the tree sample was considered as the unique factor (Dagnelie, 1998).

RESULTS

Cubic tariff

All the 105 sample trees were exploitable, this is, trees with DBH ≥ 30 cm. Measures regarding the diameters, the percentages of the slope, the thickness of the bark were collected on all the 105 trees. Table 1 shows the data recorded and the corresponding volume per tree. The thickness of the bark was measured using the "Tarrière de Pressler". The average thickness of the bark is 12.03 ± 4.6 mm. This value is comprised between that of unexploited (15.84 ± 3.7 mm) and exploited (8.56 ± 1.9 mm) *P. africana* trees observed respectively in Gang Daba and Mbabo. The means of the thickness of the bark between the un-exploited and the exploited sites, illustrated in Table 2 are significantly different ($p < 0.05$).

The relation diameter of the tree at breast height – volume of the stem bark is illustrated in Figures 2, 3, and 4 for Gang Daba, Mbabo, and the two sites together respectively. Table 3 presents for each site, the best equation which links the diameter and the stem bark of the *P. africana* tree. To examine the credibility of our equations, we compared the volume of the bark measured in the field with the volume of the bark simulated by the equations. Table 4 shows the results obtained. What ever be the site, there is no significant difference ($p > 0.05$) between the theoretical volume and the observed volume; which gives credibility to the cubage tariffs proposed for each site or for all the two sites considered together.

Cubic mass

A total of 105 samples of stem barks were collected for weighing, to determine the relation linking the volume of the bark with its mass. These samples were collected on all the same 105 trees used to yield the cubic tariff. Table 5 shows for each sample, its volume and the corresponding fresh mass in kilogram. The thickness of the bark was measured using the calliper rule. The

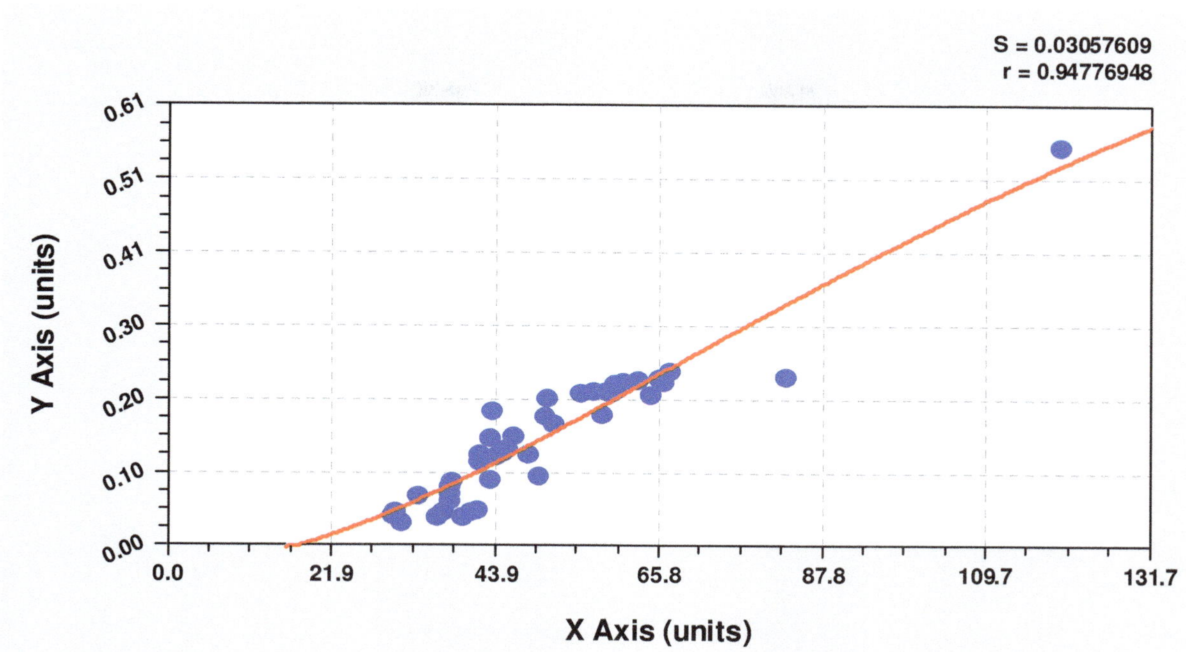

Figure 2a. evolution of the volume of the fresh bark with that of the diameter at breast high of the *Prunus* tree in Gang Daba area. MMF Model: y=(a*b+c*x^d)/(b+x^d); Coefficient Data: a = -2.85924630130E-002; b = 6.91062016528E+003; c =1.23499012989E+000; d = 1.79843264745E+000.

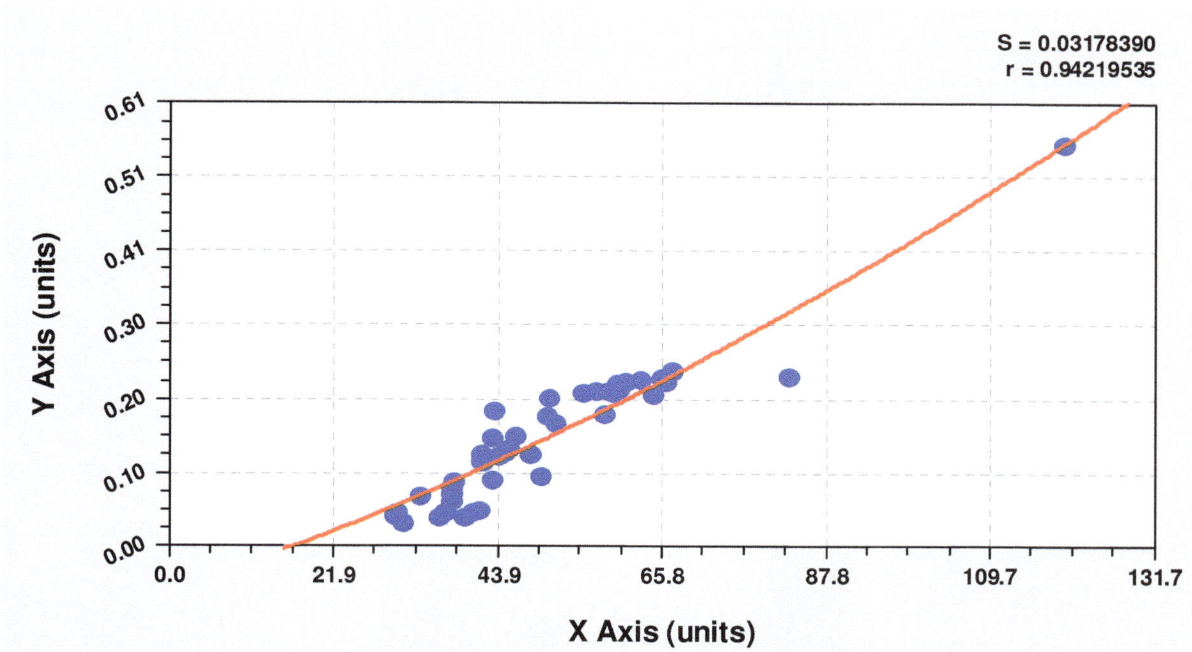

Figure 2b. evolution of the volume of the fresh bark with that of the diameter at breast high of the *Prunus* tree in Gang Daba (linear regression) Quadratic Fit: y=a+bx+cx^2. Coefficient Data: a = -5.98191756094E-002; b = 3.50369652890E-003; c =1.37448422378E-005.

average thickness of the bark is 13.01 ± 4.8 mm. This value is comprised between that of unexploited *P. africana* trees found in Gang Daba (16.99 ± 3.7 mm) and exploited *P. Africana* trees observed in Mbabo (9.40

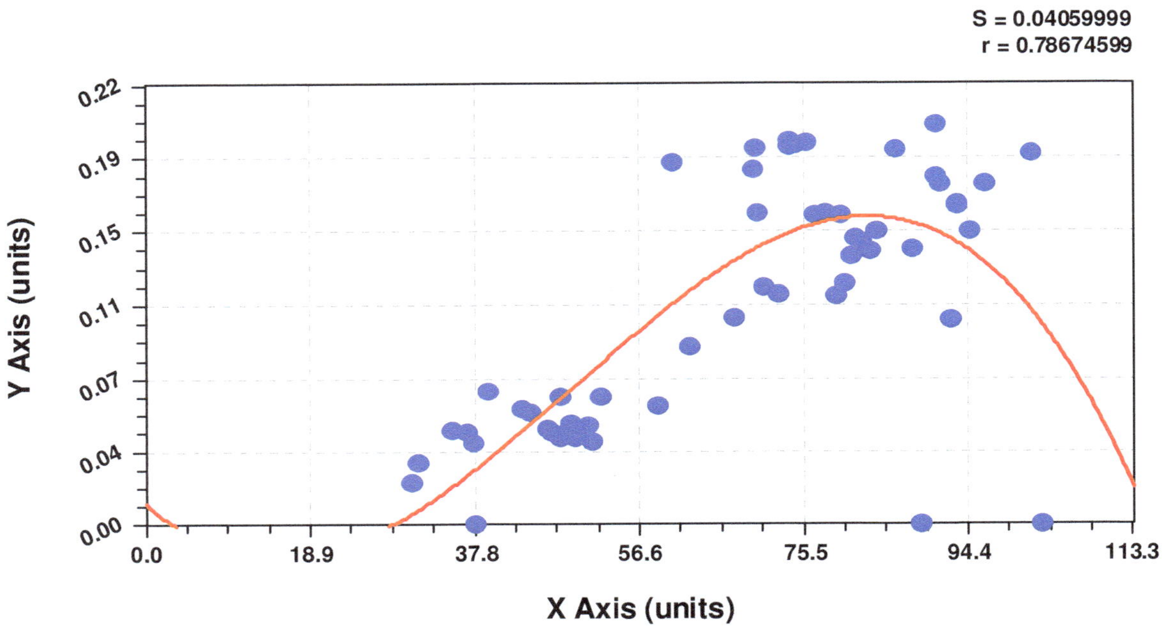

Figure 3. Evolution of the volume of the fresh bark with that of the diameter at breast high of the *Prunus* tree in Mbabo area. 3rd degree Polynomial Fit: y=a+bx+cx^2+dx^3. Coefficient Data: a = 1.10943074390E-002; b= -4.14622627475E-003; c =1.64301318294E-004; d = -1.12161703386E-006.

Table 3. Cubic tariff of *Prunus* trees in each production site in the Adamaoua region in Cameroon.

Site	Tarif	a	b	C	D
Gang Daba	Vb=(a*b+c*D^d)/(b+D^d)	-2.85924630130E-002	6.91062016528E+003	1.23499012989E+000	1.79843264745E+000
Gang Daba	Linear regression Vb=a+bD+cD^2	-5.98191756094E-002	3.50369652890E-003	1.37448422378E-005	
Mbabo	Vb=a+bD+cD^2+dD^3	1.10943074390E-002	-4.14622627475E-003	1.64301318294E-004	-1.12161703386E-006
All sites together	Vb=a/(1+b*exp(-cD))	1.79588278896E-001	5.29124992540E+002	1.45488065368E-001	

V_b is the volume of the stem bark in m^3 and D is the diameter of the tree at breast high in cm a, b, c, d, are coefficients of the equations.

± 2.07 mm). The means of the thickness of the bark between the un-exploited and the exploited sites, illustrated in Table 6 are significantly different ($p < 0.05$).

There is no significant difference ($p > 0.05$) between the two sites for what concerns the ratio mass/volume of the bark (Table 7). The average value of the ratio mass/volume obtained from the samples is 1014.80 ± 9 kg/m^3 for the two sites (Table 7). This value is comprised between that of unexploited *P. africana* trees found in Gang Daba (1014.37 ± 7.78 kg/m^3) and exploited *P. africana* trees observed in Mbabo (1015.19 ± 10.11 kg/m^3). For the two sites considered together, the equation which links the volume of the stem bark and its mass is Mb = Vb* 1014.80 kg/m^3, where Mb = mass of the fresh bark in kg and Vb = volume of the stem bark in m^3.

Mass of exploitable trees

The 105 trees were also selected to simulate the average fresh mass of *P. africana* barks. Table 8 presents for each exploitable tree, its volume and the corresponding mass in kilogram. There is a significant difference ($p < 0.05$) between the total mass of the bark of exploitable *P. africana* trees found in un-exploited and exploited sites (Table 9). It can be deduced that, an exploitable tree of *P. africana* has an average mass of 138.61 ± 79.11 Kg of stem barks. This value is comprised between that of unexploited (159.88 ± 92.55 kg) and exploited (119.27 ±

Table 4. Comparison of the two cubage tariffs (observed volume or mean 1 vs theoretical volume or mean 2) in different sites.

Parameter	Mean1	mean 2	sd1	sd2	n	F value	Pr
Gang Daba	0.1575489	0.1584337	0.09120302	0.08504862	50	0.0025	0.96
Mbabo	0.1003238	0.1115804	0.06550722	0.048937	55	1.0423	0.3096
All sites together	0.1275738	0.1338915	0.08349759	0.07214084	105	0.3442	0.5581

Table 5. Volumes of bark samples with their corresponding mass.

N°_Tree	Length (cm)	Width (cm)	Thickness (mm)	Volume (m3)	Mass (Kg)	Ratio (m/v)	Site
1	25	7.8	13.00	0.000	0.26	1025.641	Gang Daba
2	16.5	11	13.76	0.000	0.25	1001.272	Gang Daba
3	31.2	10.25	16.00	0.001	0.53	1035.804	Gang Daba
4	27.3	7.6	13.25	0.000	0.28	1018.511	Gang Daba
5	35.5	7.4	19.20	0.001	0.51	1011.134	Gang Daba
6	16	11.7	21.15	0.000	0.4	1010.285	Gang Daba
7	32	9.87	17.00	0.001	0.54	1005.721	Gang Daba
8	31	10.7	16.45	0.001	0.55	1007.979	Gang Daba
9	27	7.5	14.80	0.000	0.3	1001.001	Gang Daba
10	31.6	8.5	16.85	0.000	0.46	1016.370	Gang Daba
11	22.6	7	18.75	0.000	0.3	1011.378	Gang Daba
12	30	7.4	14.76	0.000	0.33	1007.419	Gang Daba
13	21	9.3	18.15	0.000	0.36	1015.792	Gang Daba
14	36	6.8	18.00	0.000	0.45	1021.242	Gang Daba
15	16	8.5	21.75	0.000	0.3	1014.199	Gang Daba
16	16	11.5	16.25	0.000	0.3	1003.344	Gang Daba
17	18	14.3	18.00	0.000	0.47	1014.418	Gang Daba
18	17.5	14	19.50	0.000	0.48	1004.710	Gang Daba
19	19	9.25	20.00	0.000	0.36	1024.182	Gang Daba
20	25.7	8.5	10.80	0.000	0.24	1017.268	Gang Daba
21	25.4	9	12.00	0.000	0.28	1020.706	Gang Daba
22	16.5	11	9.25	0.000	0.17	1012.583	Gang Daba
23	29.7	8.2	18.00	0.000	0.44	1003.714	Gang Daba
24	13	12.7	8.35	0.000	0.14	1015.534	Gang Daba
25	26.5	6	21.00	0.000	0.34	1018.269	Gang Daba
26	22	6.5	24.45	0.000	0.35	1001.044	Gang Daba
27	24	8.3	12.00	0.000	0.24	1004.016	Gang Daba
28	29.6	8	13.33	0.000	0.32	1013.514	Gang Daba
29	21.5	9.2	22.70	0.000	0.46	1024.485	Gang Daba
30	24	9.8	11.67	0.000	0.28	1020.408	Gang Daba
31	17.8	13	17.05	0.000	0.4	1013.847	Gang Daba
32	23	6.5	18.25	0.000	0.28	1026.252	Gang Daba
33	30	7	19.65	0.000	0.42	1017.812	Gang Daba
34	26	8	24.45	0.001	0.52	1022.328	Gang Daba
35	30	8.1	14.67	0.000	0.36	1010.101	Gang Daba
36	32.5	9	14.25	0.000	0.42	1007.647	Gang Daba
37	28.5	4.75	19.00	0.000	0.26	1010.837	Gang Daba
38	19	6	19.75	0.000	0.23	1021.334	Gang Daba
39	18	9.3	20.00	0.000	0.34	1015.532	Gang Daba
40	16.4	10.2	19.00	0.000	0.32	1006.821	Gang Daba

Table 5. Contd.

41	28.5	7.5	14.70	0.000	0.32	1018.419	Gang Daba
42	15.5	10.1	15.00	0.000	0.24	1022.038	Gang Daba
43	27	7.5	15.60	0.000	0.32	1012.979	Gang Daba
44	16	12	16.80	0.000	0.33	1023.065	Gang Daba
45	26	5.5	19.25	0.000	0.28	1016.953	Gang Daba
46	15.5	10.8	17.65	0.000	0.3	1015.362	Gang Daba
47	29.7	7	16.33	0.000	0.34	1001.266	Gang Daba
48	18.7	11.6	22.75	0.000	0.5	1013.186	Gang Daba
49	18.5	12	21.54	0.000	0.49	1024.702	Gang Daba
50	17	11.7	13.85	0.000	0.28	1016.274	Gang Daba
51	13.4	9	9.00	0.000	0.11	1013.451	Mbabo
52	27.7	8	7.15	0.000	0.16	1009.821	Mbabo
53	14	10	9.00	0.000	0.13	1031.746	Mbabo
54	19	6.5	6.45	0.000	0.08	1004.300	Mbabo
55	12.5	9	6.65	0.000	0.075	1002.506	Mbabo
56	14.5	10	8.00	0.000	0.12	1034.483	Mbabo
57	16	9.5	7.65	0.000	0.12	1031.992	Mbabo
58	15	8.5	13.65	0.000	0.18	1034.260	Mbabo
59	13.5	11	11.80	0.000	0.18	1027.221	Mbabo
60	18	9	10.70	0.000	0.18	1038.422	Mbabo
61	17	10.5	6.54	0.000	0.12	1027.934	Mbabo
62	20.1	10.6	10.00	0.000	0.22	1032.573	Mbabo
63	17.4	10.4	7.00	0.000	0.13	1026.273	Mbabo
64	14.5	9.6	11.25	0.000	0.16	1021.711	Mbabo
65	16.5	9.2	8.50	0.000	0.13	1007.518	Mbabo
66	18.3	9	9.56	0.000	0.16	1016.175	Mbabo
67	24	8.5	9.67	0.000	0.2	1014.206	Mbabo
68	16.5	7.6	12.50	0.000	0.16	1020.734	Mbabo
69	25	9.5	8.00	0.000	0.19	1000.000	Mbabo
70	18.4	8	8.65	0.000	0.13	1020.985	Mbabo
71	24.6	6.7	6.00	0.000	0.1	1011.204	Mbabo
72	23	7.5	8.00	0.000	0.14	1014.493	Mbabo
73	28	6	7.50	0.000	0.13	1031.746	Mbabo
74	16.5	8	9.00	0.000	0.12	1010.101	Mbabo
75	14.5	10.3	7.25	0.000	0.11	1015.896	Mbabo
76	18.6	8.3	11.00	0.000	0.17	1001.072	Mbabo
77	18	12	12.00	0.000	0.26	1003.086	Mbabo
78	15	9.5	10.26	0.000	0.15	1026.297	Mbabo
79	25	8.5	7.50	0.000	0.16	1003.922	Mbabo
80	17.7	8.3	12.00	0.000	0.18	1021.033	Mbabo
81	16	8.5	11.00	0.000	0.15	1002.674	Mbabo
82	21	7.6	11.00	0.000	0.18	1025.290	Mbabo
83	20.5	7.5	9.00	0.000	0.14	1011.743	Mbabo
84	20	6	9.00	0.000	0.11	1018.519	Mbabo
85	17.3	9.2	10.00	0.000	0.16	1005.278	Mbabo
86	26.5	7.5	9.00	0.000	0.18	1006.289	Mbabo
87	15	10	6.56	0.000	0.1	1016.260	Mbabo
88	20.5	8.8	12.00	0.000	0.22	1016.260	Mbabo
89	15	8	11.50	0.000	0.14	1014.493	Mbabo
90	16.5	8.9	9.50	0.000	0.14	1003.530	Mbabo
91	25	8.5	7.00	0.000	0.15	1008.403	Mbabo
92	16	12.4	10.00	0.000	0.2	1008.065	Mbabo

Table 5. Contd.

	93	20.3	8.8	10.00	0.000	0.18	1007.613	Mbabo
	94	26.7	9.4	7.00	0.000	0.18	1024.555	Mbabo
	95	16.6	8	13.50	0.000	0.18	1004.016	Mbabo
	96	17.4	11.3	8.00	0.000	0.16	1017.191	Mbabo
	97	16.5	11.5	9.40	0.000	0.18	1009.167	Mbabo
	98	18.7	7.2	10.25	0.000	0.14	1014.449	Mbabo
	99	17	7.6	13.75	0.000	0.18	1013.228	Mbabo
	100	28	8.1	7.00	0.000	0.16	1007.811	Mbabo
	101	16	9.6	12.85	0.000	0.2	1013.294	Mbabo
	102	14.7	10.2	7.30	0.000	0.11	1004.968	Mbabo
	103	20.4	8.5	8.00	0.000	0.14	1009.227	Mbabo
	104	17.8	6.7	10.00	0.000	0.12	1006.205	Mbabo
	105	20	7	12.00	0.000	0.17	1011.905	Mbabo
All sites								
N	105							
Mean				13.01	0.00	0.25	1014.80	
Sd				4.81917358	0.00012258	0.1243444	9.04330906	
Max				24.45	0.00	0.55	1038.42	
Min				6.00	0.00	0.08	1000.00	

Table 6. Comparison of the thickness of the *Prunus* bark measured using the Calliper rule between Gang Daba and Mbabo in the Adamaoua region.

Parameter	Mean	sd	n	F = 172.30	Pr < 2.2e-16 ***
Gang Daba	16.99	3.70	50		
Mbabo	9.39	2.07	55		
All sites	13.01	4.8	105		

The F value and Pr are from the comparison between Gang Daba and Mbabo. Signifance codes: 0 '***' 0.001 '**' 0.01 '*' 0.05 '.' 0.1 ' ' 1.

Table 7. Comparison of the ratio mass/volume of the *Prunus* bark measured using the Calliper rule between Gang Daba and Mbabo in the Adamaoua region.

Parameter	Mean	sd	n	F = 0.213	Pr = 0.6454
Gang Daba	1014.37	7.78	50		
Mbabo	1015.19	10.11	55		
All sites	1014.08	9.04			

59.00 kg) *P. africana* trees observed respectively in Gang Daba and Mbabo.

Comparison between the measures of the thickness of the stem barks obtained with the "Tarrière de Pressler" and those obtained with the calliper rule

Table 10 shows the results of the comparison made on the thickness of the bark measured with different materials. From the table, it can be deduced that there is no significant difference ($p > 0.05$) between the thickness of the stem bark obtained with the "Tarrière de Pressler" and that obtained with the calliper rule.

DISCUSSION

P. africana extracts that can help alleviate some prostate disorders are in high demand. The *P. africana* tree, a primary source of medicine, has become threatened as a result. Without the conservation of high yielding populations, these trees could be harvested to extinction.

Table 8. Productivity of exploitable *Prunus* trees in term of the mass of the fresh bark.

N°	DBH (cm)	Volume (m^3)	Mass (kg)	Site
1	39.49	0.041	41.371	Gang Daba
2	59.84	0.226	229.682	Gang Daba
3	36.13	0.041	41.654	Gang Daba
4	59.68	0.217	220.206	Gang Daba
5	37.05	0.049	49.849	Gang Daba
6	43.42	0.187	189.842	Gang Daba
7	31.2	0.033	33.139	Gang Daba
8	59.21	0.212	215.071	Gang Daba
9	65.89	0.233	236.056	Gang Daba
10	33.42	0.072	73.191	Gang Daba
11	58.16	0.183	185.224	Gang Daba
12	30.43	0.049	49.299	Gang Daba
13	62.87	0.229	232.514	Gang Daba
14	37.88	0.063	64.261	Gang Daba
15	41.4	0.052	52.538	Gang Daba
16	64.62	0.209	212.392	Gang Daba
17	40.64	0.048	48.504	Gang Daba
18	41.85	0.129	130.428	Gang Daba
19	51.57	0.170	172.245	Gang Daba
20	67.32	0.243	246.400	Gang Daba
21	66.53	0.229	232.461	Gang Daba
22	43.29	0.152	153.960	Gang Daba
23	82.76	0.235	238.037	Gang Daba
24	65.73	0.232	235.124	Gang Daba
25	46.28	0.154	156.030	Gang Daba
26	45.36	0.136	137.640	Gang Daba
27	37.94	0.084	85.008	Gang Daba
28	59.72	0.218	221.715	Gang Daba
29	50.45	0.180	183.066	Gang Daba
30	61.12	0.227	230.492	Gang Daba
31	41.85	0.118	119.623	Gang Daba
32	60.04	0.218	221.362	Gang Daba
33	49.66	0.099	99.959	Gang Daba
34	59.21	0.215	218.461	Gang Daba
35	119.69	0.554	562.356	Gang Daba
36	30.08	0.044	44.820	Gang Daba
37	48.19	0.128	129.545	Gang Daba
38	38.2	0.090	91.488	Gang Daba
39	58.63	0.215	217.688	Gang Daba
40	37.88	0.074	74.693	Gang Daba
41	66.37	0.228	231.195	Gang Daba
42	59.94	0.218	221.100	Gang Daba
43	43.07	0.094	95.047	Gang Daba
44	50.71	0.206	209.006	Gang Daba
45	55.26	0.214	216.693	Gang Daba
46	56.98	0.216	218.991	Gang Daba
47	30.34	0.044	44.147	Gang Daba
48	44.09	0.125	126.781	Gang Daba
49	41.95	0.119	121.147	Gang Daba
50	44.98	0.131	132.528	Gang Daba
51	46.6	0.047	47.719	Mbabo

Table 8. Contd.

52	58.76	0.061	62.050	Mbabo
53	36.8	0.047	48.108	Mbabo
54	52.3	0.065	66.266	Mbabo
55	50.8	0.051	51.909	Mbabo
56	49.43	0.045	45.257	Mbabo
57	78.05	0.159	161.318	Mbabo
58	93.14	0.164	166.754	Mbabo
59	48.83	0.052	52.410	Mbabo
60	79.68	0.158	160.316	Mbabo
61	44.15	0.057	58.205	Mbabo
62	94.54	0.150	152.039	Mbabo
63	101.73	0.189	192.045	Mbabo
64	47.59	0.045	45.391	Mbabo
65	81.08	0.137	139.364	Mbabo
66	47.65	0.066	66.507	Mbabo
67	82.25	0.144	146.519	Mbabo
68	35.24	0.048	48.970	Mbabo
69	103	0.158	160.103	Mbabo
70	90.84	0.178	180.151	Mbabo
71	31.32	0.032	32.588	Mbabo
72	96.32	0.174	177.043	Mbabo
73	39.38	0.068	69.223	Mbabo
74	51.25	0.043	43.426	Mbabo
75	62.39	0.092	93.150	Mbabo
76	67.61	0.106	107.185	Mbabo
77	86.1	0.191	194.205	Mbabo
78	69.87	0.181	184.119	Mbabo
79	38	0.029	29.910	Mbabo
80	83.3	0.140	142.365	Mbabo
81	74.71	0.195	197.537	Mbabo
82	91.17	0.174	176.207	Mbabo
83	89	0.129	130.598	Mbabo
84	60.58	0.185	188.144	Mbabo
85	37.56	0.042	42.415	Mbabo
86	70.22	0.159	161.407	Mbabo
87	75.82	0.195	198.243	Mbabo
88	81.56	0.147	148.863	Mbabo
89	74.07	0.196	198.941	Mbabo
90	92.5	0.105	106.509	Mbabo
91	72.74	0.118	119.846	Mbabo
92	30.56	0.022	22.043	Mbabo
93	93.11	0.163	165.286	Mbabo
94	74.04	0.193	196.318	Mbabo
95	43.13	0.059	60.014	Mbabo
96	90.85	0.205	207.564	Mbabo
97	79.39	0.117	118.775	Mbabo
98	80.28	0.123	125.222	Mbabo
99	48.38	0.047	47.746	Mbabo
100	70.99	0.121	123.184	Mbabo
101	88.02	0.141	142.829	Mbabo
102	46.25	0.049	49.516	Mbabo
103	70.16	0.192	194.923	Mbabo

Table 8. Contd.

104	76.91	0.158	160.314	Mbabo
105	83.94	0.150	152.557	Mbabo
All sites				
N	105	105	105	
Mean	60.02	0.14	138.61	
Sd	20.18	0.08	79.11	
Max	119.69	0.55	562.36	
Min	30.08	0.02	22.04	
Gang Daba				
N	50	50	50	
Mean	51.17	0.16	159.88	
Sd	15.66	0.09	92.55	
Max	119.69	0.55	562.36	
Min	30.08	0.03	33.14	
Mbabo				
N	55	55	55	
Mean	68.07	0.12	119.27	
Sd	20.56	0.06	59.00	
Max	103.00	0.20	207.56	
Min	30.56	0.02	22.04	

DBH: Diameter at breast high.

Table 9. Comparison of the mass of the bark of a total *Prunus* tree between Gang Daba and Mbabo in the Adamaoua region.

Parameter	Mean	sd	n	F= 7.3221	Pr= 0.007973 **
Gang Daba	159.88	92.55	50		
Mbabo	119.26	59.00	55		
All sites	138.61	79.11	105		

The F value and Pr are from the comparison between Gang Daba and Mbabo. Significance codes: 0 '***' 0.001 '**' 0.01 '*' 0.05 '.' 0.1 ' ' 1.

Table 10. Comparison of the thickness of the bark measured with different material (Tarrière de Pressler vs Calliper rule).

Parameter	Mean	sd	n	F = 2.269	Pr = 0.1335
Calliper rule	13.01	4.819	105		
Tarrière	12.02	4.66	105		

This will have a negative effect on the economic livelihoods of smallholder farmers and narrow the options for those suffering from prostate disorders. According to the World Agroforestry study published in the January-March 2012 issue of the journal Forests, Trees and Livelihoods, cultivating the tree on farms will alleviate the threat of extinction caused by unsustainable wild harvests. It takes 12 to 15 years for the tree to produce the bark that contains the prostate remedy's active ingredient (World Agroforestry Centre, 2012; Gachie et al., 2012).

Since 2008, ITTO and CITES have been working together to develop a large capacity-building program for range countries in Africa regarding tree species listed in

Appendix II of CITES. The program entitled “Ensuring international trade on CITES listing tree species is non detrimental to their conservation in the wild” aims to assist local CITES authorities in range countries in making non-detriment findings for those species. Conservation of the specimens in their natural habitat being the main target, which is not necessary the same outlined above for the World Agroforestry Centre.

Cubic tariffs are often used in forestry in three main areas including trade on forest products, forest management, and forest research, mainly oriented on forest productivity. The cubic tariff is an indispensable tool in forest management and forest inventories since it allows to quickly estimate the stand volume of trees (Pardé and Bouchon, 1988; Rondeux et al., 1991). Important things to consider in the construction of the cubic tariff is to decide on the number of entries (parameters) to use and to select the best mathematical formula (College of Forest Engineers of Quebec, 1996). The choice of the entries is guided by their simplicity and their link with the volume. Most of cubic tariffs constructed in forest management are those which yield the standing volume of the timber based on its diameter at high breast. Our cubic tariff is built from one entry, the diameter at breast heoght to be précised. It yields the volume of the stem bark of a given *P. africana* tree according to its diameter at breast height. We could also decide to use the thickness of the stem bark as another entry (parameter) for building our cubic tariff. But we found that, this parameter cannot be easy for forest officers to collect it in the field. It is more easier to measure the diameter at breast height than measuring the thickness of the tree. For a cubic tariff which concern a limit area with relatively homogenous growth conditions, the number of trees should vary from 30 to 100 (Rondeux, 1993; College of Forest Engineers of Quebec, 1996). This condition is respected in our study, since we worked on 105 trees of *P. africana* comprising 50 trees in Tchabal Gang Daba considered as an unexploited site and 55 trees in Tchabal Mbabo considered as an exploited site.

Cameroon’s *P. africana* barks are exported in two forms: the raw bark and the “powder”. Powder here is referred to the bark shavings or the ground barks. Whatever be the form, *P. africana* is exported in dried matter. The dried weight of *P. africana* barks to be exported is = 50% of Fresh weigh (Ingram and al., 2009). The minimum diameter of exploitation (MED) adopted by the Cameroon forest administration for *P. africana* barks is 30 cm at breast height. In Kenya, Gachie et al. (2012) revealed that there is a positive correlation between tree age and the extract yield. However, the trend is not linear, since at old age (> 50 cm DBH), *P. africana* trees were seen to have a drop on the crude bark extract yield. This implies that the best source of the bark is from medium-sized trees (30 to 50 cm DBH). We can therefore agree that the DBH 30 cm adopted by the Cameroon is good for maximising the extract yield. The question to ask is, if the adopted MED is enough to ensure the regeneration? Further studies are requested for defining the best MED that encompasses both the extracts yield and the regeneration concerns.

The relation diameter of the tree at breast height – volume of the stem bark varies from one site to another. In the un-exploited site (Gang Daba), this relation is almost linear, which is different to what is observed in the exploited site. We conclude that, in an un-exploited site, there is a positive correlation between tree age (diameter size) and the extract yield as suggested by Gachie et al. (2012). This hypothesis can no longer been verified in the case of an exploited site.

P. africana can survive the removal of some bark and there is the possibility, therefore, for harvesting bark without felling or killing the tree. Plantecam collectors were trained to gather the bark in the least damaging way. They were not permitted to fell the trees and they were expected to remove bark from up to only 50% of the circumference of the tree and from opposite sides of the trunk in order to prevent girdling, which could kill that specimen. Nor, were they permitted to return to the same tree for further harvesting for five years. The claim that the collection techniques described was sustainable, if properly followed, was endorsed by the botanists and conservationists working at Limbé and Buea, in the South West Cameroon. This elaborate harvesting process was a product of the species particulars characteristics (Ndibi and Kay, 1997; Marcelin et al., 2000; Page, 2003).

There is no significant difference between the thickness of the stem bark obtained with the “Tarrière de Pressler” and that obtained with the calliper rule. This means that the two materials can be used indifferently for the measurement of the thickness of the bark. Whatever be the material used, the difference between the un-exploited site and an exploited site is at least 7.3 mm. If we consider the results obtained with the calliper rule, the average thickness of the bark is 13.01 ± 4.8 mm. This value is comprised between that of unexploited *P. africana* trees found in Gang Daba (16.99 ± 3.7 mm) and exploited *P. africana* trees observed in Mbabo (9.40 ± 2.07 mm). This value is high than the 8.49 ± 2.41 mm (n = 117 trees) obtained in the mount Cameroon (Betti and Ambara, 2011), what ever be the site. The value of 9.40 mm of the Mbabo is closed to that of the Mount Cameroon, considered also as an un-exploited site (Betti and Ambara, 2011). Our findings are quite similar to those obtained in Equatorial Guinea (Sunderland and Tako, 1999). In fact, in the Bioko island, in Equatorial Guinea, Sunderland and Tako (1999) obtained an average thickness of 13.01 mm. This value was comprised between that of un-exploited (16. 21 mm) and exploited (9.94 mm) *P. africana* trees.

An exploitable tree of *P. africana* of the Adamaoua region has an average mass of 138.61 ± 79.11 kg of stem barks. This value comprise between that of un-exploited (159.88 ± 92.55 kg) and exploited (119.27 ± 59.00 kg) *P. africana* trees observed respectively in Gang Daba and Mbabo. These values represent the fresh mass for the total stem bark of a given exploitable trunk/tree. Considering that for all trees above 30 cm DBH, only two quarters of the bark are taken from the main stem up to the first branch, the average sustainable mass of *Prunus* tree in Tchabal Mbabo and Tchabal Gang Daba mountain forests will be about 69.3 kg of fresh bark per tree. This value is comprised between that of un-exploited (80 kg) and exploited (60 kg) *P. africana* trees in Ganga Daba and Mbabo respectively. Previous studies conducted in the same area using a destructive approach (cutting of trees) revealed that an exploitable *P. africana* tree harvested in sustainable manner can yield about 75 kg of fresh bark (Cunningham and Mbenkum, 1993 ; Hall et al., 2000). Sustainable manner in this paper means, as explained above, taking only two quarters of the bark from the main stem up to the first big branch. On Mount Oku in the North west region of Cameroon, Ondigui (2001) using the destructive method found that the average sustainable mass of fresh bark of *P. africana* was 55 kg/tree with four-year intervals. Results obtained in mount Cameroon using the same method suggested that the sustainable weight or mass of *P. africana* is 50 kg of fresh bark with five year intervals (Betti and Ambara, 2011). What ever be the case (exploited or un-exploited trees), the average mass of fresh bark yielded by an exploitable *P. africana* tree in Adamaoua is high than those obtained in Mont Cameroon and Mont Oku. The difference observed between different zones may be due to the environmental conditions such as soils, altitude, temperature.

Following what precedes and based on indirect method used, we suggest that for Tchabal Mbabo and Tchabal Gang Daba mountain forests, the sustainable mass/yield of an exploitable *P. africana* tree is 69.3 kg of fresh bark with at least five-year intervals. This means that to sustain *P. africana* in Adamaoua region of Cameroon, trade companies or villagers should harvest, trees of at least 30 cm of diameter at breast height, move the 69.3 kg of stem bark on the ½ opposite sides, and return at least 5 years later to move the remaining sides on the same tree, or return 10 years latter to move the same side on the same tree. Five years interval should be considered with some precautions, since quantitative studies have not yet been conducted in the Adamaoua region to monitor the rate of recovery of the stem bark after harvesting. Environmental conditions such as altitude, soils, temperature seem to have significant impact on extract yield and quality of active compounds of *P. africana*. Temperature differences may cause chilling injury in plants, which encompasses imbalances in metabolism, accumulation of toxic compounds, and increased membrane permeability (Gachie et al., 2012). Mount Cameroon is up to 4095 m and the average temperature is 22°C (Betti and Ambara, 2011; Betti et al., 2011), while Mbabo and Gang Daba may not reach even 2 500 m and the average temperature is 23°C. The high temperature observed in Adamaoua may suggest to increase the rotation period, or the time that separates two harvesting campaigns on the same tree. Further studies are required for this item.

Conclusion

The single data used for the construction of our cubic tariff is the diameter of tree at breast high. This data is well known by many foresters and is easy to collect in the field. The cubic tariff and the average mass of exploitable trees are two tools which can ease the estimation of stand bark volume for *P. africana* trees. Those two data can be used to quickly estimate the export quota of *P. africana* for Cameroon for the Tchabal Mbabo and Tchabal Gang Daba mountain forests. Our study reveals that, for these mountain forests, the sustainable mass of an exploitable *P. africana* tree is 69.3 kg of fresh bark with at least five-year intervals, this is harvesting trees of at least 30 cm of diameter at breast height, move the ½ opposite side, and return at least 5 years later to move the remaining sides on the same tree, or return 10 years latter to move the same side on the same tree. The 69.3 kg should be considered as an average, comprised between that of un-exploited (80 kg) and exploited (60 kg) *P. africana* trees in Ganga Daba and Mbabo respectively. This information is crucial for making non-detriment findings for *P. africana* barks harvested in Adamaoua mountain forests.

ACKNOWLEDGEMENTS

Authors thank all the villagers who collaborated with us in this study. The study was supported by the Join International Tropical Timber Organisation (ITTO) - Convention on International Trade in Endangered Species of Fauna and Flora (CITES) Program for Implementing CITES Listings of Tropical Tree Species, executed in Cameroon by the National Forestry Development Agency (ANAFOR).

REFERENCES

Akagou Zedong HC, Betti JL (2007). Etat des lieux sur l'exploitation de *Prunus africana* au Cameroun. Rapport de mission dans les provinces de l'Adamaoua, Ouest, Nord ouest et Sud ouest, p. 19.

Akoa A, Betti JL, Ewusi NB, Mbarga N, Akagou Zedong HC, Fonkoua C, Mala AW, Nkouna AC (2011). Preliminary report on sustainable harvesting of *Prunus africana* (Rosaceae) in the mount Cameroon. Report prepared for the National Forestry Development Agency (ANAFOR), the Cameroon CITES Scientific Authority for flora, in the frame of the project "Non-detriment findings for *Prunus africana* (Hook.f.) Kalman in Cameroon", p. 24.

Akoa A, Betti JL, Ewusi Njombe B, Mbarga N, Akagou Zedong HC, Fonkoua C, Essomba ER, Nkouna AC (2010). Preliminary report on sustainable harvesting of *Prunus africana* (Rosaceae) in the North west region of Cameroon. Report prepared for the National Forestry Development Agency (ANAFOR), the Cameroon CITES Scientific Authority for flora, in the frame of the project "Non-detriment findings for *Prunus africana* (Hook.f.) Kalman in Cameroon", p. 76.

Betti JL (2008). Non-Detriment Findings Report on *Prunus africana* (Rosaceae) in Cameroon. Report prepared for the International Expert Workshop on Non-Detriment Findings, Mexico, November 17-22, 2008. 52 p. www.conabio.gob.mx/.../TallerNDF/...CS9%20Prunus/WG1-CS9-S.pdf. 2009.

Betti JL, Ambara J (2011). Mass of *Prunus africana* stem barks on the mount Cameroon forest. Int. J. Biodiv. Conserv. 3(7): 267–279.

Betti JL, Belinga SJ, Samba D (2011). Stock of *Prunus africana* stems in the Mount Cameroon forests. Afr. J. Plant Sci. 5(13): 781-789.

College of Forest Engineers of Quebec (1996). Forestry Hand book. The Laval University Press, p. 1428.

Cunningham AB, Mbenkum FT (1993). Sustainability of harvesting *Prunus africana* bark in Cameroon: A medicinal plant in international trade. UNESCO, Paris, France.

Cunningham M, Cunningham AB, Schippmann U (1997). Trade in *P. africana* and the implementation of CITES. German Federal Agency for Nature.

Dagnelie P (1998). Theoretical and applied statistique. Volume 2 : Statistical inferences at one and two dimensions. De Boeck and Larcier (eds.). Library of Universities, Paris, Bruxelles, p. 659.

FAO (1996). Appendix 6. Forest genetics resources priorities in Africa. Report of the eighth session of the FAO Panel of experts on forest gene resources; Rome, Italy; 28 – 30 June 1993. pp. 24–25.

Fraser PJ, Healy JR, Cheek M (1996). Seedling identification. In Glyn D (eds). A strategy for the Conservation of *Prunus africana* on Mount Cameroon. Technical Papers and Workshop Proceedings, 21st and 22nd February, 1996, Limbé Cameroon. Mount Cameroon Project, pp. 1-11.

Gachie PK, Koech EK, Njunge JT, Simons AJ, Ndalut PK (2012). Variation in yield and composition of crude bark extracts of *P. africana* in different provenances of Kenya. Forest, trees and Livelihoods, DOI: 10.1080/14728028.2012.662627.

Hall JB, O'Brien E M, Sinclair FL (2000). *Prunus africana*: A Monograph. School of Agricultural and Forest Sciences Publication Number 18. University of Wales, Bangor, 104 p.

Ingram V, Owono A, Schure J, Ndam N (2009). Guidance for a national *Prunus africana* management plan, Cameroon. CIFOR, FAO.158 p.

IUCN (2001). IUCN Red List Categories and criteria : Version 3.1. IUCN Species Survival Commission. IUCN, Gland, Switzerland and Cambridge, UK.

Marcelin J, Ndam N, Bell J (2000). Medicinal tree needs sustainability cure. ITTO Newsletter 10.

National Office for Forest Development (1992). Standards study of trees. Forestry Department, MINEF. Yaoundé, Cameroon, 20 p + annexes.

Ndam N (1996). Recruitment patterns of *Prunus africana* (Hook F.) Kalkman, on Mount Cameroon: a case study at Mapandja. In Glyn D (eds). A strategy for the Conservation of *Prunus africana* on Mount Cameroon. technical papers and workshop proceedings, 21st and 22nd February, 1996, Limbé Cameroon. Mount Cameroon Project, pp. 19-34.

Ndibi B, Kay EJ (1997). The regulatory framework for the exploitation of medicinal plants in Cameroon: the case of *Prunus africana* on Mount Cameroon. Biodivers. Conserv. 6:1409–1412.

Ondigui BRP (2001). Sustainable Management of a Wild Plant Species for the Conservation of Mountan Forest Ecosystems and the Welfare of Local Communities: A Case Study of *Prunus africana* in the Mount Cameroon Area. 9 pp. In Sustainable Management of a Wild Plant Species. Proceedings of the World Mountain Symposium. Interlaken, Switzerland.

Page B (2003). The political ecology of *Prunus africana* in Cameroon. Area 35(4): 357–370.

Pardé J, Bouchon J (1988). Dendrométrie, 2e édition, Nancy. Ecole nationale du Génie rural, des eaux et des forêts, 328 p.

Rondeux J (1993). La mesure des arbres et des peuplements forestiers, Gembloux, Belgique, Les presses agronomiques, 521 p.

Rondeux J, Laurent C, Lejeune P (1991). Tarifs de cubage peuplement pour le douglas Pseudotsuga menziesii (Mirb.) Franco en moyenne et Haute-Belgique. Rev. for. Franç. XXXXIII(6): 507-513.

Sunderland T, Tako TC (1999). The exploitation of *Prunus africana* on the island of Bioko, equatorial Guinea Report for the People and Plants Initiative, WWF – Germany and the IUCN/SSC Medicinal Plants Specialist Group, June 1999.

Tchouto P (1996). *Prunus* population on Mount Cameroon. In Glyn D (eds). A strategy for the Conservation of *Prunus africana* on Mount Cameroon. Technical Papers and Workshop Proceedings, 21st and 22nd February, 1996, Limbé Cameroon. Mount Cameroon Project, pp. 12-18.

Vivien J, Faure JJ (1985). Arbres des forêts denses d'Afrique centrale. Ministère des relations extérieures, Coopération et Développement - ACCT, Paris, 551 p.

World Agroforestry Centre (2012). Unsustainable harvesting of *Prunus africana* tree threatens treatment for prostate disorders. Transformations, Issue 16.

The relative contributions of climatic elements and environmental factors to flooding in Awka urban area

E. E. Ezenwaji[1] , A. C. Okoye[2] and A. K. Awopeju[3]

[1]Department of Geography and Meteorology, Nnamdi Azikiwe University, Awka, Anambra State, Nigeria.
[2]Department of Environmental Management, Nnamdi Azikiwe University, Awka, Anambra State, Nigeria.
[3]Department of Statistics, Nnamdi Azikiwe University, Awka, Anambra State, Nigeria.

The aim of this paper was to determine the relative contributions of climatic elements and environmental factors to urban flooding in Awka urban area of Anambra State. Towards achieving this aim, 10 year (2000-2009) meteorological data of temperature and rainfall of the study area were collected from synoptic meteorological station in Awka and data on environmental factors were obtained from both the Anambra State Ministry of Environment and responses were gathered from the administered questionnaire to the respondents. Khosla's formula was employed to estimate flood discharge in the urban area and analysis was carried out with the use of multiple linear regression (MLR). Result of the regression shows that 12 variables exhibited positive signs showing that there is a direct relationship between them and flooding in the town. Also, it was found that both climatic and environmental factors contributed 60% to the explained variation, with the climatic elements alone having 33.7%. Finally, policy and planning implications of the study were discussed, while it was concluded that urgent necessary steps need to be taken to achieve goal 7 of the MDGs in the study area, which is to ensure environmental sustainability by the year 2015.

Key words: Climatic, flooding, synoptic, regression, sustainability.

INTRODUCTION

Urban areas especially those in the developing countries experience various types of disaster in most periods of the rainy season as a result of flooding. The destructive effects of floods in urban areas have been widely recognized in Nigeria (Muoghalu and Okonkwo, 1998; Odemerho, 1993, 2005; Offiong and Eni, 2008; Abaje and Giwa, 2008) and other developing countries in general (Smith, 1999; Obote, 2005; Sithole, 2009). Furthermore, the causes of such urban flooding has been linked to climate factors especially rainfall (Gray, 2005; Sadat, 2009), topography (Eze, 2008; Offiong and Eni, 2008) urban sprawl (Sule, 2004), poor urban planning (Goudie, 1981; Slamaker, 2000) and other environmental factors (Onokerhoraye, 1995). According to IPCC (1990) rapid but unchecked urban development in developing countries together with climate change have been identified as the major causes of flooding in the urban areas of the developing countries. While different views are held by researchers in the field, the problem of flooding has continued to be one of the greatest urban problems in low income countries because of the poor preventive measures adopted by the countries. The volume of water associated with each flood event in some of these urban areas is of such magnitude that it always results in the destruction of lives and property.

Most studies on urban flooding have concentrated

attention on describing the nature of urban floods, listing the causes of such floods and their destructive consequences on life and property. Some of such studies are too descriptive as they lack necessary scientific basis which therefore could not provide enough theoretical framework for scholars. However, few scholars have empirically investigated the problem of urban flood in Nigerian. They include Muoghalu and Okonkwo (1998) who employed the multiple regression analysis (MRA) to identify nine variables causing the flooding in Awka Town, and out of these three of them namely, the percentage of gravel, percentage of course sand and percentage of medium sand accounted for 92.1% of hydrologic performance of the variance explanation in the city, while Geoffrey (2001) employed time series analysis to predict flood events in Lagos Metropolis. However, there are other scholars who have done some work in the town regarding this area of study. For example, Mike and Onell (2009) isolated some climatic elements giving rise to flooding in the area which include, rainfall intensity, rainfall duration, high temperature and atmospheric humidity, and concluded that rainfall intensity and temperature were significant. Furthermore, Okoro and Basil (2012) tried to assess some environmental factors giving rise to flooding in Awka and concluded that area of impervious cover and topographical disposition of the town are significant factors. Apart from these few works known to the authors, the use of statistical techniques in the study of the problem of urban flooding has been relatively scanty. Also, the need to compare the contribution of climatic elements and other environmental factors to ascertain their relative importance to the cause of urban flooding has largely been ignored in relevant literature. To fill these gaps, this paper intends to statistically assess the relative contributions of climatic elements and environmental factors that influence flood generation in Awka and determine the relative effects of predictor variables.

MATERIALS AND METHODS

Study area

Awka town is located between latitudes 6.24°N and 6.28°N and longitudes 7.00°E and 7.06°E on the South eastern part of Nigerian (Figure 1). The study area covers 144.5 ha with a 2006 contested population of 116, 208 persons (NPC, 2006). This includes such outlying communities as Amawbia, Okpuno and Amansea which are fastly being annexed to the town by urbanization. The town's topography presents a rugged relief as it lies completely on Awka-Orlu upland.

Generally, the average height of the town range from 91 m in the western parts of the town to 160.2 m in the eastern zone, although there are local variations within the town which are drained by a number of streams. The climate is the tropical wet and dry type according to the Koppen's classification system with a clear cycle of season. The mean daily maximum temperature is usually 27°C all over the year although it could reach 34°C in March and lowest during the harmattan months of December and January. The mean annual rainfall according to the local meteorological station which has maintained climatological records since 1978 reveal a mean rainfall of about 1600 mm with a relative humidity of 80% at dawn.

Data collection

Ten (10) years (2000-2009) meteorological data of temperature and rainfall of the study area were collected from the Nigerian Meteorological Agency (NIMET) Synoptic Station at Amawbia, while data on environmental factors were obtained from questionnaire responses, field observation and published (secondary) data from the Anambra State Ministry of Environment, Awka. A total of 900 questionnaires were administered for six months, April to September, 2010 with the use of stratified sampling technique with each ward in the town forming a strata. There are nine wards in Awka town, and 100 questionnaires were administered to each of them. The data collected were used to crosscheck the veracity of environmental data collected from the Ministry. Also, *Khosla's* formula was used to estimate flood discharge in the urban area:

$$Q = P - 0.4813Tm \text{ (when } Tm > 4.5°C) \quad (1)$$

Where, Q = yearly or monthly flood discharge in cm; P = yearly or monthly rainfall in cm; Tm = mean yearly or monthly temperature in °C

Data analysis

The formular for estimating flood discharge was used to determine the flood generation in twenty flood channels in the town which was calculated to be 161.290 m^3 in six months, while we obtained the temperature and rainfall data from the NIMET station in the area. We had to carefully identify, define, describe and parametize the range of environmental and climatic factors (Table 1). The parametization was achieved by adding up the number of times a particular variable occurred in the study area during the period of data collection or record readings with respect to climatic variables.

The quantity of flood generation within the period was used as depended variable and the meteorological and environmental factors as independent variable (X_1 X_2...X_n). The data generated (dependent and independent variables) were subjected to analysis using the multiple linear regression (MLR) analytical technique. The general expression of multiple linear regression as given by Anyadike (2009) is shown below:

$$y = a + b_1 x_1 + b_2 x_2 + b_3 x_3 + \ldots b_n x_n + e \quad (2)$$

Where y = dependent variable; x_1 x_2...x_n = independent variable; a = base constant; e = error term or the proportion of the variance not explained, y the regression.

When the MLR package was used in computer calculation, it was found that the combined relationship between the 14 predictor variables (12 environmental and 2 climatic variables) and the amount of flood generated in Awka town yielded a Multiple Correlation Co-efficient (R) of 0.75 with a co-efficient of multiple determination (R^2) of 0.600. This means that the 14 predictor variables combined to explain or account for 60% of the variations in the amount of flood generation in Awka in 10 years. This, however, leaves 40.0% which the multiple regression model was unable to explain. This shows that other factors outside the climatic and environmental factors utilized in this study also account for flooding in the town. The summary of the result of the performances of the individual variables is shown in Table 2.

A similar analysis was also performed for individual years within

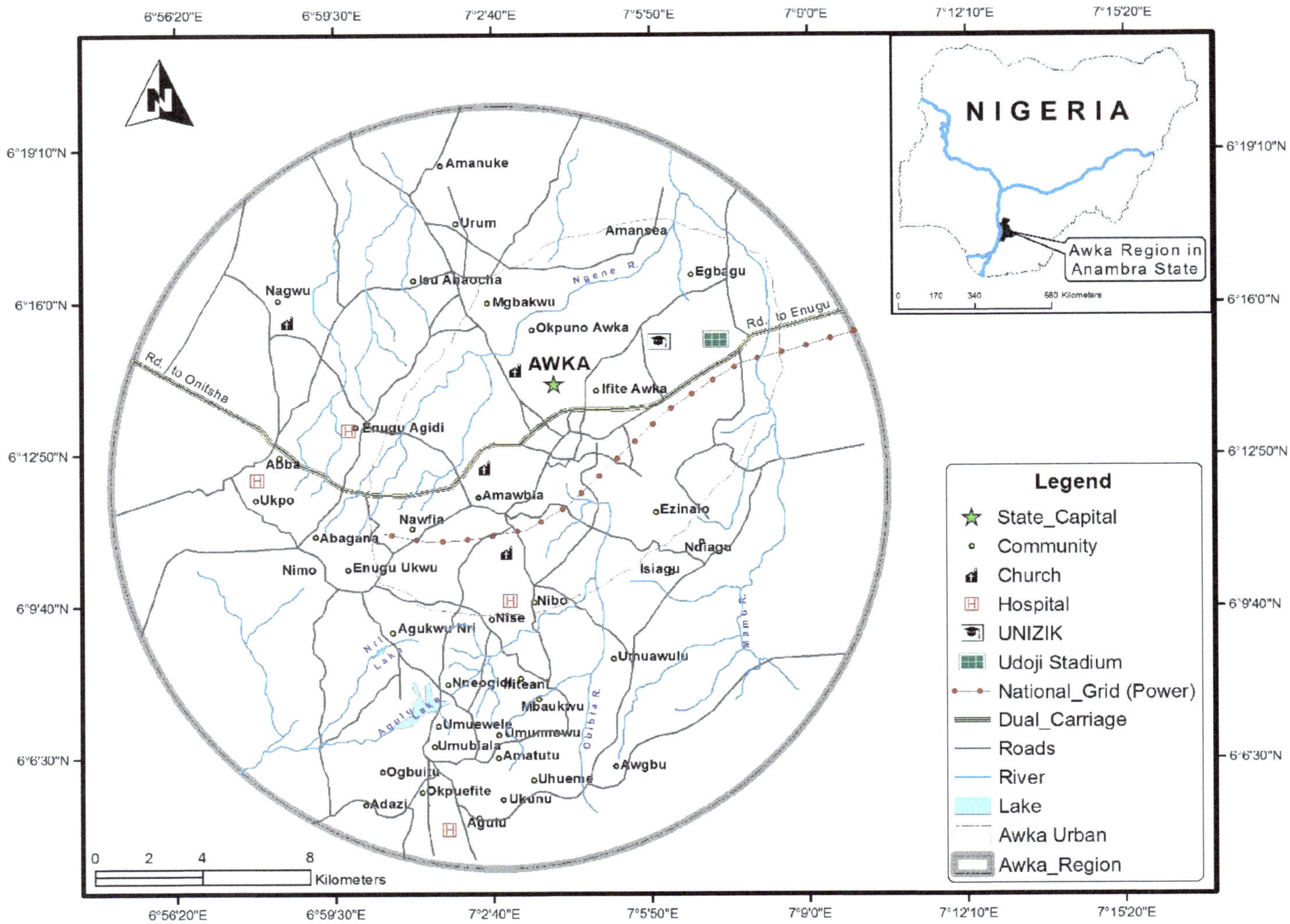

Figure 1. Map of Awka urban area, Nigeria.

Table 1. Description and parametization of climatic and environmental factors influencing flood generation in Awka town.

Name of variable	Variable label	Method of parametization
Number of flood channels used for dumping refuse	REFU	The number of flood channels used for dumping refuse during the period of data collection was determined and recorded
Number of flood channels with reverse gradient	GRAD	The identified number of flood channels with reverse gradient was recorded
Number of roads without drainage channel in Awka	DRAIN	This number was identified and recorded
Number of flood channels blocked by houses and other civil constructions	HOUS	The identified number of civil structures on the channels were recorded
The average slope of the built up area of the town	SLOP	This was calculated from the topographical map of the area and recorded
The number of urban pavements constructed within the period of Data collection	PAVE	This number was determined and recorded
Number of natural flood channels identified in the town	WATU	The number of natural flood channels were identified in the town was recorded

Table 1. Contd.

The number of Housing estates in the town	ESTA	This number was determined and recorded
Whether or not the Houses are built on sloppy land	BUIT	If the Houses were built on sloppy land, we record 1, but if not zero (0)
Number of swamp sites which have been reclaimed since 10 years	SITES	The identified number of swamp sites that were reclaimed in the last 10 years were recorded
Number of flood channels blocked by sand since 10 years	SAND	The number of flood channels blocked by sand since 10 years was recorded
Whether or not there are settlements on the identified troughs	SETT	If there are settlements in identified troughs (1) is recorded when there are no settlement on the trough (0) was recorded
Temperature	TEMP	The average temperature of the town for 10 years was calculated and recorded
Rainfall	RAIN	The average rainfall for the 10 years period was calculated and recorded

Source: Fieldwork, 2010.

Table 2. Individual contributions of independent variables in the volume of flood in the entire Awka town.

Independent variables		Standard error of Estimates	R^2	R^2 (%)	Signs	F-Statistics
X_1	TEMP	0.091	0.166	16.6	+	32.613
X_2	RAIN	0.44	0.171	17.1	+	76.114
X_3	REFU	0.007	0.019	1.9	+	4.846
X_4	GRAD	0.086	0.113	11.3	+	7.648
X_5	DRAI	0.151	0.030	3.0	+	0.029
X_6	HOUS	18.964	0.003	0.3	+	0.138
X_7	SLOP	4.322	0.022	2.2	+	2.448
X_8	PAVE	0.096	0.069	6.9	+	5.208
X_9	NATU	0.425	0.001	0.1	-	32.661
X_{10}	ESTA	4.288	0.000	0.0	+	9.049
X_{11}	BUIT	1.336	0.001	0.1	+	65.207
X_{12}	SITE	5.466	0.002	0.2	+	2.840
X_{13}	SAND	0.006	0.003	0.3	+	3.088
X_{14}	SETT	2.672	0.000	0.0	-	2.482

Source: Fieldwork, 2010.

the 10 year period (2000 to 2009) and the result is presented in Table 3.

RESULTS AND DISCUSSION

The result of the analysis in Table 2 shows that the two climatic variables TEMP (average temperature of the area) and RAIN (average amount of rainfall) contribute a total of 33.7% of the variations to flooding while environmental factors contribute 26.3%.

Out of this 26.3% contributed by environmental factors, only five variables out of twelve namely; GRAD (number of flood channels with reverse gradient) with 11.3%; PAVE (number of urban pavements constructed within the period) with 6.9%; DRAIN (number of roads without drainage channels) with 3.0%; SLOP (average slope of land in built up areas of the town) with 2.2% and REFU (number of flood channels used for dumping refuse) with 1.9%, contributed meaningfully to the observed variation in flooding. These leave the remaining seven variables with the contribution of only 1.0%. Altogether, the 14 variables (climatic elements and environmental factors) contributed 60% of the variation in flooding in Awka town. It could also be seen that 12 of the variables carried positive signs while only 2 have negative signs attached to them. The standard error of estimates of the multiple regression is also shown in the table against each variable.

Furthermore in Table 3, it could also be seen that the multiple regression model produced results for climatic elements and environmental factors for each year of the analysis. Climatic elements contributed highest in four years: 2002, 2004, 2006 and 2009, while the environmental

Table 3. Summary of the relative contributions of climatic elements and environmental factors influencing flooding in Awka town (2000-2009).

Year	Contributions of climatic elements R^2 (%)	Contributions of environmental factors R^2 (%)
2000	26.1	33.9
2001	28.9	31.1
2002	36.0	24.0
2003	29.6	30.4
2004	40.3	19.7
2005	29.9	30.01
2006	42.8	17.2
2007	29.2	30.9
2008	29.6	30.4
2009	44.3	15.7
Mean	33.7	26.3

Source: Fieldwork, 2010.

factors contributed highest in the years 2000, 2001, 2003, 2005, 2007 and 2008.

It could however, be seen that climatic elements were the greatest contributors of flooding in Awka town during the period 2000-2009. The details of this contribution will be discussed later. This finding is in agreement with Agrail (2005) that climatic factors especially rainfall is the most important factor responsible for flooding in the urban areas of most tropical countries. The standard error of estimates of the multiple regression shown in column two indicates that the predictor variables with the highest standard errors are not good indicators of the volume of flooding in Awka town. In contrast, variables like TEMP, RAIN, GRAD, PAVE, DRAIN, SLOP and REFU have low standard errors of estimates indicating that they are fairly accurate estimators of the volume of flooding in Awka.

Also, 12 of the variables in Table 2 have positive signs showing that there is a direct relationship between these predictor variables and flooding in Awka town. This means that a unit increase in any of the variables (X_1, X_2, X_3, X_4, X_5, X_6, X_7, X_8, X_{10}, X_{11}, X_{12} and X_{13}) will also lead to a unit increase in the amount of floods (Y) generated in the town. The remaining two variables (X_9 and X_{14}) have negative signs showing that there is an inverse relationship between these predictor variables and the amount of floods in Awka town. This simply means that any unit increase in any of these predictor variables will lead to a unit decrease in the amount of flooding. This result is, however, in line with natural occurrences as a unity increase in the number of natural flood channels identified in the town (X_9) will lead to the decrease in flooding.

In other words, this indicates that the presence of natural channels in the town will surely limit the incidence of flooding in the town. The findings of Ebro (2010) in Kampala, Uganda supported this result when he noted that the consumption of 1 hectare of land by urbanization led to a 3 fold increase in flooding in Southern Kampala. Again, the variable (X_{14}) whether or not, there are settlements in urban troughs and lowlands is associated with inverse relationship with flooding. This is so because it is possible that most settlements located in such areas are provided with good drainage channels.

Furthermore, we had earlier stated in Table 3, were the multiple regression analysis result of the contributions of climatic elements and environmental factors are presented that climatic elements contributed higher in 4 years. One significant thing about this contribution is that it kept increasing every year. Thus, the years when it contributed higher than the environmental factors, these continuous increases are evident. For example, in 2002 it was 32.8%, in 2004 it was 34.3%, in 2006 it was 36.6% and in 2009 it was 44.3%. What this indicates is that the effect of climatic elements on flooding in the area is increasing every year most likely because of climate change. With this trend, in the next few years, the climatic element may contribute over 50% of urban flooding in Awka town.

Conversely, the years when the environmental factors contributed highest, it does not show the kind of increase observed with that of the climatic elements as there are fluctuations in the percentage contribution in various years. Altogether, the 60% contributions of both the climatic elements and the environmental factors show that 40% of the causes of flooding in Awka town are attributed to other causes, which could be traced to poor flood management, poor urban governance and improper planning of our urban areas.

Policy and planning implications

The issue of flooding has been one of the severest urban

problems in Awka town. In 2008, the State Government approved and expended the sum of ₦133 Million for the channelization of Iyiagu flood problem (Anambra State Government, 2009). This measure has not addressed the problem of flooding in the town as virtually all the major roads are flood channels during any major flood occurrence. The flood problem in the second half of Arthur Eze Avenue towards the Expressway and that of the Post office area along Zik's Avenue is particularly worrisome. The policy and planning implications of this study is that the only measure that can solve this problem is the wholistic assessment of the entire city with a view to evolving a new master plan for urban drainage of the town.

The present piecemeal and partial planning of the flood programme in the town like the Iyiagu channelization programme earlier mentioned will not solve much problem. Presently, the master plan for the city prepared by the UN Habitat Kenya, made some provisions for addressing the problem but who knows when the government will begin to implement the plan. Even the plan document as beautiful as it may look, might not adequately solve the problem because of the poor local input into it. These issues are general to the problem, the following suggests specific measures to be undertaken, based on our findings.

Ensuring the use of climatic factors in the drainage master plan of the town

The findings in this paper clearly show that the climatic factors of rainfall and temperature are very important to the generation of flood in Awka town. Failure of the planners to recognize these vital factors will mean that the solution to the problem is not presently in sight. Muoghalu and Okonkwo (1998) suggested the incorporation of geographers, (mostly geomorphologists and meteorologists) engineers, geologist in the master planning team. According to them, only town planners and engineers are presently members of such team and they may not be able to handle the problem without the input of the other specialists mentioned.

Planning with relevant environmental factors

This study shows that it is not all environmental factors that contribute to the generation of flood in Awka. Out of the 12 environmental factors chosen, a number of flood channels with reverse gradient and the increasing rate at which pavements and concretes are being constructed in the urban area ranking highest with 11.3 and 6.9%, respectively. This particular factor should be looked at very seriously. Government should return to the old emphasis on open spaces, parks and grassy play grounds of the old. However, the construction of flood channels in the town has one basic problem which is reverse gradient, making it difficult for flood water to flow. Also, there are numerous roads without any form of flood channel instead of the gutters. This is the reason why whenever it rains, the road itself becomes a flood channel. Again, people should not be allowed to build in sloppy areas as the water that runs from their rooftops to the ground tend to accelerate very fast downhill sometime with destructive consequences. The issue of dumping refuse on flood channels needs to be addressed urgently. These factors should be seriously brought into the planning arrangement of government for the solution of flooding problems in the town.

Sensitization of the populace

The need to ensure that most environmental problems identified as important in the generation of flood are constantly put to check requiring that the populace should be constantly sensitized and educated. Considering the issue of blocking the flood channels with household and commercial refuse, government can pass information through local neighourhood leaders to solve this problem. This is the quickest and the surest way to get to the people at the grassroots. Again, government can use radio and other media outfits to get the message across to the populace.

Embarking on tree planting

This particular area needs to be seriously emphasized. This is because presently the few available trees have been cut down in the urban area mostly to give way to housing estates, roads and other human activities. Trees should be planted at the open places especially in the urban fringes where urbanization is still passive so that by the time urbanization gets to such area, they will no longer suffer the flooding problems of the urban core. In some of our old cities like Kaduna, Enugu, Kano and most northern cities, the planting of trees especially along the major streets was emphasized for the protection of the environment which includes the stopping of incidence of flooding.

Clearing the flood channels

One of the serious environmental factors generating flood in urban areas is the problem of the blockage of the flood channels. Apart from the dumping of refuse which we have been identified, the issue of the blockage of the channels by sand eroded from higher elevation areas and deposited in areas where the velocity of the flood could not carry them continues to be a recurrent problem. Government should come out with a law that makes it mandatory for all residents and shop owners to clear the gutter in front of their houses or shops so as to remove the sand and items that usually block it. The law should put stiff sanctions for all offenders.

Conclusion

We have tried in this paper to determine the relative contributions of climatic elements and environmental factors determining urban flooding in Awka town. It was seen that climatic elements contributed 33.7% out of the 60.0% generated by climatic and environmental factors. This shows that climate change may have had some serious adverse effects on the weather elements in the town. The need to isolate relevant factors and use them in the planning of urban drainage master plan needs no additional elaboration.

We have also seen that environmental factors cannot be ignored in such plan because of the contributions of such factors for urban flooding. The policy and planning implications however include the need to use relevant climate data in the planning of the towns' storm drains using relevant environmental factors in such plan, sensitization of the populace on the need to embark on measures that would reduce urban flooding, embarking on tree planting and clearing of the flood channels. There is need for government and other concerned bodies to employ our findings in this paper to reduce urban flooding in various urban areas so as to scale down the losses associated with it. Also, we need to be reminded of the fragile nature of our urban environment and as such require our protection to ensure its sustainability. This particular action is needed now than ever if we hope to achieve goal 7 of the MDGs which is ensuring environmental sustainability by the year 2015.

REFERENCES

Abaje I, Giwa PN (2008). Urban Flooding and Environmental Safety: A case study of Kafanchan Town in Kaduna State. In Mamman A. et al (eds.). Urbanization, Resource Exploitation and Environmental Stability in Nigeria. Joyce Publishers, Kaduna. pp. 64-81.

Agrail FD (2009). Rainfall and its Effects in Urban Environment. J. Atmos. Sci. 13(2):22-30.

Anambra State Government (2009). Iyiagu Channelization Programme Mimeographed.

Anyadike RNC (2009). Statistical Methods for Social and Environmental Sciences, Spectrum Publishers, Ibadan. p. 125.

Ebro E (2010). Kampala's Urban Problems. Sub-Sahara 3(2):6-13.

Eze BE (2008). Topography and Urban Expansion as Twin Factors of Street Flooding in Calabar Municipality, Cross River State in Bisong F. E. (ed.). Geography and the Millennium Development Goals. Translating Vision to Reality in Nigeria. Proceedings of the 50th Conference of Association of Nigerian Geographers (ANG), Index Book Publishers Ltd., Calabar. pp. 185-193.

Geoffrey L (2001). "Time Series" Analysis of Flood Events in Lagos, Nigeria. J. Hydrol. Afr. 8(2):20-32.

Goudie K (1981). Urban Planning in Developing Countries. Carlborne Press, New York.

Gray C (2005). Climatic Factors in the Generation of Urban Floods in Selected Sub-Sahara African Cities. Urbanization 2(2):112-121.

IPCC (1990). Climatic Change: The IPCC Scientific Assessment.

Mike JJ, Onell DB (2009). Analysis of relevant climatic elements in the study of urban flooding in Awka town, Nigeria. Clim. Environ. 5(2): 330-341.

Muoghalu LN, Okonkwo AO (1998). Effects of Urban Flooding in Awka, Capital of Anambra State, Nigeria. Environ. Rev. 2:22.

NPC (2006). The Population of Nigeria. Federal Government Gazette Vol. 96, No. 2, Government Printer, Abuja.

Obote JJ (2005). Flooding in our Cities: The case of Kampala. J. Afr. Dev. 8(2):60-69.

Odemerho FO (1993). Flood Control Failures in a Third World City: Benin City, Nigeria - Some Experimental Factors and Policy Issues. Geojournal 29:371-376.

Odemerho FO (2005). Benin City: "A case study of Urban Flood Problems'. In Onokerhoraye AG and Onita GEK (eds.). Perspectives on Development. Centre for Population and Environment.

Offiong RA, Eni DA (2007): Urban Floods: An Appraisal of the Effect of Flooding on Urban Infrastructure in Calabar Metropolis, Cross River State, Nigeria in Mamman A. B. et al (eds.). Urbanization, Resource Exploitation and Environmental Stability in Nigeria. Joyce Publishers, Kaduna. pp.75-91.

Okoro L, Basil MO (2012). A survey of environmental issues in the flooding of Awka urban. J. Environ. Anal. 3(1):15-24.

Onokerhoraye AG (1995). Benin: An African City in Transition,The Benin Social Sciences Series for Africa.

Sadat AJ (2009). Rainfall intensity and Duration as Two Factors of Flooding in Humid Tropics. J. Atmos. Sci. 13(2):10-16.

Sithole MA (2009). African Cities and their Environmental Problems. Nalsonel Press, Cape Town.

Slamaker BL (2000). Urban Development and Problems of Flooding in Nairobi. J. Urban Aff. 6(1):1110-1119.

Smith RA (1999) Effects of Urban Floods in Sub-Saharan African. Balico Press, Cairo.

Sule RO (2004). Environmental Consequences of Rapid Urbanization in Countries of the Developing World, Thumbprints International Company, Calabar.

Nature and dynamics of climate variability in the uganda cattle corridor

Nimusiima Alex[1], Basalirwa C. P. K.[1], Majaliwa J.G.M[1], Otim-Nape W.[2], Okello-Onen J.[3], Rubaire-Akiiki C.[4], Konde-Lule J.[5] and Ogwal-Byenek S.[2]

[1]Department of Geography, Geo-Informatics and Climatic Sciences, Makerere University, Uganda.
[2]Africa Innovations Institute, Kampala, Uganda.
[3]Department of Biology, Gulu University, Uganda.
[4]College of Veterinary Medicine, Animal Resources and Bio-security, Makerere University, Uganda.
[5]College of Health Sciences, Makerere University, Uganda.

The study was conducted in the districts of Nakaseke and Nakasongola stratified into four farming systems of crop dominancy, pastoralists, mixed crop and livestock and fishing. The study was guided by two research questions: (1) how do community residents perceive climate change/variability? (2) What is the trend and nature of climate variability and how does it compare with people's perceptions? Ninety eight percent (98%) of the respondents reported that the routine patterns of weather and climate had changed in the last 5 to 10 years and it has become less predictable with sunshine hours being extended and rainfall amounts being reduced. This compared well with the analyzed secondary data. Over 78% respondents perceived climate change and variability to be caused by tree cutting other than the known scientific reasons like increase in industrial fumes or increased fossil fuel use. Climate data showed that over the period 1961 to 2010 the number of dry spells within a rainfall season had increased with the most significant increase observed in the first rainfall season of March to May as compared to the season of September to November. The first dry season of June/July to August is short while the second dry season of December to February is long during the study period. The two rainfall seasons of March to May and September to November seem to be merging into one major season from May to November. Temperature data shows a significant increasing trend in mean annual temperatures with the most increase observed in the mean annual minimum temperatures than the maximum temperatures.

Key words: Climate variability, community perceptions, Uganda cattle corridor, dry spells.

INTRODUCTION

Climate variability has been identified as a major challenge hindering most developing countries from attaining Millennium Development Goals (UNDP, 2007). Climate variability manifests as floods, droughts erratic and unreliable rainfall and other extreme weather events, and has not only negative impact on natural and human systems but also creates serious developmental challenges in developing countries (AIACC, 2007).

These developing countries (most of them in Africa) are the most affected due to their dependence on climate sensitive economic sectors such as rain-fed agriculture, and their limited economic, technological and human capacities to respond to climate change and variability induced stress (Huq et al., 2004; IPCC, 2007).

Agriculture constitutes the back bone of the economies of most developing countries. In Uganda, agriculture contributes up to 22.5% of the GDP, provides employment to 66% of the total working population (UBOS, 2011). Most of this agriculture is rain-fed and therefore susceptible to climate variability and change impacts.

In Uganda, there is a strip of land stretching from the south west to the north eastern part of the country commonly known as the Uganda cattle corridor. This region has experienced contrasting climate change related conditions such as prolonged droughts, unpredictable rainfall onsets and withdraws as well as floods among other climate variability related phenomena (Kisamba, 2001). In addition to these climate variability related problems, community problems such as land disputes have also been common in this region mainly between pastoralists and crop farmers regarding the use of resources of land and water.

The nature of climate variability in this area is not well understood and how the impacts of climate variability affect the communities has not been well documented. This study therefore characterized the nature of climate variability by analyzing variations in major climatic parameters of rainfall and temperature and how they affect local communities in the central cattle corridor districts of Nakasongola and Nakaseke. The paper attempted to address two major questions: (1) How do community residents perceive climate change/variability? (2) What is the trend and nature of climate variability and how does it compare with people's perceptions?

MATERIALS AND METHODS

Study area

The study was focused on two cattle corridor districts of Nakaseke and Nakasongola in the central part of the corridor (Figure 1). The two districts had a combined population of 265,342 during 2002 population census and the projected population for 2012 was 347,600 (UBOS, 2012). The districts are composed of mainly four categories of communities: crop farmers dominated communities, livestock/pastoral dominated communities, mixed crop and livestock communities and fishing communities especially around Lake Kyoga.

The area experiences a fairly bi-modal rainfall season like most parts of the country with the first season beginning in March and ending in early June and the second season beginning in late August/early September and ends in November/December.

Selection of respondents

House-hold surveys (HHS)

In each of the two districts, communities were stratified into three strata: pastoral-majority areas; areas where pastoralists are a minority among settled agriculturalists, and areas of extensive grazing by mixed crop-livestock farmers. In each of the three (3) strata in each district, 30 villages were randomly selected using the village registers at the district giving a total of 90 villages per district. In each of the selected villages, 20 households were randomly selected, using the registers at Village Local Council offices giving a total of 1800 households for each district and 3600 households for the entire study area.

Data

Both primary and secondary data were used in this study to understand the nature and dynamics of climate variability in the two cattle corridor districts of Nakaseke and Nakasongola.

Primary data

A structured questionnaire was designed to capture the community perspectives of climate change and variability, indigenous knowledge and the coping mechanisms adapted by the communities during hash conditions. The questionnaire was pretested before implementing it in the field using trained research assistants and the entire research team.

Secondary data

This consisted of historical data mainly on rainfall and temperature from 1961 to 2010. Soroti station in the same climatological zone as the study area was selected because of its complete historical weather and climate records. This consisted of daily rainfall and temperature data from which pentad, monthly and annual means were calculated. This data was obtained from the Department of Meteorology in the Ministry of Water and Environment.

Data analysis

Primary data

The questionnaire from the HHS was coded and results entered using epi-data software which was later exported to SPSS for further analysis. Graphs and tables and cross tabulations of variables of specific interest are then presented for discussion.

Secondary data

The data was checked and quality controlled using WMO (1970) guidelines. Trend analysis on various climatic parameters mainly rainfall and temperature were used to determine the variations in pentad, monthly, seasonal and annual variations of these climatic parameters in the study area from 1961 to 2010. Other parameters including onset and cessation of rainfall seasons, dry spells with- in a rainfall season, length of both wet and dry season were derived from pentad and monthly values of climatic parameters. Correlation and regression was then applied on the derived values to determine their relationships and their temporal variations.

Onset and cessation dates of rains: A number of studies have defined onset and cessation dates of the rainfall season for different

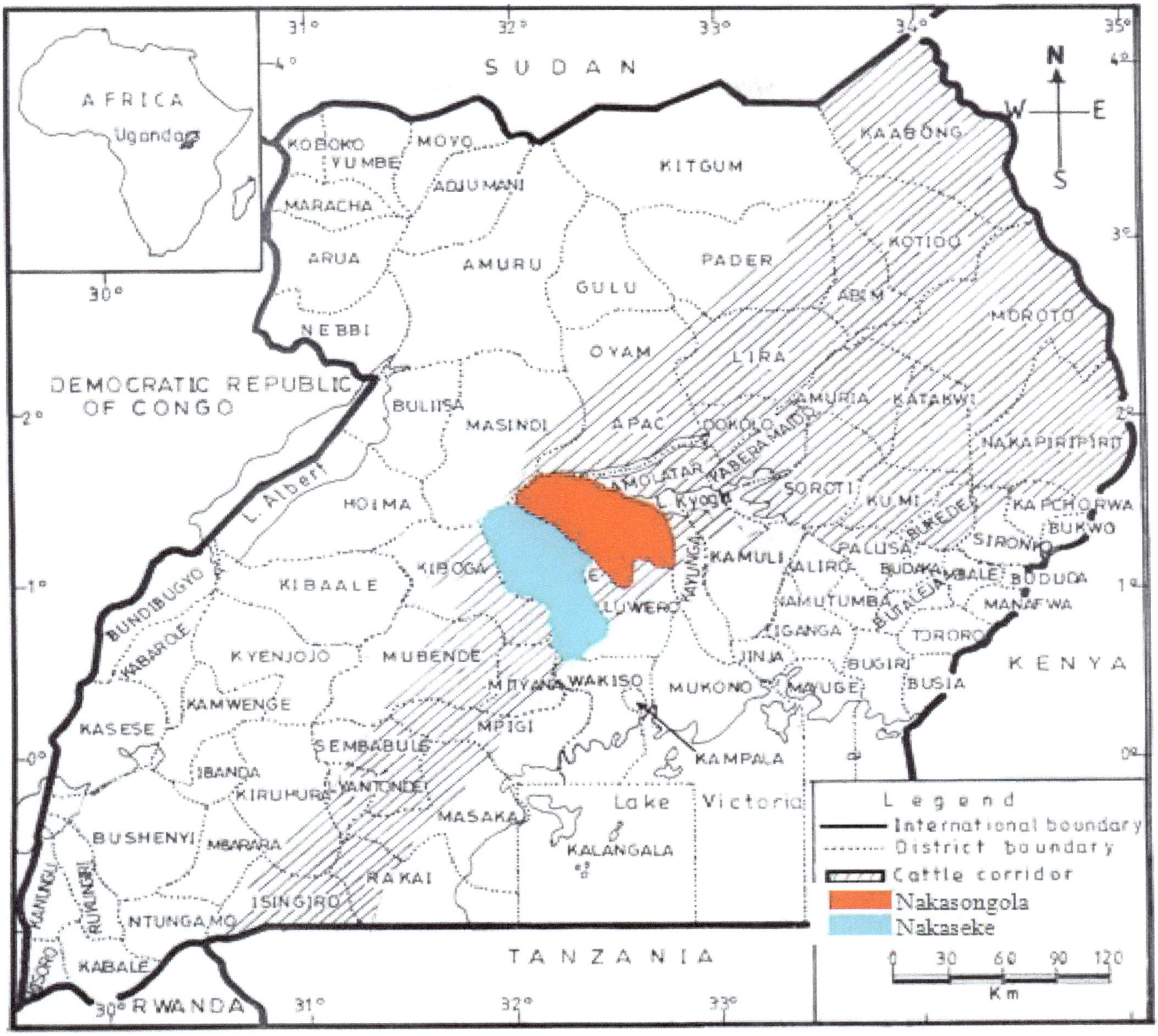

Figure 1. The Uganda cattle corridor and the districts of Nakasongola and Nakaseke.

regions across the globe. Dodd and Jolliffe (1994, 2001) in West Africa considered the onset to be the first period of 5 consecutive days in which at least 25 mm of rain falls, on condition that no dry period (seven days or more) occurs in the following 30 days. Nicholls (1984), in Austarlia defined onset as the date on which 15% of the mean annual rainfall, at a given station, has occurred In Tanzania, Mhita and Nassib (1987) defined the onset as the first week receiving at least 15 mm, after a given date determined by local climatology and agricultural practices, provided that no two week dry spells occur in the next four weeks. Marengo et al. (2001), defined the start of a rainy season using five days averages (pentads) of daily rainfall where they computed the onset as a pentad in which rainfall exceeded a given threshold, provided that the average rainfall was well below the threshold for several pentads preceding onset and well above for several pentads after onset. Prior to the report of Marengo et al. (2001), Alusa and Mishi (1973) had defined onset of a rainy season as a point when the pentad rainfall is consistently above the (1/73) of the total annual mean rainfall for a particular station. In this study, onsets and cessation dates were determined according to Alusa and Mishi (1973) and Marengo et al. (2001).

RESULTS

Community perception of climate change and variability

Community perception on changes in climate parameters

Most of the respondents (98%) interviewed perceived cli-

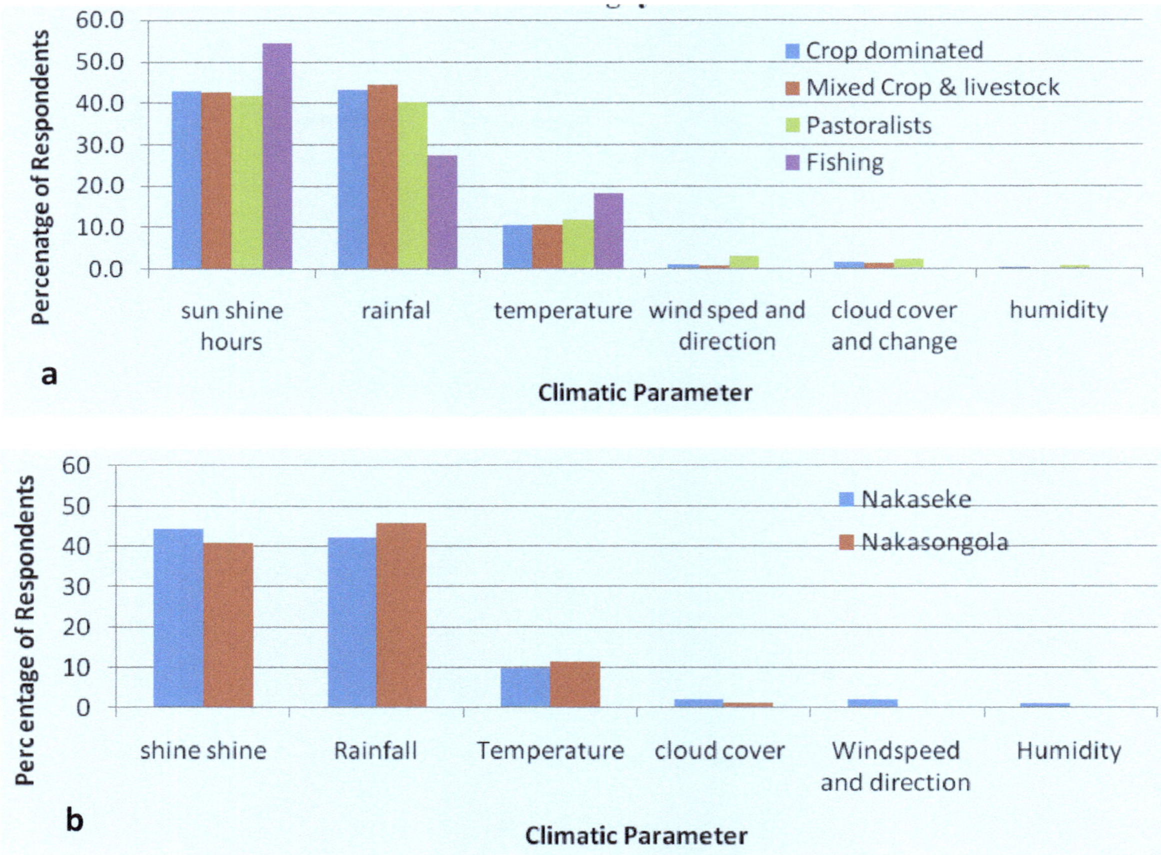

Figure 2. a. Perceived most changed parameter by communities in the different farming systems. **b.** Perceived most changed parameter by communities in Nakasongola and Nakaseke districts.

mate to have changed within the last 5-10 years as compared to 2% who either did not know or were not sure whether climate had changed (data not shown). About half of them believed sun shine hours and rainfall were the most changed climatic parameters while another 10% of them believed temperature had also changed. Very few respondents believed in changes in cloud cover, wind speed and direction and humidity (Figure 2a and b). Further, they believed (data not shown) that these changes are noticeable in increased sunshine intensity and duration; and in reduced rainfall amounts and distribution. This pattern was the same across the two districts and among the four farming systems studied (Figure 2a and b).

Community perception of causes of climate change and variability

The perceptions of communities on the causes of climate change and variability is shown in Figure 3a and b. A majority of respondents in both Nakasongola and Nakaseke districts believe that tree cutting is responsible for the changes and variability in climate (Figure 3a).

Very few of them however, believe that the perceived climate change and variability is caused by continuous cultivation, bush burning and overgrazing. Other reasons indicated by the communities were encroachment on wetlands, industrial fumes and pollution and brick making but these were suggested by very few members of the community. These perceptions were similar among communities in all four farming systems (Figure 3a).

Community perception of changes in seasonal weather

Nearly all the respondents reported changes in seasonal weather patterns during the last 5-10 years. Most of the respondents in all the four farming systems and in the two districts said the dry seasons have been the most affected as compared to the wet season (Figure 4a and b).

They reported that the length and intensity of the dry season has increased markedly as compared to what it was over ten years ago. The crop, mixed and pastoral communities said that in the past, the first dry season would fall from mid/end of December - February or early

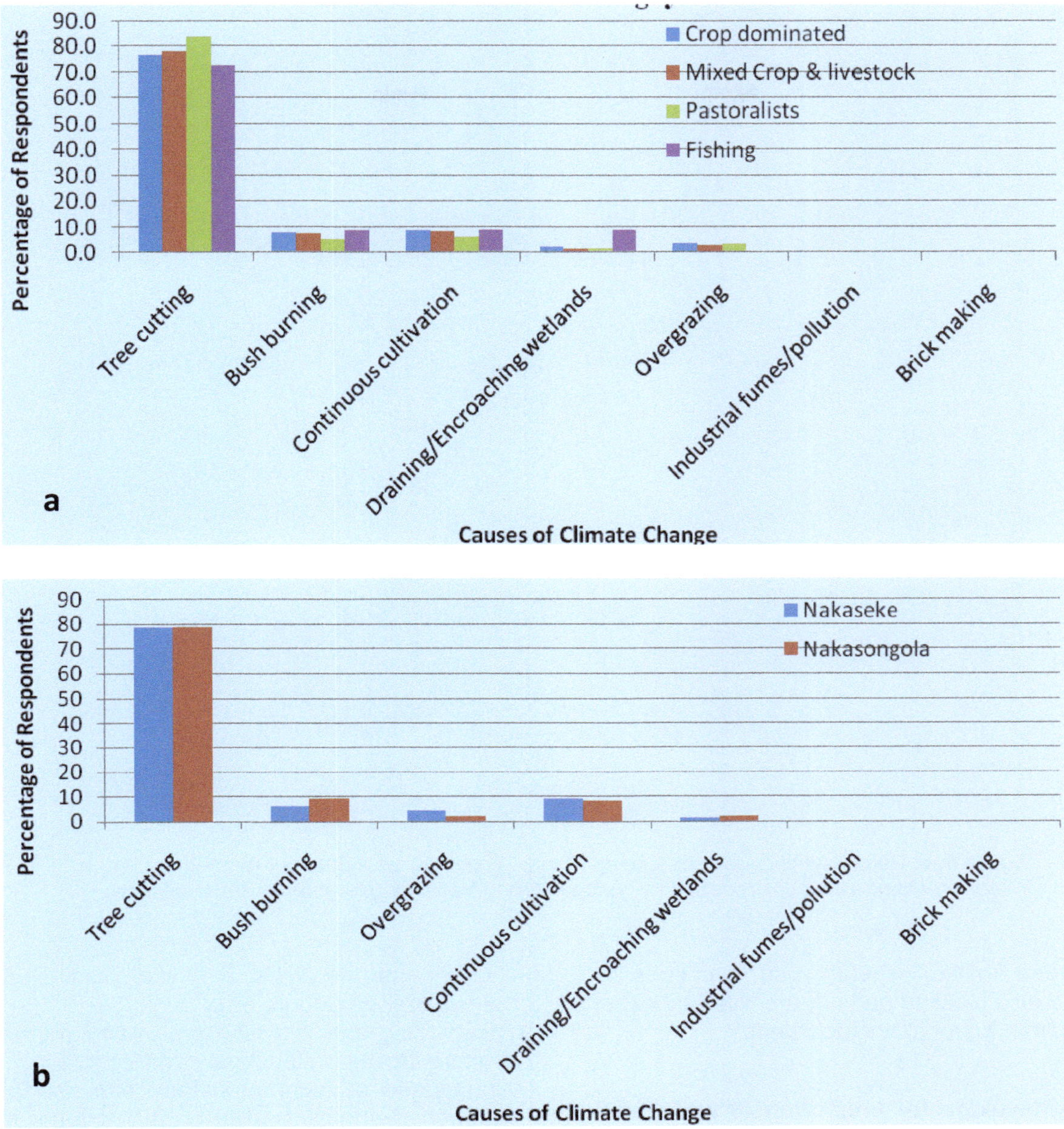

Figure 3. a. Perceived causes of climate change and variability by communities in the four farming sytems. **b.** Perceived causes of climate change and variability by communities in Nakasongola and Nakaseke districts.

March at most, with the second dry season coming around June-July. In the recent years, the farmers have been experiencing longer dry months, with the second dry season rather unpredictable and stretching from mid-November to March and sometimes early April, and the first dry season beginning at the end of June to August and that this has affected most of their operations. They lucidly recalled the years or periods when they were hit by long or severe droughts, floods, and even lightning and added that they have experienced more long droughts in the past two years and that lightning is a recent phenomenon.

Communities in all the four farming systems pointed out that they now receive less rainfall, which is also not well distributed in the villages within their parishes and is extremely unpredictable. Most of them admitted that they now receive the first rains between Mid-March to May/early June and the second rains from August/mid-September to October/early November, yet in the past 5-10 years the rains would come as early as February and go up to early June and the second season from August till the end of November.

With regard to floods, the fishing communities said they had floods in 1997. They said that when Lake Kyoga

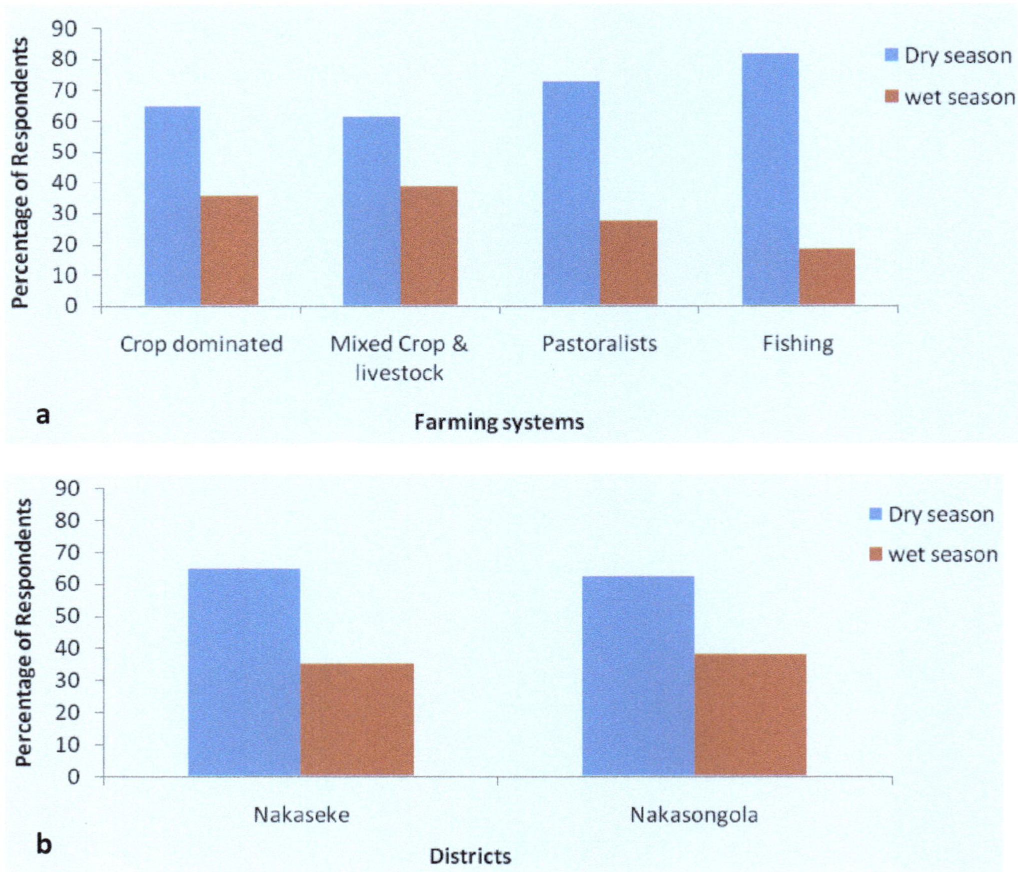

Figure 4. a. Perceived changed season reported by communities in the four farming system. **b.** Perceived changed season reported by communities in Nakasongola and Nakaseke districts.

flooded, it carried floating vegetation such as suds onto their lands and also blocked parts of the Nile and experts were called in from Egypt to unblock them.

Indigenous knowledge for prediction of wet and dry season

Indicators used by communities for prediction of onset of both wet and dry season are shown in Figure 5a and b. Increased cloud cover is the main indicator used by the communities for predicting the onset of rainy season. Other important indicators are appearance of full moon, emergence of new leaves of trees after dry season defoliation, and decrease in frequency and speed of whirling winds. Minor indicators are emergence of certain insects such as white ants (*Termitoidae),* from the ground, appearance of certain birds such as *Ebinyambala* and sudden increase in day or night temperature (Figure 5a). The indicators were the same among communities in the two districts (data not shown) and the four farming systems. However more fisher folks use increased cloud cover and emergence of new tree leaves for prediction of onset of wet season than their counterparts, the crops, crop and livestock farmers and the pastoralists (Figure 5a).

Regarding the dry season, many members of the communities use cloudless night conditions, increase in wind speed at night and early sun rise as the most common methods used to predict the coming of a dry season (Figure 5b). Other indicators are: increase in frequency and speed of whirl winds, misty conditions in the morning, increased frequency of wind storms and appearance of certain birds such as *Nyanja*. These indicators were used by communities in the two districts and in all the farming systems except the fisher folks many of whom depend on cloudless conditions and early sun rise as predictors of onset of dry seasons (Figure 5b).

Empirical evidence on climate variability and change

Rainfall

Variations in annual, seasonal, monthly and daily rain fall were considered during analysis where onset and cessation dates together with seasonal length for a period rang-

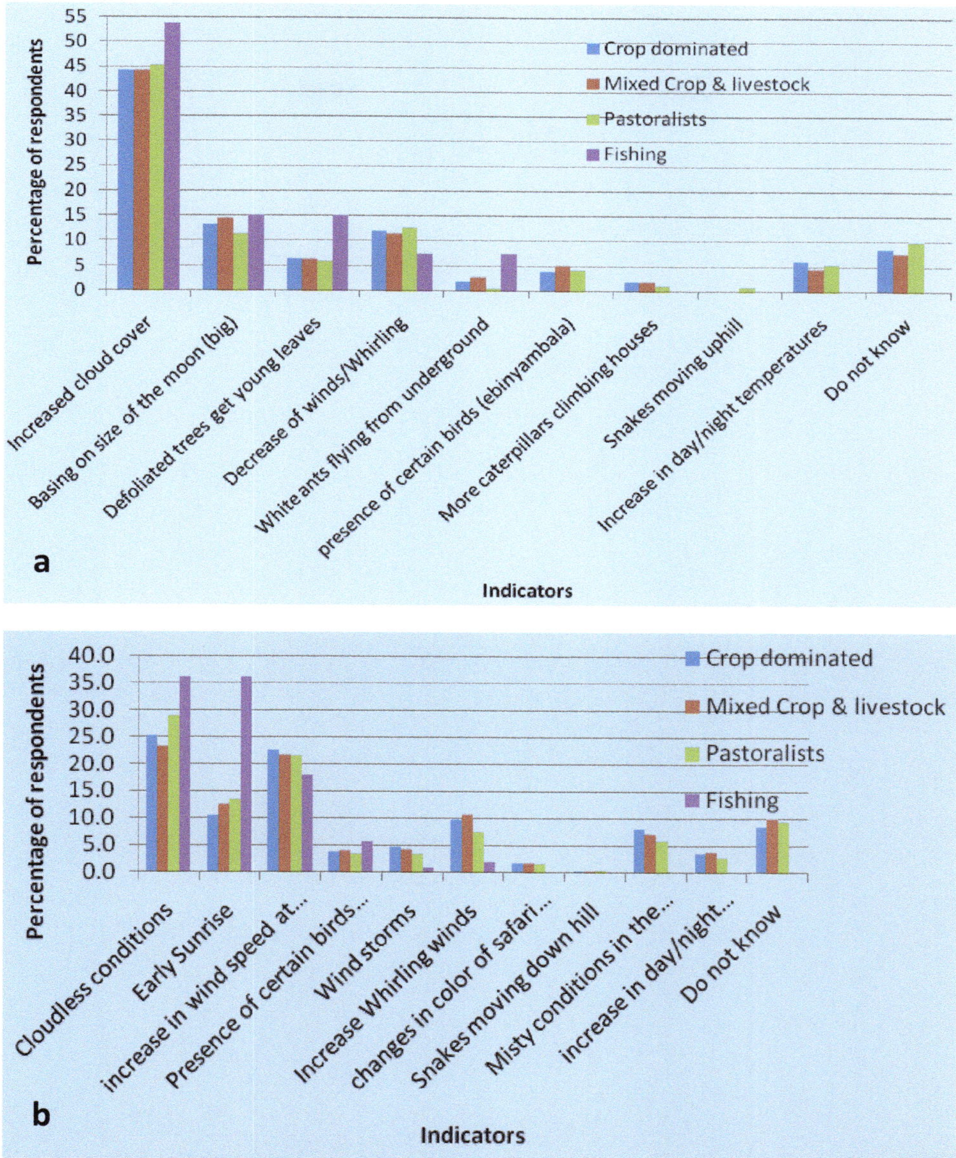

Figure 5. a. Indigenous knowledge of predicting the onset of a wet season reported by communities in the four farming systems. **b.** Indigenous Knowledge of predicting the onset of a dry season reported by communities in the four farming systems.

ing from 1961 to 2010 were deduced.

Annual and seasonal rainfall variations: Annual rainfall over the study area showed a fairly decreasing trend but the trend is not significant in the period 1961-2010 (Figure 6a). Similar non- significant trends are observed for the March to May (MAM) rainfall season (Figure 6b) and the September to December (SOND) season (Figure 6c). MAM seasonal rainfall has been variable and therefore less predictable as compared to SOND seasonal rainfall as observed in the coefficient of variation values calculated per decade from 1961/1970 to 2001/2010 (Table 1).

Monthly rainfall

Decadal monthly averages were calculated per decade from 1961/1970 to 2001/2010 (Figure 7). Monthly values show the merging of the two rainy seasons and the prolonged dry spell of December to February. The end of the first rainfall season and the start of the second rainy season is not clear. This is because the study area lies in a transition zone of the country between areas with a clear bimodal rainfall in south and areas with a unimodal rainfall in the North.

Rainfall seasonal length and onset of rains: There is a negative relationship between length of the growing sea-

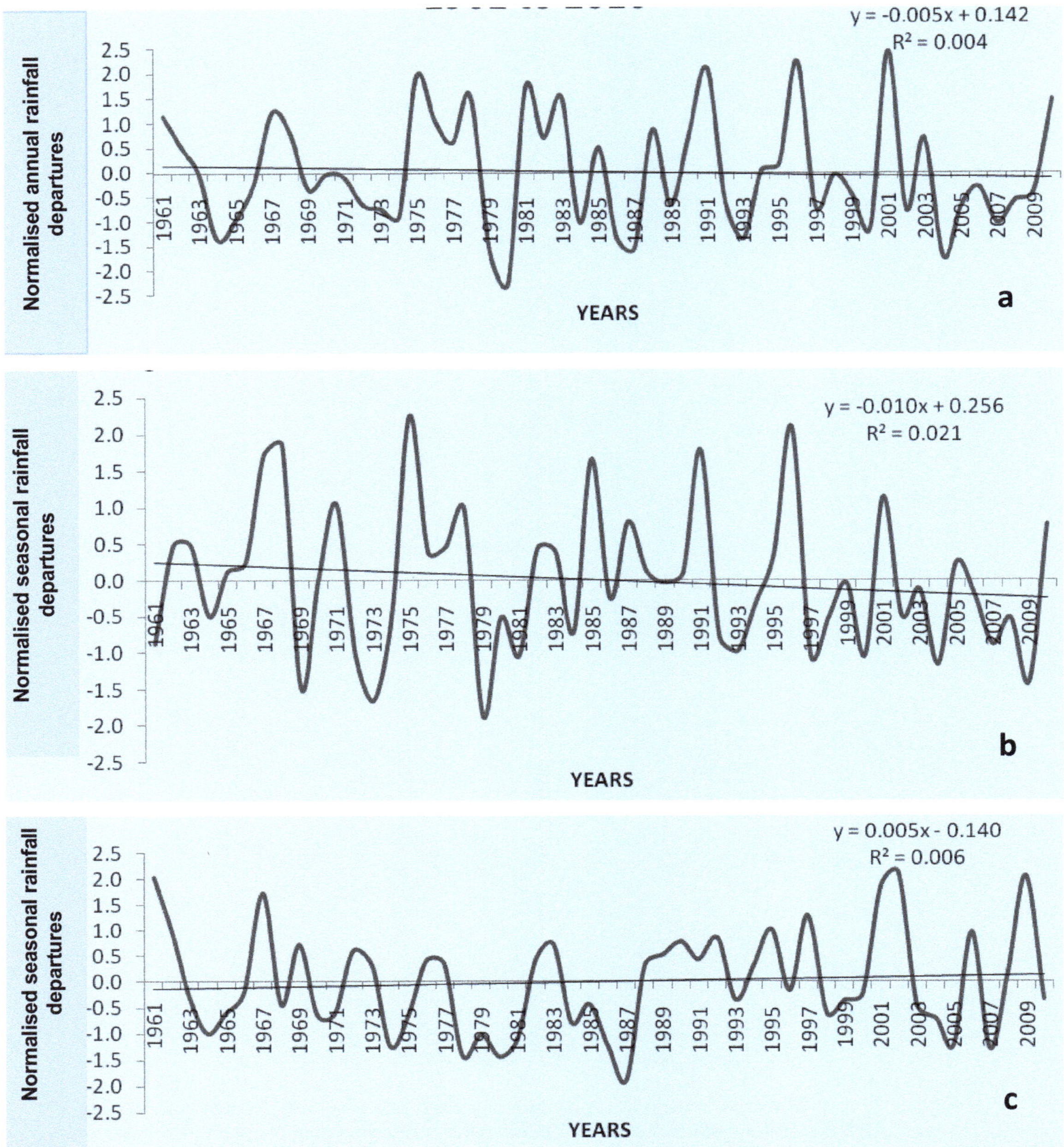

Figure 6. a. Annual rainfall departures for Soroti from 1961 to 2010. **b.** MAM seasonal rainfall anomalies for Soroti from 1961 to 2010. **c.** SOND seasonal rainfall departures for Soroti from 1961 to 2010.

Table 1. Summary of seasonal rainfall per decade from 1961 to 2010.

Period	Mean total MAM seasonal rainfall (mm)	Mean total SON seasonal rainfall (mm)	MAM standard deviation	SON standard deviation	MAM coefficient of variation	SON coefficient of variation
1961-1970	483.2	389.8	137.3	94.4	28.4	24.2
1971-1980	449.8	339.7	172	72	38.2	21.2
1981-1990	474.8	335.4	100.7	87.5	21.2	26.1
1991-2000	448.9	398.7	151.1	45.1	33.7	11.3
2001-2010	420.6	397.6	107.2	88.8	25.5	22.3

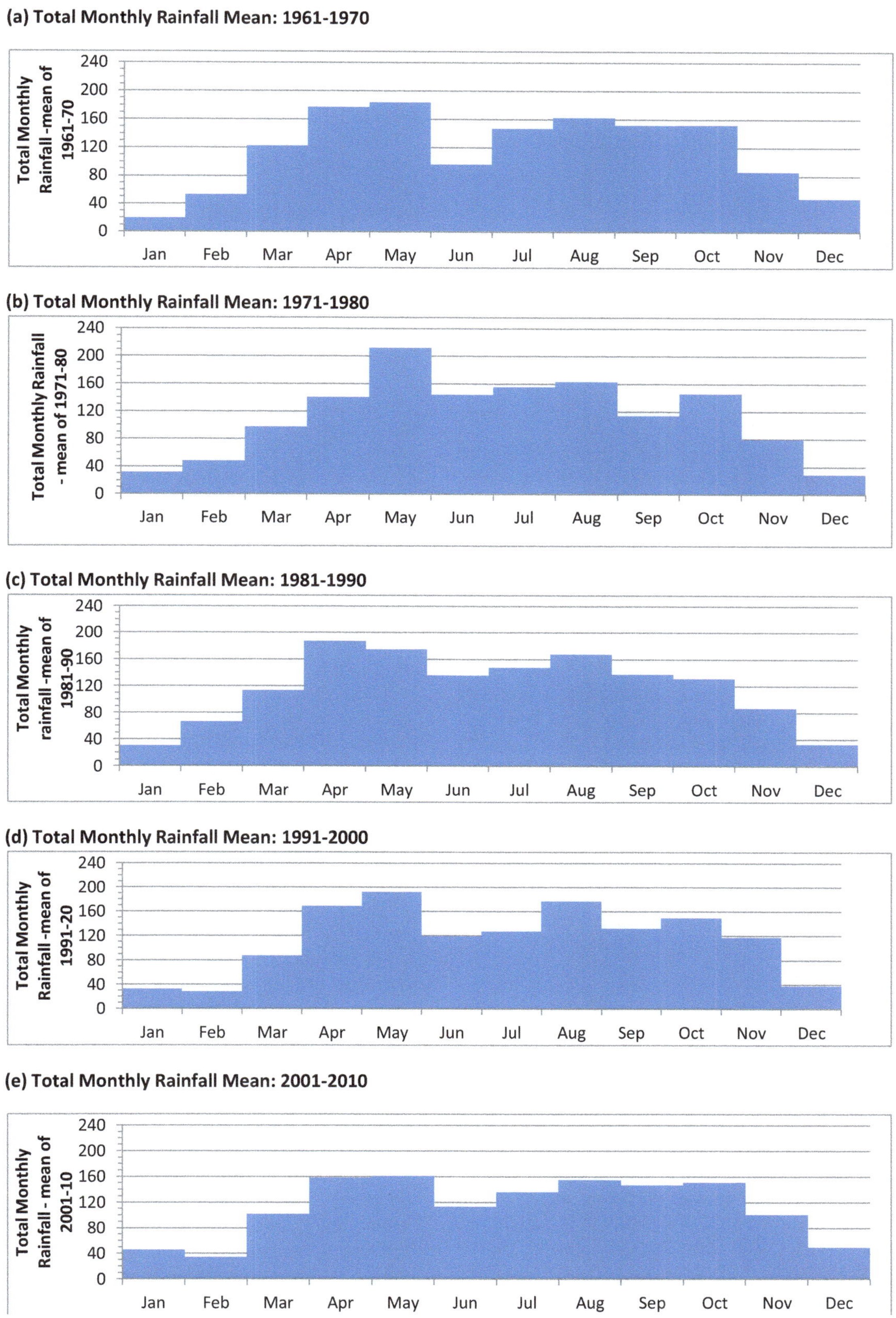

Figure 7. Total monthly distribution of rainfall in the cattle corridor from 1961/1970 -2000/2010: **(a)** Total monthly rainfall mean: 1961-1970; **(b)** total monthly rainfall mean: 1971-1980; **(c)** total monthly rainfall mean: 1981-1990; **(d)** total monthly rainfall mean: 1991-2000 and **(e)** total monthly rainfall mean: 2001-2010.

son and the start of the rains. The relationship is stronger in the second season than the first season (Figure 8a and 8b). This means early rainfall onset especially in the second season prolongs the length of the growing season and therefore farmers should take advantage of planting most crops when an early onset is predicted for the second season.

Dry spells in a rainfall season: The number of dry spells within a season has been increasing for the period 1961-2010. The increase is more pronounced in the first season than the second season (Figure 9a and 9b). This means crop failures are on the rise and hence food insecurity as reported by the community in the household surveys and focus group discussion.

Dry season length: Observations show that the first dry season (June to early August) is shortening while the second dry season (December to February) is intensifying (Figure 10). This also is in agreement with people's perceptions who reported that the dry season has been changing as compared to the wet season.

Temperature trends

There is a significant temperature increase in the study area especially in the minimum temperatures and the mean annual temperatures over a period of 17 years, from 1993 to 2010. The mean annual maximum temperatures show a non significant decreasing trend (Figure 11b), while the mean annual minimum temperatures show a significant increasing trend from 1993 to 2010 (Figure 11a) and the mean annual temperature show a fairly increasing trend (Figure 11c).

DISCUSSION

The community in general perceives climate to have changed in the last 5-10 years in the study area with sunshine and rainfall perceived to be the most changed climatic parameters in the area. They believed that the sunshine had increased in intensity as well as in duration as compared to last 5-10 years and that rainfall had decreased in amounts and distribution. The responses are similar to what was reported in other areas where such studies have been conducted (Thomas et al., 2007; Mertz et al., 2009; Gbetibouo, 2009; Rao et al., 2011) in Kenya and also reported that farmers perceived rainfall to have decreased in addition to increases in temperature. The community's response of sunshine to have changed was mainly seen as increased heat and prolonged periods of continuous sunshine (prolonged dry season). Empirical data from 1961 to 2010 shows that the total annual rainfall as well as seasonal rainfall for the two major seasons has slightly decreased although the trends are insignificant. However, from 1997 to 2010 there is a sharp decrease in the amount of rainfall received especially in the first rainfall season of March to May which is a major season in the study area. Further analysis of seasonal rainfall shows that the number of dry spells with in a rainfall season have increased especially in the first season which agrees with the community perception of un even rainfall distribution. The dry spells result in water stress which leads to crop failure and hence poor output. Empirical data also shows that early rainfall onsets results in longer rainfall seasons as compared to late onsets. This is important for agricultural planning basing on the seasonal predictions of rainfall onsets. This result compares well with Mugalavai et al. (2008) who also reported that predicted rainfall onsets determined the length of the growing season and hence the crops to grow in a particular season in western Kenya. The communities perceived the dry season to have changed most as compared to the rainy season where they had increased in intensity and duration. They reported mainly water shortages for themselves and their animals as well as shortage of food during these intensive dry seasons. However, empirical data shows that the two dry seasons have almost remained the same with the second dry season of December to February becoming more variable as compared to the June to August one.

The communities in the study area perceives tree cutting as a major cause of climate variability and change. They reported that charcoal burning and clearing of bushes for farm land are the major causes of tree cutting which has lead to the observed climatic changes in the area. Very few community members (less than 1%) however thought that industrial fumes and pollution which are major contributors of atmospheric green house gases are the causes of climate variability and change. The small number of people is attributed to the low levels of education of most of the respondents.

Observations from secondary data show that on average annual rainfall has not changed significantly but there are variations in seasonal rainfall, rainfall distribution and increasing dry spells within a rainy season. The first rainfall season of March to May is more variable as compared to the second season of September to November in the period 1961 to 2010. Minimum temperatures have been increasing at a much faster rate as compared to maximum temperatures which show a relatively decreasing trend in the period of 1993 to 2010. On average, the mean annual temperatures have been increasing in the study area since 1993 leading to increased evapotranspiration and hence water stress for their animals and crops.

ACKNOWLEDGEMENTS

The authors wish to thank International Development Research Centre (IDRC), Canada for sponsoring the

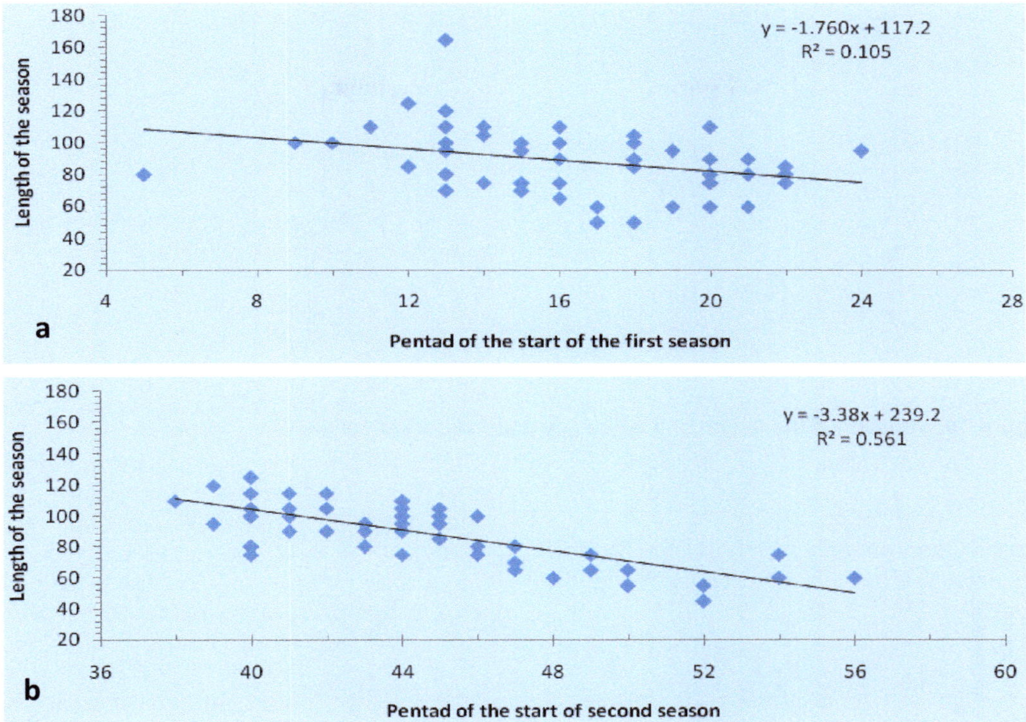

Figure 8. a. Lengthof the first season and start of the rains. **b.** Length of the second season and start of the rains.

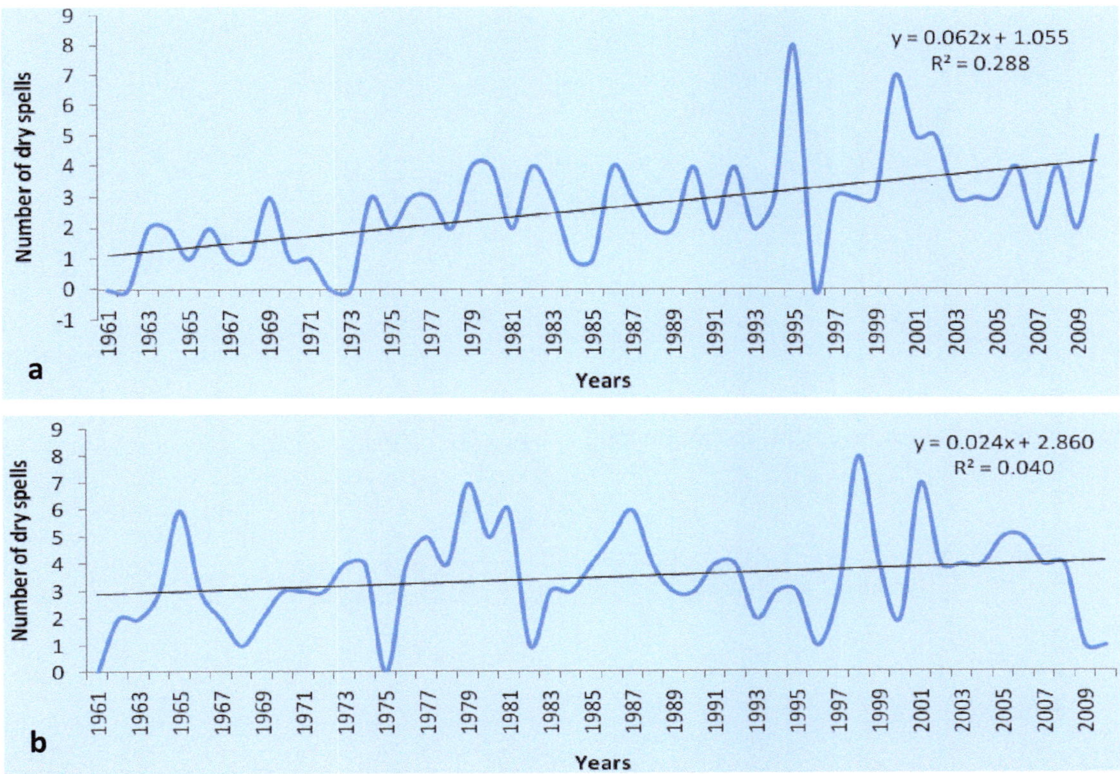

Figure 9. a. Number of dry spells in the first rainfall season. **b.** Number of dry spells in the second rainfall season.

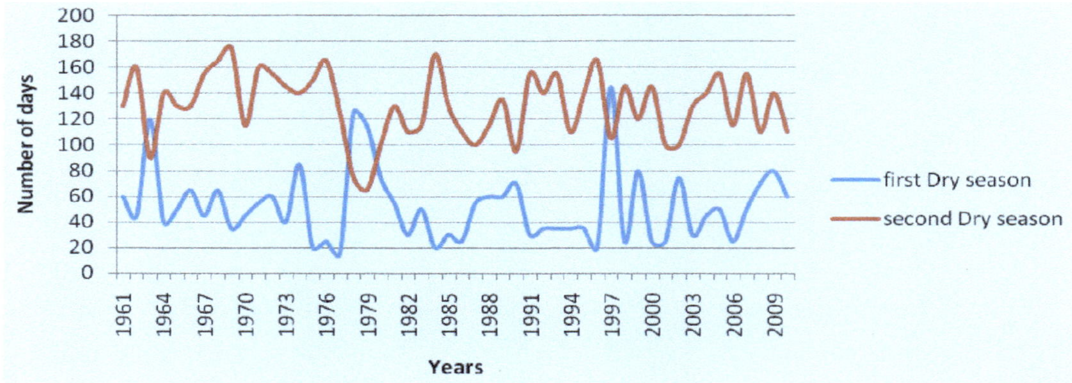

Figure 10. Variation in the length of the first and second dry seasons.

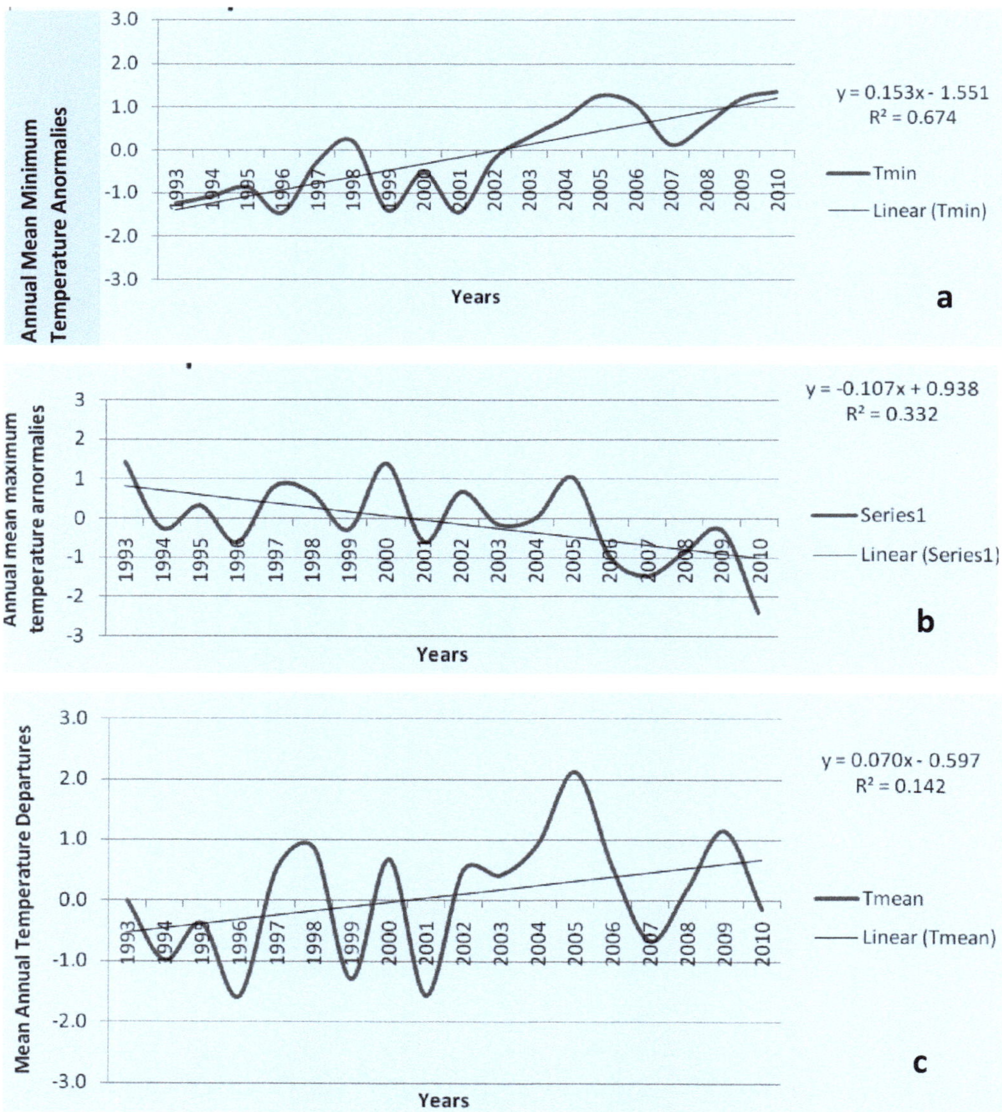

Figure 11. a. Variation of annual mean minimum temperature anormalies from 1993 to 2010. **b.** Variation in annual mean maximum temperatures anormalies from 1993 to 2010. **c.** Variation of annual mean temperature departures from 1993 to 2010.

research activities under the project “Adaptation to the Impact of Climate Variability on Food and Health Security in the Cattle Corridor of Uganda”.

REFERENCES

AIACC (2007). Climate Change Vulnerability and Adaptation in Developing Country Regions. Final Project Report, Washington DC

Alusa AL, Mushi MT (1973). A study of the onset, duration and cessation of the rains in East Africa. Reprints, Inter. Trop. Met. Meeting, Amer. Met. Soc. 3,146.

Dodd DES, Jolliffe IT (1994).Early detection of the start of the wet season in tropical climates. J. Climatol. 14:71-76

Dodd DES, Jolliffe IT (2001). Early detection of the start of the wet season in semiarid tropical climates of western Africa. Inter. J. Climatol. 21:1251-1262

Gbetibouo GA (2009). Understanding farmers' perceptions and adaptations to climate change and variability: The case of the Limpopo Basin, South Africa. IFPRI Discussion Paper No. 00849.

Huq S, Reid H, Konate M, Rahman A, Sokona Y, Crick F (2004). Mainstreaming adaptation to climate change in Least Developed Countries (LDCs). Climate Policy 4:25-43

IPCC (2007). Summary for Policymakers. In: Climate Change 2007: Impacts, Adaptation and Vulnerability. Contribution of Working Group II to the Fourth Assessmen Report of the Intergovernmental Panel on Climate Change, M.L. Parry, O.F. Canziani, J.P. Palutikof, P.J. van der Linden & C.E. Hanson, Eds., Cambridge University Press, Cambridge, UK: 7-22

Kisamba M (2001). Rangelands Management Policy in Uganda.

Mertz OC, Reenberg A, and Diouf A. (2009). Farmers' perception of climate change and agricultural adaptation strategies in Rural Sahel. Environ. Mgmt. 43(5):804-816.

Marengo JE, Liebmann B, Kousky VE, Filizola NP, Wainer IC (2001). Onset and end of the rainy season in the Brazilian Amazon Basin. J. of Climatology 14:833-849.

Mertz OC, Reenberg A, Diouf A. (2009). Farmers' perception of climate change and agricultural adaptation strategies in Rural Sahel. Environ. Mgmt. 43(5):804-816.

Mhita MS, Nassib IR (1987). The onset and end of rains in Tanzania. Proceedings of the first technical conference on meteorological research in Eastern and Southern Africa. Kenya Meteorological Department pp. 33-37.

Mugalavai EM, Kipkorir EC, Raes D, Rao MS (2008). Analysis of rainfall onset, cessation and length of growing season for western Kenya. J. Agric. Forest Met. 148: 1123-1135.

Nicholls N (1984). A system for predicting the onset of the north Australian wet-season. Inter. J. Climatol. 4:425-435

Thomas D, Twyman C, Osbahr H, Hewitson B (2007). Adaptation to climate change and variability: farmer responses to intraseasonal precipitation trends in South Africa. Climatic Change 83: 301-322

UNDP (2007). Human development report 2007/2008: Fighting climate change, Human solidarity in a divided world. UNDP.

UBOS (2012). Uganda Bureau of Statistics; statistical abstract.

Rao KPC, Ndegwa WG, Kizito K, Ayoo A (2011). Climate Variability and Change; farmer perceptions and understanding intra-seasonal variability in rainfall associated risk in semi-arid Kenya. Experimental agriculture 47(2)267-291.

14

An appraisal of the of eco-climatic characteristics in Northern Nigeria

AbdulKadir A.[1] , Usman M. T.[1], Shaba A. H.[2] and Saidu S.[1]

[1]Department of Geography, Federal University of Technology, PMB 65, Minna, Bosso Campus 920003, Niger State, Nigeria.
[2]Strategic Space Applications, National Space Research and Development Agency (NASRDA), P.M.B. 437, Garki, Abuja, Nigeria.

The character of the sub-regional eco-climatic zones has possibly changed as suggested in the dwindling fortune of agricultural productivity, documented changes in vegetation and freshwater resources and rising poverty levels especially in the hitherto productive Northern Nigeria. To determine and map the changes, if any, in these eco-climatic characteristics is therefore important for general planning. This study is using rainfall and temperature data (1950-2006) to determine factors of rainfall effectiveness; onset dates, cessation dates, hydrologic growing season (HGS), monsoon quality index (MQI) and aridity index (AI); summarizing and ranking them using numerical identifiers for the interpretation of the various moisture situations, to assess the eco-climatic characteristics in the northern Nigeria. A multi-temporal database has been developed for the factors and classes have been defined using quantitative definitions for the time series. The point data have been transformed to spatial data and subjected to geospatial analysis. The results so far reveal variability in the effects of each factor on eco-climatic zonation and it is anticipated that integrating these derived climatic parameters and remotely sensed (vegetation dynamics) data using GIS techniques, is crucial for water management and attainment of food security.

Key words: Eco-climatic, moisture effectiveness, surface, principal component analysis, environmental quality.

INTRODUCTION

Globally, increased rainfall variability, climate change and increased human activities have being and will continue to adversely modify the natural resource and the environment as a whole. Climate exerts a strong influence on relevant socio-economic sector and people's livelihood in sub-humid and semi arid zones of northern Nigeria where the drying condition have being on the increase. These changes are occurring at unprecedented rate thereby drastically reducing bio-productivity of the physical environment. This uncoordinated and undocumented process of change is a treat to food security and apparent poverty level typical of northern Nigeria. Environmental degradation also contributes significantly to food insecurity (FAO, 1996). Desertification of the northern region has being severely altering the traditional climatic and eco-climatic zones resulting to recurring crop failure, giving rise to sharp increase in the prices of agricultural products, hunger, sickness, malnutrition, starvation and livestock death in extreme cases.

The geographical location of the northern Nigeria is generally prone to drought, wind and water erosion, thereby promoting geo-environmental degradation. This region is a marginal land, where crops experience moisture stress and often fail mainly as a result of the effect of the repeated prolonged drought. Although agriculture is the major occupation in this region, desertification, repeated drought, crop failure and southward migration of the inhabitant signal is the fact that food security is threatened.

The escalating variability in rainfall distribution, deforestation and degradation of natural resources in the zone has significant eco-climatic and socio economic impacts. Understanding of the nature and future climate change has led to a realization that significant future impacts are inevitable and increased efforts towards understanding the process of adaptation to the threatened impacts are required (Adger, 2001). These cumulative effects of rapid depletion of natural resources in the region, through over exploitation and variability in the distribution pattern of rainfall have drastically affected the quality of the biotic and abiotic components in the zones.

Accordingly, this constitutes potential treat for the attainment of sustainable develop-ment. Peter (1998) states that a wide range of ecological and human crises result from inadequate access to, and the inappropriate management of the resources. Conse-quently, adequate information on the eco-climatic charac-teristics is vital in proffering solutions, combating the problem of food security in the entire northern Nigeria and a pathway for the attainment of sustainable socio-economic develop-ment in the region.

Rainfall is the principal nourishment for the terrestrial eco-system at the same times the requirement for plant and animal growth; consequently, it is a primary determining factor for eco-climatic and agro-climatic zonation. The eco-climatic and agro-climatic maps used in the zone are still the traditional maps base on conventional rainfall totals and mean temperature values. Adefolalu (1986b), Usman (2000), Zeewdu and Peter (2004) and Srikantha and Uditha (2004) affirmed the need for additional rainfall characteristics in identifying plant moisture stress other than rainfall amount. Furthermore, rainfall totals are on the increase while rainy days are declining (Tennant and Bruce, 2002). Therefore, defining moisture condition using rainfall amount may be grossly inadequate and inefficient to capture the effect on eco-climatic zonation as might be expected using other characteristics of rainfall; onset, cessation dates, hydrological growing season, moisture quality and aridity index.

Then, challenges of erratic, recurring rainfall variability have adversely affected moisture quality in the zone thereby triggering aridity. David and Chasca (2005) con-sidered climate change as 'special' amongst livelihood disturbing factors in the developing world. In Nigeria, history has proved that drastic changes in rainfall pattern have effect on agricultural productivity, disappearance of the famous groundnut pyramid in this belt follow rainfall fluctuation of 1970s.

Lázaro et al. (2001) mentioned that in order to understand the behaviour of ecosystems in semi-arid areas, rainfall must be analyzed over time. These have necessitated the appraisal of the eco-climatic characteristics other than rainfall totals and ranges whose maps has being in use over a century. Similarly, Sui-Lin and Chi-Fai (2010) identified a multifaceted strategy for the applications of science and technology, as a part of the solution, in defeating world hunger.

Also, Kufoniyi and Akinyede (2004) concluded that the fuel driving the engine of growth and sustainable development of any nation is the nation's access to reliable and adequate geospatial information. Conse-quently, the present study generate, compare and integrates spatial variability of eco-climatic characteristic to derived moisture effectiveness map of the zone, as pathway towards the attainment of food security and sustainability of the physical environment in northern Nigeria.

Study area

Northern Nigeria is located between Longitudes 3° and 15° East and Latitudes 9° and 14° North. Northern Nigeria is a political definition use to describe all states that are fully located or partially located (middle belt) in the northern portion of the country. The states located in this zone are Sokoto, Kebbi, Zamfara, Kastina, Kano, Jigawa, Yobe, Borno, Gombe, Bauchi, Kwara, Plateau, Adamawa, Niger, Federal Capital Territory (FCT), Nassarawa, Taraba, Kogi and Benue.

The climate is characterized by alternate wet and dry seasons in response to the changes in pressure patterns, the rainy season in this region is associated with late onset and earlier cessation, the onset and cessation are also characterized by destructive storms which destroy life and property. The seasonal and latitudinal variations affect diurnal and seasonal temperature ranges, the highest maximum air temperature is recorded in the northern part usually areas north of lat 9° and occur in March /April and minimum temperatures are recorded in December/January North of lat 9°N.

The general relief of this belt is between 300 to 900 m, except the Niger-Benue trough, Sokoto and Chad Basins that are below 300 m. Northern Nigeria is dominated by savanna vegetation types; Guinea, Sudan and Sahel savanna, the density of trees and grasses decrease northwards res-ponding to climatic conditions. Agriculture is the most dominant economic activity in the region.

Table 1. Multi-temporal data summary and interpretation scheme.

Factor classification	Interpretation basis (C, HGS, AI,)	Interpretation basis(Ø & MQI)
1	5&4 ≥ 70%,1,2,&3 ≤ 30% ,	1&2 = 0, 3 ≤ 20 , 4 ≥ 40 & 5 ≥ 40
2	5&4 ≥ 50,3 ≥ 40 , 1&2 ≤ 20	1, 2 &3 ≥ 40 , 4 ≥ 40 & 5 ≤ 20
3	5&4 ≥ 20, 3 ≥ 50 , 2 &1 ≤ 30	1&2 ≥ 20, 3 ≥ 50 , 4 &5 ≤ 30
4	5, 4 &3 ≥ 40 , 2 ≥ 40 & 1 ≤ 20	1&2 ≥ 50,3 ≥ 40 , 1&2 ≤ 20
5	5& 4 = 0, 3 ≤ 20 , 2 ≥ 40 & 1 ≥ 40	1&2 ≥ 70%,3,4,&5 ≤ 30%

METHODOLOGY

Daily rainfall and temperature records (1950-2006) for 12 selected meteorological stations: Minna, Zaria, Yola, Maiduguri, Bauchi, Sokoto, Yelwa, Kano, Jos, Gausau, Nguru, and Katsina were collected. These data were used to derive vital rainfall characteristic and moisture quality in the belt: onset, cessation, hydrological growing season (HGS), moisture quality and aridity index in zones. These were determined using intra-seasonal rainfall monitoring index (IRMI-) Usman and Abdulkadir (2012), moisture quality index (MQI-) Usman, 1999) and potential evapotranspiration (ET$_{O}$ Hargreaves and Samani, 1985). The computed derived parameters were summarized, analyzed and classified using SPSS software. Numerical identifiers were adapted for the interpretation of the various moisture effectiveness classes (MECs): very good, good, fair, poor and very poor zones (1, 2, 3, 4 and 5). Multi-temporal database was developed for the five factors (AI, C, HGS, MQI and Ø). Classes were defined for each factor using multi-temporal data summary and interpretation scheme for quantitative definitions of time series data, 1950-2006 (Table 1).

A geospatial database was developed using the above interpretation scheme for visualizing each factor's rainfall effectiveness: onset (Ø), cessation dates (C), hydrologic growing season (HGS), aridity index (AI) and monsoon quality index (MQI). In addition, a geo-reference base map for northern Nigeria was digitized and the point data were transformed to spatial data. All the point data (multi-temporal) for all the factors were transformed into X, Y and Z (longitude, latitude and factor) using Kriging geo-statistical defaults in SURFER 8 to produce an accurate grid of the entire factors. These grids were converted to ASCII XYZ data and imported into Idrisi Andes where they were converted to IDRISI vector point files. The vector point files were interpolated to full surface layers for the five factors. These surfaces were them exported to ArcGIS 9.2 through Idrisi Andes Geotiff conversion option. These were subjected to further geo-spatial analysis using spatial analyst module. The base area map was used to extract the study area from the surfaces and the surface maps were reclassified using moisture effectiveness classes (MECs) to derived moisture effectiveness zones (MEZ). Principal component analysis was used to compress the data, identify the variance and relationship between the eco-climatic factors, finally the relation was used to integrate and reclassified the variance in the eco-climatic characteristic using supervised classification.

RESULTS

The derived eco-climatic characteristics (AI, C, HGS, MQI and Ø) maps of the northern Nigeria reveal climatic variability typical of the rainfall distribution in the region. López et al. (2008) observed that water availability is discontinuous and highly variable. These factors captured moisture stress of the sub-humid and dry ecosystem at different levels, such that the quality of the biological and economic resources uncertainty are apparent. These surface maps depict the spatial variability that characterized the ecological zone, five or four MECs were identified for the entire factors and the moisture effectiveness generally decreases north wards. In addition, there is variability in the number MECs captured; AI, MQC and HGS captured five while onset and cessation captured four classes each.

The surface index maps were quantitatively reclassified, to depict the spatial extent of the eco-climatic factors moisture effectiveness zones, five or four MEZs were identified for all the factors and the moisture effectiveness generally decreases north wards. In addition, there is variability in the number of MEZs captured; AI, HGS, MQI capture five MEZs while onset and cessation captured four classes each. Onset map uncover the late effective onset of rain that characterized the region, as there was no earliest onset rainfall zone while cessation does not indicate extremely early rainfall cessation in the zone. Furthermore, these derivative maps show the extent of each MEZ for each factor; the worst classes are dominant to extreme north while the best are to the south and the central parts are mostly moderate as evident in the derived eco-climatic maps (Figures 1 to 5).

Aridity index map recognized five moisture effectiveness zones across the belt; lowest, low, high, very high and extremely high.

MQI portrayed five MEZs: extremely deficient, very deficient, deficient, adequate and abundant moisture effectiveness zones. The hydrological growing season unveils five moisture effectiveness zones: longest, long, short, very short and extremely short hydrological growing season.

Real monsoon onset effectiveness map depicted four moisture zones: extremely late, very late, late and early indicating the severity of rainfall onset across the region. Similarly, cessation moisture effectiveness map identified four moisture effectiveness zones: latest, late, early and very early cessation zones. Onset map reveals the late effective onset of rain that characterized the whole belt, as there was no best onset moisture effectiveness class (very good), particularly the northeast and northwest have very poor rainfall-related onset. Cessation shows there was no zone with very poor cessation moisture effectiveness zone, signifying that cessation is less stressful across the zone. Furthermore, these derivative maps confirms increase moisture stress northwards; the

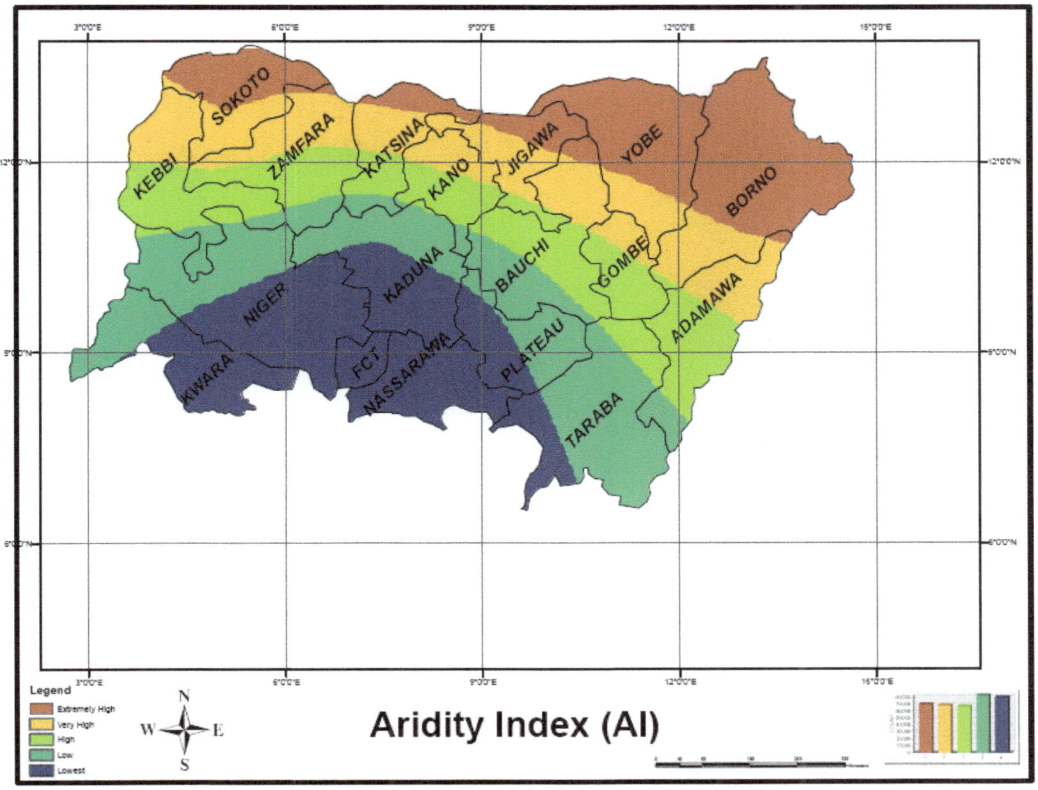

Figure 1. Aridity index map.

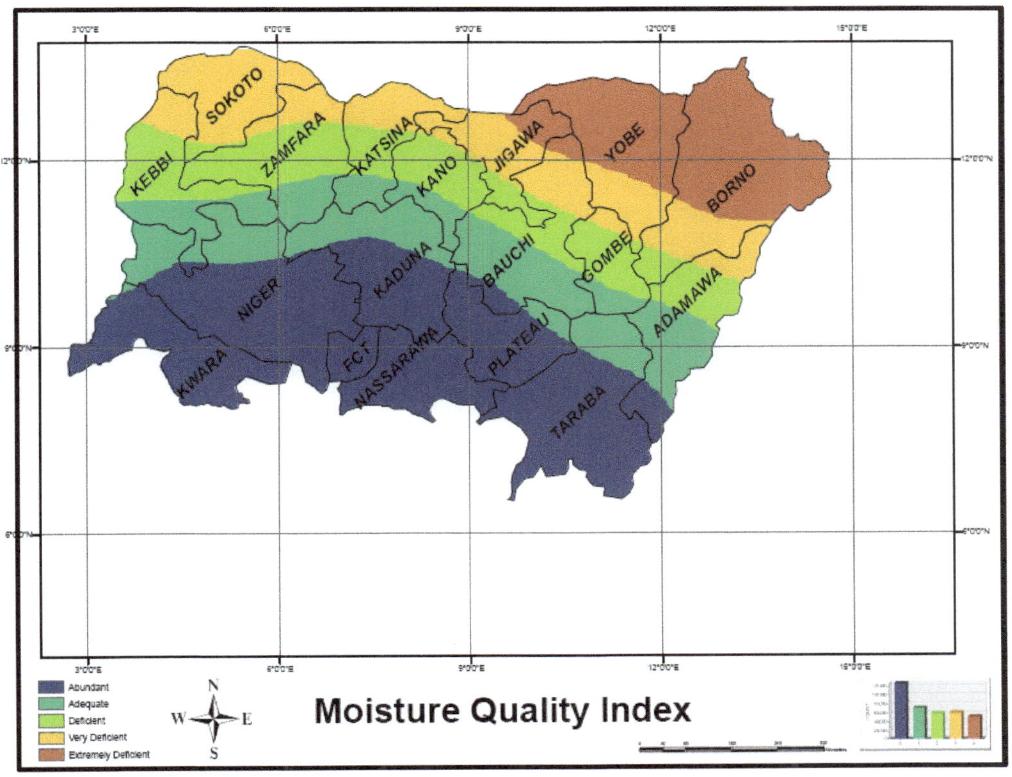

Figure 2. Moisture quality index map.

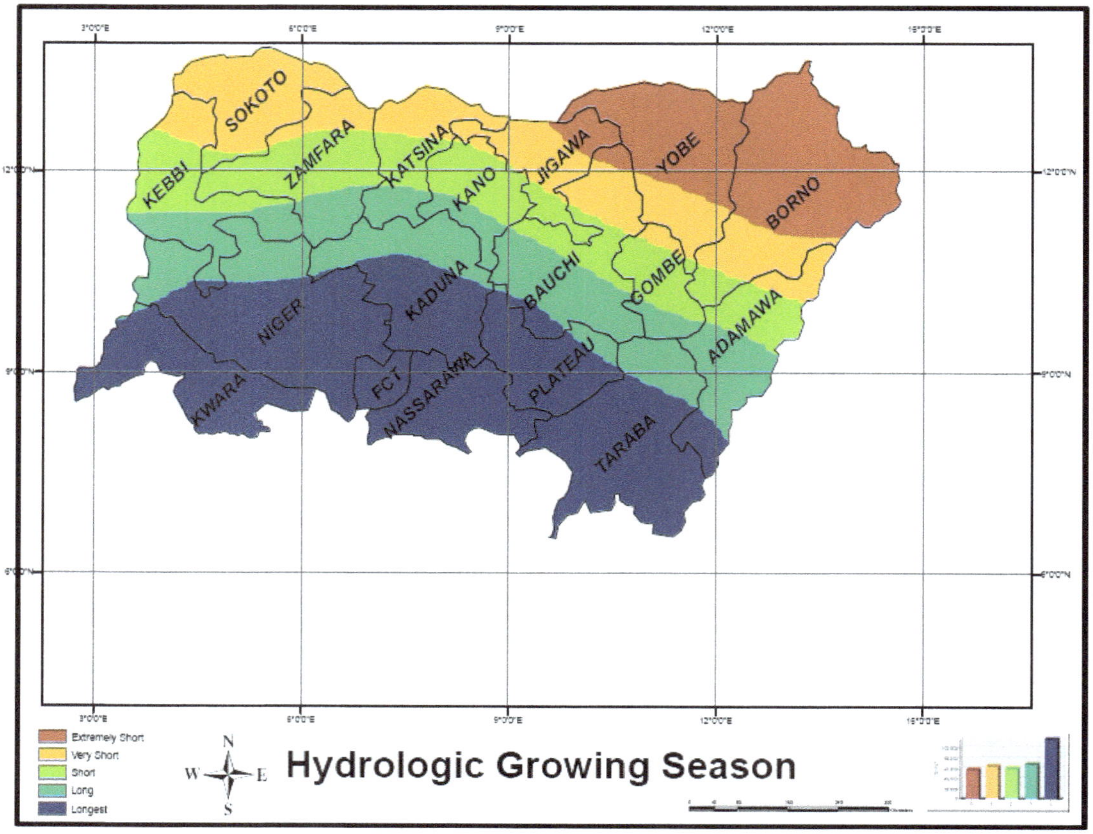

Figure 3. Hydrologic growing season map.

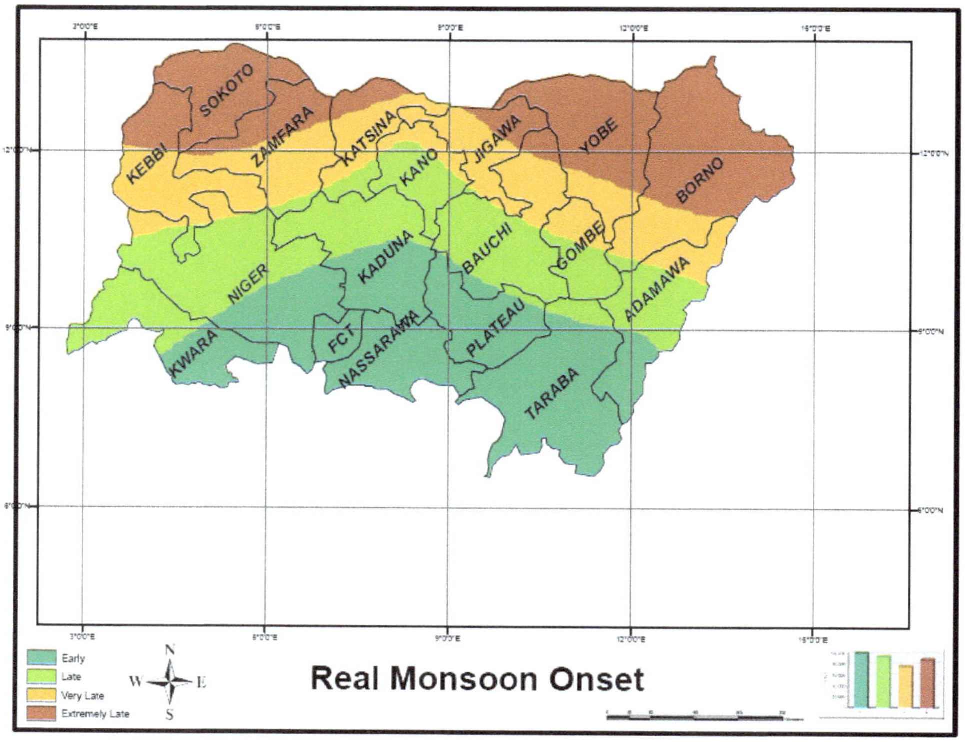

Figure 4. Onset map.

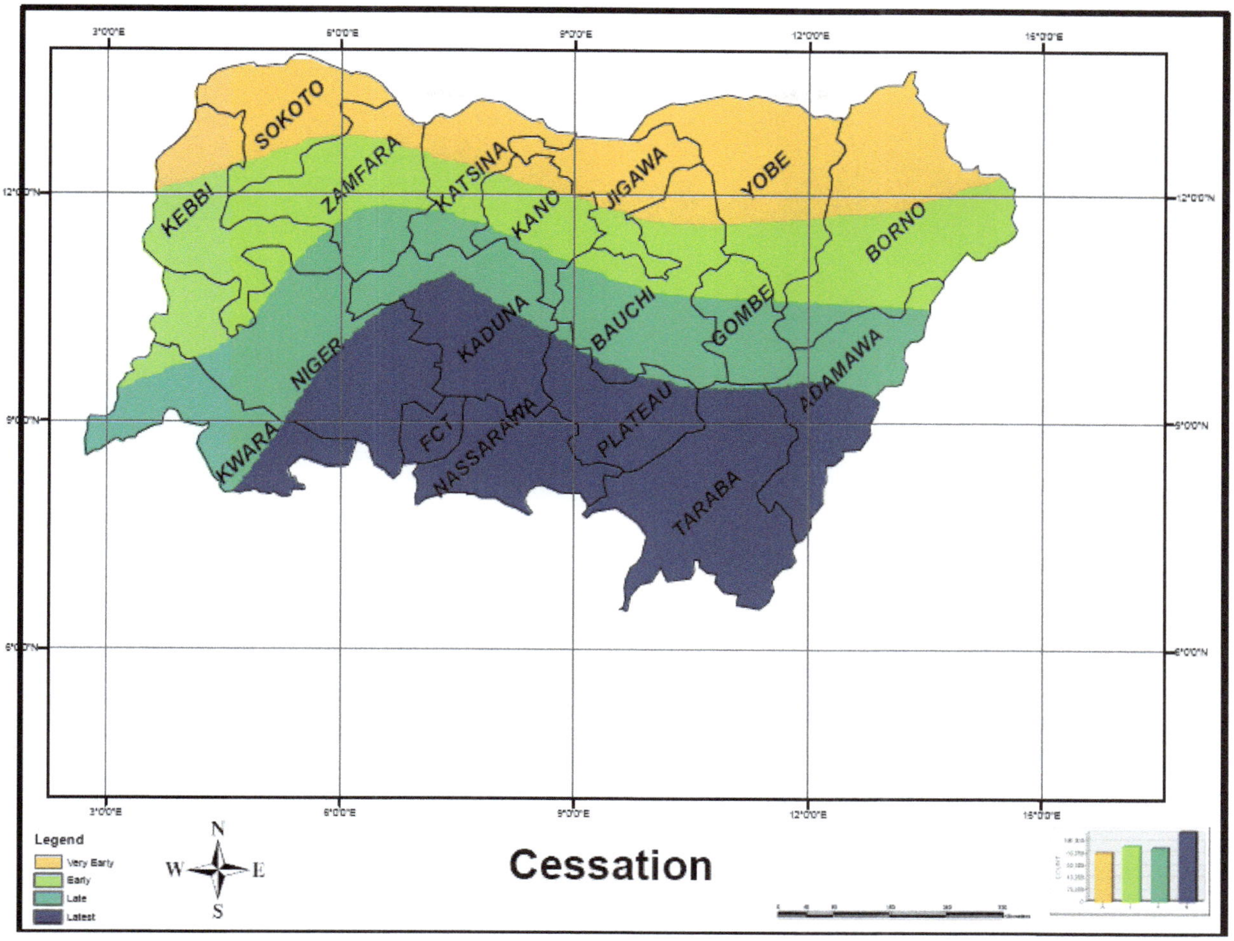

Figure 5. Cessation map.

worst classes are dominant to extreme north, while the best are to the south and the central parts are mostly moderate.

Principal component analysis

Principal component analysis (PCA), identified the union that exists between the two extreme MEZs (the best and worst) although, the intersection shows the relationship that exists between all the factors as indicated in the overlap of the intermediate classes. This signal the fact that additional techniques will be essential for the identification and delineation of these middle classes as apparent from PCA (Figure 6). To resolve this, The PCA was reclassified using supervised classification which uses union and intercessions of the various factors MEZs as an essential fact for integrating and reclassifying the factors and derives moisture effectiveness map (Figure 7).

Moisture effectiveness map

The intersections of the various factor's MEZs serve as basis for integration and reclassification of the PCA to derive moisture effectiveness maps (MEM). Supervised classification of the PCA identified five MEZs: extremely deficient, very deficient, deficient, adequate and abundant moisture zones as evident in derived moisture effectiveness map Figure 7.

The MEM identified five moisture quality zones, extremely deficient, very deficient, deficient, adequate and abundant. Abundant moisture zone comprises, FCT, Nassarawa and southern part of Kwara, Niger, Plateau, Taraba and Kaduna State. Adequate moisture zone encompasses Northern Kwara, Niger, Plateau, Taraba and Kaduna State and southern Bauchi and Adamawa state. Deficient moisture zone covers Northern Bauchi and Adamawa state, southern Gombe, Kebbi, Zamfara and Kano State. Very deficient moisture zone include northern Gombe, Kebbi, Zamfara and Kano State and southern Sokoto, Katsina, Jigawa, Yobe and Borno state. The extremely deficient moisture zone comprises the extreme north. The moisture effectiveness decreases progresssively northwards (the lower latitude are wetter than the higher ones as evident in Figure 7); thereby providing fundamental information for moisture effectiveness adaptation.

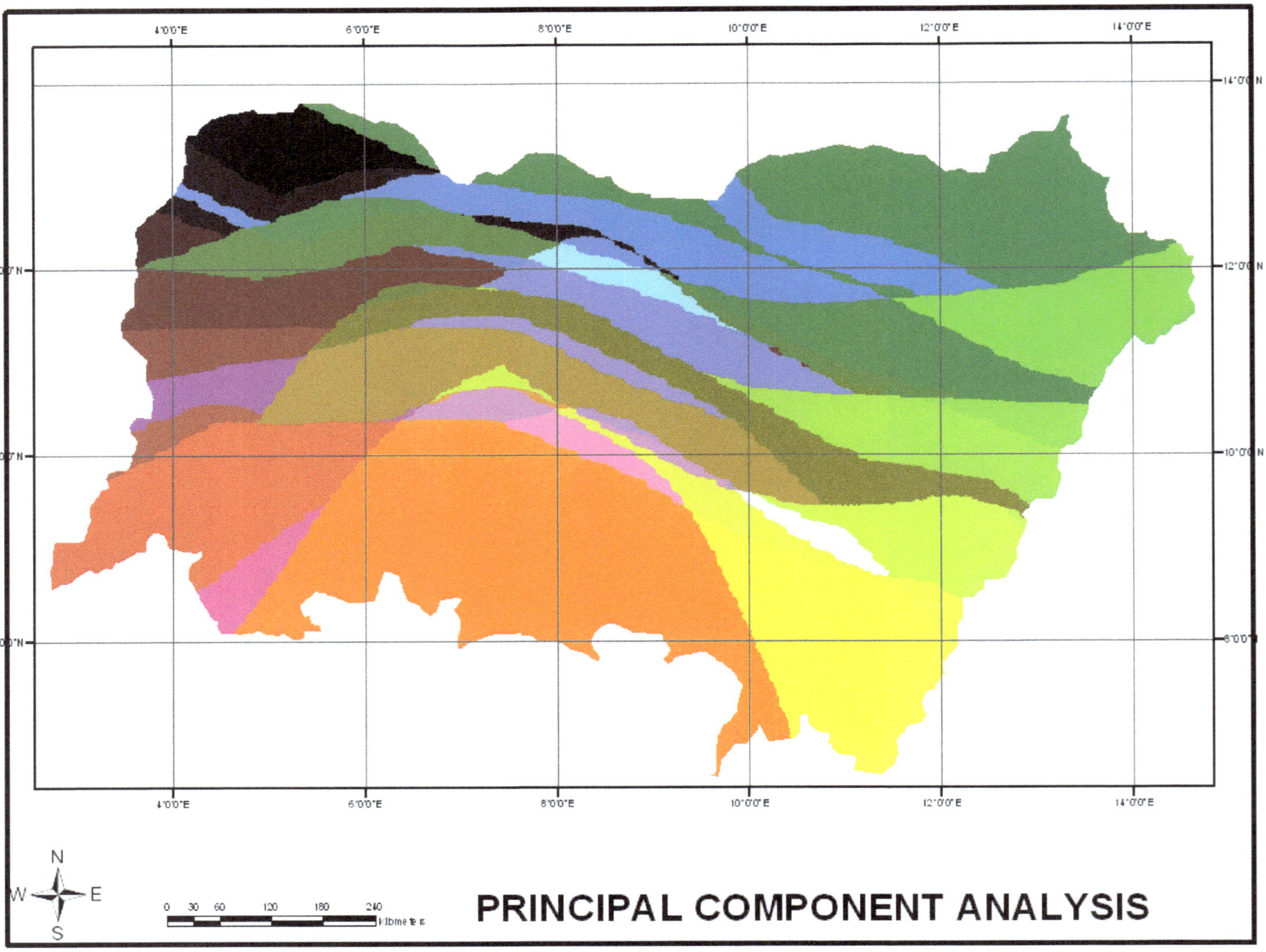

Figure 6. Principal component analyses.

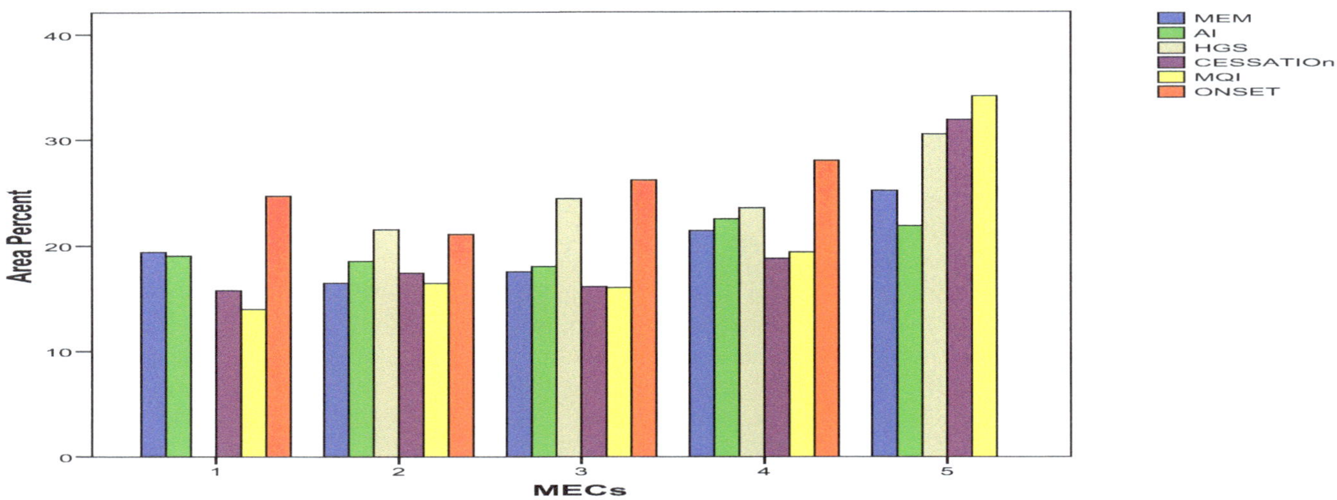

Figure 7. Eco-climatic factors moisture effectiveness zone and MEM comparison.

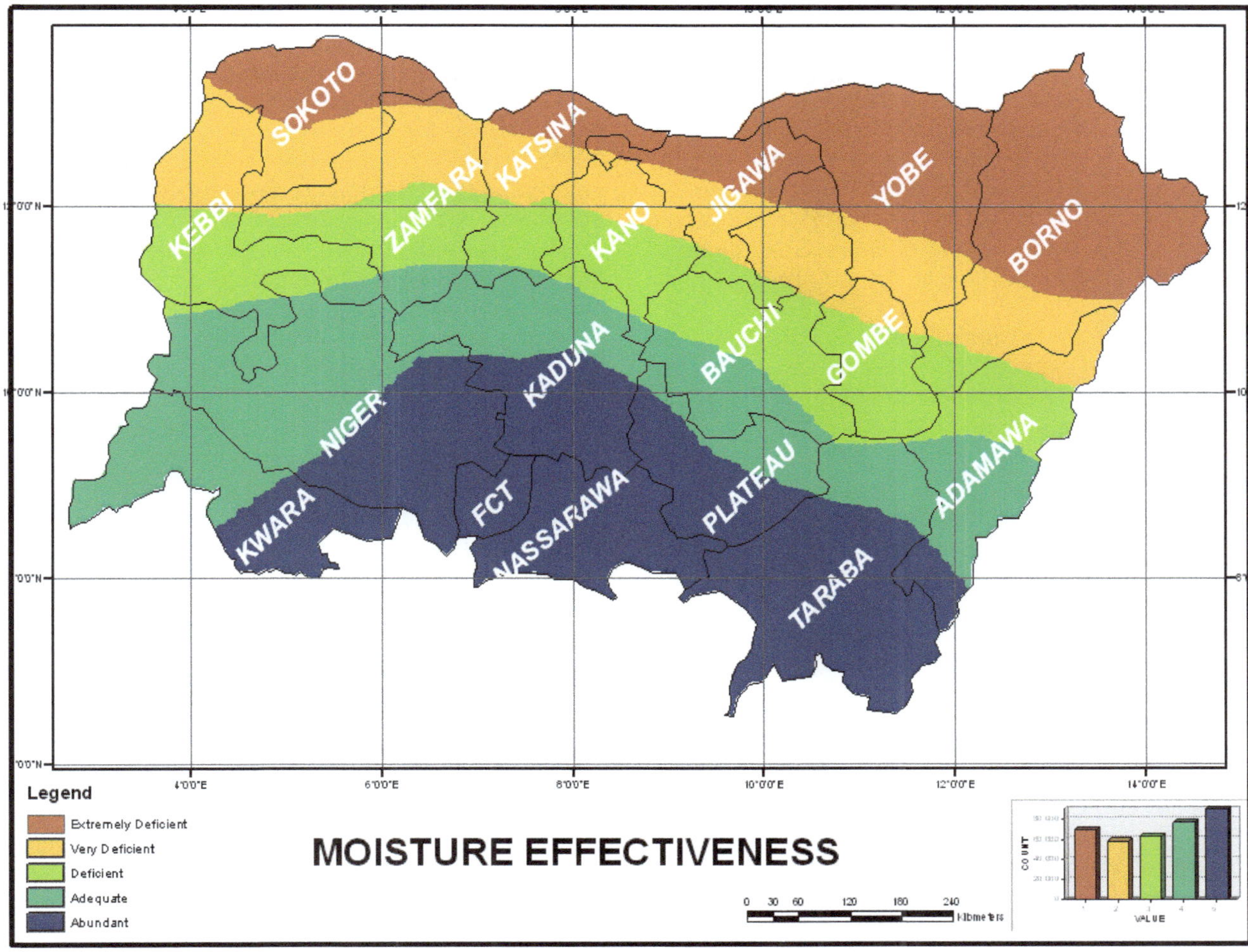

Figure 8. Moisture effectiveness map (MEM).

Table 2. Pearson correlation relationships between moisture effectiveness and eco-climatic factors.

Eco-climatic factor	Relationship with MEM
Aridity index	0.8
Cessation	0.9
Hydrologic growing season	0.9
Onset	0.8
Moisture quality index	0.9

Furthermore, comparing the spatial extend of moisture effectiveness zones and eco-climatic factors using multiple graphs (Figure 8), unveil the spatial variability associate with the entire factors moisture effectiveness Zones and the moderating effect of MEM. This justified the need for integration and classification in smoothing the variability that characterized the moisture effectiveness as well as providing adequate information on the state of the environment. Understanding the state of the environment is crucial for the attainment of sustainable development; this will enhance the sustainability of biotic and the abiotic resources thereby providing optimal environment for sustainable agriculture.

An investigation of the relationship that exists between MEM and entire factors of eco-climatic characteristics map shows that there is strong positive relationship (≥0.8) between the factors and the derived eco-climatic map (Table 2). Also, the regression analysis confirms that the MEM is a good predictor of the independent variable since R square change is 0.9, thatis, 90% of MEM can be explained by the factors. Hence, a pathway for achieving adequate information to be applied in developing sustainable practices as climatic variation is fundamental to environmental quality and food security in the zone.

DISCUSSION

The factors, moisture effectiveness indices and maps (1 to 5) illustrate the spatial extents of each factor moisture effectiveness classes throughout northern Nigeria and the challenges for the sustainability of the physical environment. The physical consequences of moisture deficit and steady deterioration of the physical environment has shown the need to provide accurate information designed towards the achievement of sustainable development. With this result, the entire northern Nigeria is vulnerable to crop failures as vast proportion of the region is characterized by varying levels of deficient moisture quality, late onset, early retreat of rainfall and shorter hydrologic growing season that were hitherto obvious. Srikantha and Uditha (2004) identified decline in the numbers of rainy days.

These are disastrous to agriculture, as delayed onset often leads to late planting of crops, while premature cessation and shorter HGS leads to wilting and dryness of the crops before maturity and consequently, poor yield. Onset, cessation and length of the hydrologic growing season (HGS) are more critical to plant (Adefolalu, 1986a). High aridity index and moisture quality reduce bio-productivity and all these endanger food security. Generally, the entire surface map affirms that the eco-climatic features are extremely variable and potentially associated with different forms of economic and environmental disruption (stress). The scenario calls for identifying reliable and adequate information on the state of the environment, using derived remotely sense data in mitigating water-related emergencies, providing safe access to water and combating desertification which is typical feature of Northern Nigeria.

The PCA spelt out the union between the features of the best and worst MEZs while the intersection between the various classes defining the relationship between the various factors. The lack of union particularly between the intermediate MEZs and the intersection between these classes also signals the fact that all the factors have certain unique features and have some relationship as evidenced in the overlap of the MEZs. Leonard and Felix (1998) showed that temporal and spatial characteristics are vital in change detection and mapping. Thus, this necessitates the use of other available techniques for smoothing this variability to provide adequate information on the state of the environment for enhanced rural livelyhood. Generally, MEZs do not vary in the same way. Specifically, very early onset does not necessarily imply very good moisture quality or aridity index. As a result, integration of all the factors with biophysical state of the environment (derived from satellite images) using empirical relations should be the basis for improved management of the vital resources (land and water). This will give an insight into specific management actions that will minimize the progressive ecological degradation and enhance food security in the region and the county at large.

The derived MEM indicates that moisture effectiveness across the region decreases progressively northwards as expected (Figure 7). Underscoring why the extreme north is characterized by extremely deficient moisture and very deficient moisture zone, this extends to about lat. 12°N and lat. 10°N in the northwest and northeast, respectively. The integration of the eco-climatic characteristics using GIS provide fundamental information for moisture stress and adaptation strategy in the region, is a signal by the strong positive correlation between the factors and MEM along with a regression analysis returning R^2 to 90%. Since climatic variation is fundamental to environmental quality and food security, such information will enhance the sustainability of biotic and the abiotic resources and provide an optimal environment for sustainable agriculture. Furthermore, the general positive relationships that exist between the factors and moisture effectiveness map confirm that all these factors are fundamental in the identification of soil moisture stress in the zones.

The eco-climatic factors moisture effectiveness maps visualized illustrates the spatial extents of each factor moisture effectiveness zones throughout the northern Nigeria and the challenges for the sustainability of the physical environment. By this result, the entire northern Nigeria is vulnerable to crop failures as vast proportion of the region is characterized by deficient moisture quality, late onset, early retreat of rainfall and shorter hydrologic growing season that were hitherto obvious. These are disastrous to agriculture: high aridity index and poor moisture quality reduce bio-productivity and all these endanger food security. Generally, the entire surface map affirms that the eco-climatic features are extremely variable and potentially associated with different forms of economic and environmental disruption (stress). The scenario calls for the need for adoption of space technologies for identifying reliable and adequate information on the state of the environment.

Conclusion

Surface interpolation of the derived parameters (onset, cessation, hydrologic growing season, aridity index and vegetation index) offered fundamental information for visualizing and understanding the eco-climatic characteristic. The plotted and classified eco-climatic characteristics provide adequate information on the moisture effectiveness in the zone. Principal components analyses compress the factors into single maps and reveal the natural variability that are associates with the factors, its reveals the complex relationships that exist among the factors. A point worthwhile is that the entire surfaces of eco-climatic characteristics that can be incorporated into ecological consideration in the resource management strategy of the northern Nigeria. The integration of the factor unveils the union and intersection that exists between the moisture effectiveness zones. These derived eco-climatic indices are indispensable for identifying, delimiting and describing the state of the environment in quantitative terms. Finally, the study confirmed that dynamics in the eco-climatic parameters moisture effecttiveness zone is fundamental for the identification of the current state of the environment in Northern Nigeria.

Consequently, necessitating the integration of derived remotely sense data for sustainable agriculture practices and other socio-economic in northern Nigeria. Principally, remotely sense data are fundament for the attainment of accurate, up-to-date and adequate information on the state of the environment. The technical procedure for doing this and the expected outcome are illustrated to demonstrate the potential for filling the gap created by outdated, inadequate and inaccurate information inherent in the use of existing classical climatic zones.

ACKNOWLEDGEMENT

Our appreciation goes to Nigeria Meteorological Agency (NIMET) and Environmental Management Program, Federal University of Technology Minna for providing the pre-requisite data for the research.

REFERENCES

Adefolalu DO (1986a). Further aspects of Sahelian drought as evident from rainfall regime of Nigeria. Meteorol. Atmos. Phys. 36(3-4):277-295.

Adefolalu DO (1986b). Rainfall Trends in Nigeria. Theor. Appl. Climatol. 37: 205-219.

Adger W (2001). Scales of governance and environmental justice for adaptation and mitigation of climate change. J. Int. Dev. 13: 921-931.

David SGT, Chasca T (2005). Adaptation to Climate Change: Perspectives Across Scales. Glob. Environ. Chang. Part A. 15(2):115-124.

FAO (1996). Rome Declaration on World Food Security and World Food Summit Plan of Action. World Food Summit held in Rome Italt http://www.fao.org/docrep/003/w3613e/w3613e00 14/7/2013.

Hargreaves GH, Samani ZA (1985). Reference crop evapotranspiration from temperature. Appl. Eng. Agric. 1(2):96-99.

Kufoniyi O, Akinyed JO (2004). Mainstreaming Geospatial Information for Sustainable National Development in Nigeria. Proceeding ISPRS Congress,Istanbul,Turkey,http://www.isprs.org/proceedings/XXXV/congress/comm4/papers/360.pdf 14/07/13.

Lázaro R, Rodrigo L, Gutirrez FD, Puigdefabregas J (2001). Analysis of a 30-year rainfall record (1967-1997) in semi-arid SE Spain for implications on vegetation. J. Arid Environ. 48 (3): 373 - 395

López BC, Holmgren M, Sabate' A, Gracia CA (2008). Estimating Annual Rainfall Threshold for Establishment of Tree Species in Water-limited Ecosystems Using Tree-ring Data. J. Arid Environ. 72(5):602-611.

Peter HG (1998). Water in Crisis: Paths to Sustainable Use. Ecol. Appl. 8:571-579.

Srikantha H, Uditha R (2004). Monitoring rainfall trends to predict adverse impacts-a case study from Sri Lanka (1964-1993). Glob. Environ. Chang. Part A. 14 (1):71-81.

Sui-Lin N, Chi-Fai C (2010) Issues and challenges in defeating world hunger. Trends Food Sci. Technol. 21(11):544-557.

Tennant WJ, Bruce CH (2002). Intra- Seasonal Rainfall Characteristics and their importance to the Seasonal Prediction. Int. J. Climatol. 22(9):1033-1048.

Usman MT (1999). An operational Index for Assessing Inter-annual, Rainfall, Variability and Agricultural droughts over the Nigeria Sahel. Afr. Climatol. Res. Ser. 3:22-32

Usman MT (2000) An Instability Index for monsoon onset characterization over the Nigerian Sahel. Environ. Rev. 3(2):367-379

Usman MT, Abdulkadir A (2012). On Determining the 'Real' Onset Date of Seasonal Rains in the Semi-Arid and Sub-Humid Areas of West Africa. Natural Hazard (NHAZ) (Online Jan. 2013 on SpringerLink:) DOI: 10.1007/s11069-012-0514-9.

15

Antimicrobial activity of earthworm (*Eudrilus eugeniae*) paste

K. Vasanthi[1], K. Chairman[2] and A. J. A. Ranjit Singh[2]

[1]Department of Zoology, Sri Parasakthi College for Women, Courtallam, M. S. University, Tirunelveli, Tamilnadu- 627 802, India.
[2]Department of Zoology, Sri Paramakalyani College, Alwarkurichi, M. S. University, Tirunelveli, Tamilnadu- 627 412, India.

Earthworm plays a major role in the proper functioning of the soil ecosystem. It acts as scavenger and helps in recycling of dead and decayed plant material by feeding on them. Earthworm increases the soil fertility and is often referred to as a farmer's friend. Earthworms have been used in medicine for various remedies. The paste prepared from earthworm, *Eudrilus eugeniae* was tested for antibacterial, antifungal activities. For the antimicrobial screening, four species of bacterial isolate and two species of fungal isolates were selected. The bacterial cultures were used for antimicrobial testing maintained on nutrient agar slant and the fungal strains were maintained on Sabouraud dextrose agar slant at 4°C. Minimum inhibitory concentration (MIC) was determined using micro dilution broth method. Earthworm paste at a dose of 100 µl was able to inhibit the growth of bacteria of *S. aureus* at a maximum level as compared to other bacteria; the growth of fungal *Candida albicans* was much inhibited. The MIC results indicated that earthworm paste at a dose of 200 µl inhibited the bacterial growth. These studies may lead to the formulation of new antimicrobial drug. The antimicrobial activity of the paste was determined by an agar diffusion method using well and disc, the study clearly indicates that the paste contain a good antibacterial potential and the bioactive compounds to inhibit the growth of bacteria and fungi. Hence earthworm paste (EP) has a good potential to develop a new drug.

Key words: *Eudrilus eugeniae*, antimicrobial activity, fungal and bacteria strains.

INTRODUCTION

Earthworms have been used in medicine for various remedies since 1340 AD (Hossam et al., 2012). Earthworm has been recognized in oriental medicine as anti-inflammatory, analgesic and antipyretic agent (Prakash and Gunasekaran, 2010). It shows anticancer effect by preventing excess glucose uptake (Balamurugan et al., 2009). Microorganisms are known to play a major role in soil characteristics, invertebrates are believed to act as regulators of antimicrobial activity. Earthworms live in an environment filled with various kinds of pathogens. Physiologically and evolutionally speaking, earthworm survival in such an environment must have favoured the development of efficient defense mechanisms against various environmental pathogens during the course of evolution, including the production of certain anti-microbiological substances, especially active proteins and enzymes (Wenli et al., 2011). Earthworm surface excreta were found to have potent antimicrobial activity. It is also having anti-coagulatory or fibrinolytic activity which results in the facilitation of blood circulation (Cooper and Balamurugan, 2010). The earthworm has been suspected to contain proteases which dissolve the fibrin

Table 1. Antibacterial activity of crude earthworm paste in well assay method.

S/N	Organism	Zone of inhibition in concentration (mm)		
		50 µl	100 µl	Tetracycline (30 µl)
1	*Escherichia coli* (NCIM-2065)	16.33±0.33	22.66±0.33	24.12±0.26
2	*Salmonella abony* (NCIM-2257)	15.33±0.33	19.66±0.33	21.66±0.23
3	*Bacillus subtilus* (NCIM-2063)	17.66±0.33	20.33±0.33	23.42±0.17
4	*Staphylococcus aureus* (NCIM-2079)	10.33±0.33	15.33±0.33	18.43±1.21
5	*Klebsiella pneumonia* (Nosocomial isolate)	11.33±0.33	23.66±0.33	25.26±0.26

P-value = 1.22E-24.

Table 2. Antibacterial activity of crude earthworm paste in disc assay method.

S/N	Organism	Zone of inhibition in concentration (mm)		
		50 µl	100 µl	Tetracycline (30 µl)
1	*Escherichia coli* (NCIM-2065)	11.33±0.33	13.66±0.33	17.66±0.33
2	*Salmonella abony* (NCIM-2257)	11.33±0.66	11.33±0.66	17.66±0.33
3	*Bacillus subtilus* (NCIM-2063)	10.33±0.33	12.66±0.33	19.66±0.33
4	*Staphylococcus aureus* (NCIM-2079)	10.00±0.00	17.66±0.33	26.66±0.33
5	*Klebsiella pneumonia* (Nosocomial isolate)	10.33±0.33	13.66±0.33	17.33±0.33

P-value = 5.33E-13.

clots or anticoagulants which selectively interfere with the intrinsic pathway of blood coagulation cascade. Medicinal properties of earthworm had also been described (Shobha and Kale, 2007). Antimicrobial potency of *Eudrilus eugeniae* extracts on certain plant pathogens were studied (Bauer et al., 1966). Hence, in the present study, the paste prepared from the earth worm, *E. eugeniae* was tested for antimicrobial potential.

MATERIALS AND METHODS

Collection of earthworms

Fully matured Earthworms, *E. eugeniae* (Kinberg) were collected from the stock culture, Department of Zoology, Sri Paramakalyani College, Alwarkurichi, Tirunelveli, Tamilnadu, India.

Preparation of earthworm paste

The *E. eugeniae* (Kinberg) were washed with running tap water and then fed with wet blotting paper for 18 to 20 h to clear their gut. The gut cleared worms were again washed with distilled water. The worms were kept in plastic troughs, covered tightly with polythene cover, and exposed to sunlight for 3 days to kill them. Mucus and coelomic fluid that oozed out digested the dead worms forming a brown coloured paste earthworm paste (EP) (Balamurugan et al., 2007).The earthworm paste were filtered and the filtrates obtained were condensed in water-bath at 35°C. The crude paste obtained was diluted in 10% DMSO for evaluation of antimicrobial activity.

Culture media

The media used for bacterial culture was nutrient agar/broth while Sabouraud's Dextrose-Agar (SDA) was used for fungal cultures.

Microbial assay (disc and well method)

The antimicrobial activity of the earthworm paste was determined by an agar-diffusion method using well and discs. The details of microbes used for the testing is given in Table 1 and Plates 2 and 3.

Test microorganisms for antimicrobial studies

For the antimicrobial screening, four species of bacterial isolate and two species of fungal isolates were selected. The bacterial and fungal strains were obtained from National Collection of Industrial Microorganism (NCIM), Pune, India. *Escherichia coli* (NCIM 2065) *Salmonella abony* (NCIM 2257), *Bacillus subtilis* (NCIM 2063) and *Staphylococcus aureus* (NCIM 2079) strains were used. *Candida albicans* (NCIM 3102), *Aspergillus niger* (NCIM 501) fungi were used as test organisms. The nosocomial isolate and local isolate strains were received from Department of Microbiology, Aasan Memorial College, Chennai-100.

The bacterial cultures were used for antimicrobial test maintained on nutrient agar slant and the fungal strains were maintained on Sabouraud dextrose agar slant at 4°C. The fresh bacterial cultures were obtained by growing the test organisms at 37°C, for 24 h while fungi were grown at 28°C for 48 h.

Antibacterial assay

Antibacterial activity was determined against certified strains of *E. coli* (NCIM 2065) *S. abony* (NCIM 2257), *B. subtilis* (NCIM 2063), *S. aureus* (NCIM 2079), *K. pneumoniae* (Nosocomial isolate), using a modification of agar diffusion assay method (Shobha and Kale, 2007) filter-paper disc of 6 mm diameter, and 6mm sterile metal borer were used (Table 2). Microorganisms were inoculated on

Table 3. Antifungal activity of crude earthworm paste in disc assay method.

S/N	Organism	Zone of inhibition in concentration (mm)	
		50 µl	100µl
1	*Candida albicans* (NCIM-3102)	12.33±0.33	16.33±0.33
2	*Aspergillus niger* (NCIM-501)	11.00±0.57	13.33±0.33
3	*Aspergillus flavus* (Local isolate)	10.33±0.33	15.00±0.57
4	*Penicillium notatum* (Local isolate)	10.66±0.33	14.33±0.33
5	*Trichophytum rubrum* (Nosocomial isolate)	0.00±0.00	12.33±0.33
6	Nystatin (30 µl)	14.33±0.33	26.66±0.33

P-value = 1.97E-16.

nutrient broth (Himedia, Bombay) during 24 h at 37°C. The inoculate absorbance was established between 0.08 and 0.10 AU (equivalent to 0.5 McFarland 10^8cfu/ml) adding sterile nutrient broth, before incorporating bacteria (λ = 625 nm). Bacterial strains were seeded on Muller-Hinton agar. The sterile discs were impregnated in the seeded agar. In the well assay method, wells were made using the sterile metal borer (6 mm). The sterile disc and well was loaded with the crude earthworm paste (EP) at a concentration of 50 and 100 µl. A disc of tetracycline (Tet) (30 µg/disc/well) was used as a positive control. The plates were incubated at 37°C for 24 h. All the assays were carried out in triplicate. The diameter (mm) of the growth inhibition zone was measured with standard zone reader scale (Himedia, Bombay) and mean diameter was recorded (Tables 2 and 3).

Antifungal assay

Antifungal activity was carried against *C. albicans* (NCIM 3102), *A. niger* (NCIM 501) *Trichophyton rubrum* (Nosocomial isolate), *A. flavus* (Local isolate), *Penicillium notatum* (Local isolate). The microorganisms were inoculated on Sabouraud dextrose broth (Hi-media, Bombay) during 24 h at 25°C. The inoculate absorbance was established between 0.08 and 0.10 AU (equivalent to 0.5 McFarland 10^8 cfu/ml) adding sterile Sabouraud dextrose broth, before incorporating fungi (λ = 530 nm). Fungal strains were seeded agar Sabouraud dextrose agar with 4% glucose. The sterile discs were impregnated in the seeded agar. In the well assay method, wells were made using the sterile metal borer (6 mm). The sterile disc and well were loaded with the crude earthworm paste (EP) at a concentration of 50 and 100 µl. Nystatin (100CFU/disc/well) antifungal agent was used as positive control. The plates were incubated at 25°C for 48 h. All the assays were carried out in triplicate. The diameter (mm) of the growth inhibition zone was measured with standard zone reader scale (Himedia, Bombay) and mean diameter was recorded.

Minimum inhibitory concentration (MIC)

Minimum inhibitory concentration (MIC) was determined using micro dilution broth method. The earthworm paste prepared (100 mg/ml) were diluted to four different concentration (50, 100, 150 and 200 µl/ml).To find out the MIC, three strains of bacteria were used *(E. coli, K. pneumoniae* and *S. aureus*). The bacterial strains were prepared in broth. The broth culture suspension of each bacterium (0.5 ml) with an optical density of MacFarland 0.5 x 10^7-10^8 CFU/ml was added to test tubes containing different concentrations of earthworm paste. To the control test tube, the earthworm paste was not added, only sterile distilled water was used. The inoculated test tubes were incubated at 37°C under aerobic conditions. After 24 h, the turbidity was evaluated. The MIC is the lowest concentration of EP that inhibited the growth of the organism completely for all the tests including control, and triplicates were maintained.

RESULTS

Earthworm paste prepared from *E. eugeniae* was tested for antibacterial and antifungal activities. For antimicrobial assay, five strains of bacteria viz. *E. coli, S. abony, B. subtilus, S. aureus* and *K. pneumoniae* were used for antibacterial assay, two concentration of the EP 50 and 100 µl were used. Of the five bacteria tested, the growth of bacteria was well inhibited by EP. The disc diffusion assay and well assay indicated a dose dependent effect of the EP to inhibit the growth of bacteria. EP at a dose of 100 µl was able to inhibit the growth of *S. aureus* at a maximum level (17.66 ± 0.33 mm), *S. abony* was less inhibited when compared with others (11.33 ± 0.66 mm). In the well assay, *K. pneumoniae* was highly inhibited (23.66 ± 0.33 mm).

The present study clearly indicates that EP contains a good antibacterial potency and the active compound in it has to be explained.

Anti fungal assay was also carried out using five different strains of fungi. The fungal strain was inhibited by EP in a dose dependent manner. The high dose of EP (100 µl) showed highest anti fungal potential. Of the five strains tested, the growth of *C. albicans* was much inhibited (16.33 ± 0.33 mm). The inhibitory potential of the EP was less in the case of *T. rubrum* (12.33 ± 0.33 mm).

To find out the minimum inhibitory concentration (MIC) of the earthworm paste, three bacterial strains were used *Viz., E. coli, K. pnemoniae* and *S. aureus*. The MIC results indicated that earthworm paste at a dose of 200 µl inhibited the bacterial growth (Table 4). The optical density (OD) was less. This indicated that EP at a dose of 200 µl is a minimum concentration to inhibit the growth of the selected bacteria.

The present study clearly indicates that the earthworm paste has bioactive compounds to inhibit the growth of

Table 4. Antibacterial activity (MIC) tested for the crude earthworm paste using optical density method.

S/N	Bacterial culture	Concentration of crude earthworm paste	OD value
1	*Escherichia coli* (NCIM-2065)	Control	19.66±0.33
		50 µl	15.00±0.57
		100 µl	14.00±0.57
		150 µl	13.00±0.57
		200 µl	10.66±0.66
2	*Klebsiella pneumoniae* (Nosocomial isolate)	Control	15.00±0.57
		50 µl	13.33±0.66
		100 µl	9.33±0.33
		150 µl	7.66±0.33
		200 µl	7.33±0.66
3	*Staphylococcus aureus* (NCIM-2079)	Control	16.33±0.33
		50 µl	10.66±0.88
		100 µl	9.66±0.88
		150 µl	8.66±0.66
		200 µl	8.33±0.33

P-value = 0.000886; TNTC - Too numerable to count.

bacteria and fungi. Hence, EP has a good potential to develop a new drug.

DISCUSSION

Khomyakov et al. (2007) Suggested that antimicrobial agents of earthworms digestive fluid are formed in the earthworm body but not by the soil microorganisms entering their digestive tract. They have observed that the digestive fluid of earthworm show the same antimicrobial activity after feeding on soil and sterile sand, and partial sterilization of the gut with streptomycin does not lower the antimicrobial activity. Mendez et al. (2003) observed that the guts of earthworms, *Onychochaeta borincana* in sterile soil contain the same microorganisms as the guts of individuals that have not been submitted to the cleansing treatment. Antimicrobial activity in the guts of earthworms derived from metabolites of symbiotic bacteria from the gut walls is possible.

The antimicrobial activity of *Eisenia foetida* coelomic fluid directed against Gram-positive and negative bacteria was analyzed. The gut extracts of earthworms have antibacterial and antifungal activity (Shobha and Kale, 2008). The new bacterial strain with antimycobacterial activity has been isolated from the midgut of *Dendrobaena veneta* (Annelida) (Marta et al., 2010). Earthworm species have rich diversity. Based on their living environments, it is rational to think that there are effective anti-infective agents in earthworm's skin. Cho et al. (1998) identified the first antimicrobial peptide (lumbricin I) from the earthworm, *Lumbricus rubellus*. Lumbricin I is considered as a proline-rich antimicrobial peptide containing 62 amino acids including proline (15%) with a molecular weight of 7231 Da. Lumbricin I showed antimicrobial activity *in vitro* against a broad spectrum of microorganisms without hemolytic activity (Cho et al., 1998). Recently, another two antimicrobial peptides (PP1 and OEP3121) have been identified from earthworms of *Pheretima tschiliensis* and *E. foetida,* respectively (Wang et al., 2003). The antimicrobial peptide, lumbricin-PG was identified from skin secretions of the earthworm, *Pheretima guillelmi* (Wenliang et al., 2011). Engelmann et al. (2004) and Balamurugan et al. (2008) found that earthworm coelomic fluid contains biologically active molecules and leukocytes that participate in phagocytosis, encapsulation and killing of HeLa, HEp-2, PC-12 and PA317 cells *in vitro*. Presumably, earthworms synthesize and secrete several effective modulators of innate immune responses such as antibacterial molecules, cytotoxic proteins and cytokines.

Conclusion

The burrowing and casting activities of earthworms contribute to the activity of soil micro organisms and nutrient enriched earthworm casts are good media supporting microbial growth. The microbial community in the gut of earthworms was predominantly Gram-negative bacteria. The present study reported that vermicompost is rich in microbial populations and diversity, particularly fungi and bacteria. It is possible that earthworms can be used not only in environmental monitoring, but also in the acquisition of novel molecules for human therapeutic purposes. This study may thus lead to formulation of new natural antimicrobial agent and thus may be found to be beneficial in future prospects for mankind.

ACKNOWLEDGEMENTS

The support extended by the UGC New Delhi and the encouragement given by the Principal of Sri Parasakthi College for Women, Courtallam, Tirunelveli, Tamilnadu, India are gratefully acknowledged.

REFERENCES

Balamurugan M, Parthasarathi K, Cooper E, Ranganathan LS (2007). Earthworm paste (*Lampito mauritii*, Kinberg) alters inflammatory, oxidative, haematological and serum biochemical indices of inflamed rat. Europ. Rev. Med. Pharmacol. Sci. 11:77-90.

Balamurugan M, Parthasarathi K, Cooper EL, Ranganathan LS (2008). Hypothetical mode of action of earthworm extract with hepatoprotective and antioxidant properties. J. Zhejiang Univ. Sci. B. 9(2):141-147.

Balamurugan M, Parthasarathi K, Cooper EL, Ranganathan LS (2009). Anti-inflammatory and anti-pyretic activities of earthworm extract-*Lampito mauritii* (Kinberg). J. Ethnopharmacol. 121(2):330-332.

Bauer AW, Kirby WMM, Sherries JC, Tuck M (1966). Antibiotic susceptibility testing by a standardized disc diffusion method. Am. J. Clin. Pathol. 45:493-496.

Cho JH, Park CB, Yoon YG, Kim SC(1998). Lumbricin I, a novel proline-rich antimicrobial peptide from the earthworm: purification, cDNA cloning and molecular characterization. Biochim. Biophys. Acta 1408:67-76.

Cooper EL, Balamurugan M (2010). Unearthing a source of medicinal molecules, Drug Discov. Today 15(21-22):966-972.

Engelmann P, Kiss J, Cooper EL, Nemeth P (2004). Earthworm leukocytes kill HeLa, HEp-2, PC-12 and PA317 cells in vitro. J. Biochem. Biophys. Methods 61:215-228.

Hossam El-Din Mohamed Omar, Zedan Z. Ibraheim, Nasser A. El-Shimy, Rouwaida S. Ali (2012). Anti-inflammatory, antipyretic and antioxidant activities of the earthworms extract. J. Biol. Earth Sci. 2:1.

Khomyakov NV, Kharin SA, Nechitailo, TYu, Golyshin PN, Kurakov AV, Byzov BA, Zvyagintsev DG (2007). Reaction of microorganisms to the digestive fluid of earthworms. Microbiology 76:45-54.

Marta J, Fiołka Mirosław P, Zagaja Tomasz D, Piersiak Marek Wróbel, Jarosław Pawelec (2010). Gut bacterium of *Dendrobaena veneta* (Annelida: Oligochaeta) possesses antimycobacterial activity. J. Invertebr. Pathol. 105:63-73.

Mendez R, Borges S, Betancourt C (2003). A microscopical view of the intestine of *Onychochaeta borincana* (Oligochaeta: Glossoscolecidae). Pedobiologia 47:900-903.

Parthasarathi KM, Balamurugan, Ranganathan LS (2008). Influence of vermicompost on the physico - chemical and biological properties in different types of soil along with yield and quality of the pulse crop - black gram. Iran J. Environ. Health. Sci. Eng. 5(1):51-58.

Prakash M, Gunasekaran G (2010). Gastroprotective effect of earthworm paste (Lampito mauritii, Kinberg) on experimental gastric ulcer in rats. Eur. Rev. Med. Pharmacol Sci.14(3):171-176.

Shobha SV, Kale R (2007). Antimicrobial potency of earthworm, Eudrilus eugeniae on certain plant pathogens. Administrator

Shobha SV, Kale RD (2008). In *vitro* studies on control of soil-borne plant pathogens by earthworm *Eudrilus eugeniae* exudates. Green Pages,2008. http://www.eco web.com/editorial/080106.html.

Wang X, Wang XX, Zhang, Y (2003). An antimicrobial peptide of the earthworm Pheretima tschiliensis: cDNA cloning, expression and immunolocalization. Biotechnol. Lett. 25:1317-1323.

Wenli Li, Chong Wang, Zhenjun Sun (2011). Vermipharmaceuticals and active proteins isolated from earthworms, Pedobiologia 54S:S49-S56.

Wenliang Li, Sisi Li , Jian Zhong, Zhu Zhu, Jingze Liu, Wenhong Wang (2011). A novel antimicrobial peptide from skin secretions of the earthworm, *Pheretima guillelmi* (Michaelsen*)* Peptides 32:1146-1150.

Effect of remediation on growth parameters, grain and dry matter yield of soybean (*Glycine max*) in crude oil polluted ultisols in Ogoni Land, South Eastern Nigeria

Ayolagha G. A.* **and Peter K. D.**

Department of Crop/Soil Science, Faculty of Agriculture, Rivers State University of Science and Technology, Port Harcourt, Nigeria.

The effectiveness of cow dung, poultry manure, NPK (mineral fertilizers) and municipal waste compost which were the easily accessible materials in the remediation of crude oil polluted soils in Ogoni, Rivers state was assessed using soybean as a test crop. A simple factorial field experiment arranged into a randomized complete block design with six treatments was used. Poultry manure gave the highest emergence of 79 and 74.62% at both seasons followed by NPK, cow dung, municipal waste compost and control. The highest plant height of 11.48 and 19.18 cm at four and six weeks after planting (WAP); and 12.70 and 24.70 cm at 4/6 WAP for both seasons were obtained in plots treated with poultry manure. NPK (15:15:15) fertilizer gave the largest leaf area (12.70 and 12.95 cm^2) and (14.43 and 14.56 cm^2) at 4 and 6 WAP for both seasons, respectively followed by poultry manure (11.80 and 11.93 cm^2) and (13.72 and 12.8 cm^2. Poultry manure gave the highest grain yield per hectare of 2,800 and 3,875 kg ha^{-1}) for both seasons. NPK had the highest dry matter yield at both seasons (8,500 and 8,000 kg ha^{-1}), followed by poultry manure (7,000 and 7,500 kg ha^{-1}). Since there were no significant differences in leaf area between NPK, poultry manure and cow dung treatments, the results indicate that poultry manure is the best remediation material for crude oil polluted soils in which soybean is planted.

Key words: Crude oil, polluted soils, soy bean, remediation material, poultry manure, NPK fertilizers, grain yield, plant height, leaf area, Niger delta.

INTRODUCTION

Remediation practices are those treatments or technologies used to restore a polluted or contaminated soil and water (Odu, 2006). Remediation is very necessary in Ogoni land as a result of crude oil spillage; a regular occurrence in the area (Legbosi, 2005). Crude oil spillage is a regular occurrence in the Niger Delta region of Nigeria where over 80% of the crude oil is produced. Dublin-Green et al. (1998) reported that from 1979- 1997, the Nigeria petroleum industries spilled 5334 barrels of oil into the Niger Delta.

Ogoni land is located in the south-eastern part of the lower Niger Delta region and is endowed with abundant natural resources including oil and gas. Ogoni land thus, plays host to five flow stations and 96 oil wells (Vikoo, 2003). Despite the suspension of oil exploration and exploitation activities in Ogoni land since 1993, the area is yet to be spared from oil spillage due to the presence of oil pipeline cris-crossing the entire Ogoni landscape and some of which are trans-Niger tank lines that transport oil to shipment terminal in Bonny Island. No other post-spill remediation responses have been embarked on beyond extinguishing the fire and plastering of the leaking pipes (Hart, 1997). There is need for remediation of the coastal plain sands of Ogoni land using less expensive and easily affordable remediation material such as cow dung, poultry manure, municipal waste compost with soybean as a test crop (Ebie and Ayolagha, 2006).

*Corresponding author. E-mail: gaskinayolate@gmail.com.

Soybean as a leguminous plant has a symbiotic relationship with microorganisms within the soil especially the symbiotic Nitrogen fixing bacteria which promote plant growth and development (Amakiri, 2000).

Soybean has been recognized as a leading tropical grain legume which provide cheap and balanced diet (Ogundipe et al., 1989) with proximate analysis of 40% protein, 32% carbohydrate, 20% edible oil, 5% mineral and 3% fibre.

Municipal waste compost (MWC) of oil polluted soils planted to soybean in Gokana Ogoni increased the levels of K, Na, and Ca; reduced total hydrocarbon content (THC) and seed germination counts; and gave no significant difference in other plant growth parameters (Ayolagha et al., 2000).

Application of N to soybean in the tropic has given variable result depending on such factors as rhizobial strains (Handerson et al., 1987; Somado and Sahrawat, 2007; Chianu et al., 2011). But no information has been obtained on the comparative effects of poultry manure, cow dung, MWC and NPK fertilizers on the remediation of crude oil polluted soils planted to soybean in Ogoni land (coastal plain sands) under natural field conditions. It is the objective of this research to investigate the effect of remediation materials on growth and yield of soybean planted to crude oil polluted soils of the coastal plain sands of Ogoni land, Rivers State.

MATERIALS AND METHODS

The experiment was conducted on a farm in Baa-Lueku in Nyokhana district of Khana local Government Area of Rivers State (Plate 1). It could be located at the following geographical co-ordinate: Latitude 4°40' 5''N and 4° 43' 19.5''N and Longitude 7° 22' 53.7''E and 7° 27' 9.8''E. The soil of the study area is the well-drained costal plain sands deposit (Ultisol). The experimental site has been fallow after some years of intensive cultivation of maize, Okra, yam, cocoyam cassava pumpkin and garden egg without history of soybean cultivation.

The rainfall distribution of the study area ranges from 2000 to 3000 mm (Oyegun and Ologunorisa, 2002) per annum in a bimodal from with average temperature of about 28 to 35°C (Oyegun and Ologunorisa, 2002) depending on the season of the year.

The experiment involves six treatments with four replicates.

These gave a total of twenty four plots in randomized complete block design. Each plot of 2 × 2 m sizes was polluted with 2.4 L of fresh Bonny light crude oil with specific gravity of 0.835 which is equivalent to one percent pollution (Elf, 2000). The no pollution and control plots were not polluted. The crude oil was measured into watering can and spread on each plot and allow to fallow for two weeks before application of remediation materials. The remediation materials used were cow dung, poultry manure, NPK and municipal waste compost. The rate of application were cow dung 10,000 kg ha^{-1}, poultry manure 10,000kg ha^{-1}, NPK 250 kg ha^{-1} and municipal waste compost 10,000kg ha^{-1} in that order (Ethirveerasingam et al., 1985). The chemical composition of the organic manures are presented in Table 1. The following chemical composition: pH, % Org. C, %N, P, Ca and Na show that poultry manure had higher levels than cow dung and MWC in that order.

However, the reverse holds through for Mg and K. The C:N ratio increased from poultry manure, cow dung to MWC. Two weeks after crude oil pollution of each plot, remediation materials were carefully applied except the control and no pollution plots. The control plot was polluted while the non pollution plot was not, however, both were not remediated. The remediation materials were tilled into the soil after being spread and left fallow for another two weeks before planting. After four weeks of planting, the second round of remediation materials were applied in the plots except the control and no pollution plots (Ayolagha et al., 2000). The rate of application of remediation materials was similar to those of the first as stated above.

Emergence of soybean plant seven days after planting

The number of soybean plants that emerged 7 days after planting were observed and estimated in each plot and the percentage emergence calculated. All plots were polluted and amended with remediation materials except the control and no- pollution plots. The control plot was polluted but not amended while the non-pollution plot was neither polluted nor amended. Emergence was determined by the formula:

$$\frac{\text{Number of seeds emerged}}{\text{Number of seeds planted}} \times \frac{100}{1}$$

Plant height 4 and 6 weeks after planting

Plant height per plot was taken at 4 and 6 weeks after planting. The height of soybean plants per plot was measured using measuring tape from the base of the plant to the collar of the last leaf in the plant.

Leaf area 4 and 6 weeks after planting

The leaf area of soybean plant 4 and 6 weeks after planting was determined using the punch techniques. Five leaves were taken from the total leaves per plot and punched. The punched leaves were weighted. The fresh weight of the total leaves per plot was also taken for analysis to relate weight with length using the following formula:

$$\frac{\text{Area of leaf disc}}{\text{Weight of punched leaves}} \times \frac{\text{Total leaf area}}{\text{Total fresh weight of the leaves}}$$

Grain and dry matter yield 12 weeks after planting

The grain yield was assessed after 12 weeks of planting by weighing the dry grains in kilogram per hectare (kg ha^{-1}). The dry matter yield was assessed by oven drying the whole plant at 80°C for 48 h to a moisture content of 14% and determining the dry weight using top weighing balance (Ebie and Ayolagha, 2006).

All data from the field were analyzed using statistical procedure for the analysis of variance (ANOVA) (Steel and Torries, 1960). The new Duncan multiple range test (DMRT, 1955) was used to compare the mean and to analyze the least significant differences (LSD).

RESULTS AND DISCUSSION

Effect of remediation materials on emergence of soybean plant seven days after planting (DAP)

The number of soybean plant that emerged at 7 DAP is

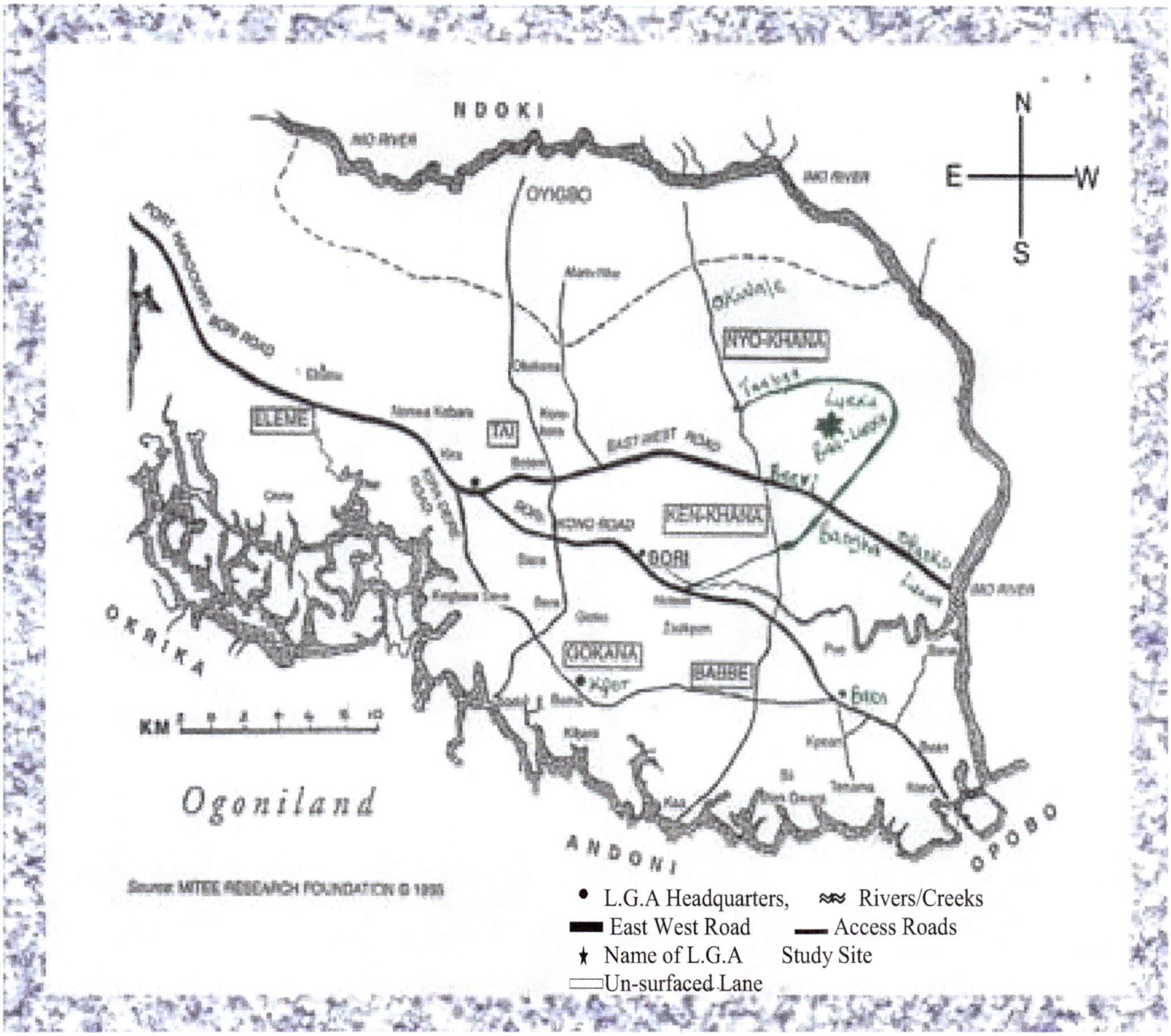

Plate 1. Map of Ogoni.

Table 1. Chemical properties of organic manure used as remediation materials.

Chemical composition of remediation materials	Remediation material		
	Poultry manure	Cow dung	MWC
pH (H_2O)	6.14	5.25	5.20
Org. C. (%)	3.26	3.25	3.24
Total N (%)	0.37	0.30	0.17
Avail. P. (mg kg^{-1})	0.65	0.44	0.64
Ca (mg kg^{-1})	5.80	3.87	3.86
Mg (mg kg^{-1})	1.26	2.43	2.43
K (mg kg^{-1})	1.98	3.61	3.21
Na (mg kg^{-1})	8.00	7.78	7.82
C :N ratio	8.81	10.83	19.06

MCW, Municipal waste compost.

Table 2. Effect of remediation materials on emergence of soybean seedlings.

Treatment	Emergence	7 DAP (%)
	Early planting	Late planting
No pollution	65.38	64.60
Control (pollution without amendment)	56.50	60.00
Cow dung	69.37	67.92
Poultry manure	79.00	74.62
NPK	73.43	68.32
Municipal waste compost	66.25	65.32
Overall means	68.32	65.70

DAP, Days after planting; MWC, municipal waste compost; NPK, nitrogen, phosphorous and potassium fertilizer.

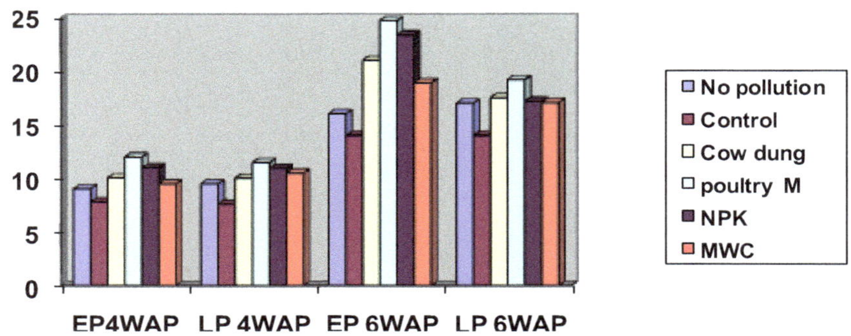

Figure 1. Effects of remediation on soybeans plant height (cm). EP, Early planting, LP, late planting; WAP, weeks after planting; NPK, nitrogen, phosphorus and potassium fertilizer; MWC, municipal waste compost.

crude oil polluted soil treated with poultry manure (79 and 74%) and decreased in the following order NPK (73 and 68%), Cow dung (69 and 67%), municipal waste compost (66 and 65%) and no pollution (65 and 64%) (Table 2). The control (pollution with no amendment) had the least emergence of 56 and 60%. However, the mean value for both seasons was not significantly different from one another.

The low seedling emergence in the control was because crude oil soaked some seeds and damaged them. This is in support of the findings of Kalio (2003) that for a seed to germinate effectively, it must be viable or alive. He reported that the crude oil soaked cotyledons or embryo lead to poor germination. The result also supported the findings of Horne (1978) that heavy crude oil pollution lead to poor germination due to poor soil aeration. He pointed out that the increased C:N ratio and soil temperature lead to poor seedling emergence.

The highest seedling emergence in plot treated with poultry manure and NPK was because these manure and fertilizers were easily dissolvable and absorbable (Awodum, 2007). He added that this promote microbial degradation of hydrocarbon because poultry manure had a lot of micro organisms and at the same time help to bind the soil particles to improve the soil structure thereby improving the soil water and temperature regimes. Cow dung and especially MWC were not as highly dissolvable as poultry manure hence gave lower seedlings emergence. The lower C:N ratio in poultry than the other manures enhanced the higher emergence. Early planting had higher seedling emergence than late planting to support the findings of Ebie (2006) who reported that high rainfall at late planting season affected the rate of seed germination at $p < 0.05$.

Effect of remediation on soybean plant height 4 and 6 weeks after planting

The result of remediation on soybean plant height is presented in Figure 1. Poultry manure had significant vertical growth measurement across the different seasons. Poultry manure also gave plant height of (11.48 and 12.70 cm) 4 weeks after planting at early and late planting and (24.70 and 19.18 cm) 6 weeks after planting at early and late planting respectively. This was followed by NPK (10.98 and 11.58 cm) 4 WAP at both early and late planting and (23.28 and 17.05 cm) 6 WAP for both early and late planting; followed by cow dung, municipal waste compost and no pollution in that order. The control gave the least vertical growth.

Poultry manure gave the highest plant height in

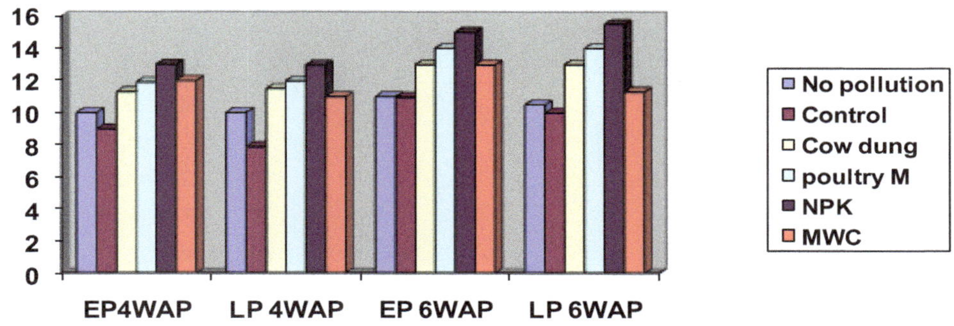

Figure 2. Effects of remediation on soybeans leaf area (cm^2). EP, Early planting; LP, late planting; WAP, weeks after planting; NPK , nitrogen; phosphorus and potassium fertilizer; MWC, municipal waste compost.

conformity with the report of some researchers (Alasiri and Ogunkeyede, 1999; Awodinn, 2007). They stated that poultry manure was easily dissolvable and absorbable and had lower C:N ratio than cow dung and MWC, as such soybean quickly absorbed the nutrient leading to fast growth rate. Soybean is a leguminous crop, hence fixed N into the soil with a resultant decrease in the C:N ratio. Nitrosomonas convert the N to Nitrite while Nitrobacter mineralized the Nitrite to Nitrate (NO_3^{-1}). This easily absorbable Nitrate enhanced growth rate of soybean. The lower C:N ratio of poultry manure enhanced better mineralization of soil minerals, improvement of soil structure, aeration and hence greater productivity and growth of soybean. The higher C:N ratio in cow dung and MWC might have lead to immobilization of minerals and lower productivity. Plant height were higher at 6 weeks than at 4 weeks after planting due to climatic advantages. The higher plant heights observed in plots treated with poultry manures and NPK were because of their quicker dissolvability and absorbability that promotes easy plant uptake for their growth (Awodinn, 2007).

At late planting, the growth rate of soybean in no pollution plot improved to (10.65 cm) because non application of remediation material increased the rate of nodulation to enhance Nitrogen fixation. This is in conformity with the findings of some researchers (Amakiri, 2000; Singh and Saxena, 1977; Mansimba and Mondibaye, 1996). The increase in root nodules could also be caused by their ability to adapt to the environmental stress created by the non-application of remediation materials. This is in line with the observations of (Giller and Wilson, 1993; Chinke et al., 2000) that the Rhizobia species have the ability to adapt to the environmental stress.

Effect of remediation on leaf area of soybean plant

The results of the effect of remediation on crude oil polluted soils as indicated by leaf area of soybean are presented in Figure 2. The results showed that leaf area generally was highest in NPK plot followed by that of poultry manure. The general trend in a decreasing order is as follows: NPK, poultry manure, municipal waste compost, cow dung, No pollution and control plot both in early and late plantings at 4 and 6 weeks after planting. The leaf area at 4 WAP ranged from 9.24 to 12.96 cm^2 for NPK during early planting, and 12.70 cm^2 for NPK at late planting. The leaf area at 6 WAP ranged from 10.48 cm^2 for control to 14.43 cm^2 for NPK at early planting and 10.69 cm^2 for no pollution plot to 14.56 cm^2 for NPK at late planting respectively.

NPK (15:15:15) plot gave higher leaf area followed by poultry manure probably because of their capacity to release nutrients to plants faster due to higher N content and lower C:N ratio (Awodinn, 2007). Micro organisms had to use up the low N in the cow dung and MWC as nutrient before breaking down their carbon. This process reduced the capacity of these materials for remediation of the oil polluted soils.

Effect of remediation on soybean grain yield (kg ha^{-1})

The result on soybean grain yield per hectare is presented in Figure 3. The results showed that at early planting poultry manure had the highest grain yield per hectare (2,800 kg ha^{-1}). This was followed by NPK fertilizer (2,220 kg ha^{-1}), municipal waste compost (2,000 kgha^{-1}), cow dung (1,631 kgha^{-1}) and no pollution (1,575 kgha^{-1}) respectively. The lowest grain yield per hectare (1,044 kg ha^{-1}) was observed in the control. The easy dissolvability and absorbability of poultry manure and NPK which promoted other growth parameters also enhanced increase in yield at early planting, during late planting, poultry manure also had the highest grain yield of (3,875 kg ha^{-1}) followed by NPK (2,687kgha^{-1}) and municipal waste compost (2,190kg ha^{-1}). Cow dung and no pollution plots had similar grain yield of 2,000 and 1,780 kg ha^{-1} respectively. Results of yield after various treatments were significantly different from one another at

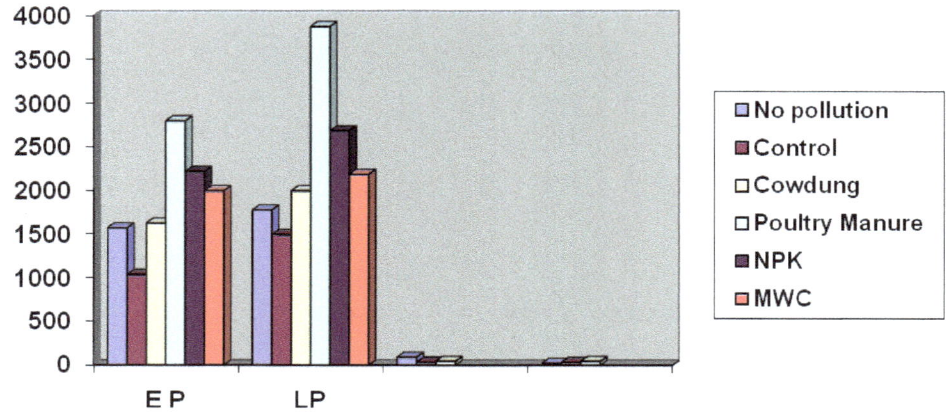

Figure 3. Effect of Remediation on soybean grain yield per hectare. EP, early planting; MWC, municipal waste compost; LP, late planting; NPK, nitrogen; phosphorus and potassium fertilizer.

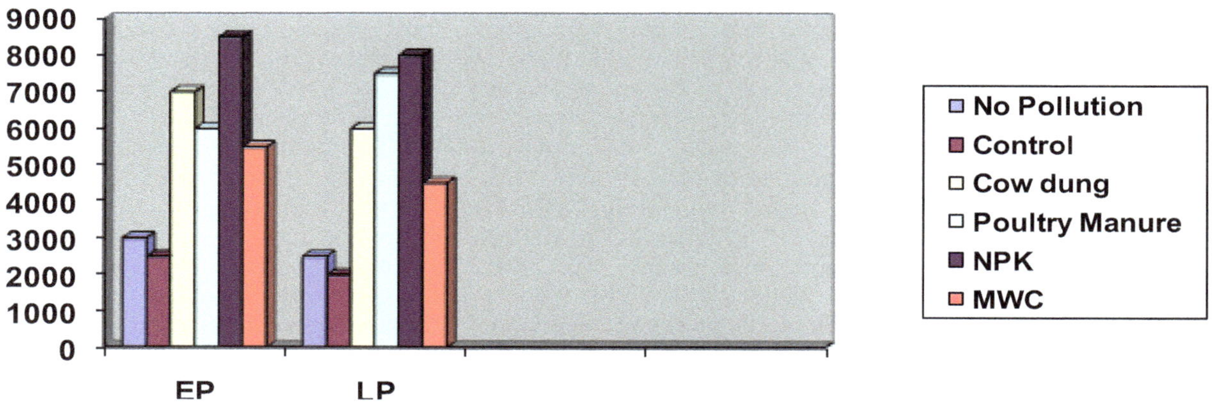

Figure 4. Effects of Remediation on soybean Dry matter yield (kg ha^{-1}). EP, early planting; LP, late planting; MWC, municipal waste compost; NPK, nitrogen; phosphorus and potassium fertilizer.

$P<0.05$. The control had the lowest grain yield of 1,500 kg ha^{-1}.

The lowest grain yield observed in the control plot was as a result of the non application of remediation materials. This agrees with Raymond et al. (1976) that crude oil has negative effect on the soil because of the high carbon content. Highly increased soil organic carbon is detrimental to plant growth and development due to higher C:N ratio, while N fixation by rhizobium in the soybean nodules and lower C:N ratio increase grain yield.

When compared with the findings of Ebie and Ayolagha (2006) on maize grain yield, it was discovered that poultry manure had (2,039 and 2,109 kg ha^{-1}) at early and late planting seasons respectively followed by cow dung (510 and 1,615 kg ha^{-1}) and NPK (1,500 kg ha^{-1}). This showed that soybean had better grain yield in crude oil polluted soils amended with similar remediation materials than maize. It was also observed that Ebie and Ayolagha (2006) worked on maize planted on the Inceptisols in the Meander Belt geomorphic regions of Bayelsa State while the present research is on soybean planted to crude oil polluted Ultisols in coastal plain sands of Ogoni Rivers State all in Niger Delta.

Effect of remediation on soybean dry matter yield (kg ha^{-1})

The effect of the various remediation materials on soybean dry matter is presented in Figure 4. It showed that soybean plant treated with NPK had the highest dry matter yield of (8,500 and 8,000 kg ha^{-1}) at both early and late planting seasons. Poultry manure had the second position (7,000 and 7,500kg ha^{-1}) followed by cow dung (6,000 and 7,000 kg ha^{-1}) and municipal waste compost

(5,500 and 4,500 kg ha^{-1}) at early and late planting respectively.

Soybean having the highest dry matter yield in NPK plot might be because of NPK's easy dissolvability and absolvability, which enhanced quick up take for proper growth and development of the soybean plant. However, Ebie and Ayolagha (2006) observed that poultry manure gave the highest dry matter of (13,187 and 12,969 kg ha^{-1}), followed by cow-dung (11,145 and 10,052 kg ha^{-1}) and NPK (8,229 and 7,239. 5 kg ha^{-1}) for early and late seasons, respectively. The higher dry matter in maize plants than soybean might be as a result of the structural or morphological differences in both plants. The dry matter yield of soybean treated with poultry manure improved to 7,500 $kgha^{-1}$ at late planting due to climatic advantage and its ability to bind soil particles to form a good soil structure, total pore spaces and hence aeration. The higher moisture retentive capacity of poultry manure might also have contributed to the improved dry matter yield at late planting.

Conclusion

The various remediation materials used for this study improved the different growth parameters of soybean such as seeding emergence 7 DAP, plant height and leaf area 4 and 6 weeks after planting at both early and late planting. Generally, the highest performance in terms of percentage emergence and plant height was recorded in plot treated with poultry manure, followed by NPK, cow dung and municipal waste compost. NPK had the highest leaf area in both seasons the least growth parameters were recorded in the no pollution and controls. Poultry manure also gave the best grain yield in both early and late planting seasons. Poultry manure gave the second best dry matter yield at both plantings. All the above showed that poultry manure is the best remediation material on crude oil polluted soils planted to soybean in the Ultisols of Ogoni land, Rivers State in the Niger Delta.

REFERENCES

Alasiri KO, Ogunkeyede SA (1999). Effect of Different Levels of Poultry Manure on Seed Yield of Okra. Soil Sci. Soc. 25:102.

Amakiri MA (2000). Microbes Moving the World Forward in The New Millennium; Inaugural Lecture delivered on 29 Nov. 2000 at the Rivers State University of Science and Technology, Nkpolu-Oroworukwu, Port Harcourt 2000, pp. 50-51.

Awoddinn AA (2007). Effect of Poultry Manure on the Growth, Yield and Nutrient Content of Fluted Pumpkin *(Telfaria occidentalis* wook F.) Asian J. Agric. Res. 1(2):67-73.

Ayolagha GA, Kio-Jack FS, Isirimah NO (2000). Remediation of Crude Oil Polluted Soils using Municipal Waste Compost for Soybean Production in Niger Delta, In Olaolu Babalola (ed) Soil Science Society of Nigeria proceedings Ibadan Oct. 30 - Nov.3rd 2000. pp.161-169.

Chianu JN, Nkonya EM, Mairura FS, Chianu JN, Akinnifesi FK (2011). Biological nitrogen fixation and socioeconomic factors for legume production in sub-saharan Africa: a review. Agron. Sustainable Dev. J. 31(1)139-154

Chinke NM, Iuafor ENO, Sanginga N, Chude VO (2000). Influence of Phosphorus and Molybdenum on the infection of Cowpea by Striga. Nig. J. Soil Sci. 12:58-66.

Dublin-Green WF, Nwankwo JN, Irechukwu DO (1998). Effective Regulation and Management of HSE Issues in the petroleum History, S.P.E International conference on Health Safety and Environment in Oil and Gas Exploration and Production, Caracas, Venezuela, 7-10, June 1998. Paper No SPE 40/26.

Duncan DB (1955). Multiple Range and Multiple F-Test. Biometrics 11:25-40.

Ebie SJ, Ayolagha GA (2006). Remediation of Crude Oil Polluted Soil using Organic and Inorganic Fertilizers for maize product in Yenagoa, Bayelsa State. M.Sc Thesis Department, of Crop/Soil Science, Faculty of Agriculture, Rivers State University of Science and Technology, Port Harcourt Rivers State Nigeria. pp. 6-20.

Elf (2000). Percentage pollution derivation form Crude Oil, Obagi Elf Operation Area OML Institute of Pollution Studies, Rivers State University and Science and Technology, Port Harcourt 58. pp. 1-3.

Ethirveerasingam N, Isirimah NO, Loganathan P, Ebong UU, Zuofa K, Daniel- Kalio LA, Ossom EM (1985). Guide to Crop Production in Rivers State. UNESCO Project. Faculty of Agriculture, Rivers State University of Sci. Tech. Port Harcourt, Nigeria. pp.1- 45.

Giller KE, Wilson KJ (1993). Nitrogen fixation in Tropical Cropping System. CAB International Walling Ford USA.

Harderson G, Danson SKA, Zapata F (1987). Biological Nitrogen fixation in Field Crops. In Hand Book of Plant Science in Agriculture. B.R. Christie (ed) CRC Press Inc. Boca Raton, USA. pp.165-192.

Hart SA (1997). Effect of Crude Oil Pollution on the Niger Delta. An MSc Seminar paper Presented to the Department of Zoology, University of Port Harcourt, Nigeria, 2 September.

Legbosi SP (2005). The Impact of Oil Spills on Biological Diversity. The Ogoni Experience 2005, Published by Sarobery Publisher Nigeria Ltd Port Harcourt. pp. 8-26.

Mansimba GR, Mondibaye R (1996). Effect of Inoculation and N-Fertilization on Soybean Grown on Congo Soil. Biol. Agric. Hortic. 13:197-204.

Odu CTI (2006). Environmental Degradation, Restoration of Pollution Sites. A lecture delivered during the 32nd world Environmental day celebration at Port Harcourt, State Government, Ministry of Environment at the Alfred Diete Spiff Civic Centre Moscow road on Monday 5th June, Port Harcourt Rivers State, Nigeria.

Ogundipe HO, Dashell KE, Osho SM (1989). Soymilk yield and Quality as Affected by Soybean Varieties and Processing Technique. Trop. Legumes Bull. p. 38.

Oyegun CU, Ologunorisa YE (2002). Weather and Climate. In The Land and People of Rivers State: Eastern Niger Delta. Alagoa, EJ, Derefaka AA (eds) Onyoma Research Publications, Choba, Rivers State, Nigeria.

Singh BB, Saxena MC (1977). Phosphorus Fertilization of Soybean. Indian J. Agric. Sci. 43(9-10) 925-929.

Somado EA, Sahrawat KL (2007). Biomass and nitrogen accumulation by green manure legumes as affected by phosphate rock in submerged soils. Afr. J. Agric. Res. 2(1):1-6.

Steel GD, Torrie JH (1960). Principle and Procedure of Statistic, MGGraw - Hill Book Co. Inc. New York, Toronto, London. pp. 312.

Vikoo B (2003). "The Mogho People and the Development of Modern Ogoni Nation: Spring Field, Owerri, Nigeria Paper 1.

A comparative study on Pb^{2+} removal efficiencies of fired clay soils of different particle size distributions

Dejene A. Tessema and Dessalew D. Alemayehu

Chemistry Department, College of Natural Sciences, Jimma University, P.O.Box 387, Jimma, Ethiopia.

Batch adsorption studies were carried out to evaluate the Pb^{2+} adsorption capacities of three different fired clay soils with different particle size distributions. Adsorption efficiency was observed to increase with an increase in clay content. Adsorption efficiencies of the fired clay soils were also influenced by the firing temperature, the solution's acidity, initial Pb^{2+} concentration, shaking time and adsorbent dose. Adsorption of Pb^{2+} by the clay soils was observed to increase linearly from pH 1 to 5 and then remained nearly constant at higher pH values. The adsorption of Pb^{2+} was observed to increase sharply with Pb^{2+} for the lower initial concentrations (5 to 10 mg/L), while a modest increase in Pb^{2+} adsorption was found from 10 to about 50 mg/L, a concentration at which maximum removal was attained. The Langmuir and Freundlich adsorption models were applied to the adsorption data and the Langmuir model represented the adsorption process better than the Freundlich model, with correlation coefficients (R^2) ranging from 0.97 to 0.99. The adsorption capacities (*qm*) calculated from Langmuir for the adsorption of Pb^{2+} by S_{3F}, S_{2F} and S_{1F} were found to be 17.3, 15.9, and 11.2 mg/g, respectively.

Key words: Adsorption, fired clay soils, isotherm, adsorption kinetics, sand, silt, particle size.

INTRODUCTION

Lead is known to be one of the most toxic contaminants, even at very low concentrations. Inorganic lead is a general metabolic poison and enzyme inhibitor (like most of the heavy metals). One of the most insidious effects of inorganic lead is its ability to replace calcium in bones and remain there to form a semi-permanent reservoir for long-term release of toxin long after the initial absorption.

Lead contamination exists in aqueous waste streams as a result of industrial processes such as the manufacture of storage batteries, pigments, leaded glass, fuels and photographic materials (Bhattacharyya and Gupta, 2006). Conventionally, lead is removed from waste water by one or more of precipitation, adsorption with activated carbon, ion exchange, membrane separation, and oxidetion and reduction technologies (Wang et al., 2003). Due to the fact that these waste water purification methods require high capital, high operating costs and often generate chemical sludge, which itself is a disposal problem, the feasibility of using various low-cost locally available adsorbents has been investigated for the removal of Pb^{2+} ions from waste water (Bulut and Baysal, 2006; Malakootian et al., 2009; Li et al., 2007; Qaiser et al., 2009). Clays are among the various locally available materials that are used for the removal of Pb^{2+} and other various heavy metals from aqueous solutions (Njoku et al., 2011; Talaat et al., 2011; Mbadcam et al., 2011; Inglezakis et al., 2007; Sar et al., 2007; Nassem and Tahir, 2001; Shi et al., 2009; Eba et al., 2010). In the soil, the variation in size of the particles and the proportionate amounts of fine and coarse materials, that is, the relative proportions of sand, silt and clay, known as the soil texture, determine to a large extent both its physical and chemical properties (McBride, 1994).

Clayey soils are mainly composed of the fine textured fractions, clay and silt, and are used for the production of bricks by firing the aforementioned to temperatures ranging from 900 to 1200°C (Ahmad et al., 2008). The phase composition of raw brick clays and fired bricks are generally complex due to the use of highly heterogenous soil. The major phases found in the raw soil materials, however, included quartz, calcite, chlorite and albite. In underfired bricks, illite is the most persistent of the clay minerals (Ahmad et al., 2008).

The use of fired clay for the removal of non-metallic and metallic pollutants has been reported in the literature (Seneveratne and Priyantha, 2009; Moges et al., 1996; McKay et al., 1985; Priyantha and Bandaranayaka, 2010; Priyantha et al., 2009; Renu et al., 2008; Bina et al., 2012). Bricks, however, are produced from clay soils that may differ widely in their mineral composition and hence textural classification. The variation in the textural composition of soils which the bricks are produced from may result in a difference in their heavy metal removal efficiencies and predicting the efficiency of a soil to adsorptively remove a heavy metal without taking its textural composition into account may be a misguided generalization.

Therefore, the main objective of this study was to evaluate and compare the ability of fired clay chips prepared in the laboratory from clay soils of different textural compositions to remove Pb^{2+} ion from aqueous solutions. In this work, the influence of clay soil firing temperature, adsorbent dosage, solution pH and shaking time on the adsorption of Pb^{2+} ions from aqueous solution by the fired clay soils were investigated. The well-known Langmuir and Freundlich isotherm models were also used to examine fit with the equilibrium data. The pseudo-first-order, pseudo-second-order and the intra-particle diffusion kinetic models were also applied to evaluate the nature of the sorption process.

MATERIALS AND METHODS

Clay soil collection and characterization

Clay soils were collected from ten different small scale brick production sites around Jimma town (South West Ethiopia). The particle size distributions of the clay soils were determined using the standard method developed by Sheldrick and Wang (1993). Three clay soils (S_1, S_2 and S_3) with wider difference in clay content were selected among the ten for further adsorption study. The proportion of clay was higher than those of silt and sand in the order: $S_3 > S_2 > S_1$.

The cation exchange capacity (CEC) and specific surface area (SSA) of the fired clay soils were determined by using the ammonium acetate (Priyantha et al., 2009) and the Sear's methods (Sears, 1956), respectively. The pH point of zero charge (pHpzc) of each of the adsorbents was measured by the potentiometric titration method using NaCl as the supporting electrolyte (Chapman, 1965).

Clay soil firing

For each of the three clay soil samples the firing temperature that produces a fired clay soil with the highest adsorption efficiency was determined by transferring about 200 g of each of S_1, S_2 or S_3 into separate crucibles and firing in a muffle furnace (3 h) at one of six different temperatures: 200, 300, 400, 500, 600, 700°C. Thus, one sample from each soil type was fired at each of the six firing temperatures and, a total of 18 samples were fired. The fired clay soils were then ground using a mortar and pestle and allowed to pass through a 0.5 mm sieve. Then, the Pb^{2+} adsorption efficiency of each of the fired clay soil samples was tested using the fraction ≤0.5 mm. The adsorption efficiency studies revealed that, the Pb^{2+} removal efficiencies of fired S_1 and S_2 increased with firing temperature between 200 and 400°C and decreased at higher firing temperatures. Fired S_3 samples, on the other hand, showed an increase in Pb^{2+} removal efficiency with firing temperature between 200 and 500°C and a decline at firing temperatures higher than 500°C. Based on these findings the samples fired at 400°C for S_1 and S_2, and the sample fired at 500°C for S_3, were used for subsequent batch adsorption studies.

For the batch adsorption experiments, 100 mL of 50 mg/L Pb^{2+} solution was transferred into a 250 mL conical flask and pH of the solution was adjusted to a desired value depending on the parameter to be studied. The required quantity of fired clay adsorbent was added and shaken on a thermostatic shaker for a predetermined contact time. The solutions were filtered before the measurement of residual Pb^{2+} concentration. For each parameter measured adsorption experiments were carried out in triplicates and averages of the results obtained from the triplicate experiments were reported.

Effect of pH

To study the pH effect on the adsorption of Pb^{2+} by each of the fired clay soils (S_{1F}, S_{2F} and S_{3F}), 100 mL of 50 mg/L Pb^{2+} solution was transferred into 250 mL Earlenmeyer flask containing four grams of the adsorbent. The pH of the solution was adjusted at the desired value by using 0.1 M NaOH or HCl and the mixture was shaken for 2 h on a mechanical shaker at 298 ± 2 K. The pH meter was calibrated with buffer solutions (pH 4.01 and 7.01) before each measurement. Finally, the solution was centrifuged, filtered and the residual metal ion concentration was determined by FAAS.

Effect of contact time

Contact time is one of the vital parameters for the successful application of adsorbents for the removal of metal ions. To study the effect of contact time, 100 mL, 50 mg/L of Pb^{2+} solution was transferred into a 250 mL volumetric flask containing four grams of the adsorbent. pH of the solution was adjusted to 5 using 0.1 M NaOH and HCl and the mixtures was shaken at 200 rpm for 2 h. The supernatant solution was collected in 30 min interval and the Pb^{2+} content of the solution was determined by FAAS.

Effect of adsorbent dose

To study adsorbent dose effect 100 mL, 50 mg/L of Pb^{2+} solution was transferred into each of seven 250 mL volumetric flasks containing one to seven grams of either adsorbent, in one gram

Table 1. Textural composition, CEC, SSA and pHpzc values.

Property	Mean ±SD		
	S_1	S_2	S_3
Clay (%)	39.98±0.02	58.20±0.12	82.01±0.31
Silt (%)	50.26±0.04	32.98±0.33	13.92±0.26
Sand (%)	9.84±0.11	8.82±0.29	4.11±0.03
CEC (meq/100 g)	29.9±0.58	51.7±0.47	69.61±0.61
	S_{1F}	S_{2F}	S_{3F}
SSA (m^2/g)	39±3.2	51.8±5.64	71±5.85
pHpzc	3.0±0.12	2.4±0.10	1.9±0.11

S_1, S_2 and S_3 = untreated soils; S_{1F}, S_{2F} and S_{3F} = fired clay soils.

interval. The pH of each solution was adjusted at 5 by using 0.1 M NaOH and HCl and the mixtures were shaken at 200 rpm for 2 h at 298 ± 2 K. Finally, the solutions were centrifuged, filtered and the residual metal ion concentrations determined by FAAS.

Effect of initial concentration

To study the effect of initial concentrations of Pb^{2+}, each of the seven 100 mL Pb^{2+} solutions having concentrations of: 5, 10, 20, 30, 40, 50 and 100 mg/L were transferred into seven separate 250 mL Erlenmeyer flasks each containing four grams of either of the three fired clay soil adsorbents. The pH of each solution was adjusted at 5 by using 0.1 M NaOH and HCl and, the mixtures were shaken at 200 rpm for 2 h at 298 ± 2 K. Finally, the solutions were centrifuged, filtered and the residual metal ion concentrations determined by FAAS.

Reagents and solutions

All chemicals used in this study were analytical grade. Deionized water was used for the preparation of stock standard solution of lead and all subsequent dilutions. A stock standard solution of Pb^{2+} (500 mg/L) was prepared by dissolving 0.80 g of $Pb(NO_3)_2$ (FINKEM) in 1000 mL deionized water. Calibration standards of the desired concentrations were prepared by serial dilution of the stock solution. Hydrochloric acid (Reideal Dehaen 37%) and/or sodium hydroxide (Fischer Scientific) solutions (0.1 N) were used throughout the experiments to adjust the pH of solutions.

Instrumentation

A NovAA 300 flame atomic absorption spectrometry (FAAS) with deuterium background signal corrector was used for the analyses of the metal ions in sample and standard solutions. An air/acetylene flame was used as an energy source for the atomization step. Before carrying out the actual analysis, the FAAS operating parameters were optimized by measuring the absorption signals of Pb^{2+} standard solutions and all parameters were adjusted at the optimum value during analysis. Ammonium chloride solution (20%) was used to minimize chemical interference.

RESULTS

Clay soil characterization

The particle size distributions and some physico-chemical properties of the selected clay soils were determined (Table 1). The clay content of S_3 was 1.41 and 2.06 times higher than those of S_2 and S_1, respectively. That of the S_2, on the other hand, was 1.46 times higher in clay content than that of S_1. The CEC and SSAs of the untreated soils follow the same order as their clay contents while their pHpzc was observed to decrease with an increase in clay content. Table 2 depicts the surface areas of the clay soils fired at different temperatures.

Table 2. SSAs of the clay soils fired at different temperatures.

Firing temperature (°C)	SSA (m^2/g)		
	S_{1F}	S_{2F}	S_{3F}
200	24±2.7	31±2.4	49±3.2
300	31±4.1	38±4.4	55±4.2
400	39±3.2	51.8±5.6	66±5.5
500	36±4.9	45.6±5.3	71±5.9
600	33±3.5	41±3.2	59±3.2

From the table, we can clearly observe that the surface areas of fired S_1 and S_2 increase in the temperature range of 200 to 400°C and decrease at temperatures higher than 400°C. On the other hand, the surface area of fired S_3 was found to increase in the temperature range of 200 to 500°C and decrease at temperatures higher than 500°C. The surface areas of the fired clay soils, although very low relative to those of activated carbon adsorbents (500 to 1400 m^2/g) which are used for various adsorption purposes, are fairly high relative to other adsorbents such as wheat bran (8.65 m^2/g) (Bulut and Baysal, 2006), natural clay (26.90 m^2/g) (Njoku et al., 2011) and vanadium mine tailing (42.39 m^2/g) (Shi et al., 2009), and cottonseed hull charred at 350 (4.7 m^2/g) and 650°C (34 m^2/g) (Uchimiya et al., 2011).

Effect of clay soil firing temperature on adsorption

The increase in the adsorption efficiencies of the fired clay soils with temperature in the range 200 to 400/500°C and the subsequent decline can be explained in terms of surface area (Table 2). With an increase in firing temperature more and more water molecules that are adsorbed in the clay matrix will be driven out, leaving porous structures in their place. This results in an increase in the surface areas of the fired clay soils. Temperatures above a certain optimum, however, destroy the clay structure and cause a decrease in surface area (Mortland, 1954). The surface area data in Table 2 confirms this fact. The reason for the higher optimum firing temperature of S_3 relative to those of S_2 and S_1 can be explained in terms of the higher water holding capacity of S_3. Due to the higher clay content of S_3 than S_2 and S_1 it may contain more water molecules

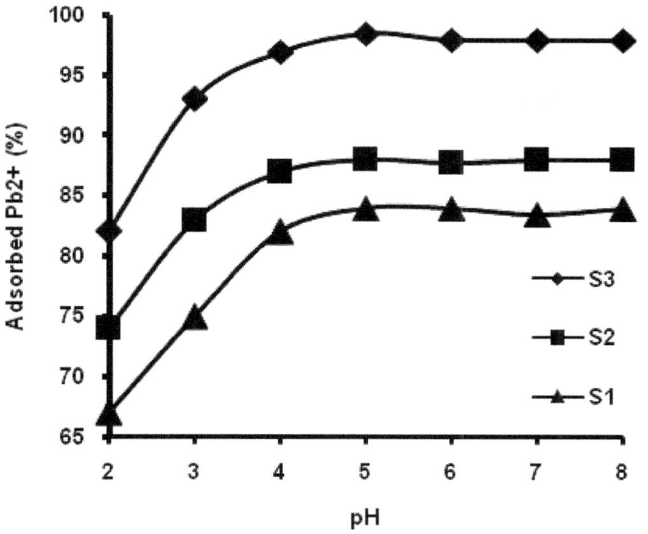

Figure 1. Adsorption of Pb^{2+} ions onto the fired clay soils at different solution pH (C_o = 50 mg/L, volume = 100 mL, adsorbent dose = 4 g, shaking time = 120 min, T = 298±2 K).

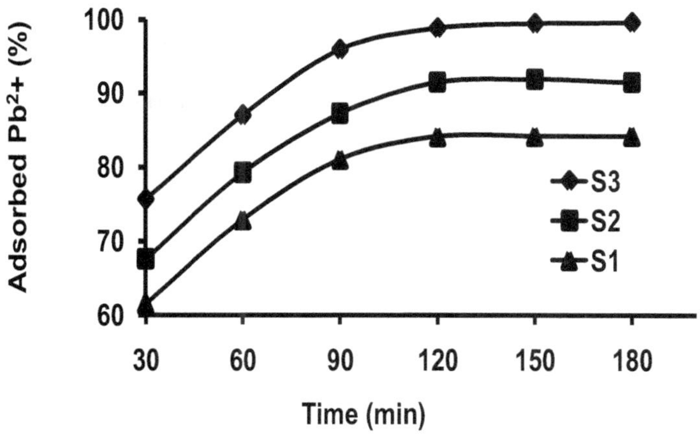

Figure 2. Removal of Pb^{2+} ions by the fired clay soils at different contact time (Co = 50 mg/L, volume = 100 mL, pH = 5, adsorbent dose = 4 g, T = 298±2 K).

that are strongly bound in its matrix. As a result, driving out the water molecules in S_3 may require higher temperature than those in S_2 and S_1.

Effect of pH

The effect of pH on the adsorption efficiency of Pb^{2+} was studied in the pH range of 2 to 10 (Figure 1). Not surprising, the results indicated that adsorption of Pb^{2+} was influenced by the pH of the solution. The adsorption of Pb^{2+} by all the clay-based soils in this study was observed to increase linearly from pH 2 to 4. At pH 5, adsorption increased slightly and remained nearly constant from then until pH 10. This phenomenon is known to be true for metal ions such as Pb^{2+} that have a strong tendency of undergoing hydrolysis (McBride, 1994).

Figure 1 clearly shows that, the Pb^{2+} adsorption efficiency of S_{3F} was higher than both S_{2F} and S_{1F} at all pH values studied. In turn, the adsorption efficiency of S_{2F} was higher than that of S_{1F}. The minimum adsorption efficiency for all the three clay soils was recorded at pH 2 and the maximum at pH 5. The minimum percent Pb^{2+} adsorbed by S_{3F}, S_{2F} and S_{1F} were, 82, 74 and 66%, respectively; while the maximum were 99, 89 and 83%.

Effect of contact time

The influence of contact time for the adsorption of Pb^{2+} on the clay soils was investigated by successively increasing the contact time from 30 to 180 min in 30 min intervals (Figure 2). The quantity of Pb^{2+} adsorbed onto the clay soils was found to increase steadily with time and equilibrium was attained at about 120 min in all the clay soils.

The increase in adsorption was rapid between 30 to 90 min and relatively slower until it appeared to reach constant state after 120 min of shaking time. After equilibrium, the adsorption rate was observed to remain nearly constant. The apparently high rate of adsorption in the first 90 min could be attributed to the fact that the available adsorption sites were sufficient when compared with the density of Pb^{2+} adsorbed at the beginning. With further increase in time, the adsorption sites become saturated gradually and the uptake rate will be controlled by the rate at which the adsorbate (Pb^{2+}) is transported from the exterior to the interior sites of the adsorbent particles, so the adsorption became much slower (Talaat et al., 2011). Finally, all possible adsorption sites will be occupied and from then on, a dynamic equilibrium in which Pb^{2+} ions adsorbed onto the clay surfaces are balanced by Pb^{2+} ions leaving the clay surfaces. As a result, no change shall be observed in the level of Pb^{2+} adsorption.

Effect of adsorbent dose

The percent removal of Pb^{2+} was studied by varying the adsorbent dose in the range of 1 to 6 g (Figure 3). It is clearly shown in Figure 3 that the adsorption of Pb^{2+} increased with an increase in the quantity of the clay adsorbent from 1 to 4 g and remained nearly constant at adsorbent quantities higher than 4g. As a result, an adsorbent dose of 4 g in 100 mL of adsorbate solution was used for the study of other parameters.

The maximum adsorptions for S_{1F}, S_{2F} and S_{3F} were found to be 83, 89 and 99%, respectively. The adsorption capacity of S_{3F}, the soil with the highest clay content, was

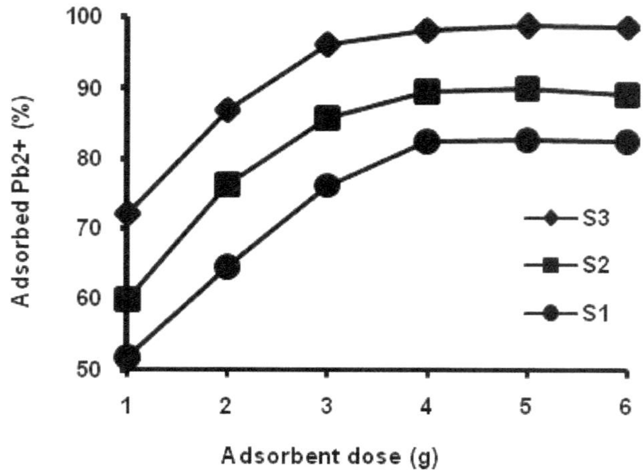

Figure 3. Adsorption of Pb^{2+} as a function of adsorbent dose (T = 298 ±2 K, C_o = 50 mg/ L, volume = 100 mL, pH = 5, shaking time = 120 min).

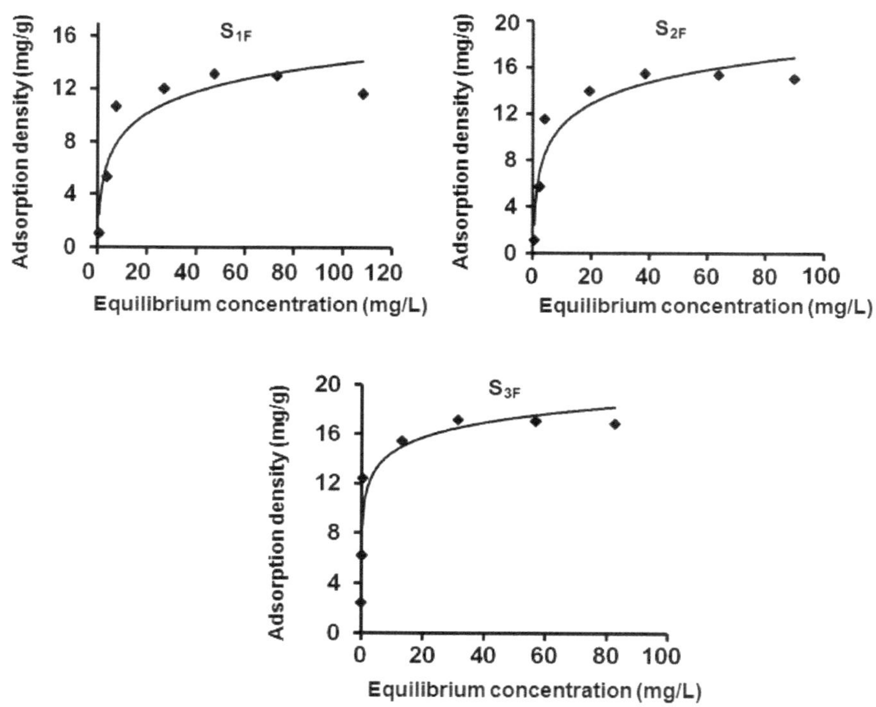

Figure 4. Adsorption isotherms of Pb^{2+} in S_{1F}, S_{2F} and S_{3F} at various equilibrium concentrations (adsorbent dose = 4 g, shaking time = 120 min, pH = 5 and T = 298 ±2 K).

observed to be about 10 and 20% higher than those of S_{2F} and S_{1F}, respectively. S_{2F}, whose clay content was 1.46 times greater than that of S_{1F} had an adsorption capacity about 6% higher than S_{1F}.

Effect of initial metal ion concentration

The results obtained for the effect of adsorbate concentration on the adsorption efficiencies of the fired clay soils are plotted as adsorption density (*qe*) versus bulk equilibrium concentration (*Ce*) (Figure 4). The isotherms indicate a sharp rise in the adsorption of Pb^{2+} for the lower (5 to 10 mg/L) $Pb2^{+}$ concentrations followed by a slower increase of up to about 50 mg/L. From this concentration onwards, the adsorption density was observed to remain constant.

Adsorption isotherms of the types shown in Figure 4 are extreme cases of the L-type (Langmuir) isotherm and are categorized under H-type (McBride, 1994). Such isotherms are indicative of a very strong adsorbate-adsorbent interaction (chemisorptions).

Adsorption isotherms

The equilibrium equation, commonly known as adsorption isotherm, is a basic requirement for the design of an adsorption system. Such an isotherm provides information on the capacity of the adsorbent or the amount required for removing a unit mass of pollutant under the operating conditions. Freundlich and Langmuir equations were early adsorption models and are still frequently used.

The two models differ from each other in their assumption concerning the existence of a finite adsorption capacity. The Langmiur equation includes a term for the maximum amount of metal that can be adsorbed; hence, it predicts a decrease in incremental adsorption, that is, in the $\Delta qe/\Delta Ce$ ratio at sufficiently high *qe* (Payne et al., 1998). The adsorption equilibrium data in the current study was applied to the Langmuir model using the linearized form of the isotherm expressed by equation (1) (Langmuir, 1918).

$$\frac{C_e}{q_e} = \frac{C_e}{q_m} + \frac{1}{K_L q_m} \tag{1}$$

Where, the term *Ce* is equilibrium concentration of the metal (mg/L) in solution and *qe* is the amount of the metal adsorbed per unit mass of clay soil (mg/g). q_m and K_L are Langmuir constants related to adsorption capacity (mg/g) and the energy of adsorption (L/g), respectively and were evaluated from the slope and intercept of the regression equation for the *Ce*/*qe* versus *Ce* plots, respectively.

The essential characteristics of the Langmuir isotherm can be expressed by a dimensionless constant called the equilibrium parameter, R_L, which is defined by Equation 2.

$$R_L = \frac{1}{1 + K_L C_o} \tag{2}$$

Where, Co = the initial concentration of Pb^{2+}; R_L = 0 indicates an irreversible adsorption; 0 < R_L < 1 favourable; R_L = 1 linear and, R_L >1 unfavourable.

The plots of *Ce*/*qe* versus *Ce* for the adsorption of Pb^{2+} on the three clay soils (figures not shown) were found to be linear with R^2 values ranging from 0.971 to 0.999. This shows that adsorption of Pb^{2+} on the fired clay soils conforms to the linear form of the Langmuir model in the concentration range considered in this study. Adsorbents of this type that exhibit the Langmuir isotherm behavior are supposed to contain fixed individual sites, each of which equally adsorbs only one molecule, thus forming a monolayer (McKinley and Jenne, 1991). The Langmuir adsorption maxima, *qm*, and the equilibrium constant, K_L, were found to be high for S_{3F}, 17.27 and 3.05 mg/g, respectively; followed by values of 15.86 and 1.99 mg/g for S_{2F} and 11.63 and 0.68 mg/g for S_{1F}. The orders of both *qm* and K_L for the three clay soils were in accordance with the order of their clay contents, that is, the higher the clay content of a soil, the higher were its *qm* and K_L values. The higher *qm* and K_L values associated with soils of higher clay content indicate that the maximum quantity of Pb^{2+} that can be adsorbed, the affinity of the binding sites and the energy of adsorption of soils of high clay content for Pb^{2+} was high relative to soils of low clay content. The calculated R_L values in all cases were found to be less than 1, indicating a favourable adsorption condition.

Using the *qm* and SSA values of each of the clay soils, the maximum adsorption density of Pb^{2+} ions on the clay soils were calculated in terms of number of Pb^{2+} ions adsorbed per unit area of soil. Accordingly, adsorption densities of 2.91, 1.17, and 0.87 Pb^{2+} ions per nm^2 area were obtained for S_{3F}, S_{2F} and S_{1F}, respectively. This reveals that the soil with the highest clay content, S_{3F}, did not only possess a higher surface area than S_{2F} and S_{1F}, but was also able to bind more Pb^{2+} per unit area. This can be attributed to the existence of a higher number of sites on the surface of S_{3F} that are preferred by Pb^{2+} ions for adsorption.

The equilibrium data in the current study were applied to the logarithmic form of the Freundlich model given by Equation 3 (Freundlich, 1906).

$$\log q_e = \log K_F + \frac{1}{n} \log C_e \tag{3}$$

Where, *qe* (mg/g) is the solid-phase concentration of Pb^{2+} which is in equilibrium with the liquid-phase concentration *Ce* (mg/L), K_F (mg/g) and *n* are Freundlich constants related to adsorption capacity and adsorption intensity, respectively. K_F and 1/*n* were determined from the plot of log *qe* versus log *Ce*, log K_F being the intercept and 1/*n* the slope of the isotherm. The equation obviously does not predict an adsorption maximum; that is, *qe* would be expected to increase as long as *Ce* increases.

The adsorption pattern of Pb^{2+} on the clay soils were found to fit well with the Freundlich model with R^2 values of 0.968, 0.966 and 0.969 for S_{3F}, S_{2F} and S_{1F}, respectively (figures not shown). Adsorbents that follow the Freundlich isotherm equation are assumed to have a heterogeneous surface consisting of sites with different adsorption potentials, and each type of site is assumed to adsorb molecules, as in the Langmuir equation. The Freundlich adsorption capacities, K_F, for S_{3F}, S_{2F} and S_{1F}, were found to be 1.92, 1.42 and 1.34 mg/g, respectively, indicating that the adsorption capacity of the soil was

Table 3. Rate constants and coefficient R^2 of models of pseudo first and pseudo second order rate equations used to describe adsorption of Pb^{2+} on the studied fired clay soils.

Adsorbent	Pseudo 1st order			Pseudo 2nd order		
	qe	k_1	R^2	qe	k_2	R^2
S_{1F}	1.206	0.015	0.937	1.909	0.0898	0.999
S_{2F}	1.328	0.021	0.925	1.913	0.102	0.998
S_{3F}	1.418	0.031	0.923	1.948	0.157	0.998

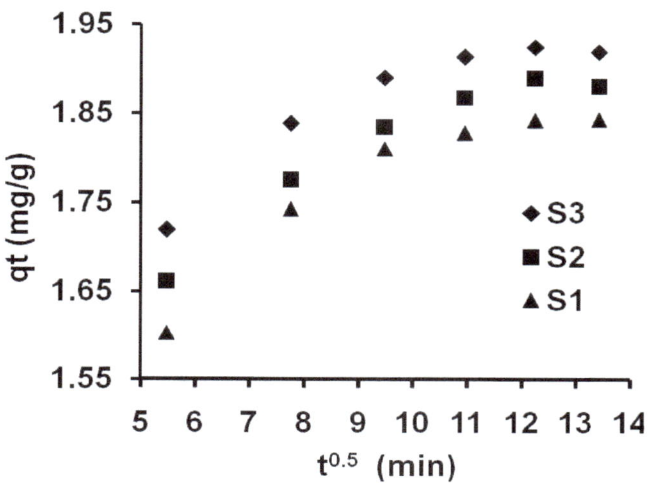

Figure 5. Intra-particle diffusion curve for the adsorption of Pb^{2+} ions on S_{3F}, S_{2F} and S_{1F}.

directly correlated with its clay content. Moreover, the Freundlich coefficients (1/*n*) were found to be below 1 (0.77, 0.71 and 0.53 for S_{3F}, S_{2F} and S_{1F}, respectively) indicating that the adsorptive removal of Pb^{2+} on the clay soils under the studied conditions was feasible.

Adsorption kinetics

Adsorption kinetics describes the solute uptake rate, which in turn governs the residence time of adsorption reaction and is one of the important characteristics in defining the efficiency of adsorption. In order to investigate the kinetics of adsorption of Pb^{2+} by the clay soils and the potential rate-controlling steps, Lagergren-first-order (Lagergren, 1898) and pseudo-second-order kinetic models (Ho and McKay, 1999) given by Equations 4 and 5, respectively, were used.

$$\log(q_e - q_t) = \log(q_e) - \frac{k}{2.303}t \quad (4)$$

$$\frac{t}{q_t} = \frac{1}{kq_e^2} + \frac{1}{q_e}t \quad (5)$$

where *qe* and *qt* are the solid-phase concentration at equilibrium and at time *t*, respectively.

Correlation coefficients and agreement between experimental adsorption data (q_{exp}) and model predicted values of the amounts of Pb^{2+} ions adsorbed (*qe*) were used to support the validity of the kinetic models. A satisfactory model shows good linearity of plots, expressed by a high R^2 value, and small deviation between *qe* and q_{exp}. The pseudo-first-order rate constant, k_1 (min^{-1}), pseudo-second-order rate constant, k_2 (g/mg/min) and, the adsorption capacity, *qe,* were computed from the slopes and intercepts of the plots (Table 3).

The R^2 values of the pseudo-second-order model were higher than those of the Lagergren-first-order model, indicating that the pseudo-second-order model was a comparatively better-fit to describe the adsorption process of Pb^{2+} onto the clay soils. The agreement between *qe* and q_{exp} values for the pseudo-second-order and pseudo-first-order models were found to be significant at $p < 0.01$. This result also supports the fact that the pseudo-second-order kinetic model was a better fit. The pseudo-second-order rate constant of S_{3F} (0.157 g min^{-1} mg^{-1}) was found to be higher than those of S_{2F} (0.102) and S_{1F} (0.0898). This could be an indication of the relatively faster rate at which equilibrium in S_{3F} is attained relative to S_{2F} or S_{1F}.

The amount of Pb^{2+} adsorbed per unit mass of adsorbent at time *t*, *qt*, as a function of the square root of time, $t^{0.5}$, was plotted using the Intra-particle Diffusion Model which is based on the theory proposed by Weber and Morris (1963) and given by Equation 6 (Figure 5).

$$q_t = k_p t^{0.5} \quad (6)$$

where, k_p (mg g^{-1} $min^{0.5}$) is the intra-particle diffusion rate constant.

The plot should be linear if the intra-particle diffusion is involved in the adsorption process. If the lines pass through the origin then the intra-particle diffusion is the rate-controlling step; if not, it is indicative of some degree of boundary layer control and this further shows that the intra-particle diffusion is not the only rate-limiting step, but that other kinetic models may also control the rate of adsorption, all of which may be operating simultaneously (Yazid and Maachi, 2008).

The plots in Figure 5 show an initial rising portion which rises linearly and then a plateau at higher $t^{0.5}$ values, which arises due to a difference between the initial and final stages of mass transfer (Panday et al., 1986). All three plots had non-zero intercepts indicating the existence of some degree of boundary layer control. Therefore, intra-particle diffusion was not a suitable mechanism to fully explain and predict the adsorption of the Pb^{2+} ion in this study. Similar results were reported by other researchers (Panday et al., 1986; Weber and Morris, 1963; Gok et al., 2008).

DISCUSSION

Adsorption efficiencies of S_1 and S_2 fired at 400°C were higher than those fired at lower or higher temperatures. While for S_3, firing temperature of 500°C was the optimum. The observed increase in adsorption efficiency up to 400/500°C and the subsequent decline can be explained in terms of surface area. With an increase in firing temperature more and more water molecules that are adsorbed in the clay matrix will be driven out, leaving porous structures in their place. This results in an increase in the surface areas of the fired clay soils. Temperatures above a certain optimum, however, destroy the clay structure and cause a decrease in surface area (Mortland, 1954). The surface area data in Table 2 confirms this fact. The reason for the higher optimum firing temperature of S_3 relative to those of S_2 and S_1 can be explained in terms of the higher water holding capacity of S_3. Due to the higher clay content of S_3 than S_2 and S_1 it may contain more water molecules that are strongly bound in its matrix. As a result, driving out the water molecules in S_3 may require higher temperature than those in S_2 and S_1.

Adsorption of Pb^{2+} by the clay soils was observed to increase from pH 2 to 5 and then levelling off after that. This could be explained as follows: at lower pH values (pH< PZC), where there are excess H_3O^+ ions in the solution, a competition exists between the positively charged H_3O^+ and Pb^{2+} ions for the available sites on the negatively charged clay surface (Eba et al., 2010). Meanwhile, the positive H_3O^+ ions that have already occupied the negative charges on the clay surface will inhibit the approach of Pb^{2+} ions to the clay surface and reduce or minimize further adsorption of Pb^{2+} ions. As a result, the level of Pb^{2+} adsorption could be predicted to be low. At higher pH values (pH > PZC), the clay surfaces become negatively charged due to deprotonation of functional groups, favouring Pb^{2+} uptake by the clay soils. Due to this, adsorption goes on increasing with an increase in pH. At about pH 5, the limited adsorption sites will be fully occupied and no further increase in Pb^{2+} ion adsorption was observed. Similar observations were reported by other researchers (Chantawong et al., 2001; Kaya and Oren, 2005).

With an increase in adsorbent dose, a corresponding increase in the removal of Pb^2 was observed until an optimum dose was reached. The initial increase in adsorption could be explained by an increase in the available active sites for the binding of Pb^{2+}. An increase in the number of active sites after equilibrium was reached, however, had no apparent effect on further increasing the removal of Pb^{2+}.

The amount of Pb^{2+} adsorbed was found to decrease with an increase in the equilibrium Pb^{2+} ion concentration in solution after an initial nearly-linear increase. This could be related to the well known "loading effect", which is conventionally interpreted as - the number of sites highly preferred by a given metal ion are a small portion of the total number of sites (Jenne, 1998; Payne et al., 1998). The initial sharp rise observed in adsorption followed by the relatively slow increase may be explained as follows: at lower equilibrium Pb^{2+} ion concentrations, surface coverage could be low and the availability of sites that are not occupied by the metal adsorbate may not be a limiting factor. As a result, incremental adsorption is expected to be independent of adsorption density. With an increase in the equilibrium Pb^{2+} ion concentration, the linearity of the increase in adsorption density with the equilibrium Pb^{2+} ion decreases due to an increase in the surface coverage.

The ratio of adsorbed Pb^{2+} to that of equilibrium concentration, however, generally decreases as the equilibrium concentration increases after only a small percentage of the adsorption capacity has been utilized. This could be due to the fact that only fractions of the total number of sites have high specificity for the adsorbate. Similar results were reported in other studies (Bulut and Baysal, 2006; Li et al., 2007; Qaiser et al., 2009; Njoku et al., 2011; Inglezakis et al., 2007; Sar et al., 2007).

Conclusion

In this study, the relative ability of fired clay soils to remove Pb^{2+} from aqueous solution was investigated. The fired clay soils showed removal efficiencies ranging from 83 to 99% where the highest removal efficiency was recorded for the soil with the highest clay content. In general, an average of 0.4% increase in adsorption capacity was achieved for a 1% increase in clay content. This may be attributable to the higher SSA and CEC possessed by the soil with higher clay content relative to the one with lower clay content. The affinity for Pb^{2+} ion, confirmed by the *qm* values determined from the Langmuir adsorption isotherm, was highest for the high-clay soil, S_{3F} (17.27 mg/g) followed by the two lower clay-content samples, S_{2F} (15.68 mg/g) and S_{1F} (11.63 mg/g). The isotherms can be better described by the Langmuir model (R^2 0.971 to 0.999) than by the Freundlich model (R^2 0.966 to 0.969). The kinetic studies conducted using the Weber and Morris equation showed that the adsorption mechanism involves intra-particle diffusion but it was not the fully operative mechanism in the adsorption of Pb^{2+} by the three clay soils. The pseudo-second-order kinetic model was found to be a better fit for the adsorption of Pb^{2+} by the clay soils.

The findings of the current study in general revealed that the Pb^{2+} (and possibly other metal ions) removal efficiencies of fired clay soils/bricks differ depending on the textural compositions of the clay soils from which the fired clays/bricks were prepared. Fired clay soils/bricks prepared from raw clay soil of higher clay : silt : sand ratio can remove Pb^{2+} better than those which were prepared from a lower ratio. Therefore, the selection of clay soils with the highest clay content relative to its silt and sand contents before firing for the purpose of adsorptive removal of Pb^{2+} will be necessary for efficient use of the fired clay soil adsorbents.

REFERENCES

Ahmad S, Iqbal Y, Ghani F (2008). Phase and Microstructure of Brick-Clay Soil and Fired Clay-Bricks from Some Areas in Peshawar Pakistan. J. Pak. Mater. Soc. 2:33-39.

Bhattacharyya KG, Gupta SS (2006). Adsorption of chromium(VI) from water by clays. Ind. Eng. Chem. Res. 45:7232-7240.

Bina R, Raaz M, Chauhan AK, Bhaskar NS (2012) Defluoridation of contaminated water employing brick powder as an adsorbent. Int. J. Sci. Nat. 3:78-82

Bulut Y, Baysal Z (2006). Removal of Pb(II) from wastewater using wheat bran. J. Environ. Manage. 78:107-113.

Chantawong V, Harvey N, Bashkin VN (2001). Adsorption of lead nitrate on Thai kaolin and ball clay. Asian J. Energy Environ. 2:33-48.

Chapman HD (1965). Cation-exchange capacity. in: Methods of Soil Analysis, Part 2: Chemical and microbiological properties (C.A. Black, Editor), American Society of Agronomy: Madison. pp. 891-901.

Eba F, Gueu S, A-Mvongbote AE, Ondo JA, Yao BK, Ndong Nlo J, Biboutou RK (2010). Evaluation of the absorption capacity of the natural clay from Bikougou (Gabon) to remove Mn (II) from aqueous solution. Int. J. Eng. Sci. Technol. 2:5001-5016.

Freundlich HMF (1906). Über die adsorption in lösungen. Zeitschrift für Physikalische Chemie. Leipzig. 57A:385-470.

Gok O, Ozcan AS, Ozcan A (2008). Adsorption kinetics of naphthalene onto organo-sepiolite from aqueous solutions. Desalination 220:96-107.

Ho YS, McKay G (1999). Pseudo-second order model for sorption processes. Process Biochem. 34:451-465.

Inglezakis VJ, Stylianou MA, Gkantzou D, Loizidou MD (2007). Removal of Pb(II) from aqueous solutions by using clinoptilolite and bentonite as adsorbents. Desalination 210:248-256.

Jenne EA (1998). Adsorption of metals by geomedia: data analysis, modeling, controlling factors, and related issues. in: *Adsorption of Metals by Geomedia* (E.A. Jenne, editor), Academic Press: San Diego. pp. 1-73.

Kaya A, Oren AH (2005). Adsorption of zinc from aqueous solutions to bentonite. J. Hazard. Mater. 125:183-189.

Lagergren S (1898) About the theory of so-called adsorption of soluble substances. K. Sven. Vetenskapsakad. Handl. 24:1-39.

Langmuir I (1918). The adsorption of gases on plane surfaces of glass, mica and platinum. J. Am. Chem. Soc. 40:1361-1403.

Li Q, Zhai J, Zhang W, Wang M, Zhou J (2007). Kinetic studies of adsorption of Pb(II), Cr(III) and Cu(II) from aqueous solution by sawdust and modified peanut husk. J. Hazard. Mater. 141:163-167.

Malakootian M, Nouri J, Hossaini H (2009). Removal of heavy metals from paint industry's wastewater using Leca as an available adsorbent. Int. J. Environ. Sci. Technol. 6:183-190.

Mbadcam JK, Anagho SG, Ndinsami J, Kammegne AM (2011). Kinetic and equilibrium studies of the adsorption of lead (II) ions from aqueous solution onto two Cameroon clays: Kaolinite and smectite. J. Environ. Chem. Ecotoxicol. 11:290-297.

McBride MB (1994). Environmental Chemistry of Soils, Oxford University Press, USA.

McKay G, Otterburn MS, Sweeney AG (1985). Fullers earth and fired clay as adsorbents for dye stuffs equilibrium and rate studies. Water Air Soil Pollut. 24:147-151.

McKinley JP, Jenne EA (1991). Experimental investigation and review of the "solids concentration" effect in adsorption studies. Environ. Sci. Technol. 25: 2082-2087.

Moges G, Zewge F, Socher M (1996). Preliminary investigations on the defluoridation of water using fired clay chips. J. Afr. Earth Sci. 21: 479-482.

Mortland MM (1954). Specific Surface Area and its Relationship to Some Physical and Chemical Properties of Soil. Soil Sci. 78:343-347.

Nassem R, Tahir SS (2001). Removal of Pb(II) from aqueous/acidic solutions by using bentonite as an adsorbent. Water Resour. 35:3982-3986.

Njoku VO, Oguzie EE, Obi C, Bello OS, Ayuk AA (2011). Adsorption of Copper(II) and Lead(II) from Aqueous Solutions onto a Nigerian Natural Clay. Aust. J. Basic Appl. Sci. 5:346-353.

Panday KK, Prasad G, Singh VN (1986). Mixed adsorbent for Cu(II) removal from aqueous solutions. Environ. Technol. Lett. 50:547-554.

Payne TE, Lumpkin GR, Waite TD (1998). Uranium(VI) adsorption on model minerals: controlling factors and surface complexation modelling. In: Adsorption of Metals by Geomedia: Variables, Mechanisms and Model Applications (E.A. Jenne, editor). Academic Press, San Diego. pp. 75-97.

Priyantha N, Bandaranayaka A (2010). Optimization of parameters for effective removal of Cr(VI) species by burnt brick clay. J. Natl. Sci. Foundat. Sri Lanka. 38:107-112.

Priyantha N, Senevirathna C, Gunathilake P, Weerasooriya R (2009). Adsorption behavior of fluoride at normal brick (NB)—water interface. J. Environ. Prot. Sci. 3:140-146.

Qaiser S, Saleemi AR, Umar M (2009). Biosorption of lead(II) and chromium(VI) on groundnut hull: Equilibrium, kinetics and thermodynamics study. Electronic J. Biotechnol. 12:1-17.

Renu S, Shiv PR, Kaushik CP (2008). Defluoridation of Drinking Water using Brick Powder as an Adsorbent. Asian J. Chem. 20: 5818-5826

Sar A, Tuzen M, Soylak M (2007). Adsorption of Pb(II) and Cr(III) from aqueous solution on Celtek clay. J. Hazard. Mater. 144:41-46.

Sears G (1956) Determination of specific surface area of colloidal silica by titration with sodium hydroxide. J. Anal. Chem. 28:1981- 1983.

Seneveratne C, Priyantha N (2009). Correlation between firing temperature and defluoridation capacity of brick clay. Int. J. Glob. Environ. 9:239-248.

Sheldrick BH, Wang C (1993) Particle size distribution. In: Soil Sampling and Methods of Analysis (M.R. Carter, editor), Lewis Publishers, U.S.A. pp. 499-511.

Shi T, Shiguo J, Ying C, Yinghong W, Changming D, Huilin G, Zhuochao W (2009). Adsorption of Pb(II), Cr(III), Cu(II), Cd(II) and Ni(II) onto a vanadium mine tailing from aqueous solution. J. Hazard. Mater. 169: 838-846.

Talaat HA, El-Defrawy NM, Abulnour AG, Hani HA, Tawfik A (2011). Evaluation of Heavy Metals Removal Using Some Egyptian Clays. 2nd International Conference on Environmental Science and Technology, 37-42.

Uchimiya M, Lynda H, Wartelle K, Klasson T, Fortier CA, Lima IM (2011). Influence of pyrolysis temperature on biochar property and function as a heavy metal sorbent in soil. J. Agric. Food Chem. 59: 2501-2510.

Wang YH, Lin SH, Juang RS (2003). Removal of heavy metal ions from aqueous solutions using various low-cost adsorbents. J. Hazard. Mater. 102:291-302.

Weber WJ, Morris JC (1963). Kinetics of adsorption on carbon from solution. J. Sanit. Eng. Div. ASCE 89: 31-39.

Yazid Y, Maachi R (2008). Biosorption of lead (II) ions from aqueous solutions by biological activated dates stems. J. Environ. Sci. Technol. 1:201-213.

18

Rainfall variability and rubber production in Nigeria

Mesike C. S. and Esekhade T. U.

Rubber Research Institute of Nigeria, P.M.B. 1049, Benin City, Edo State, Nigeria.

The role of rainfall in plant could not be overemphasized because rainfall determines the amount of moisture present in the soil which is ultimately made available to plants. The aim of this paper was to determine the variability of rainfall and its effect on rubber production in Nigeria. Towards achieving this aim, time series data from 1971 to 2009 were collected on mean annual rainfall and total annual rubber production. The data was analyzed using descriptive statistics as well as correlation analysis. The results showed that high rubber production was recorded in 1991 while production was low in 1980 and 1983. Results also show that there is an inverse relationship between rubber production and rainfall. The study recommended that protective water proof containers should be used for collection of latex during raining season so as to prevent washing away of latex by rain.

Key words: Rainfall, variability, trend, correlation, rubber production.

INTRODUCTION

Climate plays a dominant role in agriculture having a direct impact on the productivity of physical production factors, for example the soil's moisture and fertility. Adverse climate effects can influence farming outputs at any stage from cultivation through the final harvest. Even if there is sufficient rain, its irregularity can affect yields adversely if rains fail to arrive during the crucial growing stage of the crops (Smith and Skinner, 2002; Molua and Lambi, 2007; Rudolf and Hermann, 2009).

Generally, there are many factors influencing crop production and these include soil, climate and diseases among others. In relation to climate, rainfall is the dominant controlling variable in tropical agriculture since it supplies soil moisture for crops and grasses for animals. According to Ayoade (1983), agriculture largely depends on climate to function. Hence, precipitation, solar radiation, wind, temperature, relative humidity and other climatic parameters affect and solely determine the global distribution of crops and livestock as well as their productivity.

Rainfall, among other factors, has always dictated how land is used in one way or another and it also affects the humidity condition of the atmosphere. Rainfall determines the vegetation cover of a particular geological zone and crop distribution. Heavy rainfall facilitates the growing of tree crops like cocoa, rubber, oil palm possible in the rainforest. Some of the attributes of rainfall that are important to crop production are the time of onset of the raining season, total amount of rainfall, distribution, number of rainy days and duration of rainfall as well as the time of its cessation (Akintola, 1995). Furthermore, rainfall determines the amount of moisture present in the soil which is ultimately made available to plants. Water plays a vital role in the growth of plant and it provides the medium through which nutrients are carried through the plant. According to Olaoye (1999), regular occurrence of drought as a result of erratic rainfall distribution and/or cessation of rain during the growing season reduce Nigeria's capability for increased crop production. Sdoodee and Rongsawat (2012) concluded that high rainfall tended to decrease tapping days per year. From the results, it was suggested that climate change and climate variability in Songkhla province tends to reduce latex yield because of an increase of rainfall leading to a reduction of tapping days.

Rain-fed crop production is a dominant mode of

Table 1. Analysis of rainfall data from 1971 - 2009.

Variable	Coefficient	Std. error	t-statistic	Prob.
Constant	318.89	37.41685	8.522643	0.0000
DR(-1)	- 0.376412	0.159985	- 2.352793	0.0242
Mean (mm)			231.41	
Standard deviation (mm)			27.31	
Maximum rainfall (mm)			283.05	
Minimum rainfall (mm)			189.02	
Trend coefficient (mm/ year)			- 0.38	
Coefficient of variability (CV)			11.80	

Table 2. Analysis of rubber production from 1971-2009.

Variable	Coefficient	Std. error	t-statistic	Prob.
Constant	8369.54	6469.65	1.293662	0.2040
Q(-1)	0.935243	0.062823	14.88697	0.0000
Mean (tonnes)		97315.36		
Standard deviation (tonnes)		37765.84		
Maximum rubber production (tonnes)		155000.00		
Minimum rubber production (tonnes)		45000.00		
Coefficient of variability (%)		38.81		
Sum		3795299		

agricultural production in the majority of rural sub-Saharan Africa (Cooper et al., 2008). The reliance of the agricultural sector on natural rainfall places it at a serious risk of shrinkage due to the inter-annual rainfall variations. Climate change impact assessments done by the Intergovernmental Panel on Climate Change (IPCC) (2007) and Buddenhagen et al. (1992) concluded that rain fed agriculture in Africa risks negative impacts due to climate change. Rain fed agricultural production in Africa in general is projected to be reduced by up to 50% by 2020 (IPCC, 2007). With the declining rainfall trends in most of the sub-Saharan Africa, agricultural production is most likely to decline (Droogers et al., 2001; Nnyaladzi, 2009).

Rubber is very important in the economy of Nigeria and contributes significantly to Nigeria export trade. Natural rubber thrives well in humid climate with well-distributed rainfall of 1800 to 2000 mm on a well-drained soil. The rain must be evenly distributed through the year and with not more than one dry month. Ideally, the number of rainy day should range from 100-150 days (Watson, 1989). Production statistics indicated that Nigeria has a total of 247,100 hectares of land under natural rubber cultivation where over 70% are owned by small scale farmers (Delabarre and Series, 2000). Rubber cultivation in Nigeria is mainly rainfed. The rain being generally seasonal, regulate the production of planting materials and tapping activities. For this reason we studied the rainfall variability in line with rubber production in Nigeria.

MATERIALS AND METHODS

The empirical analysis covers the period between 1971 and 2009. Secondary data used for the analysis were obtained from Nigeria Meteorological Agency (NIMET), Central Bank of Nigeria (CBN) publications, such as Annual Reports and Statements of Accounts, and the Statistical Bulletin. Rainfall data used for this study were obtained from NIMET and the data related specifically to Delta state where over 70% of Nigeria rubber is produced was used. Data on quantity of rubber produced in Nigeria were obtained from CBN publications.

The data collected were analyzed with the use of descriptive statistics to analyze the trend of rainfall and rubber production in Nigeria. Correlation analysis was used to analyze the relationship between rainfall and rubber production.

RESULTS AND DISCUSSION

Trend analysis of rainfall

Rainfall shows a decreasing trend with the minimum value for the period (189.02 mm) recorded in 1977 and maximum value for the period (283.05 mm) recorded in 1999 (Figure 1). The mean and standard deviation values of rainfall are 231.41 and 27.31 mm, respectively (Table 1). This indicates that rainfall has a large variability of 11.8% with time. The rainfall trend coefficient of -0.38 mm per year is significant at 5%. This indicates a decreasing trend of rainfall in the region. The coefficient of the graph is DR = 318.89 - 0.38 t. Where: DR is the annual mean rainfall (mm) and t is the time (years). The trend equation had a negative slope of 0.38 indicating that over the time period of 1971 to 2009, the annual mean rainfall decreased by 0.38 mm per unit change in time.

Trend of rubber production in Nigeria

Figure 2 shows the trend of rubber production in Nigeria. It could be observed that the highest rubber production of 155,000 metric tonnes was recorded in 1991 and the lowest productions of 45,000 tonnes were recorded in 1980 and 1983. This however, represents 4.1% of the total production during the period. There were steady increases in rubber production in 1987-1991. This period however fall within the first decade of economic deregulation. Table 2 shows that the mean and standard deviation value of rubber production are 97315.36 tonnes and 37765.84, respectively and this indicated that rubber production has a large variability of 38.81% over time.

Correlation analysis

The result of the correlation analysis showed that rainfall

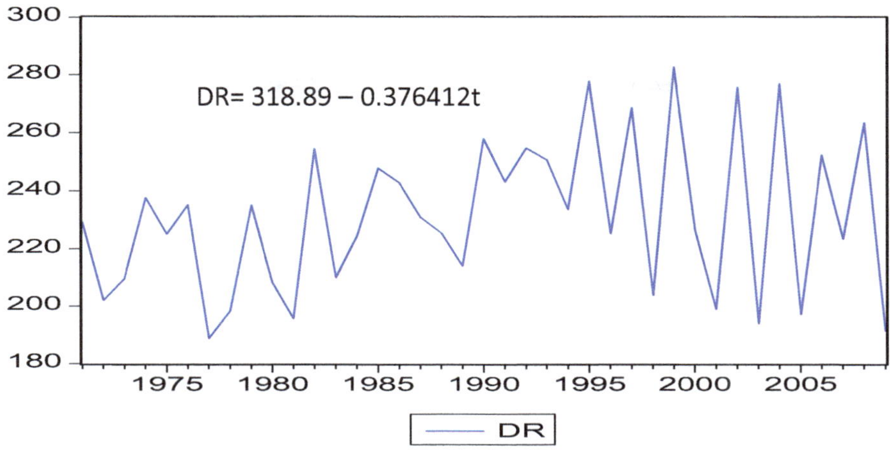

Figure 1. Trend of rainfall (1971 – 2009).

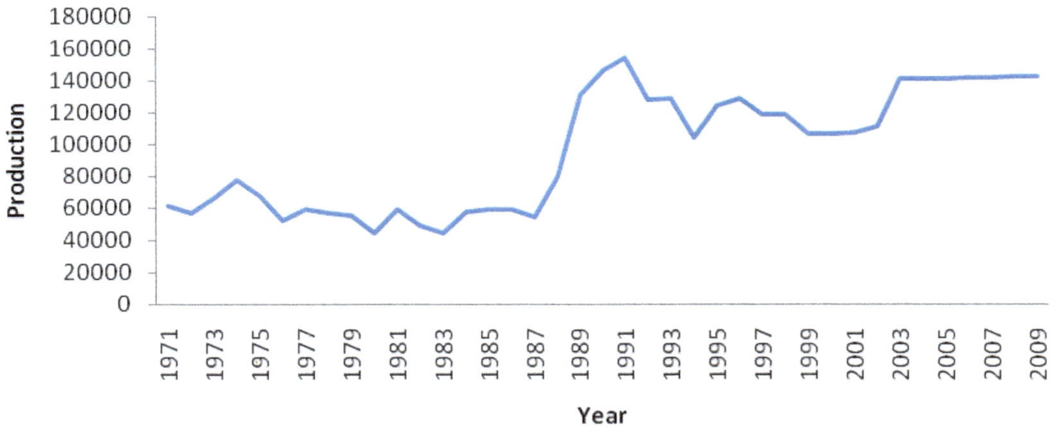

Figure 2. Trend of rubber production in Nigeria.

is inversely correlated with rubber production (-0.132). This result is line with Perera and Ranasinghe (2013) that recorded an inverse relationship between rainfall and rubber yield in Sri Lanka. The negative relationship between rubber and rainfall might be due to the fact that excess rainfall may affect tapping operation thus reducing latex yield.

Conclusion and recommendation

The study has revealed that there was high variability and decreasing trend of rainfall during the period (1971-2009). High rubber production of 155,000 tonnes was recorded in 1991 while the lowest production of 45,000 tonnes was recorded in 1980 and 1983. Rubber production is inversely correlated with rainfall. The study therefore recommend that protective water proof containers should be used for collection of latex during raining season so as to prevent washing away of latex by rain.

REFERENCES

Akintola FO (1995). Rainfall Characteristics and Cocoa production in Nigeria. In Adegeye and Ajayi (eds.). Cocoa Revolution in Nigeria. Proceedings of a National Seminar on Revolutionizing Nigeria's Cocoa Industry, University of Ibadan.

Ayoade JO (1983). Introduction to Climatology for the tropics. Spectrum book Ltd, Ibadan.

Buddenhagen IW, Gibson RW, Sweetmore A (1992). Better cultivars for resource poor farmers. In: Proceedingsof the seminar on crop protection for resource poor farmers UK: The Netherlands NRI.

Cooper PJM, Dimes J, Rao KPC, Shapiro B, Shiferaw B, Twomlow S (2008). Coping better with current climatic variability in the rain-fed farming systems of Sub-Saharan Africa: An essential first step in adapting to future climate change? Agriculture, Ecosystems and Environment, 126: 24-35.

Delabarre MA, Series JB (2000). Rubber: The Tropical Agriculturalist. CTA Macmillan Education Ltd London pp. 4-11.

Droogers P, Seckler D, Makin I (2001). Estimating the potential of rain-fed agriculture: Working paper 20. Colombo, Sri Lanka: International

Water Management Institute.
IPCC (2007). Climate Change (2007): Synthesis Report PCC Plenary XXVII. Valencia, Spain
Molua EL, Lambi CM (2007). Economic Impact of Climate change on agriculture in Cameroon. Policy Research paper No 4364 World Bank, Washington, D. C. pp. 51-55.
Nnyaladzi BN (2009) Rainfall trends in semi-arid Botswana: Implications for climate adaptation policy. Appl. Geogr. **30**:483-489.
Olaoye O (1999). Developing drought tolerant varieties for the Savanna agro-ecologies of Nigeria in 25th year Commemoratives Publications of Genesis Society of Nigeria. pp. 173-182.
Perera KKE, Ranasinghe EMS (2013).The effect of rainfall variability on rubber cultivation: A case study of Madurawala AGA division in Kalutara District, Sri Lanka. Proceedings of the Annual Research Symposium, University of Colombo, October 3-4.
Rudolf W, Hermann W (2009). Climate risk and farming Systems in Rural Cameroon. Institute of Development and Agricultural Economics. University of Hannover, Germany pp. 21-24.
Sdoodee S, Rongsawat S (2012). Impact of Climate Change on Smallholders' Rubber Production in Songkhla Province, Southern Thailand. Proceedings International and National Conference for the Sustainable Community Development of "Local Community: The Foundation of Development in the ASEAN Economic Community (AEC)" February 16-19, pp 81-86.
Smith B, Skinner M (2002). Adaptation options in Agriculture to climate change: A typology, mitigation and Adaptation Strategies for Global Change. Afr. J. Agric. Res. Econs 3(5): 78-82.
Watson GA (1989). Climate and soil in Rubber, New York: Longman Scientific and Technical. pp. 125-164.

Recovery and characterization of poly(3-Hydroxybutyric acid) synthesized in *Staphylococcus epidermidis*

Darshan Marjadi[1] and Nishith Dharaiya[2]

[1]Department of Biotechnology, Shree Ramkrishna Institute of Computer Education and Applied Sciences, Surat, Gujarat India.
[2]Department of Life Sciences, Hemchnandracharya North Gujarat University, Patan, Gujarat, India.

Polyhydroxyalkanoates (PHA) are biodegradable polyesters accumulated intracellularly as energy resources by bacterial species. In this study, fermentation process for production of PHA is carried out using sesame oil as carbon source. We studied recovery of poly(3-hydroxybutyric acid) (PHB) from *Staphylococcus epidermidis* by sodium hypochlorite digestion method. Recovered PHB sample was estimated by UV spectrophotometer. PHB from *S. epidermidis* was characterized and by these findings, we examined purified PHB by differential scanning calorimeter (DSC), a thermo gravimetric analyzer (TGA), thin layer chromatography (TLC) and infrared spectroscopy (IR). The results of our analysis of PHB while comparing with commercial source suggest that in DSC melting temperature of PHB was 173.36°C, TGA thermo grams of PHB sample was at 296.91°C, on TLC plate; Rf value was calculated as 0.71 and finally IR spectrum of the compounds showed characteristics bands for the groups CH, C=O and C-O, indicating the presence of PHB in the production medium.

Key words: Polyhydroxyalkanoates (PHA), poly(3-hydroxybutyric acid) (PHB), *Staphylococcus epidermidis.*

INTRODUCTION

Often, research and media attention on the renewable bioproduct industry is focused specifically on fuel alternatives. This is logical since diminishing fossil fuels supply nearly 80% of the global energy demands, and it is predicted that the current demand will increase by 56% by 2040 (U.S. Energy Information Administration, 2013). Currently, though, the renewable energy industry is one of the two fastest-growing industries globally, increasing at a rate of 2.5% per year (U.S. Energy Information Administration, 2013). By 2015, the demand for biodegradable plastics is estimated to reach 1.1 million tons (Metabolix, 2013a). To make environment free from plastics is one of the major interests to both decision makers and plastic industries (Chen, 2009). Poly(3-hydroxyalkanoic acid) (PHA) is a biodegradable polymer material that accumulates in numerous microorganisms under unbalanced growth conditions (Ribera et al., 2001). Poly β-hydroxybutyrate (PHB) is biopolymer that can be used as biodegradable plastics is the most common natural microbial PHA (Singh and Parmar, 2013). In terms of molecular weight, brittleness, stiffness, melting point and glass transition temperature, the PHB homopolymer is comparable to some of the more common petrochemical derived thermoplastics such as polypropylene (Sayed et al., 2009). Although PHB was found to be at advantage comparing to non-biodegradable plastics; its application is inevitably limited due to high production costs.

PHB is microbial polyester produced by many bacteria and stored in their cell in the form of granules, about 0.5 μm in diameter. β-hydroxybutyrate is connected by ester linkage and form PHB (Prasanna et al., 2011). PHB possesses only *R (alkyl group)* side chains (and lacks*S (Sulfur)* side chains) and hence reported as biodegradable materials (Anderson and Dawes, 1990; Saito et al., 1996; Jung et al., 2001); for example, vulcanized rubber. PHB is an intracellular product; the method applicable for its effective separation from other biomass component is complex and expensive.

Number of different methods for the recovery of PHB has been suggested. There are some of the known, effective methods for separation of PHB from bacterial cell: Physical method using a bead mill (Kunasundari and Sudesh, 2011), extraction method uses an organic solvent (Ibrahim and Steinbuchel, 2009), enzyme method (Kathiraser et al., 2007).

Finally, there must be a method which allows consistent recovery of the polymer with high purity. There may be a different requirement of purity of biopolymer which depends on its intended application and which ultimately designs the recovery method of PHB extraction.

In this work we studied PHB recovery from *Staphylococcus epidermidis* which possesses tendancy to utilize sesame oil and it has been reported previously that plant oils are desirable feed stocks for PHA production because they are also inexpensive in comparison with other carbon sources, such as sugar (Akiyama et al., 2003). Hence, here *S. epidermidis,* a known PHA-producing bacterium is utilized for study of PHB production and recovery of PHB from *S. epidermidis* worthy of investigation. PHB was recovered through a dispersion of a sodium hypochlorite solution and chloroform. In this paper we also described characterization and determination of native PHB like granules which recovered using various organic solvents.

MATERIALS AND METHODS

Cultivation of bacteria

S. epidermidis, isolated from edible oil contaminated sites and was used for PHB production (Marjadi and Dharaiya, 2011). *S. epidermidis* was grown and maintained in a modified mineral salts medium (MSM) (Marjadi and Dharaiya, 2011).

The production of PHB or copolymer was carried out by two-stage cultivation (Hartmann et al., 2010). First stage organisms were cultivated in the nutrient broth medium without any nutrient limitation, at 37°C and 150 rpm for 24 h. In second stage, after incubation, 2 ml of culture was taken to inoculate according to their dry cell weight (Marjadi and Dharaiya, 2011) the flask containing 200 ml of sterile production medium, and all the isolates were first grown for 72 h. at 37°C with shaking at 150 rpm in a carbon - rich MSB medium containing sesame oil (1% w/v) as a sole carbon source and cells accumulating PHB were cultivated in 250 ml of modified mineral salts basal medium (MMSB) as described by Marjadi and Dharaiya (2012).

Production and storage of PHB-containing biomass

To recover and characterizePHB produced in *S. epidermidis a*fter fermentation, the cell broth was concentrated bycentrifugation at 4,000 RPM for 15 min at 25°C, washed twice with distilled water,and then freeze dried. The resulting cell powder was stored at 4°C until they used further.

PHB recovery

After 96 h of incubation at 37°C, 10 ml of culture was taken into clean polypropylene centrifuge tubes which had been previously washed thoroughly with ethanol and hot chloroform to remove plasticizers. In each tube, 5 drops of formaldehyde were added in order to stop all biological activity and then centrifuged at 8000 rpm for 15 min. The supernatant was discarded and collected pellet was washed twice with 5 ml cold water and 2 ml cold hexane (Loba®) twice to remove hydrophobic residual oil (Kahar et al., 2004). The remaining pellet was dried in oven until obtaining the constant weight. The dried pellet was treated with the original volume of culture medium (10 ml) of 30 % (v/v) sodium hypochlororite (NaOCl) (Loba®) and the mixture was incubated at 37°C for 1 h. After incubation, the mixture containing the lipid granules was centrifuged at 6,000 rpm for 15 min and was washed with water, and then with 5 ml 96% cold acetone (Loba®) and followed by ethanol (Loba®) (1:1). The precipitates thus formed were allowed to dry to obtain PHB crystals.

PHB extraction

Extracted PHB crystals were re-dissolved in 5 mg in 5 ml (Sayyed et al., 2009) of chloroform in a test tube in water bath at 100°C for 20 min and filtered through Whatman No.1 filter paper and chloroform was evaporated by pouring the solution on sterile glass petri plate and then kept at 4°C in deep fridge. After some time, powder was collected from petri plates by slowly scratching for further analysis (Kuniko et al., 1989; Bowker, 1981; Ishizaki and Tanaka, 1991).

Estimation of PHB concentration

PHB concentration was estimated as suggested by Law and Slepecky (1961). Extracted PHB powder was transferred to clean test tube (Figure 1) and 10 ml of concentrated H_2SO_4 was added to the tube which was capped and heated for 20 min at 100°C in a water bath. PHB crystals were converted in to crotonic acid by dehydration (Aslim et al., 2002).The resultant brown colourcrotonic acid solution was cooled, and after thorough mixing, a sample was transferred to a quartz cuvette and the absorbance was measured at 235 nm in UV Spectrophotometer against a sulfuric acid blank. Standard curve of pure PHB (Sigma®, USA) was prepared by the modified method as suggested by Slepecky and Law (1960).

Characterization of PHB

The chemical structure and the thermal properties of PHB were

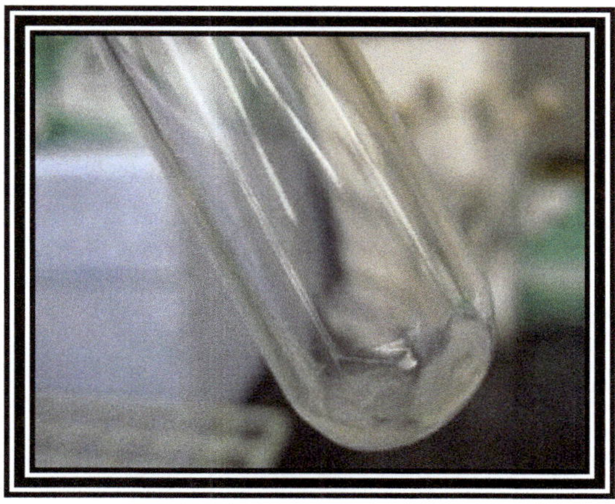

Mixture of PHB crystals and chloroform

Figure 1. Test tube containing PHB crystals and Chloroform.

used as parameters for qualitative analysis of PHB. Characterization and determination of native PHB like granules involved precise measurements to analyze their physical properties and were characterized mainly by four methods in the present study: TLC, IR, DSC and TGA.

Thin layer chromatography (TLC)

TLC was carried out in a glass plate (10 × 5 cm^2) coated with silica (3 g/15 ml of chloroform), prepared using a spreader. About 50 µl of propanolysed organic phase which involve propanolysis of PHB in a tightly sealed vial (10 ml) to which 2 ml of dichloroethane and 2 ml of a solution of propanol-hydrochloric acid (4:1 [vol/vol]) (Panda et al., 2008) of sample was loaded on the TLC plate and allowed to run in the solvent system consisting of ethyl acetate and benzene (SRL®) (1:1) mixture for 40 min. The plate was left to dry after run and for staining 50 ml of iodine solution (Hi-media®) was vaporized in water bath at 80 to 100°C. TLC plate was kept over the beaker containing iodine solution for 5-10 min in order to get it saturated with iodine vapour. The Rf values of the spots were calculated using standard formula and compared with the standard chart (Rawte and Mavinkurve, 2002).

Infrared spectroscopy (IR)

IR analysis of PHB-like granules was performed using a commercial customer service, Aarti Industries, Tarapur, India. Briefly, extracted sample and standard PHB from Sigma® was separately made in to solid pellet by making an intimate mixture of a powder sample with potassium bromide for IR analysis. The relative intensity of transmitted light was measured against the wavelength of absorption on the region 800 to 4000 cm^{-1}using IR double beam spectrophotometer (Shimandzu®). IR spectra of samples were measured at ambient condition.

Differential scanning calorimetry (DSC)

DSC analysis of PHB-like granules was performed using a commercial customer service, Center of Excellence, Vapi, India. Briefly, differential scanning calorimetry was used to characterize the melting temperature (Tm) of samples which was done in a range of 30 to 450°C air at 10°C air /min. The melting temperature (Tm) and melting enthalpy (ΔH) were determined from DSC endothermal peaks.

Thermo gravimetric analysis (TGA)

TGA analysis of PHB-like granules was performed using a commercial customer service, Center of Excellence, Vapi, India. Briefly, Thermo gravimetric analysis was used to determine the decomposition temperature (Tdecomp.) of PHB. Ten milligrams of PHB film were folded into a platinum tray and subjected to a heating rate of 20°C air/min from ambient to a final temperature of 500°C air.

RESULTS AND DISCUSSION

Lipid inclusion granules were stained black whereas the bacterial cytoplasm was stained pink in color confirming the presence of lipid inclusion granules inside the bacterial cell. PHB production was found to be influenced by the utilization of carbon from sesame oil.

PHB was isolated from the production medium by solvent extraction technique. The sodium hypochlorite digestion process enables in the digestion of cells and release of the PHB granules outside the cells for easy extraction of PHB. As present work involves utilization of edible oil as carbon source, extraction of PHB with a pre treatment of hexane helps in efficient removal of edible oil from fermentation broth.

After achieving constant weight of cell biomass at 105°C after 24 h in an oven, the biomass is treated with the hypochlorite solution, based on the fact that it can dissolve nearly all components of cell except PHB granules (Yu et al., 2006). The solvent extraction is widely used to recover PHB with high purity. Sodium hypochlorite breaks the cell wall of bacteria and facilitates elimination of Non-PHB Cellular Material (NPCM) resulted in the lysis of cells without affecting the PHB (Jacquel et al., 2008). The solvent system consisting of

Table 1. IR spectrum of sample and standard PHB.

Sample	Peak region	Comment
	1635	Carbonyl group (C=O)
	3097	Methine groups (CH)
PHB Sample	1089	Ester group (C-O)
	3578.55	Intramolecular H bond
	3415.06	H bond
	1673	Carbonyl group (C=O)
	2928	Methine groups (CH)
PHB (Sigma)	1076	Ester group (C-O)
	3330.13	Intramolecular H bond
	3417.70	H bond

Table 2. Thermal properties of PHB

Sample	n (°C)	Δ*H* (J g-1)
PHB Sample	173.36	69.54
PHB (Sigma)	172.40	65.70

Tm: melting temperature, ΔH: melting enthalpy of the sample

Table 3. Initial and maximum decomposition temperatures evaluated from TGA.

Sample	Ti (°C)*	T max (°C)
PHB Sample	208.98	296.91
PHB (Sigma)	201.16	287.60

Ti: Initial thermal decomposition, *T max*: Maximum thermal decomposition

1:1 mixture of ethanol and acetone washings serves to remove water which interferes with the extraction of the polymer into chloroform. It also proves to be specific and efficient for lysing the NPCM without affecting PHB. The present system also assists in extraction of cell lipid content and other molecules (except PHB) (Rawte and Mavinkurve, 2002).

The conversion to crotonic acid by hot concentrated sulfuric acid proved to be about one-third time more sensitive than alkaline hydrolysis of PHB extraction and analysis (Marjadi and Dharaiya, 2012). The protocol that was most efficient in determining PHB that is, β-elimination of crotonic acid; a brown colored compound affected by concentrated sulfuric acid was described by following reaction.

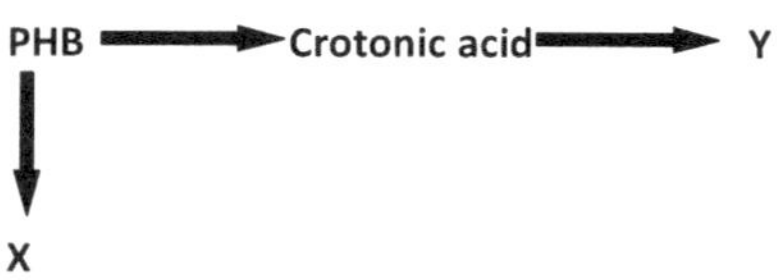

Reaction 1

Where, X refers to other degradation products of PHB and Y refers to degradation products of crotonic acid (Huang and Reusch; 1996).

The amount of PHB in the extracted samples was determined with UV spectrophotometer at 235 nm with reference to the standard graph of 3-hydroxy butyric acid (Data not shown here).

Thin layer chromatography (TLC)

As per the procedure described earlier, when the TLC plate sprayed with iodine vapour, PHB appeared as greenish-black spot surrounded by brown colour on white background. The comparison between standard and sample was performed using solvent system of ethyl acetate and benzene on TLC plate; Rf value (0.71) indicated the presence of PHB in the production medium by comparing with standard PHB.

Infrared spectroscopy (IR)

IR spectrum of the compounds were recorded in the range of 800-4000 cm^{-1} and showed characteristics bands for the groups CH, C=O and C-O (Sindhu et al., 2011). The methine groups (CH) gave strong band in the range of 1360-1416 and 2914-3097. These frequency values were higher than the normal values because of polymerization.

The carbonyl group (C=O) gave strong band in the range of 1636-1673. These frequency values were lower than the normal value because of polymerization. The (C-O) group showed strong and broad absorption in the range of 1047-1089 (Table 1 and Figure 2).

Differential scanning calorimetry (DSC)

The thermal properties of PHB samples and commercial PHB were investigated by differential scanning calorimetry (DSC). The data are illustrated in Figure 3. Since the melting temperature of PHB is around 170-180°C (Matko et al., 2005), while that of the PHB sample is within the range (173.36°C), which is close to that of the commercial PHB (Table 2).

Thermo gravimetric analysis (TGA)

PHB sample and standard. The thermal degradation of Figure 4 and Table 3 shows the TGA thermo grams of extracted PHB proceeds by a one-step process with a maximum decomposition temperature at 296.91°C. This thermal degradation at maximum decomposition temperature of approximately 300°C is mainly associated with the ester cleavage of PHB component by β-elimination reaction (Choi et al., 2003). The temperature of 296.91°C was found to be the maximum decomposition temperature for biopolymer made with extracted PHB and it was almost similar with that of the standard PHB from

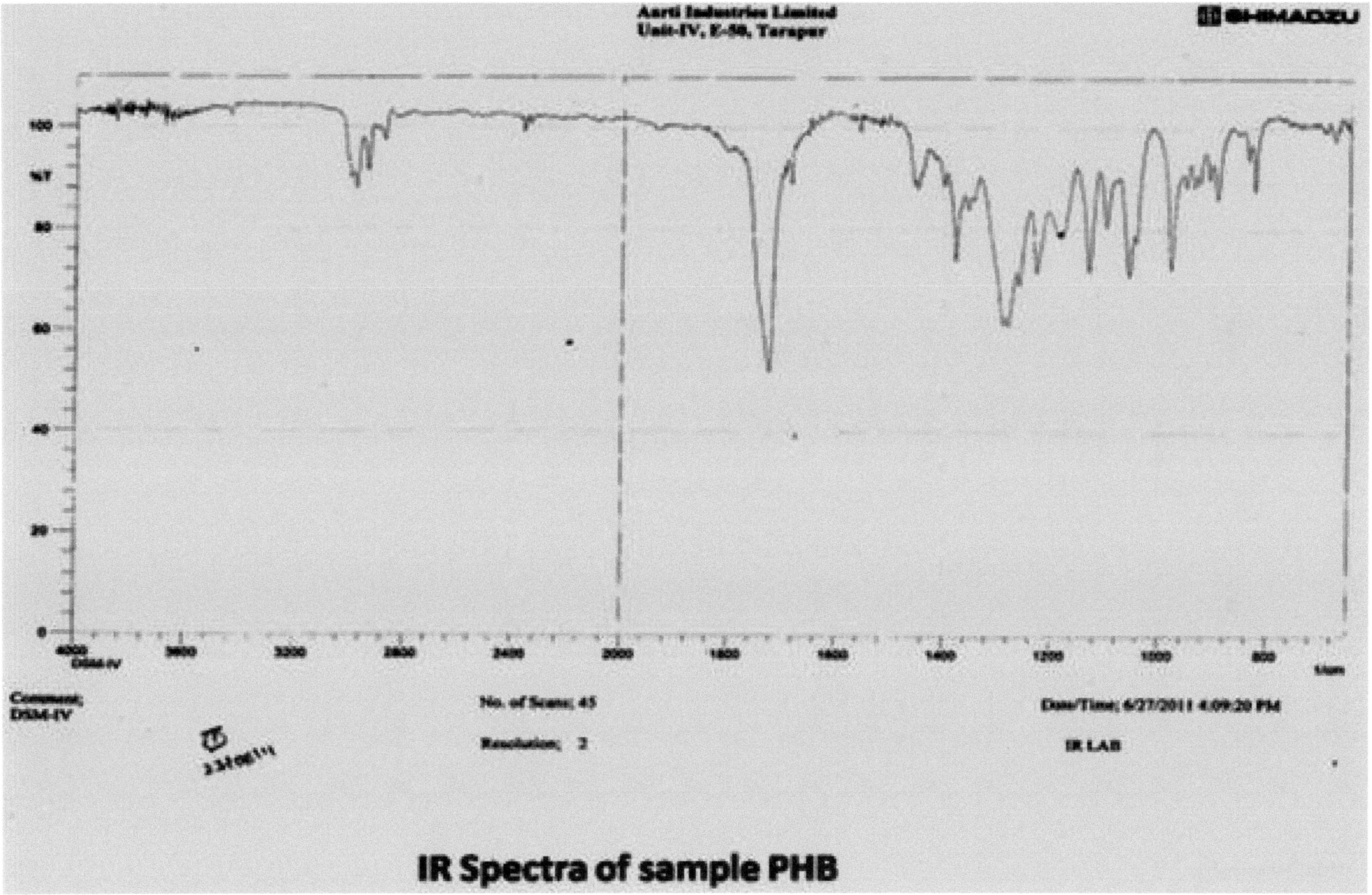

Figure 2. IR spectra comparison between sample and standard PHB.

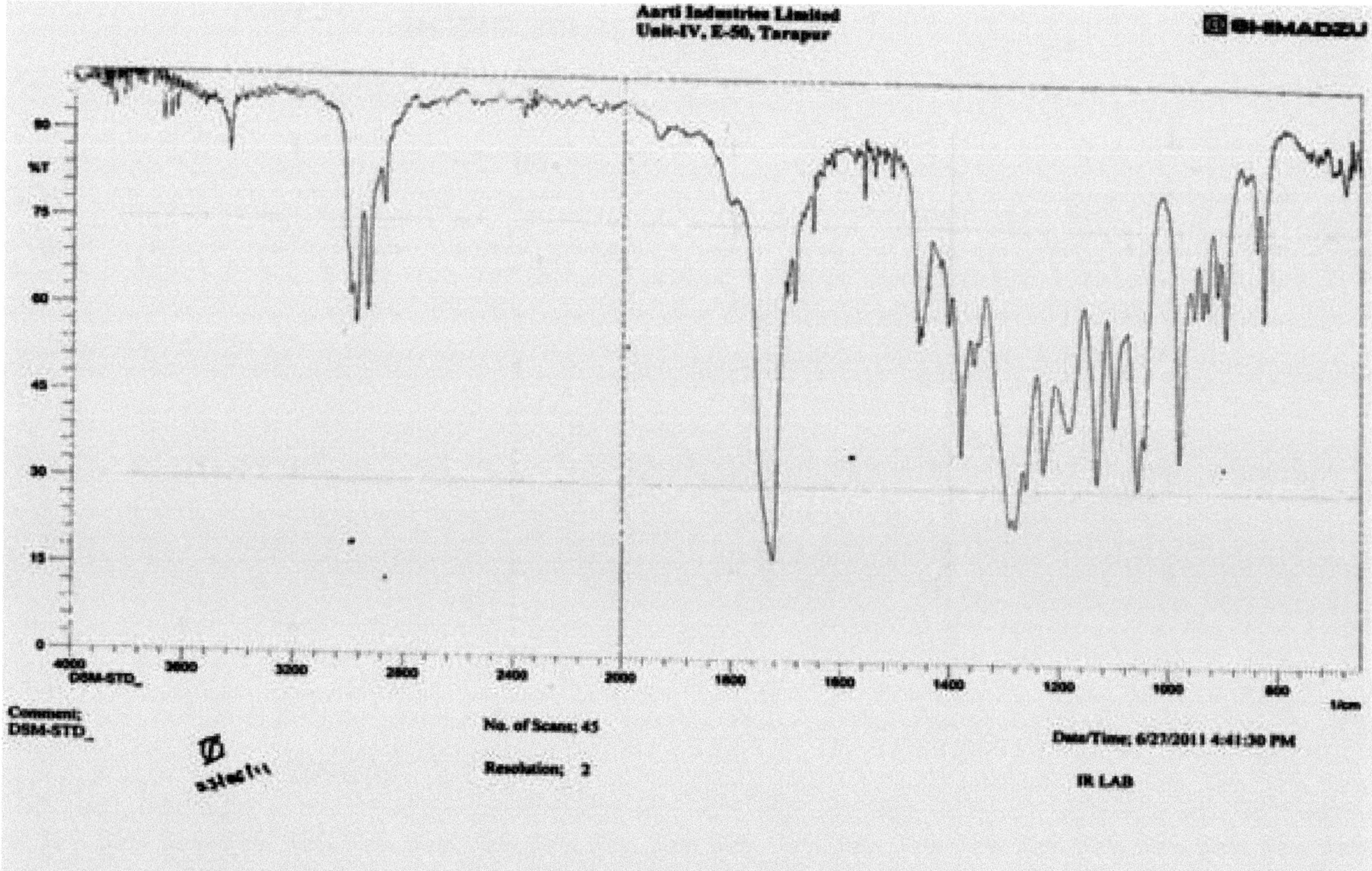

IR spectra of standard PHB

Figure 2. Contd.

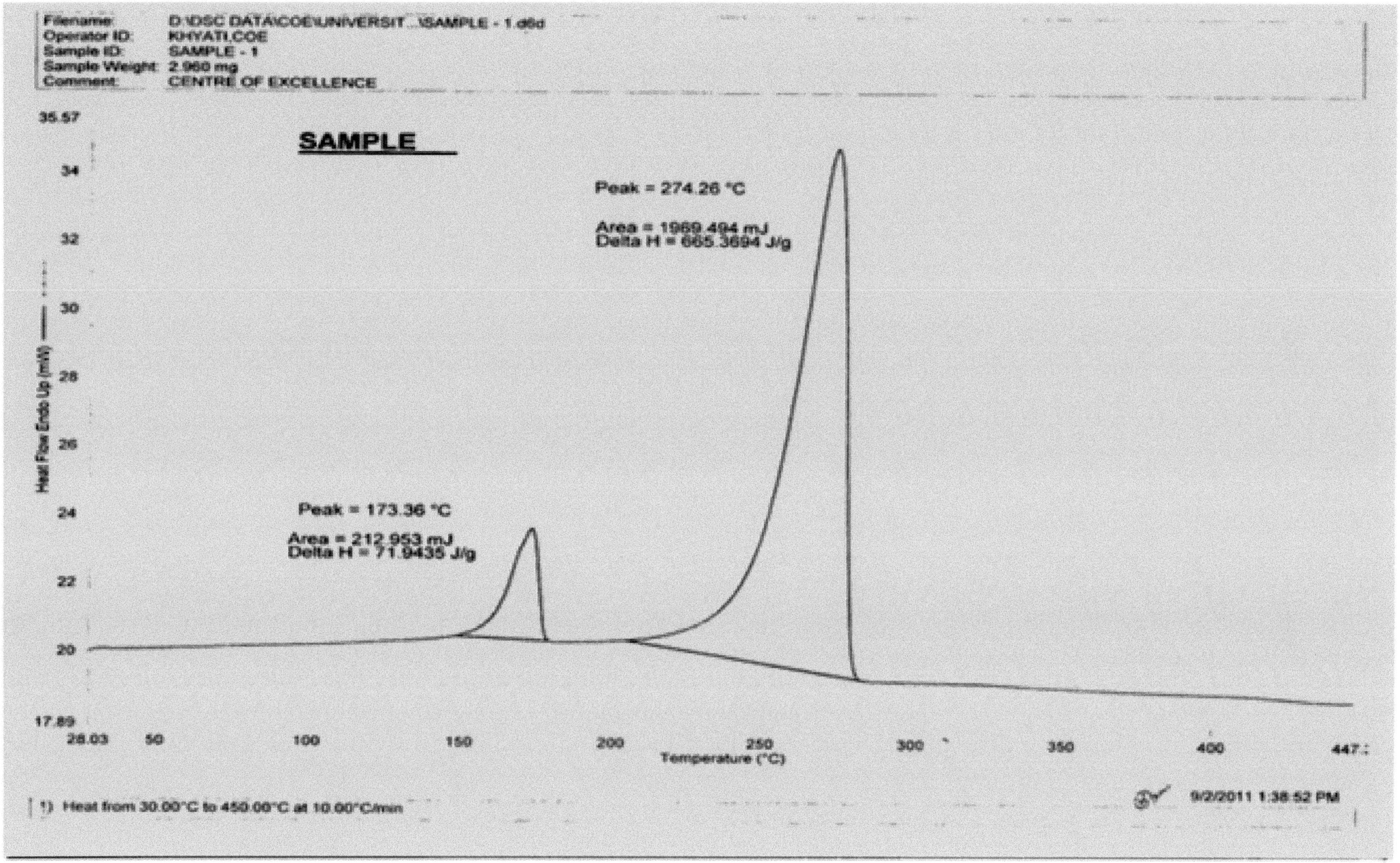

DSC thermo gram of sample PHB

Figure 3. DSC thermo gram comparison between standard and sample PHB.

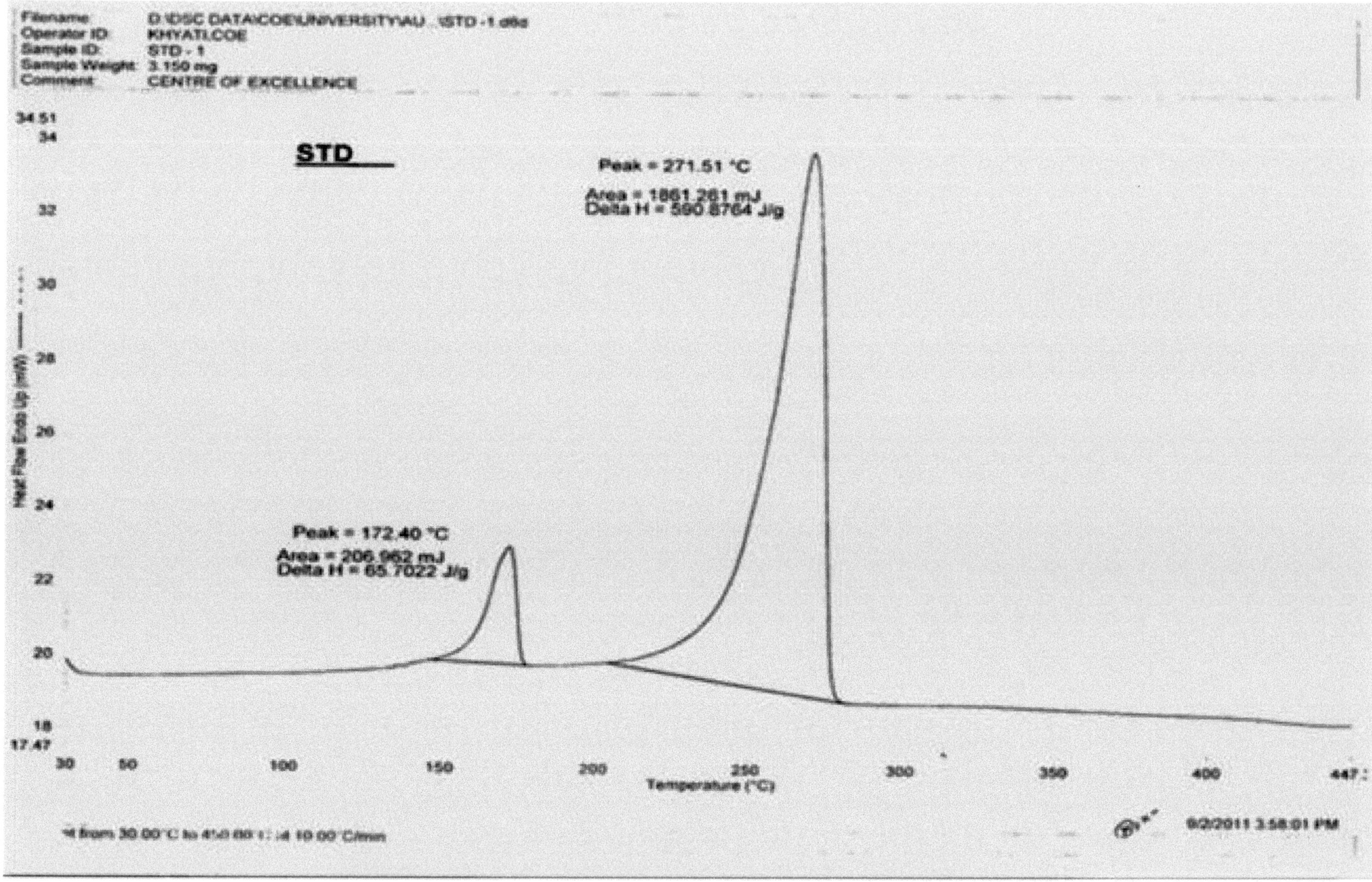

DSC thermo gram of standard PHB

Figure 3. Contd.

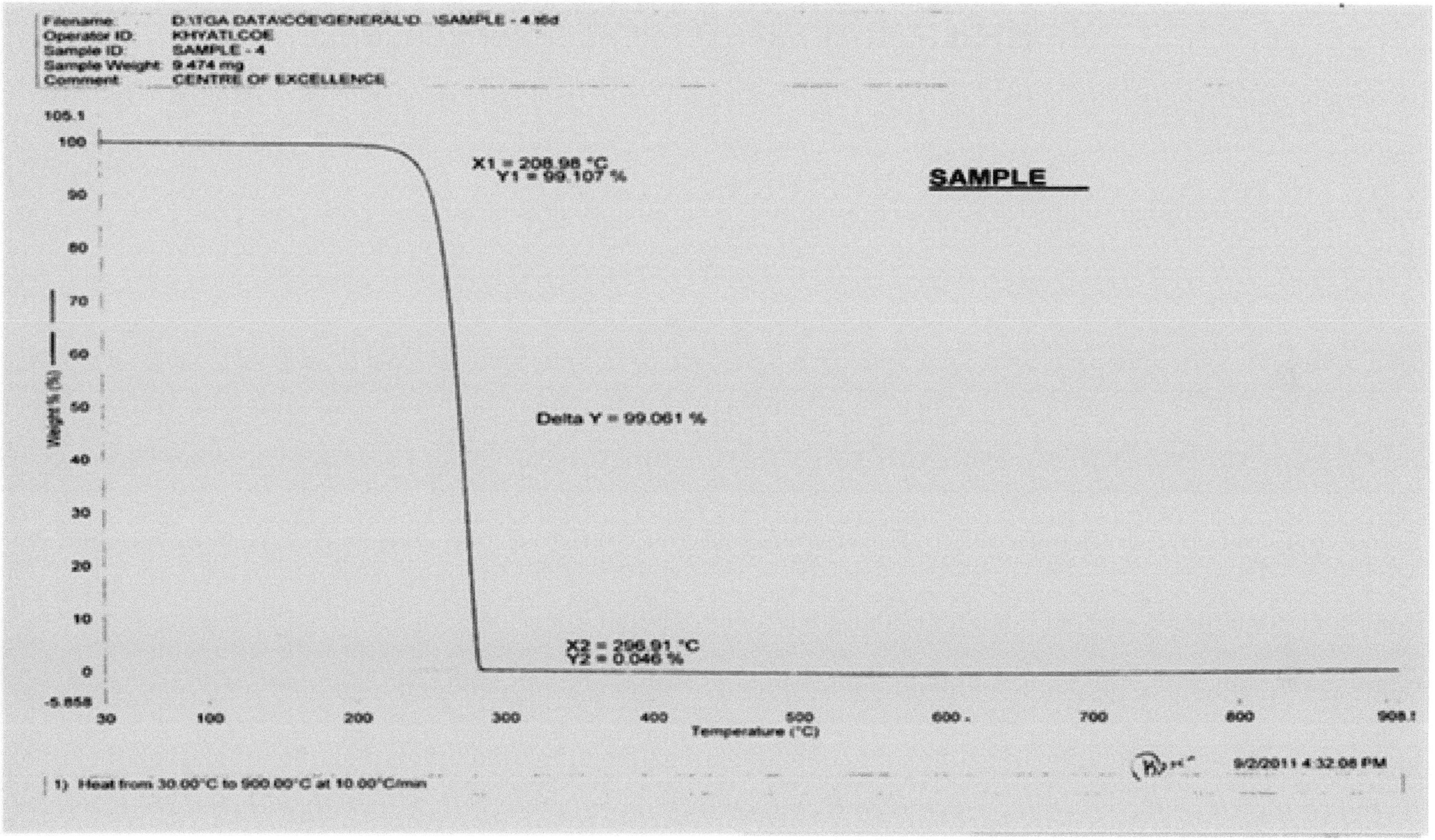

TGA thermo gram of sample PHB

Figure 4. TGA thermo gram comparison between sample and standard PHB.

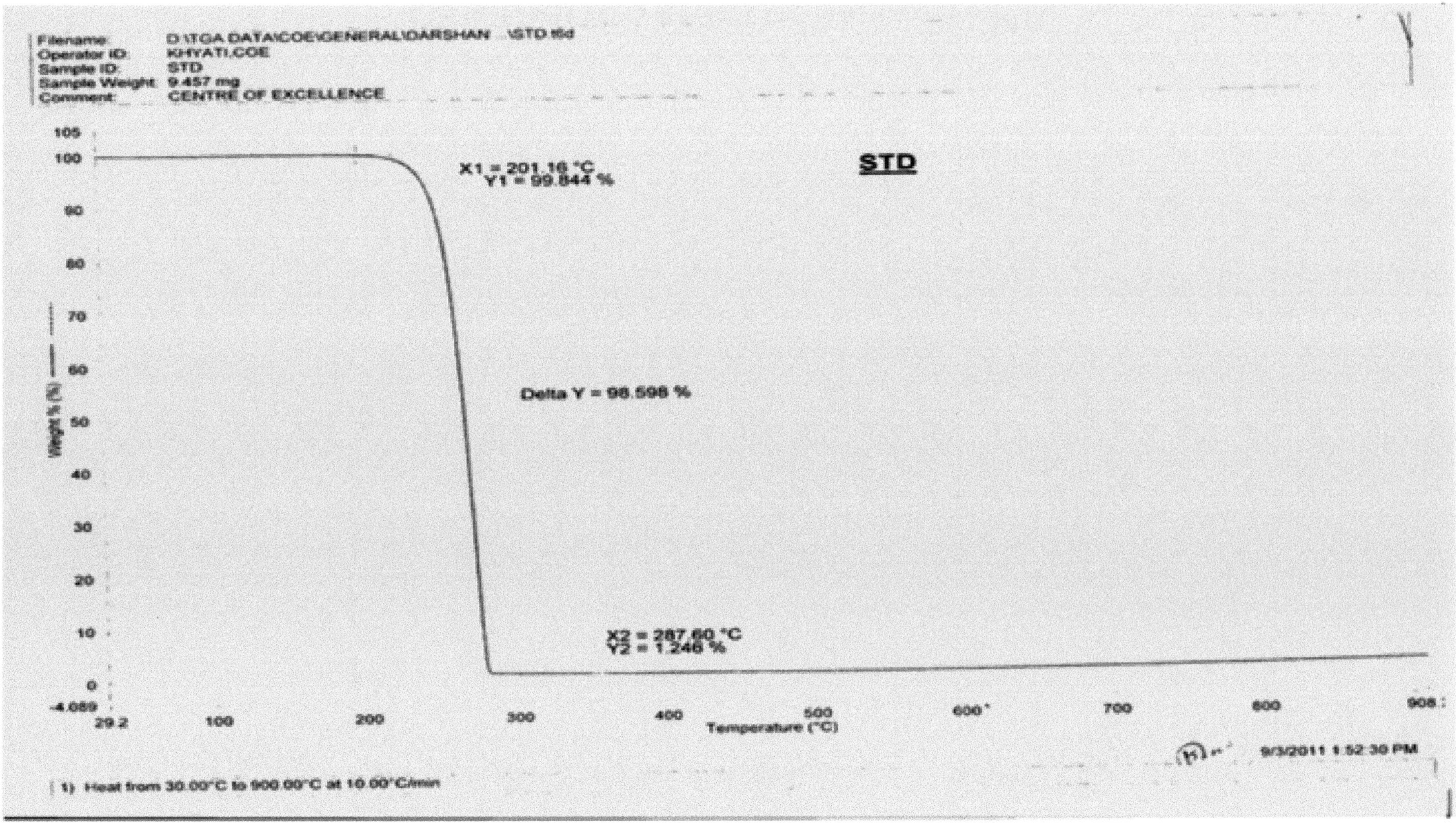

TGA thermo gram of standard PHB

Figure 4. Contd.

Sigma (287.60°C).

Further, the characterization of PHB like granules by various methodologies and their comparison with standard PHB as described earlier shows that the extracted polymer from the microbial isolate possess almost similar properties and was finally confirmed to be PHB and of good quality.

Conclusion

Polyhydroxybutyrate (PHB) was successfully produced through biosynthesis in *S. epidermidis* and recovered appropriately. The method of PHB extraction also influences the quality of polymer. Therefore, bacterial cells were blended with chloroform using high speed homogenizer for a short time to cause lower damage of PHB the molecular weight. The identical PHB sample was verified to commercial PHB and other PHB data that reported in literatures. The obtained PHB has the same thermal properties as commercial PHB with higher molecular mass (approxinatly 3.9 × 106 Da) and lower degree of crystallinity.

Conflict of Interests

The author(s) have not declared any conflict of interests.

ACKNOWLEDGEMENTS

The authors are grateful to the University Grants Commission, New Delhi, India for financial assistance through minor research project. We also acknowledge the support of Aarthi Industries, Tarapur for analysis of structure and the thermal properties of PHB. Thanks are also due to the Principal, M N Science College, Patan and the Head of Department of Life Sciences, HNG University, Patan for extending laboratory facilities.

REFERENCES

Anderson AJ, Dawes EA (1990). Occurrence, metabolism, metabolic role and industrial uses of Bacterial polyhydroxyalkanoates. Microbiol. Rev. 54:450-472.

Aslim B, Zehra N, Beta Y (2002). Determination of PHB growth quantities of certain Bacillus species isolated from soil. Turk. Electronic J. Biotechnol. 24-30.

Bowker RR (1981). Manual of Methods for General Bacteriology American Society for Microbiology Washington. D.C.

Chen GQ (2009). A microbial polyhydroxyalkanoates (PHA) based bio- and materials industry. Chem. Soc. Rev. 38(8):2434-2446.

Choi JY, Lee JK, You Y, Park WH (2003). Epoxidized polybutadiene as a thermal stabilizer for poly(3-hydroxybutyrate). II. thermal Stabilization of poly(3-hydroxybutyrate) by epoxidized polybutadiene. Fiber. Polym. 4: 195-198.

Hartmann R, Hany R, Witholt B, Zinn M (2010). Simultaneous biosynthesis of two copolymers in *Pseudomonas putida*GPo1 using a two-stage continuous culture system. Biomacromolecules 11:1488-1493.

Huang R, Reusch RN (1996). Poly(3-hydroxybutyrate) is associated with specific proteins in the cytoplasm and membranes of *Escherichia coli*. J. Biol. Chem. 271:22196-22202.

Ibrahim MH, Steinbüchel A (2009).Poly (3-hydroxybutyrate) production fromglycerol by Zobellelladenitrificans MW1 via high-cell-density fed-batch fermentation and simplified solvent extraction. Appl. Environ. Microbial. 75(19):6222-6231.

Ishizaki A, Tanaka K (1991) Production of poly-β-hydroxybutyric acid from carbon dioxide by *Alcaligenes eutrophus* ATCC 17697. J. Ferm. Bioeng. 71:254-257.

Jacquel N, Lo CW, Wei YH, Wu HS, Wang SS (2008). Isolation and purification of bacterial poly (3-hydroxyalkanoates). Biochem. Eng. J. 39(1):15-27.

Jung K, Hazenberg W, Prieto M, Witholt B (2001). Two-stage continuous process development for the production of medium-chain-length poly(3-hydroxyalkanoates). Biotechnol. Bioeng. 72:19-24.

Kahar P, Tsuge T, Taguchi K, Doi Y (2004). High yield production of polyhydroxyalkanoates from soybean oil by Ralstoniaeutropha and its recombinant strain. Polym. Degradation Stab. 83:79-86.

Kathiraser Y, Aroua MK, Ramachandran KB, Tan IKP (2007). Chemical characterization of medium-chain-length polyhydroxyalkanoates (PHAs) recovered by enzymatic treatment and ultrafiltration. J. Chem. Technol. Biotechnol. 82(9):847-855.

Kunasundari B, Sudesh K (2011).Isolation and recovery of microbial polyhydroxyalkanoates. Express Polym. Lett. 5:7.

Kuniko M, Nakamura Y, Doi Y (1989). New bacterial copolyesters produced in *Alcaligeneseutrophus*from organic acids. Polym. Commun. 29:174-176.

Law JH, Slepecky RA (1961). Assay of poly(3-hydroxybutyric acid). J. Bacteriol. 82(1):256-261.

Marjadi D, Dharaiya N (2012).Bioprospecting and Characterization of poly-β- hydroxyalkanoate (PHAs) producing Pseudomonas Spp. isolated from edible oil contaminated soil. Res. Biotechnol. 3:5.

Matko E, Tonka K, Ivka K (2005). Thermal degradation of poly(3-hydroxybutyrate) plasticized with acetyl tributyl citrate. Polym. Degrad. Stab. 90:313-318.

Metabolix (2013a). Why integrate biopolymers into your conversion process? Retrieved from http://www.metabolix.com/converters

Panda B, Sharma L, Singh A, Mallick N (2008). Thin layer chromatographic detection of poly-β-hydroxybutyrate (PHB) and poly-β-hydroxyvalerate (PHV) in cyanobacteria. Iindian J. Biotechnol. 7(2):230-234.

Prasanna T, Babu PA, Lakshmi PD, Chakrapani R, Rao CR (2011). Research Article Open Access Production of Poly (3-hydroxybutyrates) by Bacillus species isolated form Soil. J. Pharma Res. Rev.1:15-18.

Rawte T, Mavinkurve S (2002). Characterization of polyhydroxyalkanoates - Biodegradable plastics from marine bacteria. Curr. Sci. 83(5):562-564.

Ribera RG, Monteoliva-Sanchez M, Ramos-Cormenzana A (2001). Production of polyhidroxyalkanoates by *Pseudomonas putida* KT2442 harboring pSK2665 in wastewater from olive oil mills (alpechín). Electronic J. Biotechnol. 4(2):11-12.

Saito Y, Nakamura S, Hiramitsu M, Doi Y (1996). Microbial synthesis and properties of poly (3-hydroxybutyrate-co-4-hydroxybutyrate). Polym. Int. 39:169-172

Sayyed RZ, Gangurde NS, Chincholkar SB (2009).Hypochlorite digestion method for efficient recovery of PHB from *Alcaligenesfaecalis*. Indian J. Microbiol. 49:230-232.

Sindhu R, Ammu B, Binod P, Deepthi SK, Ramachandran KB, Soccol CR, Pandey A (2011). Production and characterization of poly-3-hydroxybutyrate from crude glycerol by Bacillus sphaericus NII 0838 and improving its thermal properties by blending with other polymers. Braz. Arch. Biol. Technol. 54(4):783-794.

Singh P, Parmar N (2013). Isolation and characterization of two novel polyhydroxybutyrate (PHB)-producing bacteria. Afr. J. Biotechnol. 10(24):4907-4919.

Slepecky RA, Law JHA (1960). A rapid spectrophotometric assay of alpha, beta-unsaturated acids and betahydroxy acids. Anal. Chem. 32:1697-1699.

U.S. Department of Energy (2013). How fossil fuels were formed. Retrieved from http://www.fe.doe.gov/education/energylessons/coal/gen_howformed.html

Influence of salinity on soil chemical properties and surrounding vegetation of Awe salt mining site, Nasarawa State, Nigeria

Dedeke Oyinade Aderoju[1] and Akomolafe Gbenga Festus[2]

[1]Department of Botany, Federal University, Lafia, Nasarawa State, Nigeria.
[2]Department of Botany, Federal University, Lafia, Nasarawa State, Nigeria.

This study was carried out to monitor and document the influence of salinity on soil chemical properties and surrounding vegetation of Awe salt mining site, Nasarawa State, Nigeria. The soil samples were collected randomly from the salt mining site and 10 m away from the site (control) at 5, 10, 15 and 20 cm depths, respectively, with three replications each. The soil samples were air dried, sieved and the physical and chemical properties were determined using standard methods. The exchangeable sodium percentage (ESP), cation exchange capacity (CEC) and sodium adsorption ratio (SAR) were calculated. No plant was found growing on the salt mining site (SMS) unlike the control site (CTS) which has a little diversity of plants. The soil organic carbon, organic magnesium, total nitrogen and phosphorus at SMS decreased significantly (α = 0.05) with soil depth in relation to CTS except for 10-15 cm soil depth. Soil exchangeable acidity (H^+ and Al^{3+}) at SMS significantly increased (α = 0.05) when compared with the CTS at different soil depths. Also, K^+, Na^+ and Mg^{2+} of SMS and CTS significantly increased (α = 0.05) at the different soil depths. Soil pH, electrical conductivity (EC), ESP and SAR at SMS significantly increased at different depths when compared with CTS. Awe salt mining has brought about soil nutrient imbalance at CTS with EC less than 0.2 dS/m and SAR between 0-3. Thus, it is concluded that soil at CT and SMS is a sodic soil which have reduced the diversity of plants species.

Key words: Cations exchange, mining, salinity, sodicity, Awe town.

INTRODUCTION

Soil is the world's largest terrestrial carbon (C) sink, and is estimated to contain approximately 1600 Pg of C to a depth of one metre (Eswaran et al., 1993). Salt-affected soils, as they are called, are either saline or sodic. Saline soils refer to soils with electrical conductivity above 4 dS/m and usually contain sufficient soluble salts that adversely affect the growth of most crops (Allotey et al., 2008). The soluble salts are mainly chlorides and sulphates of sodium, calcium and magnesium. Sodic soils refer to soils with exchangeable sodium percentage above 15, with sodium salts capable of alkaline hydrolysis and are mainly Na_2CO_3 (Allotey et al., 2008). These two main groups of salt-affected soils differ (physically, chemically, biologically, as well as their geographical and geochemical distribution) and therefore, require different approaches for their reclamation and agricultural utilization. Salinity in soil or water is one of the major stresses and can severely limit crop production

(Ashraf and Harris, 2004).

Salts have been a known problem for thousands of years, particularly in arid and semiarid areas where there is insufficient rainfall to leach salts from the root zone (Miller and Donahue, 1995). This problem usually arises when the total amount of salts which accumulate in the root zone is high enough to negatively affect plant growth. Excess soluble salts in the root zone restrict plant roots from withdrawing water from the surrounding soil, effectively reducing the plant available water (Bauder and Brock, 2001; USDA, Natural Resources Conservation Service, 2002). Under irrigated agriculture, salts are added continuously to the soils with each irrigation event (Qadir et al., 2005). Thus, soil salinity is a major threat to the sustainability of irrigated agriculture (Ghassemi et al., 1995). Approximately 932 million ha of land worldwide are degraded due to salinity and sodicity, usually coinciding with land available for agriculture. Of this area, salinity affects 23% of arable land while saline-sodic soils affect a further 10% (Szabolcs, 1989).

Soil water salinity can affect soil physical properties by causing fine particles to bind together into aggregates. This process is known as flocculation and is beneficial in terms of soil aeration, root penetration and root growth (Neumann, 1995). Saline soils contain sufficient soluble salts to suppress plant growth through a series of interacting factors such as osmotic potential effect, ion toxicity and antagonism, which induce nutrient imbalances (Neumann, 1995). Salinity imposes serious environmental problems that affect grassland cover and the availability of animal feed in arid and semi-arid regions (El-kharbotly et al., 2003). Salt stress undesirably affects plant growth and productivity during all developmental stages. Aslam (2006) pointed out that tree planting is the cheapest and most simple natural biological approach to controlling salinity.

Therefore, changes in salinity and sodicity affect soil physical and chemical properties, which subsequently alter nutrient cycles and decomposition processes (Wong et al. 2005). The risk of erosion is increased, while soil physical and chemical properties are altered, impacting upon aggregation and nutrient cycling as well as biotic activity. Therefore, there is a clear linkage between land management practices through their effect on salinity and sodicity (Wong et al., 2005). An excess of sodium on the cation exchange sites of fine-textured soils forms a condition in which irrigation water entering the soil is attracted to small pores with a great amount of force, resulting in soil swelling, particle slaking from aggregates, and dispersion, thus precluding drainage (Dollhopf, 2000). Upon drying, dispersed soil particles undergo a reorientation, resulting in lost of soil structure, lower hydraulic conductivity, and surface crusting that can break plant stems, inhibit germination and emergence, and slow infiltration (Dollhopf, 2000).

Soils affected by salinity, that is, those soils high in soluble salts, are characterised by rising water tables and waterlogging of lower lying areas in the landscape. Sodic soils are high in exchangeable sodium, slake and disperse upon wetting (Nelson et al., 1996). In addition, a few studies are available that unambiguously demonstrate the effect of increasing salinity and sodicity on soil C dynamics (Nelson et al., 1997; Pankhurst et al., 2001; Sarig et al., 1993). Generally, responses to salinization have been of two general kinds; engineering the environment to manage increased salt in the soil by irrigation and drainage management, or by "engineering" the plants to increase their salt tolerance. Salt tolerant plants may also ameliorate the environment by lowering the water table in salt affected soils (Pitman and Lauchli, 2002).

Nasarawa State, Nigeria is a state known for different types of solid minerals. Awe is a local government area in Nasarawa State, Nigeria where salt is been processed locally from soil and water from the river. It has an area of 2,557 km² and a population of 112,574 in the 2006 census. Despite the large area cover by salt, data on the magnitude of changes in soil chemical properties as affected by salinity remains sparse. Hence this research was conducted to monitor and documents the status of soil chemical properties of Awe salt mining site, Nasarawa State and its influence on surrounding vegetation.

MATERIALS AND METHODS

Study location

Awe town in Nasarawa state, Nigeria is located between longitudes 9°08'09, 50" E and latitudes 8°09'29, 06" N. Generally, fishing and farming are the main occupations of the people in the area. Maize, pepper and cassava are the main crops cultivated.

Plant collection and identification

Two sample plots of sizes 10 x 10 m were used for this survey in both salt mining site (SMS) and control site (CTS). The CTS was chosen to be 10 m away from the salt mining site. All the plants species at the two sites were collected, pressed, dried and identified using the herbarium of Department of Botany, Obafemi Awolowo University, Ile-Ife, Nigeria.

Soil collection and preparation

The soil in this area was mainly sandy loamy. Soil was collected randomly from the SMS at different depths using a soil auger (5, 10, 15 and 20 cm) during the rainy season. Also, soil was collected from CTS (10 m away from the SMS) at the same four different depths so as to know the changes in the chemical compositions. These collections were replicated thrice. The soil was air-dried and sieved using 2 mm mesh gauze to remove debris.

Laboratory analysis

The pH of the soil was determined in 2:1 $CaCl_2$/soil suspension using glass electrode pH meter (Crockford and Norwell, 1956). Electrical conductivity (EC) was determined using the method of

Rayment and Higgison (1992). Organic carbon (OC) and organic matter (OM) were determined by the Walkley and Black (1934) dichromate oxidation method.

Total nitrogen (N) was determined by the Kjedhal method (Jackson, 1958). Available phosphorus (P) was extracted by 0.03M NH_4F + 0.025M HCl (Bray and Kurtz, 1945) and the P in the extract were determined by colorimeter. Potassium (K) and sodium (Na), calcium (Ca) and magnesium (Mg) were extracted by 1.0 M ammonium acetate using leaching method. Micro nutrients (iron (Fe), copper (Cu), zinc (Zn) and manganese (Mn)) were extracted with Hec and determined by buck scientific atomic absorption spectrophotometer (210/211vgp).

Calculations

The exchangeable sodium percentage (ESP) of the soil was calculated using Mohsen et al. (2009) formula as follows:

$$ESP = \frac{\text{Exchangeable sodium}}{\text{Soil CEC}} \times 100$$

Where: Exchangeable acidity (EA) = H^+ + Al^{3+}, exchangeable cations (CEC) = Na + Ca + K + Mg

$$\text{SAR (sodium adsorption ratio)} = \frac{[Na]^+}{½\{[Ca] + [Mg]\}}$$

RESULTS

No plant was found growing on the SMS. However, at the CTS, there was a little diversity of plants growing. The plants and their percentage proportions are highlighted in Table 1. Two families of plant that dominated the control site are Malvaceae and Tiliaceae each having a percentage population of 28.6% out of the total population. *Azadiractha indica* is the species with the highest percentage proportion of 20% while *Senna occidentalis* is the species with the lowest percentage proportion of 8.3% as shown in Table 1 and Figure 1.

The results in Table 2 highlighted the impact of salt on soil OC, OM, total N and P. At CTS, soil OC, OM and total N ranges from 10-15cm < 15-20cm < 0-5cm < 5-10cm. Whereas at SMS, OC, OM and Total N ranges from 10-15cm > 0-5cm > 5-10cm > 15-20cm. Soil OC, OM, total N and P at SMS decreased significantly (α = 0.05) with soil depth in relation to CTS except for SMS at 10-15 cm soil depth.

Table 3 shows the effects of salinity on the exchangeable cations. At SMS, soil CEC and $Cacl_2$ had the highest value at 0-5 cm soil depth. Soil exchangeable acidity (H^+ and Al^{3+}) at SMS significantly increased (α = 0.05) when compared with the CTS at different soil depths. Also, CEC of CTS significantly increased (α = 0.05) at different soil depths.

Soil pH, EC, ESP and SAR at SMS significantly increased at different depths when compared with CTS (Table 4). The pH of the soil collected at CTS ranges from 7.4 - 8.5 at soil depth 0-5 to 15-20 cm which is slightly different from pH (8.6-8.9) of soil at SMS. At CTS, soil EC, ESP and SAR ranged from 0.0033±0.001, 2.33±1.28 and 0.308±0.239, respectively. Soil EC and SAR ranged from 0.048±0.039, 6.17±2.25 and 0.88±0.71, respectively at SMS.

DISCUSSION

The zero plant diversity at the SMS might have been a long term effect of the saline conditions of the soil which had inhibited the ability of plants to take up water and nutrients. Plants uptake water through a process of osmo-regulation. When the soil solution salinity is greater than the internal salinity of the plant, water uptake is restricted. Thus, it brings about reduction in plant growth and biomass. This work agrees with that of Munns (2002) that salt stress decreases growth in most plants, including halophytes. Also, Abari et al. (2011) observed that NaCl and KCl salt stress decreased germination percentage and speed in *Acacia oerfota* and *Acacia tortilis*. Jamil et al. (2005) stated that increasing salt concentration decreases the germination percentage and rate at an important degree and increases germination time. In another study, the effect of sea water salinity (1500, 2500 and 3500 ppm) on the growth of tomato (*Lycopersicon esculentum*) cultivars revealed that the sea water salinity delayed seed germination and reduced germination percentage especially with increasing salinity level (Hajer et al., 2006).

Therefore, it is possible over time that the plant diversities at the CTS may also be lost gradually as the soil is also being affected by salinity. This is because out of the seven plant species identified at the CTS, only three species have been found to have tolerance to salinity. *Calotropis procera* is a salt tolerant plant to a relatively high degree and is native to Nigeria. It is usually an indicator of over-cultivated or exhausted soils and prefers disturbed sandy soils (Orwa et al., 2009). This is a further indicator of the soil nutrient status of CTS. *Corchorus aestuans* and *Triumfetta rhomboidea* of the same family Tiliaceae have also been reported to have high degree of tolerance to salinity (Islam et al., 2013).

In saline soils, the soil organic carbon content is influenced by two opposing factors: reduced plant inputs and reduced rates of decomposition. This might have been responsible for the significant decrease in the soil OC, OM, Total N and P observed in this work at SMS as the soil depth increases. This trend was in reverse order at the CTS. Therefore, soil salinity decreases plant productivity and hence carbon inputs to the soil at SMS (Coleman et al., 1997).

The salinity had slightly increased the soil Na^+, K^+, Mg^{2+}, Ca^{2+} and $CaCl_2$ at CTS. This is possibly an accumulated effect over time. Low value of CEC observed at 10-15 cm soil depth at SMS could be as a result of high

Table 1. Percentage proportion and degree of salt tolerance of plants identified at control site (CTS) of Awe salt mining area of Nasarawa State, Nigeria.

Scientific name of plant	Family	Degree of salt tolerance	Proportion (%)
Azadiractha indica	Meliaceae	Zero	20.0
Calotropis procera	Asclepiadaceae	High	14.5
Corchorus aestuans	Tiliaceae	High	11.5
Senna occidentalis	Caesalpiniaceae	Zero	8.3
Sida acuta	Malvaceae	Zero	15.0
Sida corymbosa	Malvaceae	Zero	13.6
Triumfetta rhomboidea	Tiliaceae	High	17.1

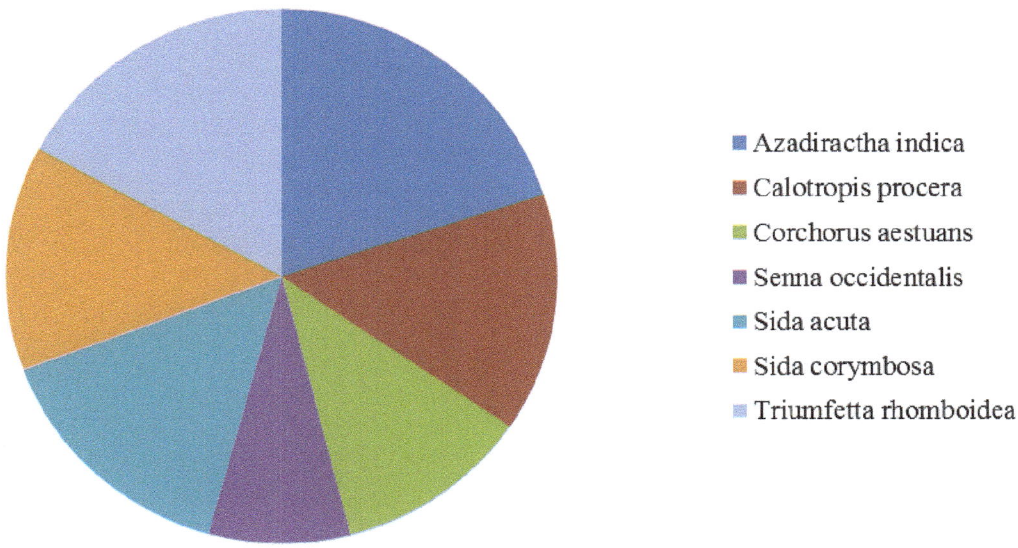

Figure 1. A chart showing the percentage proportions of the plants identified at control site of Awe salt mining area, Nasarawa State, Nigeria.

Table 2. Effect of salinity on OC, OM, total Na and P.

Depth (cm)	Location	OC (%)	OM (%)	Total N (%)	P (Ppm)
0-5	CT	1.17^c	2.01^c	0.1031^c	0.000382^a
5-10	CT	4.72^d	8.12^d	0.3530^d	0.000449^d
10-15	CT	0.20^a	0.34^a	0.0166^a	0.000438^c
15-20	CT	1.13^b	1.95^b	0.0929^b	0.000404^b
0-5	SMS	0.55^b	0.94^b	0.0495^b	0.000511^c
5-10	SMS	0.66^c	1.14^c	0.0556^c	0.000404^a
10-15	SMS	2.07^d	3.56^d	0.1695^d	0.000404^a
15-20	SMS	0.31^a	0.54^a	0.0265^a	0.000410^b
LSD (0.05)		3.87	6.66	0.293	0.0001269

CT = 10 m away from the mining site and SMS = salt mining site.

concentration of OC, OM, total N and P. It is known that increasing organic matter would increase the CEC values and subsequently decreased the ESP values. This is similar to study of Hill and James (1995) who reported

Table 3. Effects of salinity on exchangeable cations.

Depth (cm)	Location	EA (cmol/kg)	H^+ (cmol/kg)	Al^{3+} (cmol/kg)	K^+ (cmol/kg)	Na^+ (cmol/kg)	Mg^{2+} (cmol/kg)	Ca^{2+} (cmol/kg)	$Cacl_2$ (cmol/kg)
0-5	CT	1.4^b	0.20^a	1.20^c	0.42^c	0.44^d	2.53^c	7.00^a	7.10^a
5-10	CT	1.3^a	0.80^c	0.40^a	0.50^b	0.25^c	1.84^b	10.40^d	7.60^b
10-15	CT	2.5^d	1.90^d	0.60^b	0.60^a	0.20^b	2.76^d	9.50^c	7.70^c
15-20	CT	1.9^c	0.50^b	1.40^d	0.30^b	0.17^a	1.38^a	8.40^b	7.70^c
0-5	SMS	1.3^a	0.70^a	0.50^a	1.11^c	1.65^c	3.91^d	11.0^d	8.30^d
5-10	SMS	2.3^b	0.90^b	1.40^c	0.81^b	0.61^d	2.99^b	10.80^c	8.20^b
10-15	SMS	3.2^d	2.00^c	1.20^b	0.58^a	0.40^a	3.22^c	4.60^b	8.10^a
15-20	SMS	2.6^c	0.90^b	1.70^d	0.58^a	0.49^b	1.84^a	4.30^a	8.20^c
LSD(0.05)		1.356	0.328	1.155	0.309	0.737	0.754	6.80	0.5719

CT = 10 m away from the mining site and SMS = salt mining site.

Table 4. Effects of salinity on conductivity of the soil.

Depth (cm)	Location	pH_{H2O}	EC (dS/m)	ESP	SAR
0-5	CT	8.0^a	0.005^c	4.36^d	0.647^d
5-10	CT	8.5^c	0.002^b	1.95^c	0.319^c
10-15	CT	8.2^b	0.005^c	1.34^a	0.024^a
15-20	CT	8.5^c	0.001^a	1.66^b	0.243^b
0-5	SMS	8.6^a	0.110^d	9.34^d	1.910^d
5-10	SMS	8.7^b	0.034^c	4.01^a	0.073^a
10-15	SMS	8.9^c	0.021^a	4.54^b	0.640^b
15-20	SMS	8.7^b	0.026^b	6.79^c	0.884^c
LSD(0.05)		0.762	0.065	2.35	0.987

CT = 10 m away from the mining site and SMS = salt mining site.

that the addition of an organic amendment to soil is likely to increase the soil CEC simply from the additive effect of the organic matter and the high CEC associated with organic matter depending on soil pH.

Salinity limits water uptake by plants by reducing the osmotic potential and thus the total soil water potential. Salinity may also cause specific ion toxicity or upset the nutritional balance. In addition, the salt composition of the soil water influences the composition of cations on the exchange. The salinity at SMS has affected the pH and the exchangeable cations of CTS at soil depth 0-5 and 5-10. High value of EC, ESP and SAR at 0-5 cm could be as a result of high soil CEC and $Cacl_2$. Also, the higher soil pH, EC, ESP and SAR observed at SMS than CTS in this study is another proof of the effect of salinity on this soil. Therefore, soil at CTS and SMS were severely affected by salinity as the EC of both sites were < 0.2 and SAR values range between 0-3. This is confirmed by Miller and Gardiner (2007) who reported that when SAR is 0-3 and EC is < 0.2, the soil is sodic.

Conclusively, due to all the outcomes from this study, the influence of salinity on the soil in Awe salt mining area of Nasarawa State, Nigeria is sodic. This has really affected the plant diversities in the mining site and the control site. It is hereby recommended that appropriate soil reclamation technique be employed early in bringing back the soil to the form that can be suitable for agricultural production, in as much as the control soil is already affected.

REFERENCES

Abari KA, Nasr, MH, Hojjati M, Bayat D (2011). Salt effects on seed germination and seedling emergence of two Acacia species. Afr. J. Plant Sci. 5(1):52-56.

Allotey DFK, Asiamah RD, Dedzoe CD, Nyamekye AL (2008). Physico-Chemical Properties of Three Salt-Affected Soils in the Lower Volta Basin and Management Strategies for their Sustainable Utilization. W. African J. Appl. Eco. 12:163-182.

Ashraf M, Harris PJC (2004). Potential biochemical indicators of salinity tolerance in plants. Plant Sci. 166:3-16.

Aslam Z (2006). *Saline Agriculture.* In: Pakistan Agriculture Management and Development. Quraishi MAA, Zia MA, Quraishi MS (Eds.), pp. 210-221. A-One Publishers, Lahore, Pakistan.

Bauder JW, Brock TA (2001). Irrigation water quality, soil amendment and crop effects on sodium leaching. Arid Land Res. and Man. 15:101-113.

Bray RH, Kurtz LT (1945). Determination of total, organic, and available forms of phosphorus in soils. Soil Sci. 59:39-45.

Coleman K, Jenkinson D, Crocker G, Grace P, Klir J, Körschens M (1997). Simulating trends in soil organic carbon in long-term experiments using RothC-26.3. Geoderma 81:29-44.

Crockford L, Nowell R (1956). Laboratory manual of Physical Chemistry. John Wiley and Sons, New York, Expts. 31-32, 58-59.

Darmody, and W.L. Daniels (Eds.). pp. 21-40. American S. Agro., Madison, Wisconsin.

Dollhopf DJ (2000). Sampling strategies for drastically disturbed lands. In "Reclamation of Drastically Disturbed Lands". R.I. Barnhisel, R.G.

El-kharbotly A, Maghloub O, Al-subhi A, Al-halhali A (2003). Indigenous grass species with potential for maintaining rangeland and livestock feeding in Oman. Agric. Ecosyst. Environ. 95:623-627.

Eswaran H, Van den berg E, Reich P (1993). Organic carbon in soils of the world. Soil Sci. Soc. Am. J. 57:192-194.

Ghassemi F, Jakeman AJ, Nix HA (1995). Salinisation of land and water resources: Human causes, extent, management and case studies. UNSW Press, Sydney, Australia, and CAB International, Wallingford, UK.

Hajer AS, Malibari AA, Al-Zahrani HS, Almaghrabi OA (2006). Responses of three tomato cultivars to sea water salinity: Effect of salinity on the seedling growth. African J. Biotech. 5(10):855-861.

Hill RL, James BR (1995). The influence of waste amendments on soil properties, soil amendments and environmental quality. CRC Press, Inc. 0-87371-859-3/95.

Islam MS, Azam MS, Sharmin S, Sajib AA, Alam MM, Reza MS, Ahmed R, Khan H (2013). Improved salt tolerance of jute plants expressing the *katE* gene from *Escherichia coli.* Turk. J. Biol. 37:206-211.

Jackson ML (1958). *Soil Chemical Analysis*. p. 216, Prentice-Hall, Inc. Englewood Cliffs, N.J.

Jamil M, Lee CC, Rehman SU, Lee DB, Ashraf M (2005). Salinity (NaCl) tolerance of Brassica species at germination and early seedling growth. *Electron.* J. Environ. Agric. Food Chem. 4:970-976.

Miller RW, Donahue RL (1995). Soils in our environment. 7th edition. p. 323, Prudence Hall, Englewood, Cliffs, NJ.

Miller RW, Gardiner DT (2007). Soils in our environment. 9th edition, Pp. 452, Prentice Hall-Inc., Upper. Sddle River, New Jersey 07458.

Mohsen S, Majid R, Borzoo GK (2009). Prediction of soil exchangeable sodium percentage based on soil sodium adsorption ratio. American-Eurasian J. Agric. Environ. Sci. 5 (1):01-04.

Munns R (2002). Comparative Physiology of salt and water stress. Plant Cell Environ. 25:239-250.

Nelson PN, Barzegar AR, Oades JM (1997). Sodicity and clay type: influence on decomposition of added organic matter. Soil Sci. Soc. Am. J. 61:1052-1057.

Nelson PN, Ladd JN, Oades JM (1996). Decomposition of 14C-labelled plant material in a salt affected soil. Soil Biol. Biochem. 28:433-441.

Neumann PM (1995). Inhibition of root growth by salinity stress. In: Structure and Function of Roots. Baluska F, Ciamporova M, Gasporikova O, Barlow PW (Eds). pp. 299-304. Kluwer Academic Publishers, Netherlands.

Orwa C, Mutria A, Kindt R, Jamnadass R, Anthony S (2009). Agro-forestree Database: a tree reference and selection guide version 4.0. Available at http://www.worldagroforestry.org/sites/treedbs/treedatabases.asp.

Pankhurst CE, Yu S, Hawke BG, Harch BD (2001). Capacity of fatty acid profiles and substrate utilization patterns to describe differences in soil microbial communities associated with increased salinity or alkalinity at three locations in South Australia. Biol. Fert. Soils 33:204-217.

Pitman MG, Lauchli A (2002). Salinity: Environment - Plants - Molecules. pp. 3-20. Kluwer Academic Publishers, Netherlands.

Qadir LM, Noble AD, Schubert S, Thomas RJ, Arslan A (2005). Sodicity-induced Land Degradation and its sustainable Management: problems and prospects. Land Degrad. Develop. DOI: 10.1002/ldr.751

Rayment GE, Higgison FR (1992). Australian Laboratory Handbook of Soil and Water Chemical Methods. Melbourne, Inkata Press (Australian Soil and Land Survey Handbooks vol. 3)

Sarig S, Roberson EB, Firestone MK (1993). Microbial activity-soil structure: response to saline water irrigation. Soil Biol. Biochem. 25:693-697.

Szabolcs I (1989) *Salt affected soils.* Pp. 274, Boca Raton, CRC Press.

USDA, Natural Resources Conservation Service (2002). Soil Conservationists. Salinity Management Guide - Salt Management. Available at http://www.launionsweb.org/salinity.htm.

Walkley A, Black IA (1934). An Examination of Degtajaroff method for Determining Soil Organic Acid Filtration Method. Soil Sci. 37:29-38.

Wong VNL, Greene RSB, Murphy BW, Dalal R, Mann S (2005). Decomposition of added organic material in salt-affected soils. IN ROACH, I. (Ed.) Cooperative Research Centre for Landscape Environments and Mineral Exploration Regional Regolith Symposia, 2nd-4th November. Canberra, CRC LEME.

21

Microbiological quality and safety of some selected vegetables sold in Jimma town, Southwestern Ethiopia

Alemayehu Dugassa[1], Ketema Bacha[2] and Tsige Ketema[3]

Jimma University, College of Natural Sciences, Ethiopia.

Raw vegetables are major vehicles for the transmission of foodborne infections. In Ethiopia, there is a habit of consuming raw vegetables as salad, particularly tomato (*Solanum lycopersicum* L.), cabbage (*Brassica oleracea* L.), carrots (*Daucus carota* L.), lettuces (*Lactuca sativa* L.) and green peppers (*Capsicum annuum* L.) without adequate treatment. The objectives of this study were to evaluate the microbiological loads, assess the prevalence of some food borne pathogens, and investigate the antimicrobial susceptibility of pathogens. A total of 180 vegetable samples were purchased from different sites and markets of Jimma town and analyzed for their microbial loads following standard microbiological methods. In addition, antibiotic resistance pathogens and prevalence of Salmonella and *Staphylococcus aureus* were also determined. Ninety percent of vegetable samples had aerobic mesophilic counts of ≥ 5 log^{10} CFU g^{-1}. Similarly, 82.2, 92.8 and 97.8% of samples had coliform, Enterobacteriaceae and lactic acid bacteria counts of ≥ 4 log^{10} CFU g^{-1}, respectively. However, most of staphylococci and aerobic spore counts varied between 2 - 3.9 log^{10} CFU g^{-1}, but greater than 74% of yeasts and molds were counted ≤ 2.9 log10 CFU g^{-1}. The aerobic mesophilic flora of the vegetable samples was dominated by *Bacillus* spp. (22.3%) followed by Staphylococcus spp. (17.7%). Salmonella and *S. aureus* were isolated from 23 (12.8%) and 18 (10%) vegetable samples, respectively. All of Salmonella and *S. aureus* isolates showed resistance to ampicillin and penicillin G, respectively. However, they were 100% sensitive to ciprofloxacin and gentamicin. Lettuce had high microbial load and Salmonella were most prevalent in lettuce but *S. aureus* were more prevalent in green pepper. Most of the pathogens were multiple drugs resistant. The use of food grade chemicals to kill pathogens and reduce the microbial load before consumption is recommended.

Key words: Drug resistance, prevalence, pathogens, raw vegetables.

INTRODUCTION

Vegetable is the tender plant part which is not sweet and may be flavored or spiced with condiments before consumption. The consumption of minimally processed food or raw vegetables has been increased tremendously due to their nutritive values in human dietary. Vegetables can be used as salad mixes, side dishes or ingredients in the meals. Fresh products that contain complex and colorful blends incorporating a wide variety of vegetable mixes and flavors would especially benefit from sales in all market segments. In general, as consumers continue to lead a healthy lifestyle, there are broad product development opportunities in this category. Currently, supermarkets and the food service outlets are the primary retail outlets for these products (Amoah et al., 2009). Thus despite their nutritional and health benefits, outbreaks of human infections associated with the consumption of fresh or minimally processed vegetables have increased in recent years (Beuchat, 2002).

Since vegetables are produced in a natural environment, they are vulnerable to contamination by human pathogens. The majorities of diseases associated with fresh vegetables are primarily those transmitted by the fecal oral route, and therefore, are a result of contamination at some point in the process (Johnston et al., 2005). Vegetables could be contaminated with bacterial pathogens from human or animal sources including *Salmonella*, *Shigella*, *Escherichia coli* O157:H7, *Listeria monocytogenes*, *Staphylococcus aureus* and *Campylobacter*, and resistance pathogens to different antimicrobials (Al-Binali et al., 2006; Simon et al, 2007; Allende et al., 2008; Elhariry, 2011). As the result, vegetables have been associated with outbreak of foodborne disease in many countries.

The presence of antibiotic resistance both in normal flora and pathogenic microorganisms in fresh vegetables may contribute to horizontal spreading of resistance between different isolates, species and genera. The presence of resistance gene on transferable elements facilitates distribution of resistance and wide spread use of antibiotics allows direct selection or co-selection of resistance (Heuer and Smalla, 2007). Therefore, the presence of antibiotic resistant bacteria in fresh vegetables constitutes an additional concern for consumer safety (Aarestrup et al., 2008; Walsh and Fanings, 2008).

Plate count of aerobic mesophilic microorganisms found in food is one of the microbiological indicators for food quality (Aycicek et al., 2004). These organisms reflect the exposure of the sample to any contamination and in general, the existence of favorable conditions for multiplication of microorganisms. Food borne bacterial pathogens commonly detected in fresh vegetables were coliform bacteria, *S. aureus* and *Salmonella* spp. (Tambekar and Mundhada, 2006).

Coli forms are commonly used bacterial indicator of sanitary quality of foods and water and considered as an indicator of microbial pollution and they are common inhabitant of animal and human guts (Tortora, 1995). The presence of these bacteria poses a serious threat to public health with outbreaks arising from food and water that has been contaminated by human or animal feces or sewage. *S. aureus* is the third most common cause of confirmed food poisoning in the world and the illness is due to the ingestion of preformed enterotoxin produced in foods (Acco et al., 2003).

Ethiopia has highly diversified agroecological zones which are suitable for the production of various types of vegetables. Vegetables are mainly grown by traditional farmers in home gardens. About 27% of the vegetable species recorded from home gardens in Ethiopia were consumed as raw or cooked (Asfaw, 1997). Particularly, in the urban parts of the country eating of raw vegetables becomes more common. Vegetable farmers around Jimma town supply vegetables to the local market but the market place of Jimma town is not well organized. Vegetables are sold in front of shops besides with other goods and on street by street vendors.

In addition, vegetables can be stored in poor quality containers and house before sell for at least one day. This can increase potential contamination of vegetables with animals and human's feces, soil, dusts and other postharvest contaminants (Al-Binali et al., 2006). Contamination of vegetables are of special concern, because it is likely to be consumed raw, without any type of microbiologically lethal processing, thus posing a potential food safety problem.

The present study was under taken to examine the microbiological quality and safety of fresh vegetables particularly tomato (*Solanum lycopersicum* L.), cabbage (*Brassica oleracea* L.), carrot (*Daucus carota* L.), lettuce (*Lactuca sativa* L.), and green pepper (*Capsicum annuum* L.) samples collected from different sites (Kochi, Agip and Merkato) and markets (shops and street venders) of Jimma town, assess the growth potential of standard strains and evaluate the drug resistance ability of *Salmonella* spp and *S. aureus* isolated from these vegetables.

Materials and Methods

Description of the study area

The study was conducted in Jimma town, which is located at 353 km south west of Addis Ababa (Figure 1). The town's geographical coordinates are approximately 7°41' N latitude and 36° 50'E longitude. From a climatic point of view, abundant rainfall makes this region one of the best watered of Ethiopian highland areas, conducive for agricultural production (Alemu et al., 2011).

Study design and study population

The cross sectional study design was used. The sampling sites were Kochi, Agip, and Merkato. The study periods covered from September, 2011 to May, 2012.

Sampling techniques

A simple random sampling technique was used to address representative of the whole population.

Data collection

As the study has survey and experimental parts, data were collected using structured questionnaires.

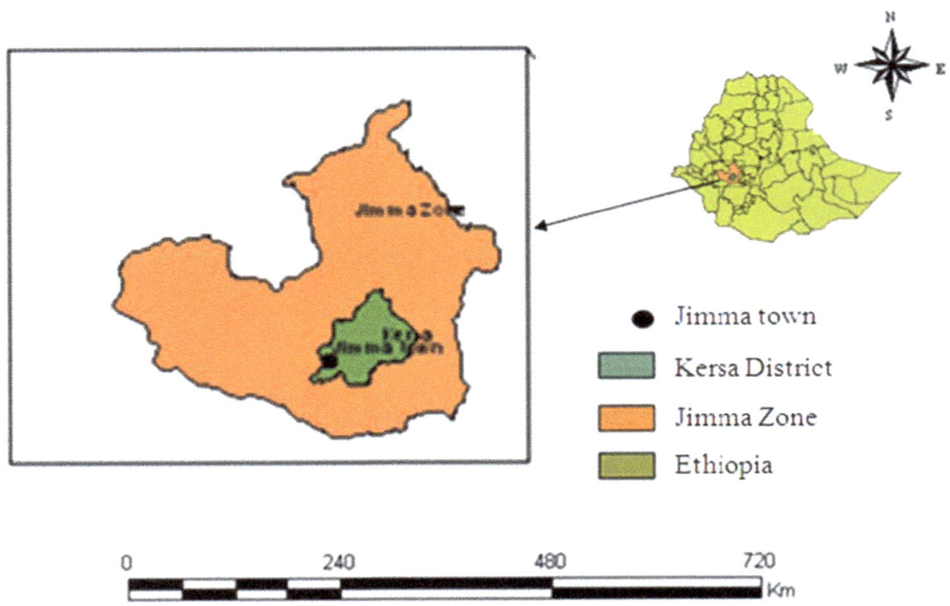

Figure 1. Map of the study area.

Collection of samples

A total of 180 fresh vegetable samples were purchased at different sampling days from local markets of Jimma town, southwest Ethiopia. The samples consisted of 36 each of tomato (*S. lycopersicum* L.), cabbage (*B. oleracea* L.), carrots (*D. carota* L.), lettuces (*L. sativa* L.), and green peppers (*C. annuum* L.). All samples were collected using sterile plastic bags aseptically and immediately brought to the Postgraduate and Research Laboratory of Biology Department, Jimma University, for analysis. Microbiological analysis was conducted within 3 h of sample collection.

Sample preparation

For sample preparation, 25 g samples were aseptically removed from each sample, shredded into approximately 2 - 3 cm pieces using a sterile stainless steel knife and vigorously shaken in 225 ml of sterile 0.1% (w/v) bacteriological (buffered) peptone water (Oxoid) for 3 min separately to homogenize the samples (Shalini, 2010).

Microbiological enumeration

The homogenate from sample preparation in buffered peptone water was used for the following procedures.

Total aerobic mesophilic count

Total viable aerobic mesophilic count of all vegetables samples were determined by plate count using standard plate count agar (PCA) (Oxoid) medium (Shalini, 2010).

Total coliform count

A 0.1 ml of homogenate from 10^{-1} - 10^{-3} dilution was pipetted and spread on violet red bile agar (VRBA) (Oxoid). Red to pink colonies were counted after incubating plates at 32□C for 18 - 24 h (Spencer et al., 2007).

Enterobacteriaceae count

All purple colonies were counted on MacConkey agar (Oxoid) as members of Enterobacteriaceae after incubation for 24 h at 32°C (Spencer et al., 2007).

Staphylococci count

Mannitol salt agar (MSA) (Oxoid) was surface plated with 0.1ml of the homogenate from 10^{-1} - 10^{-2} and incubated at 32°C for 36 h. Then, golden yellow color colonies were aseptically picked and purified (Acco et al., 2003).

Aerobic spore count

Bacterial spores were counted after heating the suspension of vegetable samples for 10 min in water bath at 80°C and incubation at 32°C for 36 to 72 h (Acco et al., 2003).

Lactic acid bacteria (LAB) count

To count LAB, 0.1 ml of 10^{-1} - 10^{-3} dilution of homogenate was spread on de Mann Rogosa Sharpe (MRS) agar (Oxoid) media and incubated at 37°C for 48 h in anaerobic condition using anaerobic jar (Oxoid) (Pal et al., 2005).

Yeasts and molds counts

The yeasts and molds count of all vegetables samples were determined by direct plate count using potato dextrose agar (PDA) supplemented with 0.1 g Chloramphenicol. The plates were incubated at 25- 28°C for three to five days (Spencer et al., 2007).

Microbial analysis

For microbial analysis, 15 - 20 colonies with different morphology and color were picked randomly from countable plate count plates and were purified by repeated plating and characterized to the family and genus level using the following tests.

Cell morphology

These were carried out by Gram staining techniques and observing under microscope using oil immersion objective. Schefer fulton endospore staining techniques were used to identify the presence or absence of endospore (Krieg, 1981).

KOH-test (test on lipopolysaccharide)

A colony was aseptically picked from the surface of plate count agar plates using an inculcating loop and stirred in the KOH solution for 10 s to 2 min (Gregerson, 1978).

Oxidation Fermentation (O/F) test

This test is used to assess the ability of the isolate to utilize glucose and determine the metabolic way they used as well (that is by fermentation or oxidation) (Hugh and Leifson, 1953).

Catalase test

Catalase taste was carried out after young colonies flooded with a 3% solution of hydrogen peroxide (H_2O_2) (Chelikani et al., 2004).

Cytochrome Oxidase test

This test was conducted following the method outlined by Kovacs (1956).

Detection of pathogens

Salmonella

For detection of *Salmonella*, 25 g vegetable samples were added to 225 ml buffered peptone water, vigorously shaken and the suspension was incubated at 37°C for 24 h for metabolic recovery and proliferation of cells (Deza et al., 2003). From this, 1 ml of culture was transferred into separate tubes each containing 10 ml of Selenite Cystein Broth. The broth was incubated at 37°C for 24 h. After secondary enrichment, culture from enrichment broth was separately streaked on plates of Xylose Lysine Desoxycholate (XLD) (Oxoid) medium. Pink colonies with or without black centers from selective medium was picked, purified and tested biochemically (Cheung et al., 2007).

Staphylococcus aureus

For detection of *S. aureus*, golden yellow colonies from MSA during staphylococci count were picked, purified and preserved. Coagulase test was done by two ways: slide coagulase test and tube coagulase test (Cheesbrough, 2006).

Antimicrobial susceptibility testing of the isolated pathogens

This was investigated on Mueller Hinton Agar (Oxoid) plates following the standardized disk diffusion techniques. The antibiotic discs were placed on the medium by using forceps and incubated at 35°C for 18 h and the zones of inhibition was measured manually with a transparent ruler. The results of the antimicrobial susceptibility were interpreted based on the guidance of National Committee for Clinical Laboratory Standards (CLSI, 2007).

For this tests, Ampicillin (AMP), (10 µg/ml); Chloramphenicol(C), (30 µg/ml); Ciprofloxacin (CIP), (5µg/ml); Gentamicin(CN), (10 µg/ml); Kanamycin (K), (30 µg/ml); Nalidixic acid(NA),(30 µg/ml); Norfloxacin (NOR), (10 µg/ml); Streptomycin (S),(10 µg/ml) and Tetracycline (TE), (30 µg/ml) were used for *Salmonella* and Penicillin G (P), (10 µg/ml); Erythromycin(E), (15 µg/ml); Clindamycin (DA), (2µg/ml); Chloramphenicol(C), (30 µg/ml); Ciprofloxacin (CIP), (5µg/ml); Gentamicin(CN), (10 µg/ml); Kanamycin (K), (30 µg/ml); Streptomycin (S),(10 µg/ml) and Tetracycline (TE), (30 µg/ml) were used for *S. aureus*. The reference strains, *S. aureus* ATCC 25923 and *Escherichia coli* ATCC 25922, sensitive to all the drugs were used in this study.

Statistical analysis

Coefficient of variation (% CV) was calculated and significance of variation in microbial counts within the vegetable samples was analyzed. Mean values of the microbial counts of various vegetable samples from different sites and markets were compared using one way ANOVA and the significance of difference between groups were considered at 95% confidence interval ($p < 0.05$). In addition, the data obtained from the respondents were analyzed by SPSS version 16.

Ethical consideration

Ethical clearance was obtained from Research Review and Ethical committee of College of Natural Science, Jimma University.

RESULTS

Socio-demographic characteristics

A total of 90 farmers and vegetable venders were interviewed. A significant number of the respondents were females (60%) (Table 1). Forty percent of the respondents were within an age group of 30 to 39 years. With respect to the educational status, about 34.4, 32.2, 16.7 and 16.7% of the farmers or sellers attended secondary school, elementary school, capable of reading and writing, and illiterate, respectively (Table 1). Occupationally, the respondents (55.6%) were vegetable sellers and 44.4% were farmers (Table 1).

General vegetable farm and management conditions

The general characteristics of farm and management conditions of vegetables sold in Jimma town are summarized in Table 2. Vegetables sold in Jimma town were 100% cultivated in traditional farming methods by rural farmers. The preferred cultivation seasons of the

Table 1. Socio-demographic characteristics of vegetable farmers and sellers or venders, Jimma town, south western Ethiopia, 2011/12 .

Characteristic		**Number of respondents(n=90)**	
		Frequency	**Percent (%)**
Sex	Male	36	40.0
	Female	54	60.0
Age	20-29	24	26.7
	30-39	36	40.0
	40-49	19	21.1
	> 50	11	12.2
Education status	Illiterate	15	16.7
	Read and write	15	16.7
	Elementary school	29	32.2
	Secondary school	31	34.4
Occupation	Farmer	40	44.4
	Vegetable sellers	50	55.6

Table 2. General vegetable farm and management conditions, Jimma town, Southwestern Ethiopia, 2011/12.

Characteristic	**Respondents(n=90)**	
	Frequency	**Percent (%)**
Methods of cultivation		
Traditional	90	100
Water source of irrigation		
River	57	63.3
Well	33	36.7
To increase fertility of farm		
Inorganic fertilizers	66	73.3
Animal manure	24	26.7
Harvesting equipments		
Sac	37	41.1
Hand basket	29	32.2
Storage place before selling		
In store room	56	62.2
On the floor in vegetable farm	34	37.8
Transporting containers		
Sac	58	64.4
Plastic bags	32	35.6
How long do you store before sell		
1 day	5	5.6
2 days	29	32.2
3 days	36	40.0
More than 3 days	20	22.2
Consumption habit		
Without heat treatment	49	54.4
With heat treatment	30	33.3
With food grade chemicals	11	12.2

vegetables were found out to be during dry season (41.1%) using irrigation. The water sources for irrigation were river (63.3%) and well (36.7%). A large number of vegetable farmers (73.3%) were used inorganic fertilizers

Table 3. Mean microbiological counts (log CFU g^{-1}) of selected vegetables purchased from shops and vended markets, Jimma town, southwestern Ethiopia, 2011/12.

Microbial group	Vegetables									
	Tomato (T)		Cabbage (Ca)		Carrot (Cr)		Lettuce (L)		Green pepper (G)	
	Mean ± S.D	%C.V	Mean ± S.D	%C.V	Mean ± S.D	%C.V	Mean ± S.D	%C.V	Mean ± S.D	%C.V
AMC	5.3 ± 0.7	13.2	5.7 ± 0.4	7.0	5.5 ± 0.4	7.3	6.0 ± 0.4	6.7	5.4 ± 0.5	9.3
Coliforms	3.4 ± 0.9	26.5	5.2 ± 0.5	9.6	5.0 ± 0.5	10.0	5.2 ± 0.6	11.5	4.7 ± 0.8	17.0
Enterobacteriaceae	4.5 ± 0.9	20.0	5.5 ± 0.5	9.1	5.1 ± 0.7	13.7	5.5 ± 0.6	10.9	5.0 ± 0.6	12.0
Staphylococci	2.8 ± 0.8	28.6	3.4 ± 0.6	17.6	3.5 ± 0.6	17.1	3.7 ± 0.5	13.5	3.8 ± 0.5	13.2
Aerobic Spore	3.6 ± 0.6	16.7	3.5 ± 0.4	11.4	3.7 ± 0.4	10.8	3.7 ± 0.4	10.8	3.4 ± 0.5	14.7
LAB	4.7 ± 0.3	6.4	4.5 ± 0.6	13.3	4.8 ± 0.6	12.5	4.8 ± 0.5	10.4	4.6 ± 0.3	6.5
Yeast	2.5 ± 0.5	20.0	2.5 ± 0.4	16.0	2.6 ± 0.5	19.2	2.9 ± 0.7	24.1	2.5 ± 0.5	20.0
Molds	2.1 ± 0.3	14.3	2.2 ± 0.3	13.6	2.4 ± 0.4	16.7	2.4 ± 0.4	16.7	2.2 ± 0.4	18.2

AMC, Aerobic mesophilic count; LAB, lactic acid bacteria; S.D, standard deviation; C.V, coefficient of variation.

although 26.7% were using animal manure to increase the fertility of the farm land. The vegetable farmers used different materials to harvest the produce including sack (41.1%), hand basket (32.2%) and plastic bags (26.7%). The harvested vegetables were stored at different places before selling. About 62.2% of the vegetable farmers were stored in store room. However, 37.8% of the respondents stored vegetables simply on the floor in the vegetable farms (Table 2).

Vegetables were transported from farm site to market by different means of transportation. Donkey were mostly used (35.6%) followed by horse cart (26.7%), car (23.3%) and humans back (14.4%). Sack and plastic bags were used as transporting containers while 64.4% of the respondents were used sack and 35.6% of vegetable farmers and sellers were used plastic bags. About 63.3% of the respondents were placed vegetables on the bed infront of the shop for sell. On other hand, 35.6% of vegetable sellers vended vegetables on street without using bed or plastic sheet. However, 1.1% of respondents used plastics to vend vegetables on floor. Vegetables were not availed to the consumers as soon as harvested. Therefore, 77.8% of the sellers stored vegetables for up to three days, whereas 22.2% stored for more than three days before sold to consumers. Over 54% of the respondents consumed vegetables without heat treatment. However, 33.3 and 12.2% of the respondents consumed after heat treatment and treating with food grade chemicals, respectively (Table 2).

Microbiological count of raw vegetables

The microbiological load of vegetables sampled in this study was varied with types, sites and markets. The mean microbial counts for selected raw vegetables sold in Jimma town are shown in Table 3. Accordingly, high aerobic mesophilic bacteria counts (6.0 log10 CFU g^{-1}) followed by Enterobacteriaceae (5.5 log10 CFU g^{-1}) and coliforms (5.2 log10 CFU g^{-1}). Lactic acid bacteria (LAB) were the forth dominant bacterial groups, but yeasts and molds were the least dominant (< 3.9 log10 CFU g^{-1}). The maximum aerobic mesophilic bacteria count was recorded in lettuce (7.3 log10 CFU g^{-1}) while the minimum was in carrot (3.3 log10 CFU g^{-1}) samples (Appendix A). Over all, there was significant variation among each microbial counts in tomato samples (C.V > 10%) except LAB. In tomato, lettuce, and green peppers there was significant variation (CV> 10%) within the samples in coliforms counts. The counts of Enterobacteriaceae were significantly different (C.V > 10%) in samples of tomato, carrot, lettuce, and green peppers. *Staphylococcus* spp., aerobic spore formers, and yeast counts significantly varied (C.V > 10%) within samples of all types of vegetable samples. However, LAB was not significantly varied (C.V < 10%); only in tomato and green pepper. On other hand, there was significant variation (C.V > 10%) among yeast and mold counts of all vegetable samples analyzed (Table 3). In general, there was significant variation ($p < 0.05$) between vegetable samples analyzed for various microbial groups. LAB was not significantly different ($p > 0.05$) between vegetables.

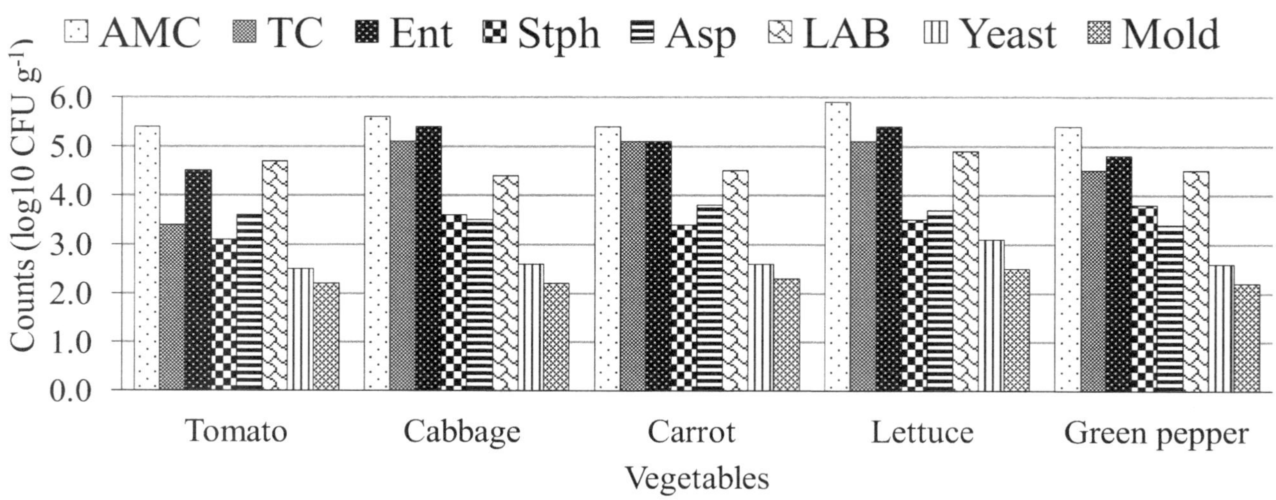

Figure 2. Microbial load of some selected raw vegetables purchased from shops, Jimma town, south western Ethiopia, 2011/12. AMC, Aerobic mesophilic count; TC, total coliforms; Ent, Enterobacteriaceae; Stph, *Staphylococcus* count; ASP, aerobic spores.

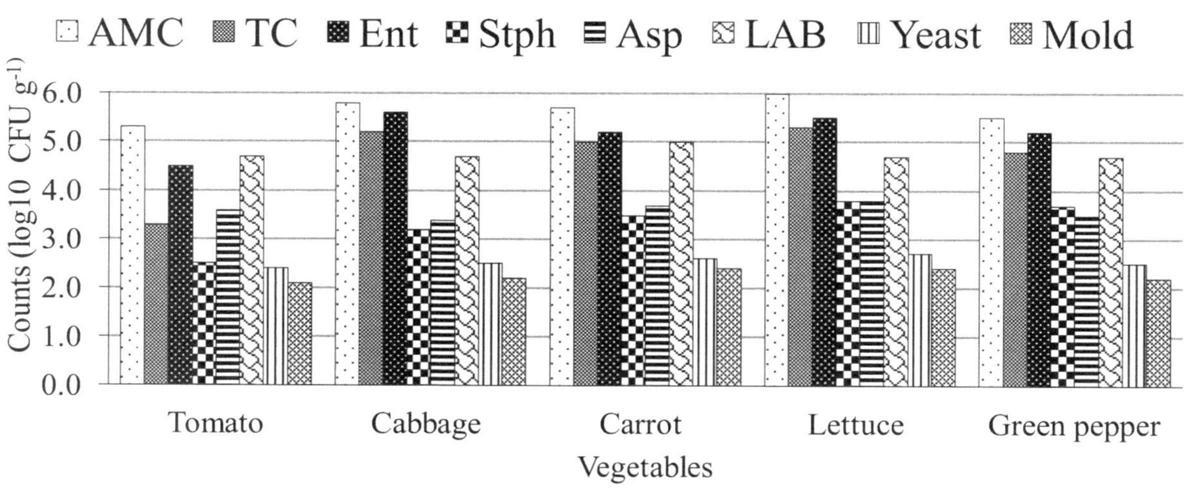

Figure 3. Microbial load of some selected raw vegetables purchased from street venders, Jimma town, south western Ethiopia, 2011/12. AMC, Aerobic mesophilic count; TC, total coliforms; Ent, Enterobacteriaceae; Stph, *Staphylococcus* count; ASP, aerobic spores.

Aerobic mesophilic counts of vegetables analyzed in this study were detected in the range of 3.5 - 6.9, 4.8 - 6.9, 3.3 - 6.1, 5.3 - 7.3, and 3.7 - 6.3 $\log^{10}$ CFU g^{-1} in tomato, cabbage, carrot, lettuce and green pepper, respectively (Appendix A). Most of staphylococci and aerobic spore counts were in the range of 2.0 - 6.5 log10 CFU g^{-1} except in cabbage. Similarly, yeast and mold counts were in the range of 1.9 - 4.2 log10 CFU g^{-1}. However, these ranges were varied based on types of markets and sites from which vegetables were purchased (Appendix A - D).

All vegetable samples purchased from shops contained higher aerobic mesophilic bacterial count than other microbial groups (Figure 2). The counts of coliforms and Enterobacteriaceae were higher in cabbage, carrot and lettuce with counts ≥ 5 log10 CFU g^{-1} (Figure 2). However, counts of the microbial groups of vegetables purchased from shops were ≥ 2 log10 CFU g^{-1}. On other hand, yeast and molds were the least dominant in all vegetables purchased from shops (Figure 2).

Similarly, the aerobic mesophilic bacteria counts of vegetables purchased from street venders were higher than others microbial groups (Figure 3). Likewise, the counts of Enterobacteriaceae and coliforms were ≥ 5 log10 CFU g^{-1} except in tomato. Staphylococci and aerobic spore counts from shops were higher than from

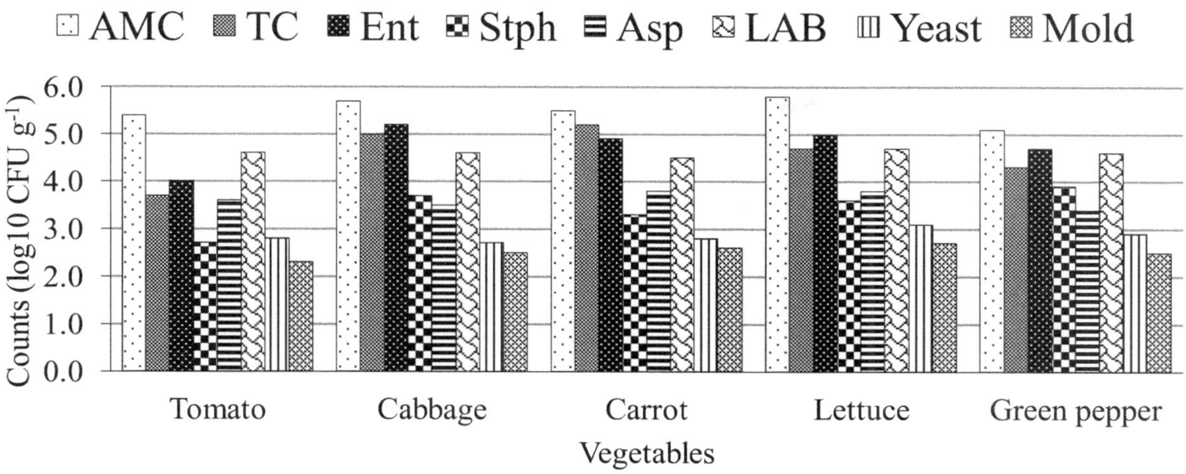

Figure 4. Microbial load of some selected raw vegetables purchased from Kochi site, Jimma town, South western Ethiopia, 2011/12.

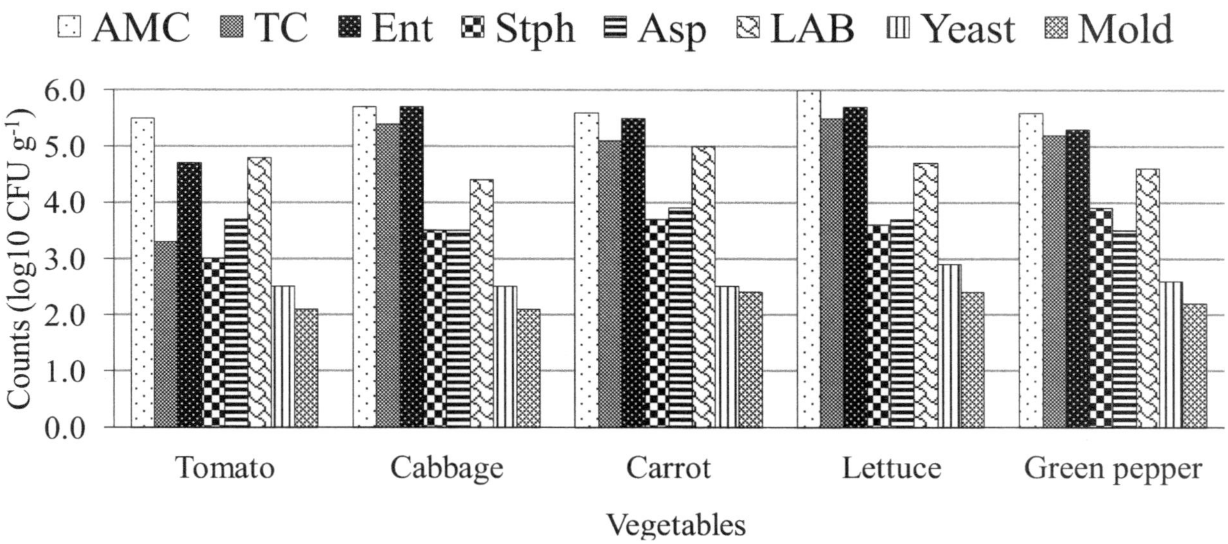

Figure 5. Microbial load of some selected raw vegetables purchased from Agip site, Jimma town, south western Ethiopia, 2011/12. AMC, Aerobic mesophilic count; TC, total coliforms; Ent, Enterobacteriaceae; Stph, *Staphylococcus* count; ASP, Aerobic spores.

street venders with counts ≥ 3 log10 CFU g^{-1} (Figures 2 and 3). But staphylococci counted < 3 log10 CFU g^{-1} in tomato purchased from street venders. Yeasts and molds counts were the lowest in vegetables purchased from street venders and shops (Figure 2 and 3). However, there was no significant variation among counts of aerobic mesophilic bacteria and coliforms among cabbage samples and LAB among tomato and green pepper purchased from different markets. However, counts of the other microbial groups were significantly varied (C.V > 10%) (Appendix A and C). On other hand, the microbial loads of vegetables were analyzed based on the three sampling sites Kochi, Agip and Merkato. Accordingly, the aerobic mesophilic count of both vegetables in Kochi sites were < 6 log^{10} CFU g^{-1} (Figure 4). However, lettuce purchased from both Agip and Merkato contained the maximum aerobic mesophilic bacteria with an average count 6 and 6.1 log^{10} CFU g^{-1}, respectively (Figures 5 and 6). Nevertheless, coliform, Enterobacteriaceae, staphylococci and aerobic spores' counts were ≥ 3 log^{10} CFU g^{-1} in all vegetables purchased from Kochi, Agip and Merkato sites (Figures 4, 5 and 6). However, staphylococci counted < 3 log^{10} CFU g^{-1} in tomato purchased from the three sites (Figures 4, 5 and

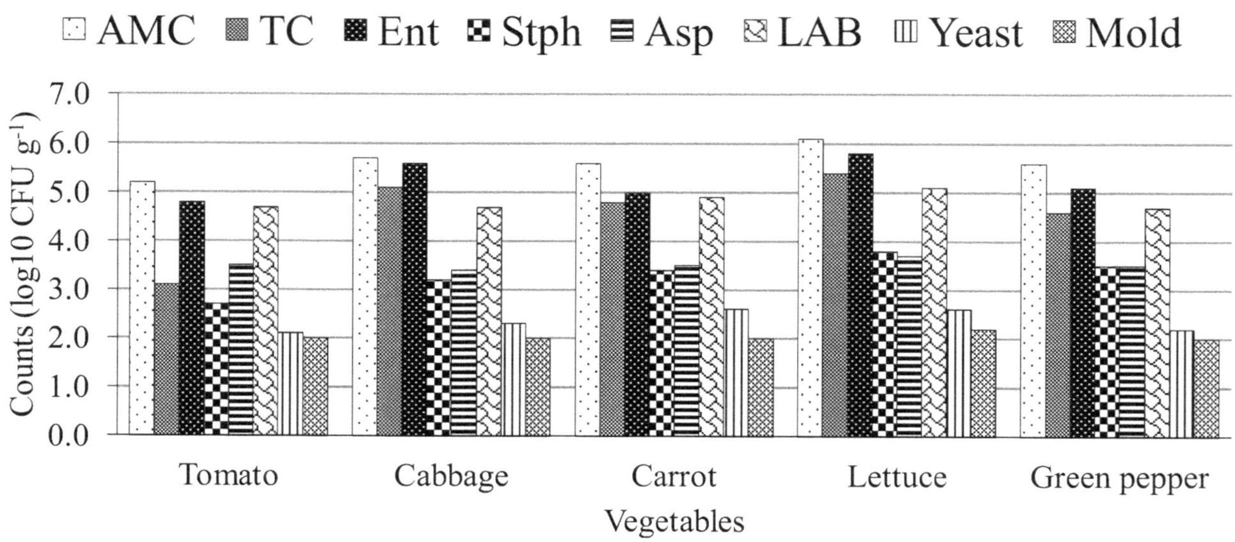

Figure 6. Microbial load of some selected raw vegetables purchased from Merkato site, Jimma town, Southwestern Ethiopia, 2011/12. AMC, Aerobic Mesophilic Count; TC, total coliforms; Ent, Enterobacteriaceae; Stph, *Staphylococcus* count; ASP, Aerobic spores.

6). LAB counts were similar among all vegetables purchased from all sites and counted ≥ 4 log^{10} CFU g^{-1}. But yeast and mold counts were ≤ 3 log^{10} $CFUg^{-1}$ in all samples from Kochi, Agip and Merkato sites. There was no significant variation in counts of aerobic mesophilic bacteria and coliforms among lettuce samples and LAB among tomato purchased from different sites (CV ≤ 10). Counts of other bacterial groups, however, varied significantly (CV>10%) among samples of both vegetable types at both sites (Appendix A - D).

The frequency distribution of different microbiological groups of raw vegetables in Jimma town is as shown in Table 4. Accordingly, 97.2% of tomato and green pepper samples had aerobic mesophilic bacteria counts between 4 - 6.9 log^{10} CFU g^{-1}. However, all of aerobic mesophilic bacteria counts of cabbage and lettuce were higher than 4 log^{10} CFU g^{-1} and 5 log^{10} CFU g^{-1}, respectively. Over 97.7, 92.7 and 82.2% of vegetable samples had LAB, Enterobacteriaceae and coliforms ≥ 4 log^{10} CFU g^{-1}, respectively. The other microbial groups of vegetables were mostly counted between 2 - 3.9 log10 CFU g^{-1}. However, 24.4 and 58.3% of the samples had yeast and mold counts below the detectable level, respectively (Table 4).

Microbial analysis of vegetables

Based on cultural, morphological and biochemical characteristics of the organisms, a total of 1476 bacterial isolates were isolated from 180 vegetable samples. A total of six bacterial genera were identified (Table 5). The number and type of microbial groups isolated from the different vegetable samples were varied (Table 5). *Bacillus* spp (22.3%) was the most frequently isolated group being present in all vegetable types sampled followed by *Staphylococcus* spp. (17.7%), Enterobacteriaceae (15.5%) , *Micrococcus* (14.3%) and *Pseudomonas* (11.6%). *Aeromonas* (9.3%) and other Gram positive (G+) bacteria (9.3%) were the least isolated (Table 5). The most dominant bacterial group isolated from tomato samples were *Bacillus* spp. (29.6%) followed by *Micrococcus* (18.5%) and *Staphylococcus* (13.4%). However, cabbage samples were dominated by Enterobacteriaceae (21.8%) followed by *Bacillus* spp. (20.2%) and *Micrococcus* (15.1%). In carrot, *Bacillus* spp. (26.7%) were the most dominant followed by *Staphylococcus* spp. (19.4%) and Enterobacteriaceae (15.3%). Similarly, *Bacillus* spp. (22.2%) were dominant in lettuce followed by *Staphylococcus* (17.3%) and Enterobacteriaceae (16.4%). On other hand, green peppers were dominated by *Staphylococcus* (21.2%) followed by *Bacillus* spp (16.4%) and *Micrococcus* (15.2%) (Table 5).

Frequency of isolation of Salmonella spp. and S. aureus

Among 180 vegetable samples analyzed 23 (12.8%) samples were positive for *Salmonella* isolates (Table 6). With regard to frequency distribution in each vegetable type, *Salmonella* isolates were highly prevalent in lettuce (16.7%). The frequency distribution of *Salmonella* in both tomato and cabbage were equal (13.9%). On other hand, *Salmonella* were isolated in 11.1% of carrot samples.

Table 4. Frequency distribution of various microbial groups in some selected vegetable samples, Jimma town, south western Ethiopia, 2011/12.

Microbial group	Sample type	Log10 CFU g-1						
		<2 (%)	2-2.9 (%)	3-3.9 (%)	4-4.9 (%)	5-5.9 (%)	6-6.9 (%)	7- 7.9 (%)
Amc	Tomato	0 (0.0)	0 (0.0)	1 (2.8)	8 (22.2)	23 (63.9)	4 (11.1)	0 (0.0)
	Cabbage	0 (0.0)	0 (0.0)	0 (0.0)	2 (5.6)	26 (72.2)	8 (22.2)	0 (0.0)
	Carrot	0 (0.0)	0 (0.0)	1 (2.8)	0 (0.0)	32 (88.9)	3 (8.3)	0 (0.0)
	Lettuce	0 (0.0)	0 (0.0)	0 (0.0)	0 (0.0)	22 (61.1)	13 (36.1)	1 (2.8)
	Green pepper	0 (0.0)	0 (0.0)	1 (2.8)	5 (13.9)	27 (75.0)	3 (8.3)	0 (0.0)
Coliforms	Tomato	6 (16.7)	3 (8.3)	18 (50.0)	8 (22.2)	1 (2.8)	0 (0.0)	0 (0.0)
	Cabbage	0 (0.0)	0 (0.0)	0 (0.0)	13 (36.1)	22 (61.1)	1 (2.8)	0 (0.0)
	Carrot	0 (0.0)	0 (0.0)	0 (0.0)	18 (50.0)	18 (50.0)	0 (0.0)	0 (0.0)
	Lettuce	0 (0.0)	0 (0.0)	0 (0.0)	14 (38.9)	21 (58.3)	1 (2.8)	0 (0.0)
	Green pepper	0 (0.0)	0 (0.0)	5 (13.9)	20 (55.6)	11 (30.6)	0 (0.0)	0 (0.0)
Enterobacteriaceae	Tomato	1 (2.8)	2 (5.6)	7 (19.4)	14 (38.9)	11 (30.6)	1 (2.8)	0 (0.0)
	Cabbage	0 (0.0)	0 (0.0)	0 (0.0)	5 (13.9)	28 (77.8)	3 (8.3)	0 (0.0)
	Carrot	1 (2.8)	0 (0.0)	0 (0.0)	12 (33.3)	23 (63.9)	0 (0.0)	0 (0.0)
	Lettuce	0 (0.0)	0 (0.0)	1 (2.8)	4 (11.1)	27 (75.0)	4 (11.1)	0 (0.0)
	Green pepper	0 (0.0)	0 (0.0)	1 (2.8)	17 (47.2)	18 (50.0)	0 (0.0)	0 (0.0)
Staphylococci	Tomato	14 (38.9)	5 (13.9)	14 (38.9)	3 (8.3)	0 (0.0)	0 (0.0)	0 (0.0)
	Cabbage	0 (0.0)	6 (16.7)	24 (66.7)	5 (13.9)	1 (2.8)	0 (0.0)	0 (0.0)
	Carrot	2 (5.6)	1 (2.8)	26 (72.2)	6 (16.7)	1 (2.8)	0 (0.0)	0 (0.0)
	Lettuce	0 (0.0)	2 (5.6)	26 (72.2)	7 (19.4)	1 (2.8)	0 (0.0)	0 (0.0)
	Green pepper	0 (0.0)	1 (2.8)	26 (72.2)	9 (25.0)	0 (0.0)	0 (0.0)	0 (0.0)
Aerobic spore	Tomato	1 (2.8)	2 (5.6)	23 (63.9)	10 (27.8)	0 (0.0)	0 (0.0)	0 (0.0)
	Cabbage	0 (0.0)	2 (5.6)	30 (83.3)	4 (11.1)	0 (0.0)	0 (0.0)	0 (0.0)
	Carrot	0 (0.0)	1 (2.8)	27 (75.0)	8 (22.2)	0 (0.0)	0 (0.0)	0 (0.0)
	Lettuce	0 (0.0)	1 (2.8)	24 (66.7)	11 (30.6)	0 (0.0)	0 (0.0)	0 (0.0)
	Green pepper	1 (2.8)	2 (5.6)	29 (80.6)	4 (11.1)	0 (0.0)	0 (0.0)	0 (0.0)
Lactic acid bacteria	Tomato	0 (0.0)	0 (0.0)	0 (0.0)	29 (80.6)	7 (19.4)	0 (0.0)	0 (0.0)
	Cabbage	1 (2.8)	0 (0.0)	1 (2.8)	30 (83.3)	4 (11.1)	0 (0.0)	0 (0.0)
	Carrot	1 (2.8)	0 (0.0)	0 (0.0)	22 (61.1)	13 (36.1)	0 (0.0)	0 (0.0)
	Lettuce	0 (0.0)	0 (0.0)	0 (0.0)	23 (63.9)	13 (36.1)	0 (0.0)	0 (0.0)
	Green pepper	0 (0.0)	0 (0.0)	1 (2.8)	31 (86.1)	4 (11.1)	0 (0.0)	0 (0.0)
Yeast	Tomato	14 (38.9)	15 (41.7)	7 (19.4)	0 (0.0)	0 (0.0)	0 (0.0)	0 (0.0)
	Cabbage	8 (22.2)	19 (52.8)	9 (25.0)	0 (0.0)	0 (0.0)	0 (0.0)	0 (0.0)
	Carrot	7 (19.4)	21 (58.3)	7 (19.4)	1 (2.8)	0 (0.0)	0 (0.0)	0 (0.0)
	Lettuce	4 (11.1)	21 (58.3)	7 (19.4)	4 (11.1)	0 (0.0)	0 (0.0)	0 (0.0)
	Green pepper	11 (30.6)	14 (38.9)	11 (30.6)	0 (0.0)	0 (0.0)	0 (0.0)	0 (0.0)
Molds	Tomato	27 (75.0)	8 (22.2)	1 (2.8)	0 (0.0)	0 (0.0)	0 (0.0)	0 (0.0)
	Cabbage	24 (66.7)	11 (30.6)	1 (2.8)	0 (0.0)	0 (0.0)	0 (0.0)	0 (0.0)
	Carrot	17 (47.2)	14 (38.9)	5 (13.9)	0 (0.0)	0 (0.0)	0 (0.0)	0 (0.0)
	Lettuce	12 (33.3)	18 (50)	6 (16.7)	0 (0.0)	0 (0.0)	0 (0.0)	0 (0.0)
	Green pepper	25 (69.4)	9 (25)	2 (5.6)	0 (0.0)	0 (0.0)	0 (0.0)	0 (0.0)

AMC, Aerobic mesophilic counts.

However, green pepper contained the least *Salmonella* isolates (8.3%) as compared to lettuce, cabbage, tomato, and carrot samples (Table 6).

Of the total 180 vegetable samples, 18 (10.0%) were positive for *S. aureus* (Table 6). *S. aureus* was prevalent in each vegetable type. In most case, the levels of prevalence were different between vegetables. However, the prevalence of *S. aureus* in both cabbage and lettuce were equal (11.1%). *S. aureus* was most frequently isolated from green pepper (13.9%) followed by cabbage

Table 5. Dominant bacteria in some selected vegetables purchased form shops and vended markets, Jimma town, south western Ethiopia, 2011/12.

Sample type	Number of isolates	Number of different bacterial isolates (%)						
		Enterobacteriaceae	*Pseudomonas*	*Aeromonas*	*Bacillus*	*Micrococcus*	*Staphylococcus*	Other Gram + bacteria
Tomato	216	24 (11.1)	26 (12.0)	20 (9.3)	64 (29.6)	40 (18.5)	29 (13.4)	13 (6.0)
Cabbage	252	55 (21.8)	35 (13.9)	22 (8.7)	51 (20.2)	38 (15.1)	36 (14.3)	15 (6.0)
Carrot	288	44 (15.3)	32 (11.1)	13 (4.5)	77 (26.7)	29 (10.1)	56 (19.4)	37 (12.8)
Lettuce	324	53 (16.4)	44 (13.6)	32 (9.9)	72 (22.2)	44 (13.6)	56 (17.3)	23 (7.1)
Green pepper	396	53 (13.4)	34 (8.6)	50 (12.6)	65 (16.4)	60 (15.2)	84 (21.2)	50 (12.6)
Total	1476	229 (15.5)	171 (11.6)	137 (9.3)	329 (22.3)	211 (14.3)	261 (17.7)	138 (9.3)

Table 6. Prevalence of *Salmonella* and *S. aureus* in raw vegetables, Jimma town, south western Ethiopia, 2011/12.

Sample type	Sample size (180)	Number of *Salmonella* positive samples (%)	Number of *S. aureus* positive samples (%)
Tomato	36	5 (13.9)	2 (5.6)
Cabbage	36	5 (13.9)	4 (11.1)
Carrot	36	4 (11.1)	3 (8.3)
Lettuce	36	6 (16.7)	4 (11.1)
Green pepper	36	3 (8.3)	5 (13.9)
Total	180	23 (12.8)	18 (10.0)

and lettuce. In carrot, the prevalence was 8.3% with least prevalence (5.6%) in tomato samples (Table 6).

Antimicrobial susceptibility patterns of Salmonella isolates and S. aureus

Salmonella isolates were most susceptible to Ciprofloxacin (100%) and Gentamicin (100%) followed by Norfloxacin (95.7%), Chloramphenicol (87%) and Kanamycin (78.3%) (Table 7). On other hand, it exhibited slight resistance to Streptomycin (43.4%), Chloramphenicol (13%) and Kanamycin (21.7%). All *Salmonella* isolates were resistance to Ampicillin (100%) and 82.6% were resistance to Nalidixic acid (Table 7).

Multiple drug resistance patterns of *Salmonella* spp and *Staphylococcus aureus*

Among nine antimicrobial drugs used in this study, both *Salmonella* and *Staphylococcus aureus* showed Multiple Drug Resistance (MDR) to seven of them (Table 8). The highest MDR by *Salmonella* isolates were noted against TE/AMP/NA (26.1%). The maximum MDR registered was resistance to six drugs with the combination of C/K/S/TE/AMP/NA although less frequent (4.3%). Similarly, a total of six MDR patterns were observed among isolates of *S. aureus*. The highest MDR were observed against TE/P/DA (50%) followed by P/DA (22.2%). The maximum MDR registered was resistance to six drugs with the combination of K/S/TE/E/P/DA (5.6%) (Table 9).

DISCUSSION

The current study revealed the possible source of

Table 7. Antibiotic susceptibility patterns of *Salmonella* isolates from raw vegetables sold in Jimma town, Southwest Ethiopia, 2011/12.

Antimicrobial agents	Disk content (µg/ml)	Resistance		Intermediate		Sensitive	
		Number	%	Number	%	Number	%
Ampicillin (AMP)	10	23	100	0	0	0	0
Chloramphenicol (C)	30	0	0	3	13	20	87
Ciprofloxacin (CIP)	5	0	0	0	0	23	100
Gentamicin (CN)	10	0	0	0	0	23	100
Kanamycin (K)	30	2	8.7	3	13	18	78.3
Nalidixic acid (NA)	30	17	73.9	2	8.7	4	17.4
Norfloxacin (NOR)	10	0	0	1	4.3	22	95.7
Streptomycin (S)	10	5	21.7	5	21.7	13	56.5
Tetracycline (TE)	30	15	65.2	0	0	8	34.8

Chloramphenicol, Ciprofloxacin and Gentamycin were the most effective drugs against *Staphylococcus aureus* and shown the same activity level (100%). Kanamycin (94.4%), Erythromycin (88.8%) and Streptomycin (83.3%) were also partly effective against *S. aureus.* On other hand, *Staphylococcus aureus* was 100% resistance to penicillin G followed by Clindamycin (88.9%) and Tetracycline (66.6%) (Table 8).

Table 8. Antibiotic susceptibility patterns of *Staphylococcus aureus* isolated from raw vegetables, Jimma town, south west Ethiopia, 2011/12.

Antimicrobial agent	Disk content (µg/ml)	Resistance		Intermediate		Sensitive	
		Number	%	Number	%	Number	%
Chloramphenicol (C)	30	0	0	0.0	0	18	100
Ciprofloxacin (CIP)	5	0	0	0.0	0	18	100
Clindamycin (DA)	2	14	77.8	2.0	11.1	2	11.1
Erythromycin (E)	15	1	5.6	1.0	5.6	16	88.8
Gentamicin (CN)	10	0	0	0.0	0	18	100
Kanamycin (K)	30	1	5.6	0.0	0	17	94.4
Penicillin G (P)	10	18	100	0.0	0	0	0
Streptomycin (S)	10	2	11.1	1.0	5.6	15	83.3
Tetracycline (TE)	30	8	44.4	4.0	22.2	6	33.3

Table 9. Multiple drug resistance patterns in *Salmonella* and *S. aureus* isolated from raw vegetables, Jimma town, southwest Ethiopia, 2011/12.

Isolate	Number of drug resisted	Drug resisted	Number of resistant isolates	Percent of resistant isolates (%)
Salmonella spp. (23 isolates)	2	AMP/NA	1	4.3
	3	S/AMP/NA	1	4.3
		TE/AMP/NA	6	26.1
	4	K/S/AMP/NA	2	8.7
		S/TE/AMP/NA	2	8.7
		K/TE/AMP/NA	1	4.3
		TE/AMP/NA/NOR	1	4.3
		C/S/TE/AMP/NA	2	8.7
	5	K/S/TE/AMP/NA	1	4.3
		C/S/TE/AMP/NA	1	4.3
	6	C/K/S/TE/AMP/NA	1	4.3
Staphylococcus aureus (18 isolates)	2	P/DA	4	22.2
	3	TE/P/DA	9	50

Table 9. Contd.

	4	S/TE/P/DA	1	5.6
	5	S/TE/E/P/DA	1	5.6
	6	K/S/TE/E/P/DA	1	5.6

AMP, Ampicillin; C, Chloramphenicol; DA, Clindamycin; E, Erythromycin; K, Kanamycin; NA, Nalidixic acid; NOR, Norfloxacin; P, Penicillin; S, Streptomycin; TE, Tetracycline.

pre- and post-harvest contaminants of vegetables. In Jimma, farmers are cultivating vegetables following traditional farming system. Farmers cultivate vegetables during rainy season, dry season and throughout the year. Most of the time, they used water from river and well as source of water for irrigation purpose. Therefore, river could be the main source for contamination of vegetables during pre-harvest in the field since it could contain sludge from different towns and villages (Aycicek et al., 2006). Pathogens from irrigation water may survive in soil and contaminate vegetable which in turn be transported to consumers with the possibility of causing diseases (Halablab et al., 2011). Other possible source of contamination could be animal manure used by farmers to increase the fertility of farm land. In addition, harvesting equipments, storage place, mechanisms of transportation to the market, placement in the market, and length of storage before selling could be the source of post-harvest contamination of vegetables (Natvig et al., 2002).

Extremely high counts of aerobic mesophilic bacteria reflect exposure of the vegetables to contaminants with the existence of favorable conditions for multiplication of microorganisms (Tortora, 1995). This study showed that the counts of aerobic mesophilic bacteria ranged between 3.3 log10 CFU g^{-1} (carrot) to 7.3 log10 CFU g^{-1} (lettuce). In contrary to this, other researchers from different countries reported a varied load of aerobic mesophilic counts in various vegetables. For instance, Chang and Fang (2007) from Taiwan, Vural and Erkan (2008) and Temiz et al. (2011) from Turkey , Eni et al. (2010) from Nigeria and Khiyami et al.(2011) from Saudi Arabia reported that aerobic mesophilic bacteria counts were between 3.3 - 8.6, 6.4 - 7.6, 6.2 - 7.1, 5.9 - 7.5 and 5 - 5.7 log10 CFU g^{-1}, respectively. Moreover, 82% of whole vegetables investigated in Spain revealed aerobic mesophilic bacteria count < 7 log10 CFU g^{-1} (Abadias et al., 2008). In the present study, 97.2% of aerobic mesophilic bacteria counts were < 7 log10 CFU g^{-1}. The difference in the counts between this study and previous reports may probably be due to difference in cultivation areas of vegetables, seasonal and climatic variation and/or difference in the microbial quality of manure and irrigation water used.

Hazard analysis and critical control point total quality management (HACCP- TQM) technical guide lines set the microbial quality standards for raw foods, whereby the food containing < 4, 4.0 - 6.7, 6.7 - 7.7 and > 7.7 log^{10} CFU g^{-1} aerobic plate count are rated as good, average, poor and spoiled food, respectively (Aycicek et al., 2006). Based on these criteria, 2.8% of each tomato, carrot and green peppers were regarded as good whereas, 97.2% were average; but, all of cabbage samples could be regarded as average in its microbial quality. About 97.2 and 2.8% of lettuce samples were rated as average and poor, respectively. Thus, the consumption of street vended vegetables without any treatment could potentially leads to certain health problem. The poor microbial quality of lettuce could be due to the use of animal manure and river water for irrigation. Lettuce is known to serve as a vehicle of foodborne pathogens and toxins of which the principal source of contamination, are the cultivation stages, processing and operation for preparation (Halablab et al., 2011). In agreement with these authors' findings, this study showed that all lettuce samples collected from different sites and markets in Jimma town had higher incidence of aerobic organisms than any other vegetable samples collected from the same location ($p < 0.05$). Accordingly, the total aerobic bacterial count on lettuce ranged from 5.3 - 7.3 log^{10} CFU g^{-1} as compared to tomato, cabbage, carrot and green pepper.

Total coliform and Enterobacteriaceae count can be considered as a hygiene quality indicator especially for fecal contamination. Their presence could indicate the pathogens might be present due to fecal contamination of human and animal origin or irrigation water. In this study, the counts of coliforms in all vegetable samples ranged from 2.0 log^{10} CFU g^{-1} (tomato) to 6.2 ^{log10} CFU g^{-1} (cabbage). In contrary, the coliform counts of salad vegetables in related study ranged from 4.3 - 4.9 log10 CFU g^{-1} (Khiyami et al., 2011). In addition, report from Zambia (Nguz et al., 2005) found coliform counts from vegetable products between 2.2 - 5.9 log^{10} CFU g^{-1} and Temiz et al. (2011) from Turkey reported that average total coliform counts of vegetables were between 3.4 - 4.9 log10 MPN g^{-1}. However, Aycicek et al. (2006) obtained a range of total count of coliforms on vegetable samples from 3.0 to 6.9 log10 CFU g^{-1}. In agreement with what was reported by Aycicek et al. (2006), the coliform counts in the current study were less than 6.9 log10 CFU g^{-1}.

Similarly, the highest counts of Enterobacteriaceae were encountered in cabbage samples collected from Agip venders (6.7 log^{10} CFU g^{-1}) and lowest from cabbage samples purchased from Kochi shops (4.1 log10

CFU g^{-1}). In related study conducted at Addis Ababa,Biniam and Ashenafi (2010) reported counts of Enterobacteriaceae at levels higher than 4 log^{10} CFU g^{-1} in lettuce and green pepper. Similar counts of Enterobacteriaceae were reported from vegetables examined in Morocco (Ibenyassine et al., 2007). Out of 28 vegetable samples collected from Spain, Abadias et al. (2008) found that 78.6% of the samples had Enterobacteriaceae counts < 5 log^{10} CFU g^{-1}. In contrast to this, 95% of Enterobacteriaceae count in current study was < 6 log^{10} CFU g^{-1}. The high coliform and Enterobacteriaceae counts in cabbage samples and other vegetables in this study could be attributed to poor hygiene of vegetable store room, market place, transporting containers, irrigation water and animal manure used by rural farmers to increase fertility of the farm land.

The contamination of vegetables with high level of *Staphylococcus* may cause *Staphylococcus* food poisoning. It has been reported that production of enterotoxin occurs when the counts of *S. aureus* reach 6 log10 CFU g^{-1} (Schelin et al., 2011). In our study, high *Staphylococcus* count was frequently counted between 3.0 - 3.9 log10 CFU g^{-1} in all vegetable samples analyzed. Accordingly, the frequency of isolation of *Staphylococcus* in this study was 38.9 and 66.7% for tomato and cabbage, respectively. In contrast to this, Biniam and Ashenafi (2010) reported over 80% of green pepper and lettuce harbored *Staphylococcus* counts ranging between 4.0 - 6.0 log10 CFU g^{-1}. The relatively low level of *Staphylococcus* count in present study could be due to short period of storage of the vegetables before sell since vegetables were brought to the market from nearby farmers living around Jimma town.

Higher bacterial spore counts from raw vegetables were found in our study than the mean aerobic spore count observed in lettuce and green pepper by Biniam and Ashenafi (2010). In other study, Ijabadeniyi et al. (2011) from South Africa reported aerobic spore count of 1.5 - 2 log10 CFU g^{-1}. In contrary, the aerobic spore count of present study was between 2.0 to 4.5 log10 CFU g^{-1}. Vegetables treated with food grade chemicals do not support the proliferation of spore forming bacteria. The presence of these bacteria at this level could indicate lack of treatment of vegetables with food grade chemicals to enhance the safety level of vegetables. However, the observed counts were not significantly high to pose health risk.

Lactic acid bacteria (LAB) are the biological basis for the production of a great multitude of fermented foods (Lasagno et al., 2002). The most important contribution of these bacteria is to preserve the nutritive qualities of the raw material and inhibit the growth of spoilage and pathogenic bacteria. This inhibition may be due to the production of many metabolites such as organic acids (lactic and acetic acid), hydrogen peroxide, diacetyl and bacteriocins (Ennahar et al., 2000; Lasagno et al., 2002).

In the present study, all vegetable samples had LAB counts < 5 log10 CFU g^{-1}. The high count of LAB are important to lower pH of the vegetables and contributes to accumulate sufficient antimicrobial metabolites to exert inhibitory effect against potential foodborne pathogens that contaminate the raw vegetables. Similarly, Abadias et al. (2008) from Spain reported LAB counts < 5 log10 CFU g^{-1} in all samples examined. Trias et al. (2008) reported the wide distribution of LAB in fresh vegetables of different origins.

Most of vegetable samples (> 74%) in the current study showed yeast and mold counts ≤ 2.9 log10 CFU g^{-1}. Contrary to our observation Meher et al. (2011) reported that the counts of yeasts and molds in carrot and tomato were below detectable level. The presence of molds in vegetables could pose the possible health problems as some may produce mycotoxins and others are known to cause allergies when they are able to produce large numbers of conidia (Seo et al., 2010).

The level of microbial contamination observed in vegetables of our study may be a reflection of poor storage conditions and how long these produce were kept before they were collected. Bacteria on storage materials may transfer to and cross contamination between produce. Different bacteria were identified and number of the bacteria isolated from each of the samples was varied. Some of the bacteria isolated in this study may be part of the natural flora of the vegetables or contaminants from various sources. *Pseudomonas* spp. and *Bacillus* spp. are part of the natural flora and are among the most common vegetable spoilage bacteria (Jay et al., 2005).

The microbial load of different vegetables was varied based on vegetable types, sites of sample collections, and market place. It was observed that level of lactic acid bacteria between different vegetables were similar ($P > 0.05$) although significant difference were observed between vegetables in other microbial counts ($P < 0.05$). Moreover, the high variability of all microbial groups within the samples of each vegetable showed the lack of uniformity in irrigation water, storage container and placement in the market before sell, consistent sanitation practices. Thus, there is an increased potential for vegetables to become contaminated with pathogenic species during production and processing as there was no system for control of microbiological safety of vegetables.

The presence of *S. aureus* and *Salmonella* spp. in vegetables are dangerous to consumers. *Salmonella* spp. was isolated from higher number of lettuce (16.7%) than other vegetable samples. This may be due to having foliar surfaces with many folds and the fragility of leaves (Aycicek et al., 2006).

In other report, too, *Salmonella* spp. was isolated from vegetables particularly lettuce samples (Rajkowski and Fan, 2008). The contamination of vegetables with human pathogen could occur during the growth of the produce using animal manure, contaminated water or cross contamination during the cutting as the cut of vegetable can harbor and support the growth of food borne pathogen due to nutrients leakage from plant cellular material (Eni et al., 2010). The presence of *Salmonella* in 25 g of sample examined is regarded as potentially hazardous to consumers, and is unacceptable for consumption (Cheung et al., 2007). In addition, *S. aureus* was isolated from higher number of green pepper (13.9%). In similar study, Eni et al. (2010) from Nigeria were reported *S. aureus* was the most frequently isolated pathogens from vegetable samples. *S. aureus* is a dangerous pathogen and one of the most causative agents of hospital infectious (nosocomial infections) in human beings. Surface of vegetables may be contaminated by this organism through human handling and other environmental factors and can be able to survive for several weeks. Thus, contamination of vegetables during distribution and handling may allow bacterial growth and subsequently production of toxins which may represent potential risk to humans. Therefore, cleaning and use of the right types and concentrations of food grade chemicals for cleaning should be practiced to make the vegetables fit for consumption. Emergence of drug resistant pathogens is one of the most serious health problems in developing countries. This happens, for instance, when antibiotics are misused or overused (Nuermberger and Bishai, 2004). In our study, all isolates of *Salmonella* spp. and *S. aureus* were resistance to Ampicillin and penicillin G, respectively. The resistance of *Salmonella* to Streptomycin, Nalidixic acid and Tetracycline in this study was lower than reported from Malaysia (Yoke-Kqueen et al., 2008) and Brazil (Geimba et al., 2005). On other hand, all *Salmonella* isolates were sensitive to Ciprofloxacin and Gentamicin (Table 7). Similarly, all *Staphylococcus aureus* were sensitive to Ciprofloxacin and Chloramphenicol (Table 8).

In agreement with our study, Meher et al. (2011) from Bangladesh were reported similar results on susceptibility of *Salmonella* and *S. aureus* to Ciprofloxacin. Most of *Salmonella* isolates (82.6%) and *S. aureus* (88.9%) were multiple drugs resistant. About 30.3% of *Salmonella* isolates were resistant to three antimicrobials, namely TE/AMP/NA and S/AMP/NA. Likewise, 50% of *S. aureus* were resistant to three antimicrobials (TE/P/DA). Such antimicrobial resistance pattern clearly indicates that isolated pathogens were more resistant to easily available and most frequently used antibiotics. Resistance of *Salmonella* and *S. aureus* isolates to specific drugs could possibly be due to dissemination of drug resistance microbes in the environment arising from the misuse of antibiotics among the general population. In other study, Akbarmehr (2012) reported that 28 % of *Salmonella* isolates were resistant to four antibiotics.

Conclusion

There was lack of awareness on feasible sanitation methods to prevent foodborne diseases associated with consumption of fresh vegetables.

The possible source of contamination of vegetables could be irrigation water, animal manure used as fertilizers and water used to wash vegetables as most sellers wash or refresh different vegetables before selling them with the same water again and again.

All samples analyzed in this study were contaminated with high microbial load. The highest microbial load was recorded in lettuce followed by cabbage and carrot which could be attributed to various preharvest and postharvest sources of contamination. However, there was significant difference in microbial load between vegetable samples.

Out of the total 180 samples of different vegetables, *Salmonella* isolates were found from 23 samples with more prevalence in lettuce than other vegetable samples. Likewise, *Staphylococcus aureus* were encountered from 18 samples with more prevalence in green pepper. This could be an indication of poor hygienic practice and frequent hand contact at the time of harvesting and in the market.

Most of *Salmonella* spp. was resistant to three antibiotics (TE/AMP/NA). Similarly, 50% of *S. aureus* was resistant to three different antibiotics (TE/P/DA).

Recommendations

To limit the introduction of pathogenic bacteria to vegetables through irrigation, the origin of irrigation water should be known. Where wells are used, such wells should be well maintained, and all irrigation sources should be monitored routinely for human pathogens.

Manure used as fertilizer should be treated by composting to eliminate pathogenic microorganisms and farmers should be educated on the need to allow sufficient amount of time between the final manure application and harvest.

Vegetable processors should be educated on the adverse effect of using untreated or polluted water for food processing as these could serve as sources of contamination.

Consumers should treat raw vegetables with food grade chemicals to kill pathogens and significantly reduce the microbial load before consumption.

Different vegetables should be stored separately before consumption to prevent cross contamination and the transfer of drug resistant bacterial pathogens.

In general, to reduce health risk associated to vegetable consumption, intervention mechanisms should be identified and the government should impose strict measures to control or at least minimize the risk of microbial contamination by implementing the Hazard Analysis and Critical Control Point (HACCP).

In the future, the effect of storage time and minimal processing on microbiological quality and safety of vegetables should be analyzed. Vegetables should reach consumers with in short period of time after harvest.

Conflict of Interests

The author(s) have not declared any conflict of interests.

ACKNOWLEDGEMENTS

I would like to express my sincere gratitude to Dr. Ketema Bacha and Tsige Ketema for their professional guidance, valuable, practical and useful comments.

I would like to thank Jimma University for funding this research. Additionally, I would like to express my thanks to Department of Biology for provision of laboratory facilities and other supports. Particularly, my thanks goes to Dr. Diriba Muleta, Head, Department of Biology for kind assistance he rendered me through providing scientific journals and in issues related to administrative matters and continuous encouragement.

I am profoundly grateful and indebted to Mulugeta Dugassa, who helped me financially starting from the beginning of this study up to the end. In addition, I am immeasurably grateful to Zenebu Gurmu and Obse Dugassa and Gelane Dugassa for their moral, financial and all round support.

REFERENCES

Aarestrup F, Wegener H, Collingnon P (2008). Resistance in bacteria of the food chain. Epidemiology and control strategies. Expert Rev. Anti Infect. Ther. 6:733-750.

Abadias M, Usall J, Anguera M, Solsona C, Viñas I (2008). Microbiological quality of fresh minimally processed fruit and vegetables, and sprouts from retail establishments. Int. J. Food Microbiol. 123:121-129.

Acco M, Ferreira FS, Henriques JA, Tondo EC (2003). Identification of multiple strains of *Staphyloccocus aureus* colonizing nasal mucosa of food handlers. Food Microbiol. 20:489-493.

Akbarmehr J (2012). Antimicrobial resistance in *Salmonella* isolated from broiler chicken carcasses. Afr. J. Microbiol. Res. 6:1485-1488.

Al-Binali AM, Bello CS, El-Shewy K, Abdulla SE (2006). The prevalence of parasites in commonly used leafy vegetables in South Western Saudi Arabia. Saudi Med. J. 27:613-616.

Alemu A, Tsegaye W, Golassa L, Abebe G (2011). Urban malaria and associated risk factors in Jimma town, south-west Ethiopia. Malar. J. 10:173-200.

Allende A, Selma MV, López-Gálvez F, Villaescusa R, Gil MI (2008). Role of commercial sanitizers and washing systems on epiphy tic microorganisms and sensory quality of fresh-cut escarole and lettuce. Post Harvest Biol. Technol. 49:155-163.

Amoah P, Drechsel P, Abaidoo RC, Abraham EM (2009). Improving food hygiene in Africa where vegetables are irrigated with polluted water. West Africa Regional Sanitation and Hygiene Symposium, Accra, Ghana.

Asfaw Z (1997). Conservation and Use of Traditional Vegetables in Ethiopia. In Traditional African vegetables, Proceedings of the International Plant Genetic Resources Institute (IPGRI), Guarino, L., Ed., International workshop on genetic resources of traditional vegetables in Africa: Conservation and use. Nairobi. pp. 57-65.

Aycicek B, Sarimeh M, Cakiroglu S (2004). Assessment of microbiological quality of meals sampled at the meal serving units of military hospital in Ankara, Turkey. Food Control 15:379-384.

Aycicek H, Oguz U, Karci K (2006). Determination of total aerobic and indicator bacteria on some raw eaten vegetables from wholesalers in Ankara, Turkey. Int. J. Hyg. Environ. Health 209:197-201.

Beuchat LR (2002). Ecological factors influencing survival and growth of human pathogens on raw fruits and vegetables. Microbes Infect. 4: 413-423.

Biniam G, Ashenafi M (2010). Microbial load, prevalence and antibiograms of *Salmonella* and *Shigella* in lettuce and green peppers. Ethiop. J. Health Sci. 20:41-48.

Chang JM, Fang TJ (2007). Survival of *Escherichia coli O157:H7* and *Salmonella entrica* serovar Typhimurium in iceberg lettuce and the antimicrobial effect of rice vinegar against *E. coli O157:H7*. Food Microbiol. 24:745-751.

Cheesbrough M (2006). District Laboratory Practice in Tropical Countries. Part 2 (2nd ed.). Cambridge: Cambridge University Press.

Chelikani P, Fita I, Loewen P (2004). Diversity of structures and properties among catalases. Cell. Mol. Life Sci. 61:192-208

Cheung PY, Kwok KK, Kam KM (2007). Application of BAX system, Tecra UniqueTM *Salmonella* test, and a conventional culture method for the detection of *Salmonella* in ready-to-eat and raw foods. J. Appl. Microbiol. 103:219-227.

Clinical and Laboratory Standards Institute (CLSI) (2007). Performance standards for Antimicrobial susceptibility Testing: Seventeenth informational supplement. M100-S17.

Deza MA, Araujo M, Garnido MJ (2003). Inactivation of *Escherichia coli* 0157:H7, *Salmonella entritidis* and *Listeria monocytogenes* on the surface of tomatoes by neutral electrolysed water. Lett. Appl. Microbiol. 37:482-487.

Elhariry HM (2011). Biofilm formation by *Aeromonas hydrophila* on green leafy vegetables: cabbage and lettuce. Foodborne Pathog. Dis. 8:80-95.

Eni AO, Oluwawemitan AI, Solomon U (2010). Microbial quality of fruits and vegetables sold in Sango Ota, Nigeria. Afr. J. Food Sci. 4:291-296.

Ennahar S, Sashihara T, Sonomoto K, Ishizaki A (2000). Class IIa bacteriocins: biosynthesis, structure and activity. Microbiol. Rev. 24: 85-106.

Geimba MP, Tondo EC, Brandelli A (2005). Antimicrobial resistance in *Salmonella enteritidis* isolated from foods involved in human foodborne outbreaks that occurred in the south of Brazil. J. Food Saf. 25:173-182.

Gregerson G (1978). Rapid method for distinction of Gram-positive from Gram-negative bacteria. Eur. J. Appl. Microbiol. 5:123-127.

Halablab MA, Sheet IH, Holail HM (2011). Microbiological Quality of Raw Vegetables Grown in Bekaa Valley, Lebanon. Am. J. Food Technol. 6:129 - 139.

Heuer H, Smalla K (2007). Horizontal gene transfer between bacteria. Environ. Biosaf. 6:3-17.

Hugh M, Leifson E (1953). The taxonomic significance of fermentative versus oxidative Gram-negative bacteria. J. Bacteriol. 66:24-26.

Ibenyassine K, Mahand RA, Karakomo Y, Anajjar B, Chouibani M, Enanaji MM (2007). Bacterial pathogens recovered from vegetables irrigated by wastewater in Morocco. J. Environ. Health 9:42-47.

Ijabadeniyi O, Legesse K, Mike V, Elna M (2011). Irrigation water as a potential preharvest source of bacterial contamination of vegetables. J. Food Saf. 31:452-461.

Jay JM, Loessner MJ, Golden DA (2005). Modern food Microbiology (7th edition). Springer science + Business Media, Inc., 233 Springer Street, New York, NY 10013, USA.

Johnston LM, Jaykus LA, Moll D, Martinez MC, Anciso J, Mora B, Moe CL (2005). A field study of the microbiological quality of fresh produce. J. Food Prot. 68:1840-1847.

Khiyami M, Noura A, Basel B, Hassan S (2011). Food borne pathogen contamination in minimally processed vegetable salads in Riyadh, Saudi Arabia. J. Med. Plants Res. 5:444-451.

Kovacs N (1956). Identification of Pseudomonas pyocyanae by the oxidase reaction. Nature 178:703.

Krieg NR (1981). Enrichment and isolation. In: Manual of Methods of Bacteriology, eds. Gerhardt, P., Murray, R.G., Costilow, R.N., Nester, E.W., Wood, W.A., Krieg, N.R. and Briggs Phillips, G., Washington DC: ASM Press. pp. 112-142.

Lasagno M, Beoletto V, Sesma F, Raya R, Font De Valdez G, Eraso A (2002). Selection of bacteriocin producer strains of lactic acid bacteria from a dairy environment. Microbiologia 25:37-44.

Meher NN, Reaz MM, Mahmudul H, Saiful I, Habibur R (2011). Prevalence of multi drug resistant bacteria on raw salad vegetables sold in major markets of Chittagong City, Bangladesh. Middle East J.

Sci. Res. 10:70-77.
Natvig EE, Ingham SC, Ingham BH, Cooperband LR, Roper TR (2002). *Salmonella* enterica serovar Typhimurium and *Escherichia coli* contamination of root and leaf vegetables grown in soils with incorporated bovine manure. Appl. Environ. Microbiol. 68: 2737-2744.
Nguz K, Shindano J, Samapundo S, Huyghebaert A (2005). Microbiological evaluation of fresh-cut organic vegetables produced in Zambia. Food Control 16:623-628.
Nuermberger EL, Bishai WR (2004). Antibiotic resistance in *Streptococcus pneumoniae*: what does the future hold? Clin. Infect. Dis. 38:363-371.
Pal V, Jamuna M, Jeevaratnam K (2005). Isolation and characterization of bacteriocin producing lactic acid bacteria from a south indian special dosa (appam) batter. J. Cult. Collect. 4:53-60.
Rajkowski KT, Fan X (2008). Microbial quality of fresh-cut iceberg lettuce washed in warm or cold water and irradiated in a modified atmosphere package. J. Food Saf. 28:248-260.
Schelin J, Wallin-Carlquist N, Cohn MT, Lindqvist R, Barker GC, Radstrom P (2011). The formation of *Staphylococcus aureus* enterotoxin in food environments and advances in risk assessment. Foodborne Infect. 2:580-592.
Seo Y, Jang J, Moon K (2010). Microbial Evaluation of Minimally Processed Vegetables and Sprouts Produced in Seoul, Korea. Food Sci. Biotechnol. 19:1283-1288
Shalini S (2010). Study on microbiological aspects of fresh fruit and vegetables (including green leafy vegetables) in and around National Capital Region (NCR). Bhaskaracharya College of Applied Sciences. Dwarka, New Delhi, India.
Simon SL, Holzapfel WH, Halasz A (2007). Biogenic amine contents and microbial contamination of leafy vegetables during storage at 5°C. J. Food Biochem. 17: 407-418.
Spencer K, John FT, Spencer AL (2007). Food Microbiology Protocols, Homana Press, Totowa, New Jersey.
Tambekar DH, Mundhada RH (2006). Bacteriological quality of salad vegetables sold in Amravati city (India). J. Biol. Sci. 6: 28 - 30.
Temiz A, Bağcı U, Özmen S (2011). Efficacy of different decontamination treatments on microbial population of leafy vegetables. Gida 36:9-15.
Tortora G (1995). Microbiology. *The Benjamin puplishing co. Ic.*, New York, USA. pp. 274-278.
Trias R, Bañeras L, Montesinos E, Badosa E (2008). Lactic acid bacteria from fresh fruit and vegetables as biocontrol agents of phytopathogenic bacteria and fungi. Int. Microbiol. 11:231-236.
Vural A, Erkan ME (2008). Investigation of Microbial quality of some leafy green vegetables. J. Food Technol. 6:285-288.
Walsh C, Fanings S (2008). Antimicrobial resistance in foodborne pathogens. Curr. Drug Targets 9:808-815.
Yoke-Kqueen C, Learn-Han L, Noorzaleha AS, Son AS, Sabrina S, Jiun-Horng S (2008). Characterization of multiple-antimicrobial-resistant *Salmonella enterica* sub sp. Enterica isolated from indigenous vegetables and poultry in Malaysia. Lett. Appl. Microbiol. 46:318-324.

Appendix A. Aerobic mesophilic bacteria and coliforms load of vegetables.

Vegetable sample	Sampling sites and markets	Microbial Counts in log10 CFU g^{-1}										P-value
		Aerobic mesophilic bacteria					coliforms					
		min	max	Mean	SD	%CV	min	max	Mean	SD	%CV	
Tomato	Kochi	3.5	6.6	5.4	0.9	16.7	2.0	5.4	3.7	0.8	21.6	
	Agip	4.5	6.9	5.5	0.5	9.1	2.0	4.6	3.3	0.9	27.3	
	Merkato	4.4	5.8	5.2	0.5	9.6	2.0	4.5	3.1	0.9	29.0	
	Shop	4.4	6.6	5.4	0.6	11.1	2.0	4.6	3.4	0.8	23.5	
	Vender	3.5	6.9	5.3	0.8	15.1	2.0	5.4	3.3	1.0	30.3	
	All sites	3.5	6.9	5.3	0.7	13.2	2.0	5.4	3.4	0.9	26.5	
Cabbage	Kochi	4.8	6.6	5.7	0.5	8.8	4.5	5.7	5.0	0.5	10.0	
	Agip	4.9	6.6	5.7	0.5	8.8	4.5	6.2	5.4	0.5	9.3	
	Merkato	5.3	6.3	5.7	0.3	5.3	4.5	5.8	5.1	0.5	9.8	
	Shop	4.8	6.5	5.6	0.4	7.1	4.5	6.2	5.1	0.5	9.8	
	Vender	5.3	6.6	5.8	0.4	6.9	4.5	5.8	5.2	0.5	9.6	
	All sites	4.8	6.6	5.7	0.4	7.0	4.5	6.2	5.2	0.5	9.6	
Carrot	Kochi	3.3	6.1	5.5	0.7	12.7	4.3	5.8	5.2	0.5	9.6	
	Agip	5.4	5.8	5.6	0.1	1.8	4.4	5.7	5.1	0.4	7.8	
	Merkato	5.3	5.9	5.6	0.2	3.6	4.3	5.4	4.8	0.4	8.3	
	Shop	3.3	6.0	5.4	0.6	11.1	4.5	5.8	5.1	0.5	9.8	
	Vender	5.3	6.1	5.7	0.2	3.5	4.3	5.6	5.0	0.5	10.0	
	All sites	3.3	6.1	5.5	0.4	7.3	4.3	5.8	5.0	0.5	10.0	
Lettuce	Kochi	5.3	6.8	5.8	0.4	6.9	4.2	5.5	4.7	0.4	8.5	
	Agip	5.4	7.3	6.0	0.5	8.3	4.6	5.9	5.5	0.5	9.1	
	Merkato	5.5	6.8	6.1	0.4	6.6	4.6	6.0	5.4	0.5	9.3	
	Shop	5.4	6.8	5.9	0.4	6.8	4.2	5.9	5.1	0.6	11.8	
	Vender	5.3	7.3	6.0	0.5	8.3	4.5	6.0	5.3	0.5	9.4	
	All sites	5.3	7.3	6.0	0.4	6.7	4.2	6.0	5.2	0.6	11.5	
Green pepper	Kochi	3.7	5.8	5.1	0.6	11.8	3.0	5.4	4.3	0.7	16.3	
	Agip	4.5	6.2	5.6	0.4	7.1	4.6	5.9	5.2	0.6	11.5	
	Merkato	4.7	6.3	5.6	0.4	7.1	3.0	5.8	4.6	0.7	15.2	
	Shop	4.5	5.8	5.4	0.4	7.4	3.0	5.8	4.5	0.7	15.6	
	Vender	3.7	6.3	5.5	0.6	10.9	3.3	5.9	4.8	0.8	16.7	
	All sites	3.7	6.3	5.4	0.5	9.3	3.0	5.9	4.7	0.8	17.0	

Min, Minimum; Max, Maximum; SD, Standard Deviation; CV, Coefficient of variation.

Appendix B. Counts of Enterobacteriaceae and Staphylococci load of vegetables.

Vegetable sample	Sampling sites and markets	Microbial Counts in log10 CFU g^{-1}										P-value
		Enterobacteriaceae					Staphylococci					
		Min	Max	Mean	SD	%CV	Min	Max	Mean	SD	%CV	
Tomato	Kochi	2.0	5.4	4.0	1.2	30.0	2.0	4.7	2.7	1.0	37.0	
	Agip	3.5	5.8	4.7	0.8	17.0	2.0	4.2	3.0	0.7	23.3	
	Merkato	3.8	6.4	4.8	0.6	12.5	2.0	3.3	2.7	0.5	18.5	
	Shop	2.0	5.8	4.5	1.0	22.2	2.0	4.7	3.1	0.8	25.8	
	Vender	2.9	6.4	4.5	0.8	17.8	2.0	3.3	2.5	0.5	20.0	
	All sites	2.0	6.4	4.5	0.9	20.0	2.0	4.7	2.8	0.8	28.6	

Appendix B. Contd

Cabbage	Kochi	4.1	5.8	5.2	0.6	11.5	2.8	5.4	3.7	0.8	21.6
	Agip	4.7	6.7	5.7	0.5	8.8	2.8	4.4	3.5	0.5	14.3
	Merkato	5.3	6.0	5.6	0.2	3.6	2.3	3.6	3.2	0.4	12.5
	Shop	4.1	6.1	5.4	0.6	11.1	3.0	4.4	3.6	0.5	13.9
	Vender	4.7	6.7	5.6	0.4	7.1	2.3	5.4	3.2	0.6	18.8
	All sites	4.1	6.7	5.5	0.5	9.1	2.3	5.4	3.4	0.6	17.6
Carrot	Kochi	2.0	5.8	4.9	1.0	20.4	2.0	4.5	3.3	0.7	21.2
	Agip	4.5	5.8	5.5	0.4	7.3	3.0	4.7	3.7	0.5	13.5
	Merkato	4.2	5.7	5.0	0.5	10.0	2.0	5.0	3.4	0.7	20.6
	Shop	2.0	5.8	5.1	0.9	17.6	2.0	4.7	3.4	0.7	20.6
	Vender	4.2	5.8	5.2	0.5	9.6	2.5	5.0	3.5	0.6	17.1
	All sites	2.0	5.8	5.1	0.7	13.7	2.0	5.0	3.5	0.6	17.1
Lettuce	Kochi	3.6	5.7	5.0	0.7	14.0	2.9	4.3	3.6	0.4	11.1
	Agip	5.3	6.3	5.7	0.3	5.3	3.1	4.2	3.6	0.3	8.3
	Merkato	5.3	6.3	5.8	0.3	5.2	2.8	5.7	3.8	0.8	21.1
	Shop	4.2	6.0	5.4	0.5	9.3	2.8	4.3	3.5	0.4	11.4
	Vender	3.6	6.3	5.5	0.7	12.7	2.9	5.7	3.8	0.6	15.8
	All sites	3.6	6.3	5.5	0.6	10.9	2.8	5.7	3.7	0.5	13.5
Green pepper	Kochi	3.4	5.6	4.7	0.6	12.8	3.2	4.8	3.9	0.5	12.8
	Agip	4.5	5.9	5.3	0.5	9.4	3.5	4.9	3.9	0.4	10.3
	Merkato	4.0	5.7	5.1	0.5	9.8	2.8	4.6	3.5	0.5	14.3
	Shop	3.4	5.7	4.8	0.6	12.5	2.8	4.9	3.8	0.5	13.2
	Vender	4.5	5.9	5.2	0.5	9.6	3.0	4.8	3.7	0.5	13.5
	All sites	3.4	5.9	5.0	0.6	12.0	2.8	4.9	3.8	0.5	13.2

Min, Minimum; Max, Maximum; SD, Standard Deviation; CV, Coefficient of variation.

Appendix C. Aerobic spore formers and Lactic Acid Bacteria load of vegetables.

Vegetable sample	Sampling sites	Microbial counts in log10 CFU g^{-1}										P-value
		Aerobic spore					Lactic Acid Bacteria					
		Min	Max	Mean	SD	%CV	Min	Max	Mean	SD	%CV	
Tomato	Kochi	2.0	4.8	3.6	0.9	25.0	4.3	5.4	4.6	0.4	8.6	
	Agip	3.0	4.5	3.7	0.6	16.2	4.5	5.4	4.8	0.3	6.3	
	Merkato	3.0	4.4	3.5	0.4	11.4	4.5	5.4	4.7	0.3	6.4	
	Shop	2.7	4.8	3.6	0.6	16.7	4.4	5.4	4.7	0.3	6.4	
	Vender	2.0	4.5	3.6	0.7	19.4	4.3	5.4	4.7	0.4	8.5	
	All sites	2.0	4.8	3.6	0.6	16.7	4.3	5.4	4.7	0.3	6.4	
Cabbage	Kochi	2.9	4.5	3.5	0.5	14.3	4.5	4.6	4.6	0.0	0.0	
	Agip	2.8	4.2	3.5	0.4	11.4	2.0	5.3	4.4	0.9	20.5	
	Merkato	3.0	3.6	3.4	0.2	5.9	4.1	5.7	4.7	0.4	8.5	
	Shop	2.8	4.5	3.5	0.4	11.4	2.0	5.3	4.4	0.7	15.9	
	Vender	2.9	4.3	3.4	0.4	11.8	4.3	5.7	4.7	0.4	8.5	
	All sites	2.8	4.5	3.5	0.4	11.4	2.0	5.7	4.5	0.6	13.3	
Carrot	Kochi	3.5	4.5	3.8	0.4	10.5	2.0	5.3	4.5	0.8	17.8	
	Agip	3.3	4.4	3.9	0.4	10.3	4.5	5.5	5.0	0.4	8.0	

Appendix C. Contd

	Merkato	2.5	3.7	3.5	0.3	8.6	4.3	5.6	4.9	0.5	10.2
	Shop	2.5	4.5	3.8	0.5	13.2	2.0	5.5	4.5	0.7	15.6
	Vender	3.3	4.4	3.7	0.3	8.1	4.1	5.6	5.0	0.5	10.0
	All sites	2.5	4.5	3.7	0.4	10.8	2.0	5.6	4.8	0.6	12.5
Lettuce	Kochi	2.8	4.8	3.8	0.6	15.8	4.5	5.3	4.7	0.3	6.4
	Agip	3.3	4.2	3.7	0.3	8.1	4.0	5.6	4.7	0.5	10.6
	Merkato	3.3	4.3	3.7	0.3	8.1	4.5	5.8	5.1	0.5	9.8
	Shop	3.2	4.8	3.7	0.4	10.8	4.5	5.6	4.9	0.4	8.2
	Vender	2.8	4.3	3.8	0.4	10.5	4.0	5.8	4.7	0.5	10.6
	All sites	2.8	4.8	3.7	0.4	10.8	4.0	5.8	4.8	0.5	10.4
Green pepper	Kochi	2.0	4.5	3.4	0.6	17.6	4.5	4.6	4.6	0.1	2.2
	Agip	2.5	4.6	3.5	0.5	14.3	4.2	5.2	4.6	0.2	4.3
	Merkato	3.0	4.5	3.5	0.4	11.4	3.5	5.5	4.7	0.6	12.8
	Shop	2.0	4.5	3.4	0.6	17.6	3.5	4.7	4.5	0.3	6.7
	Vender	2.8	4.6	3.5	0.4	11.4	4.1	5.5	4.7	0.4	8.5
	All sites	2.0	4.6	3.4	0.5	14.7	3.5	5.5	4.6	0.5	10.9

Min, Minimum; Max, Maximum; SD, Standard Deviation; CV, Coefficient of variation.

Appendix D. Yeast and Mold load of vegetables.

Vegetable sample	Sampling sites	Microbial counts in log10 CFU g^{-1}										P- value
		Yeast					Molds					
		Min	Max	Mean	SD	%CV	Min	Max	Mean	SD	%CV	
Tomato	Kochi	2.0	3.5	2.8	0.6	21.4	2.0	3.1	2.3	0.4	17.4	
	Agip	2.0	3.5	2.5	0.4	16.0	2.0	2.5	2.1	0.2	9.5	
	Merkato	2.0	2.5	2.1	0.2	9.5	2.0	2.0	2.0	0.0	0.0	
	Shop	2.0	3.5	2.5	0.6	24.0	2.0	3.1	2.2	0.3	13.6	
	Vender	2.0	3.2	2.4	0.4	16.7	2.0	2.7	2.1	0.2	9.5	
	All sites	2.0	3.5	2.5	0.5	20.0	2.0	3.1	2.1	0.3	14.3	
Cabbage	Kochi	2.0	3.2	2.7	0.4	14.8	2.0	3.1	2.5	0.4	16.0	
	Agip	2.0	3.5	2.5	0.5	20.0	2.0	2.8	2.1	0.2	9.5	
	Merkato	2.0	2.8	2.3	0.3	13.0	2.0	2.0	2.0	0.0	0.0	
	Shop	2.0	3.5	2.6	0.6	23.1	2.0	3.1	2.2	0.4	18.2	
	Vender	2.0	3.1	2.5	0.4	16.0	2.0	2.9	2.2	0.3	13.6	
	All sites	2.0	3.5	2.5	0.4	16.0	2.0	3.1	2.2	0.3	13.6	
Carrot	Kochi	2.0	4.2	2.8	0.7	25.0	2.0	3.3	2.6	0.5	19.2	
	Agip	2.0	3.1	2.5	0.3	12.0	2.0	2.8	2.4	0.3	12.5	
	Merkato	2.0	3.0	2.6	0.4	15.4	2.0	2.3	2.0	0.1	5.0	
	Shop	2.0	4.2	2.6	0.6	23.1	2.0	3.0	2.3	0.4	17.4	
	Vender	2.0	3.2	2.6	0.4	15.4	2.0	3.3	2.4	0.4	16.7	
	All sites	2.0	4.2	2.6	0.5	19.2	2.0	3.3	2.4	0.4	16.7	
Lettuce	Kochi	1.9	4.3	3.1	0.7	22.6	2.0	3.3	2.7	0.4	14.8	
	Agip	2.0	4.3	2.9	0.8	27.6	2.0	3.1	2.4	0.4	16.7	
	Merkato	2.0	3.0	2.6	0.3	11.5	2.0	2.5	2.2	0.2	9.1	
	Shop	2.0	4.3	3.1	0.8	25.8	2.0	3.1	2.5	0.4	16.0	

Appendix D. Contd

	Vender	1.9	3.4	2.7	0.4	14.8	2.0	3.3	2.4	0.4	16.7
	All sites	1.9	4.3	2.9	0.7	24.1	2.0	3.3	2.4	0.4	16.7
Green pepper	Kochi	2.0	3.4	2.9	0.4	13.8	2.0	3.2	2.5	0.4	16.0
	Agip	2.0	3.4	2.6	0.5	19.2	2.0	3.3	2.2	0.4	18.2
	Merkato	2.0	2.7	2.2	0.3	13.6	2.0	2.0	2.0	0.0	0.0
	Shop	2.0	3.4	2.6	0.6	23.1	2.0	3.3	2.2	0.4	18.2
	Vender	2.0	3.3	2.5	0.4	16.0	2.0	2.9	2.2	0.3	13.6
	All sites	2.0	3.4	2.5	0.5	20.0	2.0	3.3	2.2	0.4	18.2

Min, Minimum; Max, Maximum; SD, Standard Deviation; CV, Coefficient of variation.

22

Arsenic pollution of surface and subsurface water in Onitsha, Nigeria

A. C. C. Ezeabasili[1] , O. L. Anike[2], B. U. Okoro[3] and C. M. U-Dominic[4]

[1]School of the Built Environment, Heriot-Watt University, UK.
[2]Department of Geological Sciences, Nnamdi Azikiwe University, Anambra state, Nigeria.
[3]Department of Civil Engineering, Nnamdi Azikiwe University, Anambra State, Nigeria.
[4]Department of Industrial Engineering, University of Ibadan, Nigeria.

Study on arsenic pollution was conducted in Onitsha town in Anambra State, Nigeria to examine the pollution status of surface and subsurface water. Arsenic concentration in surface water within Onitsha metropolis ranged from 0.2001 mg/L (River Niger upstream) to 1.5883 mg/L (River Niger central drainage surface), while concentration in boreholes lies within the range of 0.0000 mg/L (22.2% of boreholes) and 1.2507 mg/L (creek road), respectively. Arsenic concentration in sampled water from river sources is above the WHO standard (0.01 mg/ L) while 87.8% of Arsenic concentration in boreholes adjacent and closest to these rivers and creek is also above the world Health Organization standard (WHO, 2011) of 0.01 mg/L. it is discovered that pollution of surface water is greater than that of the groundwater sources. Major sources of arsenic in Onitsha are refuse dumps, effluent from industries and sewage amongst other sources. Arsenic is known to be the most poisonous metal in existence on earth's surface, its production should be carefully monitored and effluents treated before discharge into the environment.

Key words: Arsenic, water, health.

INTRODUCTION

Water is one of the major means of transport of arsenic in the environment. Soil erosion and leaching have been reported to contribute to the quantity of arsenic in dissolved and suspend form. Soil erosion and agricultural runoff are large contributors to the arsenic concentration in sediments too. Arsenic in its various chemical forms and oxidation states is released into the aquatic environment by natural erosion processes, sewage, refuse and industrial discharges. On release to the aquatic environment, the arsenic species enter into a methylation/ demethylation cycle, while some are bound to the sediments or taken up by biota where they could undergo metabolic conversion to other organoarsenicals (Islam and Tanaka, 2004). A chemical factory manufacturing several chemicals including the insecticide Paris - green (acetocopper arsenite), was responsible for the contamination of wells in the southern part of Calcutta, India. Over seven thousand people consumed the arsenic-contaminated water for several years, but this fact remained unnoticed until September, 1989. A few died, and some of the victims were hospitalized, while symptoms of arsenic poisoning were evident in many families living in the area. Water samples analyzed for arsenic indicated extremely high levels of contamination, with total arsenic concentration ranging from as low as 0.002 to as high as 58 mg/L (Viaraghavan et al., 1999). According to Krishnamurthy (1992), about 1% of Americans consume drinking water that has arsenic levels of 25 ppb or more, and in Utah and California water supplies have been found to contain as much as 500 ppb; also an estimated one-in-a-thousand lifetime risk of dying from cancer induced by normal background levels of arsenic (this equals the risk estimate due to tobacco smoke and radon exposure) (Kosnett, 2005).

Inorganic arsenic is a potent human carcinogen and general toxicant. More than one hundred million people are exposed to elevated concentrations, mainly via drinking water, but also via industrial emissions (Vahter, 2008). Arsenic is metabolized via methylation and reduction reactions, methylarsonic acid and dimethylarsinic acid being the main metabolites excreted in urine. Both inorganic arsenic and its methylated metabolites easily pass the placenta and both experimental and human studies have shown increased risk of impaired foetal growth and increased foetal loss. Recent studies indicate that prenatal arsenic exposure also increases the risk of adverse effects during early childhood. There is a growing body of evidence that the intrauterine or early childhood exposure to arsenic also induces changes that will become apparent much later in life. One epidemiological study indicated that exposure to arsenic in drinking water during early childhood or in utero was associated with an increased mortality in young adults from both malignant and non-malignant lung disease.

Furthermore, a series of experimental animal studies provide strong support for late effects of arsenic, including various forms of cancer, following intrauterine arsenic exposure. The involved modes of action include epigenetic effects, mainly via DNA hypomethylation, endocrine effects (most classes of steroid hormones), immune suppression, neurotoxicity, and interaction with enzymes critical for foetal development and programming (Vahter, 2008).

The mechanism of toxicity of arsenic is through induction of oxidative stress and impairment of the antioxidant defense mechanism, leading to apoptosis and cell death (Flora et al., 2008). Arsenicosis is the effect of arsenic poisoning, usually over a long period such as from 5 to 20 years.

Drinking arsenic-rich water over a long period results in various health effects including skin problems (such as colour changes on the skin, and hard patches on the palms and soles of the feet), skin cancer, cancers of the bladder, kidney and lung, and diseases of the blood vessels of the legs and feet, and possibly also diabetes, high blood pressure and reproductive disorders. Absorption of arsenic through the skin is minimal and thus hand-washing, bathing, laundry, etc. with water containing arsenic do not pose human health risks. In China (Province of Taiwan) exposure to arsenic via drinking-water has been shown to cause a severe disease of the blood vessels, which leads to gangrene, known as 'black foot disease'. This disease has not been observed in other parts of the world, and it is possible that malnutrition contributes to its development. However, studies in several countries have demonstrated that arsenic causes other, less severe forms of peripheral vascular disease (WHO fact sheet, 2001). Chronic arsenic toxicity occurred as a large epidemic in Bangladesh (Smith et al., 2000, WHO, 2011). Ogbuagu and Ajiwe (1998) found that Onitsha North and South have the greatest concentration of Industries in Anambra state, and stand a high risk of industrial pollution. He inferred that some of the industries in the state discharge their waste into the environment with little or no treatment.

This study investigates concentration of Arsenic in surface and subsurface water within Onitsha metropolis and possible impact on populace.

Study location

Onitsha is located in Anambra State in Eastern Nigeria and lies within latitudes 5°22^1 and 6°48^1 and longitudes 6°32^1 and 7°20^1. Onitsha in Anambra State is located on the east bank of River Niger and covers an area of about 49,000 km^2. It is one of most important commercial centres in sub-Sahara African, as well as a transit city in Nigeria. It has an estimated population of about one million inhabitants (Figure 1).

The socioeconomic characteristics of Onitsha consist of about 75% labour force that is engaged in tertiary sector, such as, trading and services. The remaining 25% of labour force is engaged in manufacturing and industrial activities. However, Onitsha is a centre for the production of local goods and services, as well as, a market for the sale of foreign goods. The Onitsha main market, which is reputed as the largest market in sub-Saharan Africa, has increased the tempo of commercial activities in the city in recent times.

MATERIALS AND METHODS

Twenty seven (27) water samples (11 of which were from surface water sources and 16 from boreholes) were obtained from the Onitsha metropolitan city. Surface water sampling was conducted at constant depth of 0.5 - 1.0 m in areas where Onitsha main market/ auxiliary drains and industrial effluents enter the rivers. Borehole samples extracted were very close to the Niger river and in most cases less than 10 km away from the bank.

Research work was conducted for period of 2004-2008. Water samples were stored in sterilized 50cl PVC bottles; and acidulated prior to analysis.

Atomic Absorption Spectrometry method was used to determine concentrations of Arsenic in water samples from Onitsha. This method is based upon the absorption of radiant energy, usually in the ultra-violet and visible regions by neutral atoms in the gaseous state. Water sample were stored in 50 ml of PVC bottles, analysis was done using UNICAM 969 model Atomic Absorption

Table 1. Arsenic concentration in sampling points within Onitsha metropolis

Sampled area	Samples Identifier	Concentration (ppm)	pH
River Niger upstream surface	R. Niger Surf.	0.2001	6.1
River Niger upstream riverbed	R. Niger bed	0.9669	
Nwanaene Lake/ drain	Nwanaene	0.6058	5.4
River Niger/Otumoye surface	Otumoye surf.	1.4076	6.1
River Niger/Otumoye riverbed	Otumoye bed	1.0311	
River Niger/Creek road surface	Creek surf.	0.3203	6.1
River Niger/Creek road riverbed	Creek bed	0.6966	
River Niger Downstream surface	R. Niger Surf.	0.3409	6.1
River Niger Downstream riverbed	R. Niger bed	1.5779	
River Niger Central drainage surface	N. central surf.	1.5883	6.1
River Niger central drainage riverbed	N. central bed	0.3625	
Borehole at 20 Niger street, Onitsha	No 20 Niger	0.0188	6.1
Borehole at 2 Niger street, Onitsha	2nd Niger	0.0000	6.7
Borehole at 9 Abatete/Creek road, Fegge, Fegge	9 Abat.	1.2507	6.1
Borehole at Nigeria Mineral Water Industries, Onitsha	Nig. Min.	0.0026	6.6
Borehole at Royal Queen Table Water, Obosi	Royal Queen	0.0000	7.1
Borehole at Osemaco water, Onitsha	Osemaco	0.0000	7.1
Borehole at Oseokwodu market	Oseokwodu	0.0023	6.8
Borehole at Ivory metal industries, Onitsha	Ivory Metals	0.5140	6.8
Borehole at Akapaka forest avenue	Akapaka	0.0017	7.0
Borehole near Otumonye/ Relef market	Otumonye	0.3060	6.3
Borehole at Enda Table water, Onitsha	Enda	0.0000	6.4
Borehole at St. Joseph drinking water, Bridge Head, Onitsha	Onitsha	0.0000	6.4
Borehole at Banny table water, Nkisi layout, Onitsha	Banny	0.0000	6.8
Borehole at Onitsha South local government area	Onitsha south	0.0008	6.5
Borehole at 35a Ajasa street, Onitsha	35 Ajasa	0.0014	6.1
Borehole at Life Brewery company, Onitsha	Life brewery	0.0038	5.4

Spectrometry. Other chemical utilized were analytical grade chemicals (HNO_3, Sigma chemicals, Australia and standard heavy metal solutions). Samples for heavy metal analysis were digested using nitric acid (65% purity) as described in the APHA methods (APHA, 1999).

RESULTS AND DISCUSSION

Arsenic concentration in surface water within Onitsha metropolis range from 0.2001 mg/L (River Niger up-stream surface) to 1.5883 mg/L (River Niger central drainage surface), while concentration in boreholes falls within range of 0.0000 mg/L (22.2% of boreholes that is, 2nd Niger, Royal Queen, Osemaco, Enda, Onitsha, Banny) and 1.2507 mg/L (No. 9 Abbatete/creek road), respectively (Table 1, Figures 2 to 4). Arsenic pollution in these areas are traced to anthropogenic pollution especially industrial pollution, refuse dumps and oil/fuel dumps that get washed down to streams/rivers of these major cities on daily basis.

One hundred percent of arsenic concentration in river bodies is above the WHO standard while 87.8% of Arsenic concentration in boreholes adjacent and closest to these rivers and creek is over above the world Health Organization standard (WHO, 2011a) of 0.01 mg/L. It is inferred that river water intrusion into the adjacent soil formation is responsible for the high value of arsenic in the borehole beyond the World Health Organization standard (Al-Juboury, 2009; Igwe et al., 2007) especially as these soils have high permeability coefficients and the pH of these subsurface water are on the average below 7 making for the mobilization of metallic ion like arsenic.

In Nigeria, the World Health Organisation (WHO, 2011b) and the Nigeria Standard for Drinking Water Quality (NSDWQ, 2007) drinking water standard is 0.01 mg/L for Arsenic. Higher range of value is common both in the surface or bottom waters of heavily polluted rivers like River Niger at Onitsha, and their adjacent underground waters.

Degree of pollution of various water sources at Onitsha

In Onitsha, various kinds of pollutants are all over the places; they range from biological to chemical. The most polluted zones are the Nwangene Lake zone where all kinds of sewage and chemical effluent drain into. It is also in this place that the biggest refuse dumps are emptied. This has caused major pollution problems in the River

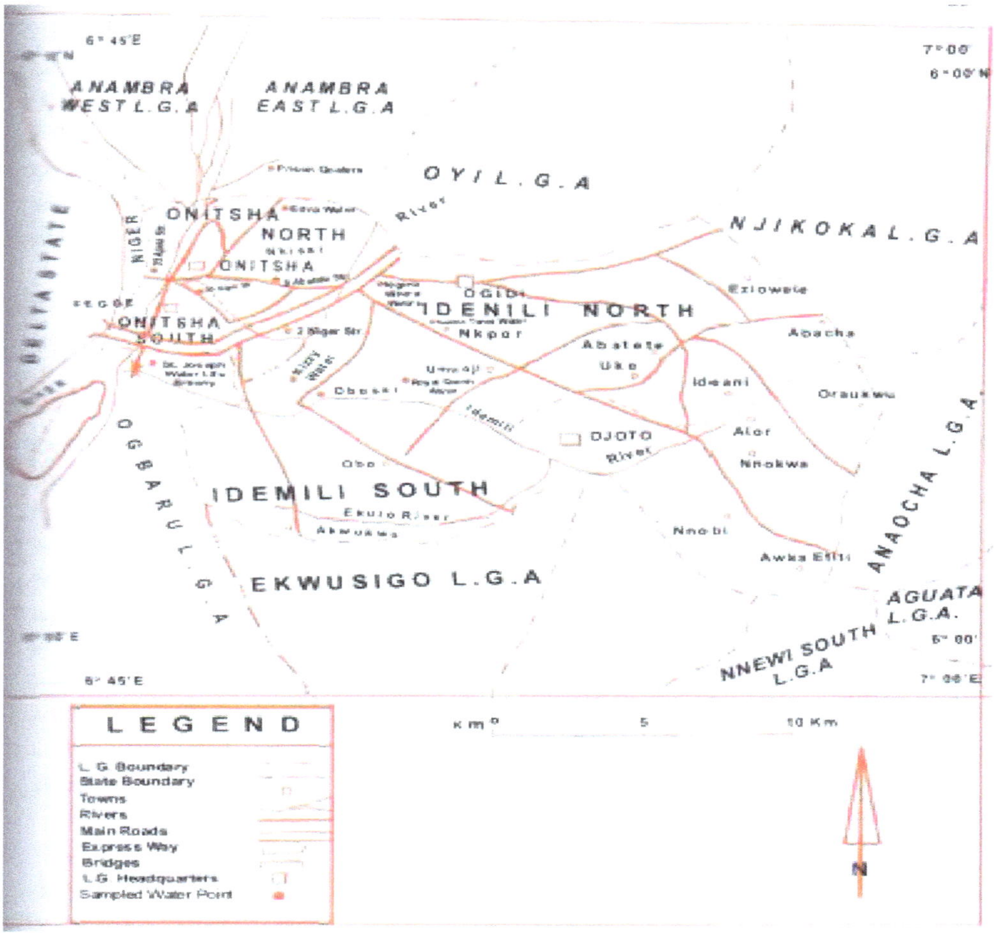

Figure 1. Map of Onitsha Metropolis showing some of the study locations (Source: Adapted from the Federal Survey, 1983).

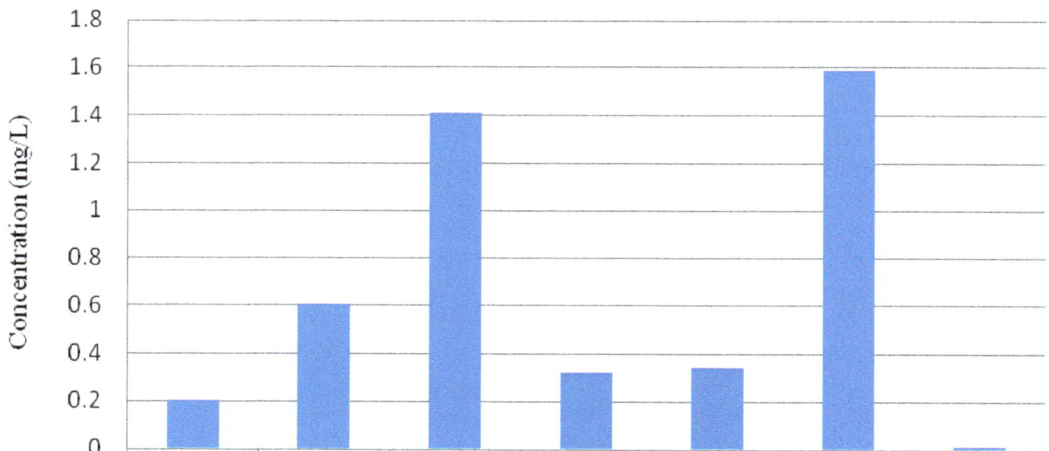

	R. Niger Surf.	Nwanaene	Otumoye surf.	Creek surf.	R. Niger Surf.	N. central surf.	WHO (2011)
Series1	0.2001	0.6058	1.4076	0.3203	0.3409	1.5883	0.01

Figure 2. Arsenic concentration in surfaces of water bodies within Onitsha metropolis.

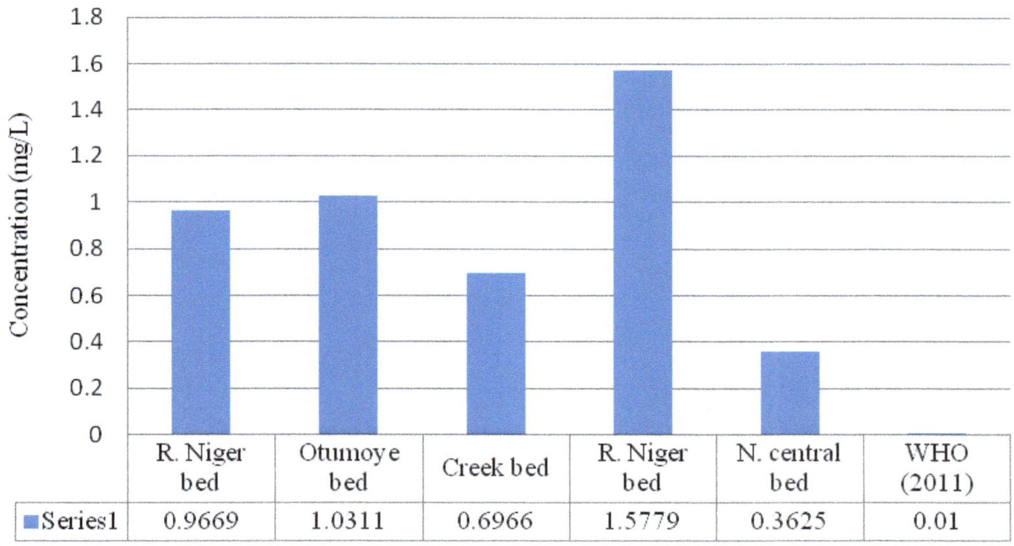

Figure 3. Arsenic concentration in river bed of surface water bodies within Onitsha metropolis.

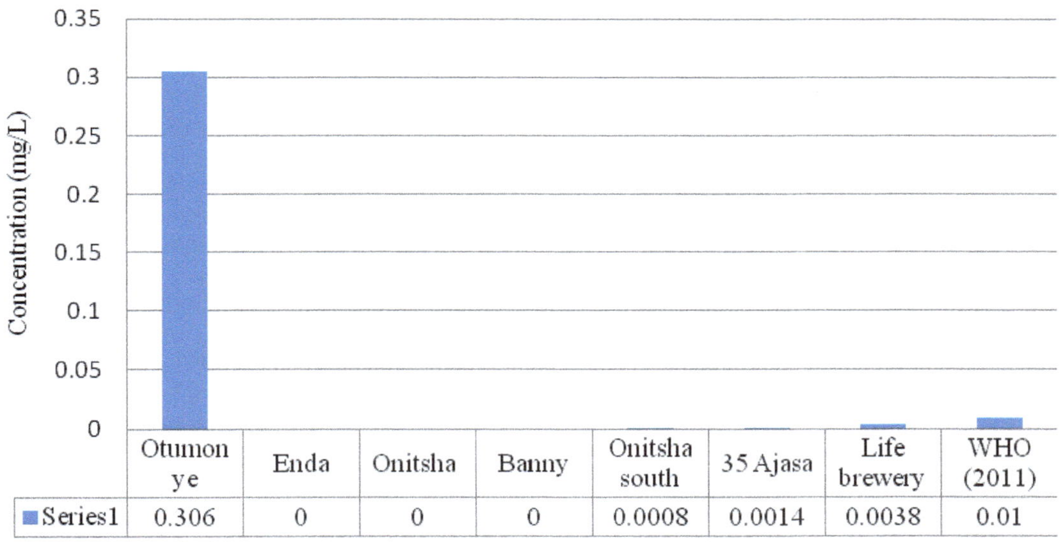

Figure 4a. Arsenic concentration in borehole water Otumoye, Enda, Onitsha, Banny, Onitsha south, 35 Ajasa and Lifebrewery.

Niger where it drains into at Otumoye major drain as can be seen from the results of the tests, but also have polluted boreholes along Creek road and Abatete streets as reported in Table 1 and Figure 4b; the pollution of water bodies in this area can be predicted by the extent of solid and liquid disposals in them. The intrusion of this polluted water is suspected to have stretched up to 100 m radius from the Nwangene Lake and River Niger. It is therefore a serious health hazard for hundreds of inhabitant in the area who use untreated borehole water. The concentration of the heavy metal is much higher than the WHO guidelines.

River Niger water is one of the major sources of water to the people of Onitsha city especially in times of scarcity. Currently all types of wastewater and refuse are disposed in the River Niger. The enormous growth of new industries and urbanization, results in pollution of the river water and the adjoining ground water. Urbanization and technological development results in generation of more and more waste effluents and is discharged into the water bodies resulting in pollution of the surface water. The water borne sewage, storm drainage from Onitsha city streets and effluents from smaller industries combined with flow from municipal outfalls in mounting

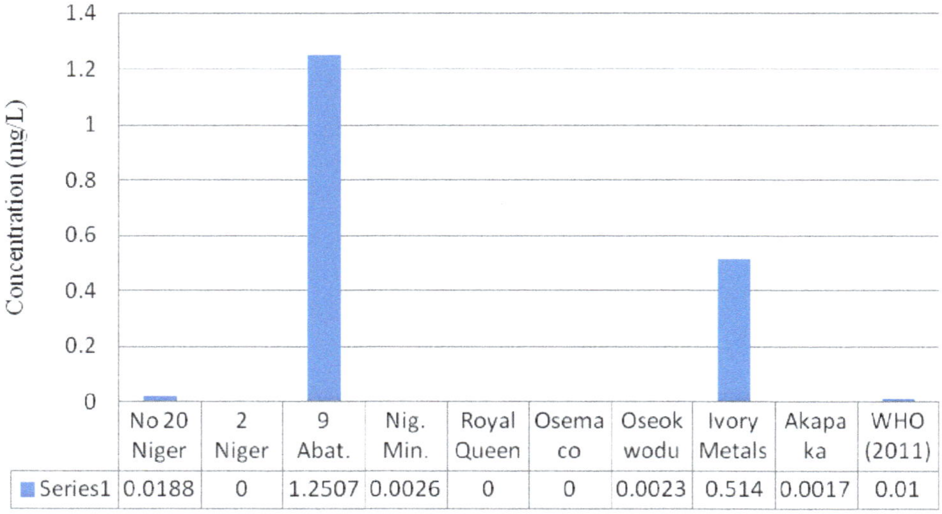

Figure 4b. Arsenic concentration in borehole water of No 20 Niger, 2 Niger, 9 Abat. Nig. Min., Royal Queen, Osemaco, Oseokwodu, Ivory Metals and Akapaka.

volumes along with agricultural runoff from the hinterland ultimately joins the surface water, which drains down the River Niger. Due to inadequacy of public water supply, usage of more ground water for domestic, industrial and agricultural purposes has become very important. Generally, the subsurface water is preferable because of its inherent advantages over surface water.

Major storm water drains in the Onitsha metropolis empties into the River Niger through the main market itself. This also impacts on the quality of water (Figure 2); most results also shows that the bed of the river where the velocity of water is near zero and where a lot of pollutants are supposed to settle are higher than the surface water, which expectedly has self cleansing effects. The general low values of arsenic in most subsurface waters is a thing of joy but it is noticed that consistent pollution activities could aggravate concentration of these dangerous metals with time especially with the rapid increase in urbanization.

Onitsha was a major battle field during the three (3) year Civil War in Nigeria. It is known that Onitsha was contaminated with a wide range of pollutants like hydrocarbons from fuel spillage and leaks, PAHs from fires, explosives, mines, bullets, smokes generating compounds which were not cleared up after the war, this adversely impacts on surface and sub-surface water quality of the environment.

Conclusions and recommendations

Reliable monitoring and surveillance methods are essential for keeping close watch on the health of the environment. Some unacceptable results were recorded for heavy metal loads in the surface and subsurface water in the area. It is obvious that if the rate of pollution is not abated, common health problems which are already found in at Onitsha would increase. The introduction of modern, sensitive and more reliable methods for testing drinking water for physical, chemical and biological parameters should be explored.

More intense training of health experts in the area in water pollution is a sine qua non. There are enough unpolluted ground water resources to satisfy the needs of the society and these could be exploited while effluents from various sources should be treated before discharge. The proper monitoring of industries through environmental impact assessment and environmental audit reports should be enforced by the Ministry of Environment in Anambra State. Information on increasing presence of heavy metals and their affect on health should be disseminated to various stakeholders, including researchers, policy makers, politicians, and donor agencies.

Conflict of Interests

The author(s) have not declared any conflict of interests.

REFERENCES

Al-Juboury AI (2009). Natural Pollution By Some Heavy Metals in the Tigris River, Northern Iraq. Int. J. Environ. Res. 3(2):189-198

APHA (1992) American Public Health Association. Standard methods for the examination of water and wastewater. 18th Ed. Washington, D.C.

Federal Surveys (1983) Catalogue of Maps and Aerial Photographs Produced by the Federal Surveys, Federal Surveys Dept, Lagos, 151 Pages.

Flora SJ, Mittal M, Mehta A (2008). Heavy metal induced oxidative stress and its possible reversal by chelation therapy. Indian J. Med. Res. 128(4):501-523.

Igwe JC, Abia AA, Ibe CA (2008). Adsorption kinetics and intraparticulate diffusivities. Int. J. Environ. Sci. Tech. 5(1)83-92.
Islam MS, Tanaka M (2004). Impacts of pollution on coastal and marine ecosystems including coastal and marine fisheries and approach for management: a review and synthesis. Mar. Pollut. Bull. 48:624-649.
Kosnett MJ (2005). Critical Care Toxicology. Diagnosis and Management of the Critically Poisoned Patient. Elsevier Mosby.
Krishnamurthy S (1992). Biomethylation and environmental transport of metals. J. Chem. Educ. 69(5):347
NSDWQ (2007) Nigerian standard for drinking water quality. Standard Organization of Nigeria. Wuse, Abuja.
Ogbuagu JO, Ajiwe VI (1998). Industrial pollution survey in Anambra. J. Environ. Contam. Toxicol. 9(5):269-275.
Smith AH, Lingas EO, Rahman M (2000). Contamination of drinking-water by arsenic in Bangladesh: a public health emergency. Bulletin of the World Health Organization. 78(9):1093-1103.
Vahter M (2008). Health effects of early life exposure to arsenic. Basic Clin. Pharmacol. Toxicol. 102(2):204-211.
Viraraghavan T, Subramanian KS, Aruldoss JA (1999). Arsenic in drinking water, problems and solutions. Water Sci. Technol. 40(2):69-76.
WHO (2011a) Water-related diseases. Available at www.who.int/water_sanitation_health/diseases/arsenicosis/en/ - accessed 22 September 2011.
WHO (2011b). Guidelines for drinking-water quality. World Health Organization Geneva Vol.1: 4 ed. ISBN 978924 1548151.
WHO Fact Sheet No. 210. (2001) Arsenic in Drinking Water. Bulletin of the World Health Organization, Prepared for World Water Day 2001. Reviewed by staff and experts from the Programme for Promotion of Chemical Safety (PCS), and the Water, Sanitation and Health unit (WSH), World Health Organization (WHO), Geneva 78(9):1096.

Assessment of heavy metals in vegetables irrigated with Awash River in selected farms around Adama town, Ethiopia

Girmaye Benti

Department of Biology, Dilla University College of Natural and Computational Science, P. O. Box.419 Dilla, Ethiopia.

A study was conducted at Melka Hida and Wonji Gefersa farms where Awash River was used for cultivation of vegetable crops to assess heavy metal contamination of vegetables. To what extent these vegetables are contaminated is not known. Three leafy vegetable samples, namely, cabbage (*Brassica oleracea Linn.*), lettuce (*Lactuca sativa*) and spinach (*Spinacea oleracea*) from Melka Hida and Wonji Gefersa farms were examined for heavy metal (Cd, Cr and Pb) contamination using atomic absorption spectroscopy. The results indicate that the heavy metals in vegetables of Melka Hida farm were higher than those of the vegetables in Wonji Gefersa farm. In all the samples analyzed, the concentration of Pb and Cd was more than the maximum limit and their levels varied from 0.31 to 0.65 and 0.21 to 0.40 mg/kg, respectively. However, the level of chromium was generally within the normal range in cabbage (0.85 and 0.29 mg/kg) and spinach (1.30 and 1.06 mg/kg) from Melka Hida and Wonji Gefersa farms, respectively, except in lettuce from Melka Hida farm, 2.4 mg/kg. The high levels of these heavy metals place the consumers of these vegetables grown within the study area at health risk with time unless an urgent step is taken by relevant agencies to address this issue.

Key words: Awash River, contamination, trace elements, vegetables.

INTRODUCTION

Vegetables constitute an important part of the human diet since they contain carbohydrates, proteins, as well as vitamins, minerals and trace elements (Abdola and Chmtelnicka, 1990). In recent years, their consumption is increasing gradually, particularly among the urban community. This is due to increased awareness on the food value of vegetables, as a result of exposure to other cultures and acquiring proper education (Thompson and Kelly, 2003). However, they contain both essential and toxic elements over a wide range of concentrations.

Rapid and unorganized industrialization and urbanization have contributed to the elevated levels of heavy metals in the urban environment in developing countries (Wong et al., 2003). As urban populations in developing countries increase, and residents seek better living standards, larger amounts of freshwater are diverted to domestic, commercial and industrial sectors, which generate greater volumes of wastewater (Asano et al., 2007; Qadir et al., 2007a). There is, however, a lack of investment capacity worldwide for construction and operation of adequate treatment facilities which threatens the quality of surface waters, soils and groundwater to which wastewater is discharged.

In addition to this, excessive application of nitrogen and other inorganic fertilizers and organic manures to these vegetables can accumulate high levels of nitrate and other anions as well as heavy metals. Heavy metals, such as cadmium, copper, lead, chromium and mercury, are important environmental pollutants, particularly in areas under irrigated with wastewater. Several studies have revealed that contamination of vegetables with heavy metals and pesticides poses a threat on consumers (D'Mello, 2003; Sharma et al., 2006; Zandstra and De Kryger, 2007).

International Water Management Institute (IWMI) (2006) suggests that at least 3.5 million ha of land is irrigated globally with untreated, partly treated, diluted, or treated wastewater. Wastewater often contains a variety of pollutants: salts, metals, metalloids, pathogens, residual drugs, organic compounds, endocrine disruptor compounds and active residues of personal care products. Moreover, the excessive accumulation of heavy metals in agricultural soils through wastewater irrigation may not only result in soil contamination but also may lead to elevated heavy metal uptake by vegetables, and thus affect food quality and safety (Muchuweti et al., 2006).

According to Bahemuka and Mubofu (1999) and Ikeda et al. (2000), humans are exposed to the risks through the consumption of vegetables contaminated with heavy metals. Most consumers are not aware of the source of the produce and the use of polluted irrigation water.

Adama and Wonji Gefersa towns are found in Adama *Woreda* of East Showa Zone, Oromia Region, with a population of 260,600 and 23,510, respectively (CSAE, 2005). Awash River is the only important river in Adama *Woreda* used for irrigating around 1132 ha of land. Among the major rivers of Ethiopia, Awash River, which flows from central highlands through Ethiopia's major industrial and agro-industrial belt, taking in a whole burden of all types of raw effluent stands as one of Ethiopia's river streams in urban areas of developed rivers (Tesfamariam, 1989). Most of the existing industries and major towns with in the upper watershed have no treatment plants for discharge of their wastes and are seriously polluting the water course (MWEE, 2010).

In addition, the Modjo River, which is highly polluted by discharging effluent from Modjo tannery industry and waste disposed from the town, is the main tributary of Awash River. Besides this, the expansion of new industries and disposal of industrial wastes to the Awash River is of great concern to the nation (Girma, 2001). Furthermore, food and beverage factories tend to discharge heavy organic pollutants and dyes from textile factories are also released into the same river. The level of toxic heavy metals in vegetables grown on irrigated Awash River is not known in the study area. Therefore, the study was designed to assess the levels of some toxic heavy metals contaminants on selected vegetables irrigated with Awash River in Melka Hida and Wonji Gefersa farms around Adama town.

MATERIALS AND METHODS

Description of the study area

This study was conducted at two wastewater irrigated vegetable growing farms, that is, Melka Hida and Wonji Gefersa that are found in Adama *Woreda*. Melka Hida is found in Adama Town Administrative Zone, Oromia Region, which is 99 km away from Addis Ababa and is located at latitude of 8° 33′ 0″ north and longitude 39° 16′ 12″ east. It has an elevation of 1620 m above sea level.

Wonji Gefersa is a town that is found in Adama *Woreda* of East Showa Zone, Oromia Region, nearby Adama Town, which is 107 km away from Addis Ababa and located at a latitude of 8° 26' 59" north and longitude of 39° 16' 48" east. It has an elevation of 1588 m above sea level and its temperature and annual rain fall is 23°C and 500-800 mm, respectively (Environmental Protection Authority, 2005). Awash River is the only important river in Adama *Woreda* used for irrigating around 1132 ha of land, which originates from the highlands of Dandi *Woreda* located west of Addis Ababa, Ethiopia, and flows along the rift valley into the Afar region, where it eventually terminates in a salty lake, Lake Abbe, found on the border with Djibout.

Study design

A cross sectional survey was conducted to assess the level of heavy metal contamination on the main leafy vegetables [lettuce (*Lactuca sativa*), cabbage (*Brassica oleracea Linn.*), and spinach (*Spinacea oleracea*)] that were grown in Melka Hida and Wonji Gefersa vegetable farms irrigated with Awash River. The samples were regularly collected in three week intervals during January 2013 - March 2013. The vegetable samples were analyzed for heavy metals, cadmium (Cd), lead (Pb), and chromium (Cr) concentrations using atomic absorption spectroscopy (AAS).

Sample collection

A total of 72 samples comprising three types of fresh vegetables (cabbage, lettuce, spinach,) were collected from Melka Hida and Wonji Gefersa vegetable farms using a random sampling technique method. Recently, matured leaves of lettuce, cabbage, and spinach were sampled at early maturity according to methods used by Fisseha (1998). All samples were collected aseptically in a sterilized

Table 1. Mean concentration of three purposively selected heavy metals in vegetables cultivated at Melka Hida and Wonji Gefersa farms in terms of mg/kg dry weight (Mean±SE).

Parameter	No. of examined sample	Site	Vegetable type				
			Cabbage	Lettuce	Spinach	Mean	Maximum limit
	12	MH	0.31 ± 0.3^{b}	0.65 ± 0.3^{a}	0.33 ± 0.3^{b}	0.43 ± 0.2^{a}	
Pb	12	WG	$0.30\pm.03^{b}$	0.40 ± 0.3^{a}	0.31 ± 0.3^{b}	0.34 ± 0.2^{b}	0.3*
		Mean	0.31 ± 0.2^{b}	0.53 ± 0.2^{a}	0.32 ± 0.2^{b}		
	12	MH	0.85 ± 0.4^{b}	2.4 ± 0.4^{a}	2.1 ± 0.4^{a}	1.78 ± 0.2^{a}	
Cr	12	WG	0.29 ± 0.4^{b}	1.33 ± 0.4^{a}	1.06 ± 0.4^{a}	0.89 ± 0.2^{b}	2.3**
		Mean	0.57 ± 0.3^{b}	1.86 ± 0.3^{a}	1.58 ± 0.3^{a}		
	12	MH	0.23 ± 0.3^{b}	0.40 ± 0.3^{a}	0.3 ± 0.3^{b}	0.31±0.2	
Cd	12	WG	0.20 ± 0.3^{b}	0.32 ± 0.3^{a}	0.22 ± 0.3^{b}	0.25±0.2	0.2*
		Mean	0.22 ± 0.2^{b}	0.36 ± 0.2^{a}	0.26 ± 0.2^{b}		

[a-b] Means with different superscript letters along the row for the same parameter in the same site do significantly differ ($P<0.05$). Pb = Lead, Cr = Chromium, Cd = Cadmium. **Source: Weigert (1991) *Source: FAO/WHO (2001).

universal container and plastic bags and transported to Dilla University for laboratory processing.

Determination of heavy metals in leafy vegetables

Vegetable samples were washed with distilled water to eliminate suspended particles. The leafy stalks were removed from all samples and these were sliced and dried on a sheet of paper to eliminate excess moisture; and then carefully dried in oven at 70°C for 24 h. 2 g of the plant material were weighed and ground in a pestle and mortar followed by wet digestion with HNO_3 and H_2O_2 in Tappi (1989) test method, as cited in Subhashini (2013) in the ratio of 3:1. The samples were digested on a hot plate at a temperature of 93°C for 4 h. Heating was done such that it did not boil and until it dried up completely to give a whitish brown dry mass. It was then cooled and the precipitate/digest mixture was extracted in acid water mixture (HCl: distilled water, ratio 3:1) and filtered through Whatman filter paper No. 42. Following this, the volume of the filtrate was made up to 50 ml. Finally, the filtrate was analysed for heavy metal content using atomic absorption spectroscopy (GBC 932AA).

Data analysis

In this study, all statistical analyses were computed using SAS software version 9.1 for heavy metal analysis. The recorded data were subjected to analysis of variance (ANOVA) to assess the effect of vegetable type and site of production on the concentrations of heavy metal contaminant in the vegetables tested. As the level of heavy metal contamination might vary with sample collection site and vegetable type, ANOVA was used to test the existence of significant difference between means. In all statistical analyses, confidence level was held at 95% and $P<0.05$ was considered as significant.

RESULTS AND DISCUSSION

Heavy metal contamination of vegetables

The concentrations of heavy metals in leafy vegetables (lettuce, spinach and cabbage) from both study sites are shown below in Table 1. Average concentration of Pb in lettuce was 0.65 and 0.4 mg/kg; spinach 0.33 and 0.31 mg/kg; cabbage 0.31 and 0.30 mg/kg dry weight each at Melka Hida and Wonji Gefersa, respectively. Lettuce showed a highly significant difference ($p<0.01$) in Pb concentration among the other vegetables and between the sites. However, all the Pb concentrations on the samples analyzed exceeded the permissible limit of 0.3 mg /kg dry weight (FAO/WHO, 2001), except cabbage at Wonji Gefersa farm. Meanwhile the cadmium (Cd) concentration in the leafy vegetables (spinach, lettuce and cabbage) analyzed mean were 0.3, 0.4, 0.23 and 0.22, 0.32, 0.20 mg/kg dry weight at Melka Hida and Wonji Gefersa sites, respectively. The concentrations of Cd in leaf vegetables had highly significant difference ($P <0.01$) between vegetable types, and there was significant difference ($p<0.05$) between sites. This might be because the plantation of industries there directly discharges their effluents into Awash River where the water is used in Melka Hida farm.

However, the concentrations of Cd in almost all samples exceed the tolerable limit of 0.2 mg/kg dry weight (FAO/WHO, 2001), except cabbage at Wonji Gefersa site. However, chromium accumulation in the analyzed leafy vegetables was within normal range of the permissible levels, except lettuce at Melka Hida farm, which is 2.40 mg/kg dry weight. There were highly significant differences ($P <0.01$) between vegetable types and between sites.

Lead

Elevated lead levels are toxic to plants and humans.

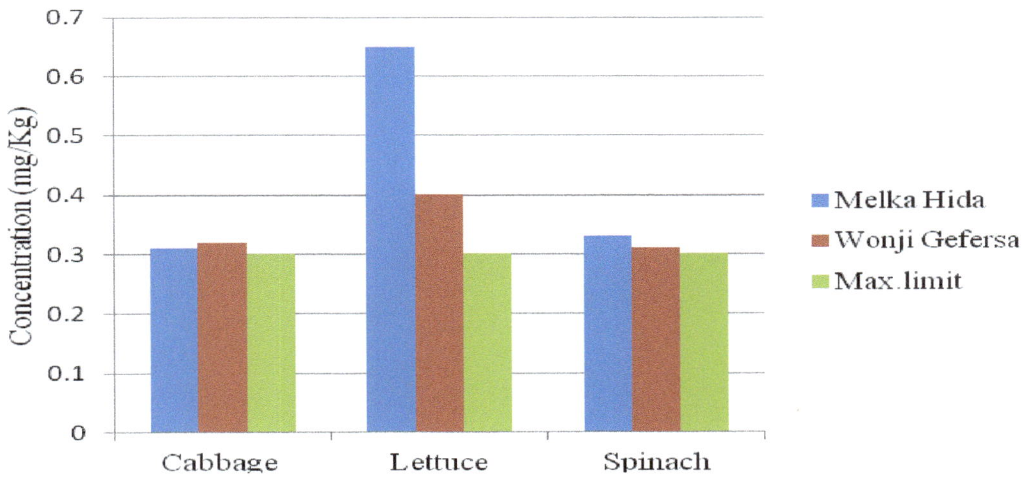

Figure 1. Lead concentration levels in vegetables of Melka Hida and Wonji Gefersa farms.

Even though many plants accumulate large amounts of lead without visible changes in their appearance or yield, this may cause a series of metabolic changes in plant such as slow growth, delayed flowering and reduction in quality. In addition, this can bring Pb into the human food chain, thereby becoming a major concern for health (Concon, 1988). Like most heavy metals, Pb can bioaccumulate overtime and exist in the body for long periods.

Thus, it is necessary to detect such metals even at very low concentrations. Heavy metal contamination of vegetables may occur due to irrigation with contaminated water (Sinha et al., 2006) and emissions of heavy metals from the industries and vehicles may be deposited on the vegetable surfaces during production, transport and marketing (Maleki and Zarasvand, 2008; Sharma et al., 2008a and 2008b) the addition of fertilizers and metal based pesticides on vegetable farm. A similar study that is conducted by Singh and Kumar (2006) and Kumar et al. (2009), reported that lead concentrations of lettuce ranged from 2.3-5.30 mg /kg sample by which super passed the maximum permissible level of Pb set by (FAO/WHO, 2001).

In respective of these results, Farooq et al. (2008) stated that Pb concentration was above toxicity level in leafy vegetables grown in vicinity of an industrial area of Faisalabad, Pakistan. In this study, the concentration of Pb levels in cabbage is least among the other vegetables (Figure 1).

This finding is similar to that of Fisseha (1998) who commented that cabbage was generally the least accumulator of heavy metals as compared to other vegetables. In the current study, the high level of Pb in all vegetables suggests that the water used for irrigation was not good for irrigation of crops in general and leafy vegetables in particular.

Chromium

Exposure to chromium may occur through breathing air, drinking water, or eating food containing chromium or even through skin contact. Exposure to elevated levels chromium leads to skin irritation, ulceration, damage to circulatory and nerve tissue which lead to health problems (Bubb and Lester, 1994). In this study, the chromium content of samples analyzed ranged from 0.29 to 2.40 mg/kg of dry weight. This shows that chromium metal levels are generally within normal range in all vegetable samples from both farms, except lettuce in Melka Hida farm. The chromium contents of lettuce at Melka Hida farm exceed maximum limit of metal concentration set by Weigert (1991) among all vegetables studied as shown in Figure 2.

In line with these results, Fisseha (1998) reported that lettuce had generally the highest concentrations of Cr, Cd, Co, Fe and Mn at Peacock (Addis Ababa) vegetable farm. Similarly, Farooq et al. (2008) reported that Pb and Cd were above toxicity level in leafy vegetables grown in vicinity of an industrial area of Faisalabad, Pakistan whereas other heavy metal (Cr) were within the permitted limits.

Cadmium

Cadmium is a non-essential element in foods and natural waters. It accumulates principally in the kidneys and liver (Divrikli et al., 2006). Various sources of environmental contamination have been implicated for their presence in

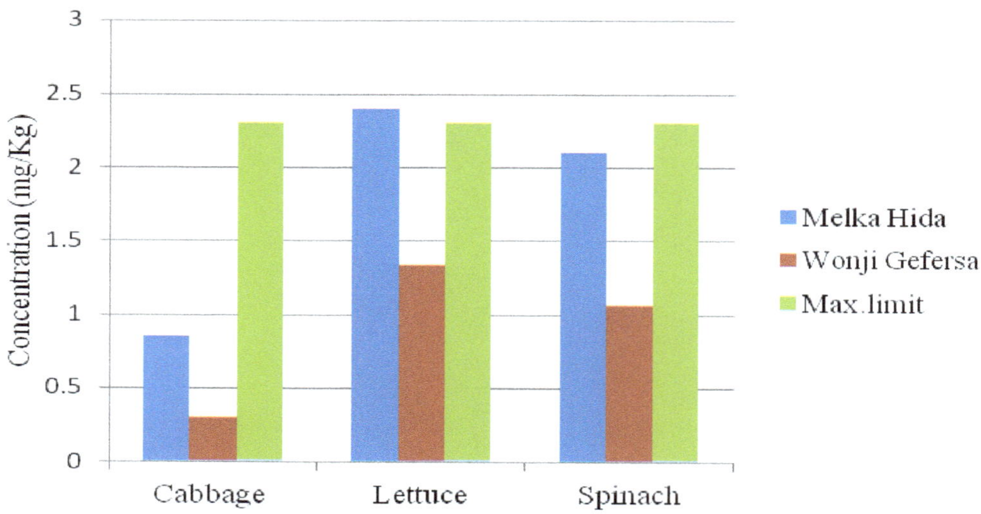

Figure 2. Chromium concentration levels of vegetables in Melka Hida and Wonji Gefersa farms.

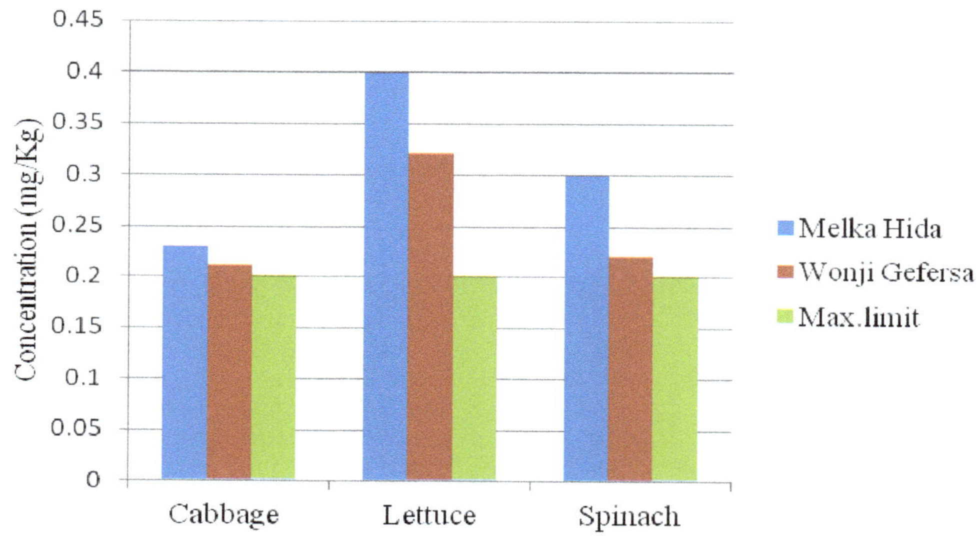

Figure 3. Cadmium concentration levels of vegetables in Melka Hida and Wonji Gefersa farms.

foods. Cadmium is the most toxic heavy metal because its bioaccumulates has a long half-life of about 30 years and may cause health disorders even at low doses (Lenntech, 2006). Toxic effects of Cd on plants include chlorosis, growth inhibition, reduction in water and nutrient uptake and crop protein synthesis (Dass et al., 1997). Cadmium is easily absorbed and translocated to shoots of food crops and may lead to chronic Cd toxicity in human (Mumba et al., 2008). However, it can also cause damage to the skeletal system and kidneys and induce cancer in humans (He et al., 2005). The level of cadmium in this study ranges from 0.20 to 0.40 (Figure 3) which exceeded the recommended limit set by FAO/WHO (2001), 0.20 mg /kg dry weight. In strong connection with this finding, an earlier study by Rahlenbeck et al. (1999) reported that metal contents of vegetables from Addis Ababa market showed that lettuce contained the highest Cd accumulation and cabbage contained the least.

Similarly, Bhatia and Choudhri (1991) reported the concentrations of Cd in edible vegetables ranged from 0.05 to 0.9 mg/kg dry weight and leafy plants such as lettuce, spinach and cabbage contains relatively higher Cd than grain or fruit plant such as apple, due to their higher transpiration. In addition, investigation done by Awashthi (2000) showed that the concentrations of Cd in spinach (4 mg/kg) and radish (2.5 mg/kg) were above the recommended level. In the present study, the accumulation of elevated concentration of Cd in lettuce and spinach might be attributed to the use of contaminated water and industrial effluents for their cultivation. The other possible reason for accumulation is that Cd is relatively easily taken by food crops and especially by leafy vegetables.

Comparing the two contaminated sites, lettuce and spinach accumulated more Pb and Cd at both sites (Melka Hida and Wonji Gefersa); while the cabbage and lettuce accumulated (Pb and Cr) more at Melka Hida farm, respectively. This shows that Melka Hida farm is, hence, more contaminated than Wonji Gefersa farm. This might be due to the fact that more industrial effluents from various industrial sources enter Awash River near the point where water is used for irrigation.

Conclusions

The study reveals that there was heavy metals contamination of fresh leafy vegetables grown in Melka Hida and Wonji Gefersa vegetable farms. In this study, lettuce exhibits higher Cr, Pb and Cd concentrations than other vegetables, whereas elevated Cd level was also exhibited by spinach. However, cabbage was found to be the least accumulator of heavy metals. Generally, the results of the present study revealed heavy metal contamination of vegetables in varying magnitude among vegetables in the study area. Hence, it poses an important public health risk. So monitoring heavy metals in plant tissues is essential in order to prevent excessive build-up of these metals in the human food chain.

Conflict of Interests

The author(s) have not declared any conflict of interests.

ACKNOWLEDEGMENTS

The author gratefully acknowledges all collaborators in this study.

REFERENCES

Abdola M, Chmtelnicka J (1990). New aspects on the distribution and metabolism of essential trace elements after dietary exposure to toxic metals. Biol. Trace Element Res. 23:25-53.

Asano T, Burton F, Leverenz H, Tsuchihashi R, and Tchobanoglous G (2007). Water reuse: issues, technologies, and applications. McGraw-Hill. pp. 345-350.

Awashthi SK (2000). Prevention of Food Adulteration, Act no. 37 of 1954. Central and State rules as amended for 1999. Ashoka Law House, New Delhi.

Bahemuka TE, Mubofu EB (1999). Heavy metals in edible green vegetables grown along the sites of the Sinza and Msimbazi Rivers in Dar es Salaam, Tanzania. Food Chemistry 66:63-66.

Bhatia I, Choudhri GN (1991). Impact of automobile effusion on plant and soil. Int. J. Ecol. Environ. Sci. 17:121-127.

Bubb JM, Lester JN (1994). Anthropogenic heavy metals inputs to low land river system, a case study, the river store. U.K, water, air and soil pollute. 78: 279-297.

Concon JM (1988). Food Toxicology: Contaminants and Additives, New York, Marcel Dekker. pp:1371

CSAE (Central Statistical Agency of Ethiopia), 2005. Census preliminary pdf-file.

D'Mello J PF (2003). Food safety: Contaminants and Toxins. CABI Publishing, Wallingford, Oxon, UK,Cambridge, MA. p. 480.

Dass P, Samantaray S, Rout GR (1997). Studies on cadmium toxicity in plants: A review. Environmental Pollution. 98:29-36.

Divrikli U, Horzum N, Soylak M, Elci L (2006). Trace heavy metal contents of some spices and herbal plants from western Anatolia, Turkey. Int. J. Food Sci. Technol. 41: 712-716

FAO/WHO (2001). Food additives and contaminants, Joint *codex alimentarius* commission, FAO/WHO. Food standards program, ALINORM 01/12A.

Farooq M, Farooq A, and Umer R (2008). Appraisal of heavy metal contents in different vegetables grown in the vicinity of an industrial area. Pak. J. Bot. 40 (5):2099-2106

Fisseha I (1998). Metal concentrations of some vegetables irrigated with industrial liquid waste at Akaki, Ethiopia. Ethiop. J. Sci. 21(1):133-144.

Girma T (2001). Land degradation: A challenge to Ethiopia. Environmental Management 27:815-824.

He Z, Jiangchuan L, Haiyan Z, Ma M (2005). Different effects of calcium and lanthanum on the expression of phytochelatin synthase gene and cadmium absorption in *lactuca sativa*. Plant Science, 168:309-318.

Ikeda M, Zhang ZW, Shimbo S, Watanabe T, Nakatsuka H, Moon CS, Matsuda- Inoguchi, N and Higashikawa K (2000). Urban population exposure to lead and cadmium in east and south-east Asia. Science of the Total Environment 249:373-384.

IWMI (2006). Recycling realities: Managing health risks to make wastewater an asset, IWMI, issue 17 papers prepared for presentation at the American Agricultural Economics Association Annual Meeting, Portland, OR, July 29August 1, 2007.

Kumar A, Sharma IK, Sharma A, Varshneg S and Verma PS (2009). Heavy metal contamination of vegetable foodstuffs in Jaipur (India). Electro. J. Environ. Agric. food chem. 8(2):96-101

Lenntech (2006). Water treatment and air purification: http://www.lenntech.com/heavy-metals.htm; accessed on January 23, 2006

Maleki A, Zarasvand MA (2008). Heavy metals in selected edible vegetables and estimation of their daily intake in Sanandaj, Iran. Southeast Asian J. Trop. Med. Public health 39:335-340

Muchuweti M, Birkett JW, Chinyanga E, Zvauya R, Scrimshaw MD and Lester JN (2006). Heavy metal content of vegetables irrigated with mixture of wastewater and ewage sludge in Zimbabwe: implications for human health. Agriculture. Ecosystem and Environment 112:41-48.

Mumba PP, Chibambo BQ Kadewa W (2008). A comparison of the levels of heavy metals in cabbages irrigated with reservoir and tap water. Int. J. Environ. Resour. 2(1): 61-64.

MWEE (2010). Water quality management. Addis Ababa, Ethiopia.

Qadir M, Wichelns D, Raschid-Sally L, Minhas PS, Drechsel P,

Bahri A, McCornick P (2007a). Agricultural use of marginal quality water opportunities and challenges. Comprehensive assessment of water. Management in Agriculture. Mc Grow Hill, New York. 3:567-70.

Rahlenbeck SI, Burberg A and Zimmermann RD (1999). Lead and cadmium in Ethiopian vegetables. Bulletin of Environmental Contamination and Toxicology. 62:30

Sharma RK, Agrawal M and Marshall FM (2006). Heavy metals contamination in vegetables grown in wastewater irrigated areas of Varanasi, India. Bulletin of Environmental Contamination and Toxicology. 77:311-318.

Sharma RK, Agrawal M and Marshall FM (2008a). Heavy metals (Cu, Cd, Zn and Pb) contamination of vegetables in Urban India: a case Study in Varanasi. Environmental Pollution.154:254-263.

Sharma RK, Agrawal M and Marshall FM (2008b). Atmospheric depositions of heavy metals (Cd, Pb, Zn, and Cu) in Varanasi city, India. Environmental Monitoring and Assessment. 142 (1-3):269-278.

Singh S, Kumar M (2006). Heavy metal load of soil, water and vegetables in peri-urban Delhi. Environmental Monitoring and Assessment. 120:71-79.

Sinha S, Gupta AK, Bhatt K, Pandey K, Rai UN and Singh KP (2006). Distribution of metals in the edible plants grown at Jajmau, Kanpur (India) Receiving Treated Tannery Wastewater: Relation with physiochemical properties of the soil. Environmental Monitoring and Assessment.115:1-22.

Subhashini V, Swamy AVVS, Hema Krishna R (2013). Pot Experiment: To study the uptake of zinc by different plant species in artificially contaminated soil. 1(2):27-33.

Tesfamariam T (1989). Water pollution and natural resources degradation. A challenge to Ethiopia Beyen D (ed) , First Natural Resources Conservation Conference,7-8 February ,1989, IAR, Addis Ababa.

Thompson HC, Kelly WC (2003). *Vegetable Crops.* 5th Edn. New Delhi: McGraw Hill Publishing Company Ltd; 199067. Türkdogan MK, Kilicel F, Kara K, Tuncer I, Uygan I. Heavy metals in soil, vegetables and fruit in the endemic upper gastrointestinal cancer region of Turkey. Environ Toxicol Pharmacol., 13(3):175-179.

Weigert P (1991). Metal loads of food of vegetable origin including mushroooms. In: Merian E, ed. Metals and their compounds in the environment: occurrence, analysis and biological relevance. pp. 458-68.

Wong CSC, Li XD, ZhangG, Qi SH and Peng XZ (2003). Atmospheric depositions of heavy metals in the Pearl River Delta, China. *Atmospheric Environment* vol. 37 issue 6 February, pp. 767-776

Zandstra BH, De Kryger TA (2007). Arsenic and lead residues in carrots from foliar applications of monosodium methanearosonate (MSMA): A comparison between mineral and organic soils, or from soil residues. Food Additives Contaminants. 24:34-42.

A precision nutrient variability study of an experimental plot in Mukono Agricultural Research and Development Institute, Mukono, Uganda

F. O. Adekayode[1] , T. Lutaaya[2], M. O. Ogunkoya[3], P. Lusembo[2] and P. O. Adekayode[4]

[1]Department of Crop, Soil and Pest Management, Federal University of Technology, Akure, Ondo State, Nigeria.
[2]National Agricultural Research Organization, Mukono Zonal Agricultural Research and Development Institute (NARO MuZARDI), Uganda.
[3]Federal College of Agriculture, Akure, Ondo State, Nigeria.
[4]Adekayode Farms and Agricultural Consultancy, Fanibi Road, Akure, Ondo State, Nigeria.

The spatial soil fertility status of a 2.5 ha experimental plot was generated by mapping the soil nutrient concentration and fertility status using GIS kriging technique. The research was conducted in Mukono Zonal Agricultural Research and Development Institute, Mukono, Uganda in October 2013. Soil samples across the experimental plot were randomly taken for laboratory analysis of nitrogen (N), phosphorus (P), potassium (K), calcium (Ca), magnesium (Mg) and the organic matter content. The mean values of N, P, K, Ca, Mg and the organic matter content were 0.16%, 13.7 ppm, 0.44 cmol/kg, 5.35 cmol/kg, 4.83 cmol/kg and 2.78% respectively. The spatial concentration of each element and the organic matter was carried out by the interpolation technique using the 3D Analyst/Raster Interpolation/Kriging Tools while the overlay operations to generate the soil fertility map was carried out using the 3D Analyst/Raster Math Tools in ArcMap. The autocorrelation analysis was carried out using the Spatial Statistics/Spatial Autocorrelation Tools. The autocorrelation analysis indicated N, Ca, Mg and organic matter to be somewhat clustered each with the Moran's 1 Index of 0.37, P was clustered with Moran's 1 Index of 0.5 while potassium pattern was neither clustered nor dispersed. The spatial soil fertility pattern reflected the distribution of nutrient concentration.

Key words: Nutrient variability mapping, Kriging technique, fertility mapping.

INTRODUCTION

The mapping of nutrient distribution in soils had been reported in previous investigations (Jobbagy and

Jackson, 2001; She and Shao, 2009; Craine and Dybzinski, 2013; Laiho et al., 2004; Salehi et al., 2013). The five major factors influencing soil formation and nutrient distribution were climate, organisms, relief, parent materials and time (Gebrelibanos and Assen, 2013). The effects of climate and organisms were explained in Jobbagy and Jackson (2000) on the distribution of soil organic carbon in the soil profile where it was observed that at a given climate, the percentage soil organic carbon in the soil was deepest in the shrublands, intermediate in grassland and shallowest in forests. The effects of parent rock and parent materials on soil chemical properties and nutrient distribution was discussed in the relationship among the six main rock types identified in south west Nigeria and the corresponding soil associations (Smyth and Montgomery, 1962; Periaswamy and Ashaye, 1982). In the report, the soils of Iwo soil association, were related to the coarse grained granite and gneisses, Ondo soil association related to the medium grained granites and gneisses, Egbeda soil association related to the fine grained biotite gneisses and schists, Itagunmodi soil association related to the amphibolites, Okemessi soil association related to quartz gneisses and schists, Mamu soil association related to sericite schists.

The spatial distribution pattern of soil nutrients and their relationships with topographic factors were reported in a research conducted by Song et al. (2011) in the Huangshui River drainage basin, China where slope curvature was observed to have significant effects on spatial distribution of nutrients and the generated digital mapping of the soil nutrients provided data support for the precise management of soil resources in the study area.

The use of remotely sensed imagery in assessing soil nutrient concentration was demonstrated by Chen et al. (2000) in the determination of soil organic carbon concentration using aerial photograph of a bare soil of a 115-ha field, located in Crisp County, Georgia. The use of geostatistical technique in mapping soil nutrient content was also demonstrated in the research by Ismail and Junusi (2009) in a Durian Orchard at Beudang, Malaysia using the Geostatistic Plus (GS++) tool to quantify the spatial nutrient content to predict nutrient values at unsampled location.

The spatial variability of soil nutrients could be conducted by either the grid-cell or grid point methods. The grid-cell method involved dividing fields into square cells and composting soil cores to give one sample per cell while the grid point method involved soil sampling at grid intersection points spaced on a square grid. Soil nutrient maps could then be generated by such methods as Delaunay triangulation, polynomial trend surface, inverse distance squared gridding, point kriging and block kriging (Wollenhaupt et al., 2013).

The spatial distribution of nutrient which employed interpolation methods were demonstrated in the spatial interpolation of soil pH across the Loess Plateau, China using the inverse distance weighted (IDW), splines, ordinary kriging and cokriging methods (Liu et al., 2013). The several other investigations conducted to assess spatial variability of nutrients had been previously reported (Sadeghi et al., 2006; Shah et al., 2013).

The objective of the research was to use the GIS kriging technique to produce precision soil nutrient concentration and fertility maps of a 2.5-ha experimental land in Mukono Agricultural Research and Development Institute Mukono, Uganda.

MATERIALS AND METHODS

Soil sampling and laboratory analysis

Soil samples to a depth of 25 cm from randomly selected locations across the 2.5 hectare experimental area were taken for laboratory analysis. Soil samples were air-dried and sieved through a 2 mm sieve and analysed for N, P, K, Ca, Mg and organic matter content following the laboratory procedures described by Carter (1993). Organic carbon was determined by oxidising soil sample with dichromate solution and later titrated with ferrous sulphate solution. The total nitrogen was determined using micro-kjeldahl method and the available P determined by the Bray P-1 method. The exchangeable cations of K, Ca and Mg were extracted by leaching 5 g of soil with 100 ml ammonium acetate at pH 7 and the potassium in the leachate determined with a flame spectrophotometer while Ca and Mg were determined with atomic absorption spectrophotometer.

Interpolation technique for spatial distribution of nitrogen, phosphorus, potassium and the organic matter content and overlay operations for production of soil fertility map

The values of the nutrients input in Microsoft Excel and saved in coma delimited (csv) file format was added as a layer on the map in ArcMap in the Projected Coordinate Systems WGS 1984 UTM zone 31N. It was added as a layer and exported to Shape file through the Data/Export Data pathway. The spatial distribution of N, P, K, Ca, Mg concentration and the organic matter content was carried out separately for each element with the 3D Analyst/Raster Interpolation/Kriging Tools while the overlay operations to generate the soil fertility map was carried out using the 3D Analyst/Raster Math/Plus Tools in ArcMap. The autocorrelation analysis was carried out using the Spatial Statistics/Analyzing Patterns/Spatial Autocorrelation Tools.

RESULTS

Table 1 shows the coordinates of the perimeter and the

Table 1. Coordinates of the perimeter and the soil sampling locations.

Coordinates of the experimental land perimeter		Coordinates of the soil sampling locations	
Easting	**Northing**	**Easting**	**Northing**
470299	42304	470320	42292
470435	42326	470363	42301
470481	42242	470420	42312
470316	42214	470324	42267
470299	42304	470329	42228
		470354	42241
		470390	42238
		470381	42259
		470439	42252
		470431	42284
		470362	42271
		470339	42280
		470400	42285
		470410	42262

soil sampling locations of the 2.5 ha experimental plot while Figure 1 shows the spatial representation of soil sampling locations. The sampling locations were randomly selected across the experimental land with each representing an area within 10 m radius. On the map part of the administrative block and the major road from the main gate are shown.

Table 2 shows the values of nutrient elements with the mean values and the organic matter content as 0.16%, 13.7 ppm, 0.44 cmol/kg, 5.35 cmol/kg, 4.83 cmol/kg and 2.78%, respectively.

Figures 2, 3, 4, 5, 6 and 7 show the spatial concentration of nitrogen, phosphorus, potassium, calcium, magnesium and the organic matter content, respectively, while Figure 8 showed the soil fertility map. The autocorrelation analysis indicated N, Ca, Mg and organic matter to be somewhat clustered with the Moran's 1 Index of 0.37, phosphorus was clustered with Moran's 1 Index of 0.5 while potassium was neither clustered nor dispersed with Moran's 1Index of 0.28. The soil fertility classes I, II and III on the soil fertility map indicated low, medium and high fertility status respectively.

DISCUSSION

The attribute data which indicated the coordinates of the perimeter and soil sampling locations in Table 1 was used to generate the spatial data in Figure 1. This corroborated the previous study by Murray and Shyy (2000) on the integration of attribute and spatial data for identifying patterns in spatial information. The previous investigation by Andrienko and Andrienko (2001) also reported on the technique of analysis of numerical data associated with area geographical objects. The attribute data had been described to contain information about the features on a map that was linked to the map and the linkage achieved by specifying variables in the attribute data set and composite association in the spatial definition that had the same values (GISTUTOR, 2001).

The spatial concentration of each of N, P, K, Ca, Mg and the organic matter content in Figures 2 to 7 was generated through interpolation method of kriging. The interpolation technique enabled predicted values to be assigned to all other locations to create continuous surface representation of the nutrient concentration. This corroborated with the research findings of Oliver and Webster (1999) in the stochastic models of spatial variation in the mapping of soil salinity in the Jordan Valley of Israel and also the herbaceous cover in semi-arid Botswana. The geostatistical interpolation technique of kriging was also used to prepare the landslide susceptibility analysis map of Kota Kinabalu in Malaysia to locate areas prone to landslides (Roslee et al., 2012). The degree of accuracy of kriging technique in the prediction of soil properties was explained in the report of Omran (2012) in the descriptive tools of semivariograms to characterize the spatial patterns of continuous and categorical soil attributes. The application of kriging technique had been premised on the principle that soil properties closer together would tend to be more alike

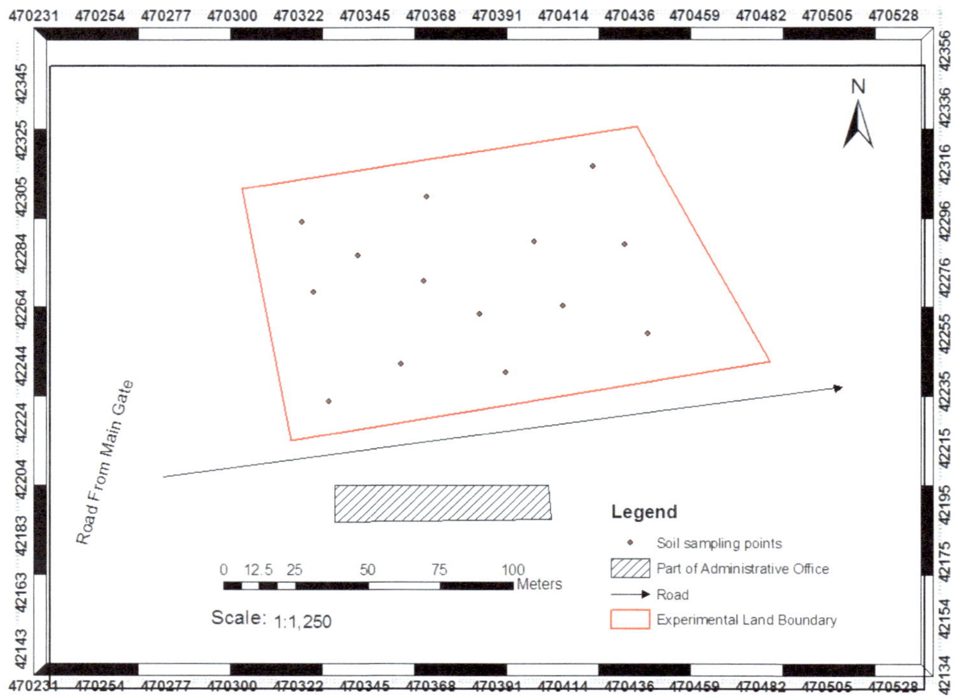

Figure 1. Experimental site with soil sampling locations.

Table 2. The values of N, P, K, Ca, Mg and the organic matter content at each sample location.

Easting	Northing	Nitrogen (%)	Phosphorus (ppm)	Potassium (cmol/kg)	Calcium (cmol/kg)	Magnesium (cmol/kg)	Organic matter (%)
470320	42292	0.16	13.4	0.36	4.48	4.88	2.78
470363	42301	0.12	11.8	0.28	3.68	4.22	2.12
470420	42312	0.14	11.2	0.21	3.24	3.86	2.12
470324	42267	0.19	15.7	0.70	8.17	5.56	3.60
470329	42228	0.18	14.8	0.65	7.49	5.48	3.46
470354	42241	0.19	15.6	0.54	6.56	5.42	3.24
470390	42238	0.18	14.9	0.71	6.52	5.52	3.42
470381	42259	0.14	13.9	0.43	5.52	4.86	2.86
470439	42252	0.13	13.2	0.34	3.86	4.14	2.26
470431	42284	0.14	11.6	0.24	4.42	4.64	2.88
470362	42271	0.15	13.8	0.42	5.58	5.24	2.88
470339	42280	0.17	14.1	0.51	6.51	5.38	2.86
470400	42285	0.15	13.6	0.32	4.40	4.12	2.14
470410	42262	0.14	13.6	0.41	4.42	4.32	2.24

than the distant points and the interpolation was a prediction made within the spatial extent of the measured locations.

The spatial concentration of the nutrient elements from the autocorrelation analysis which indicated N, Ca, Mg and organic matter to be somewhat clustered each with

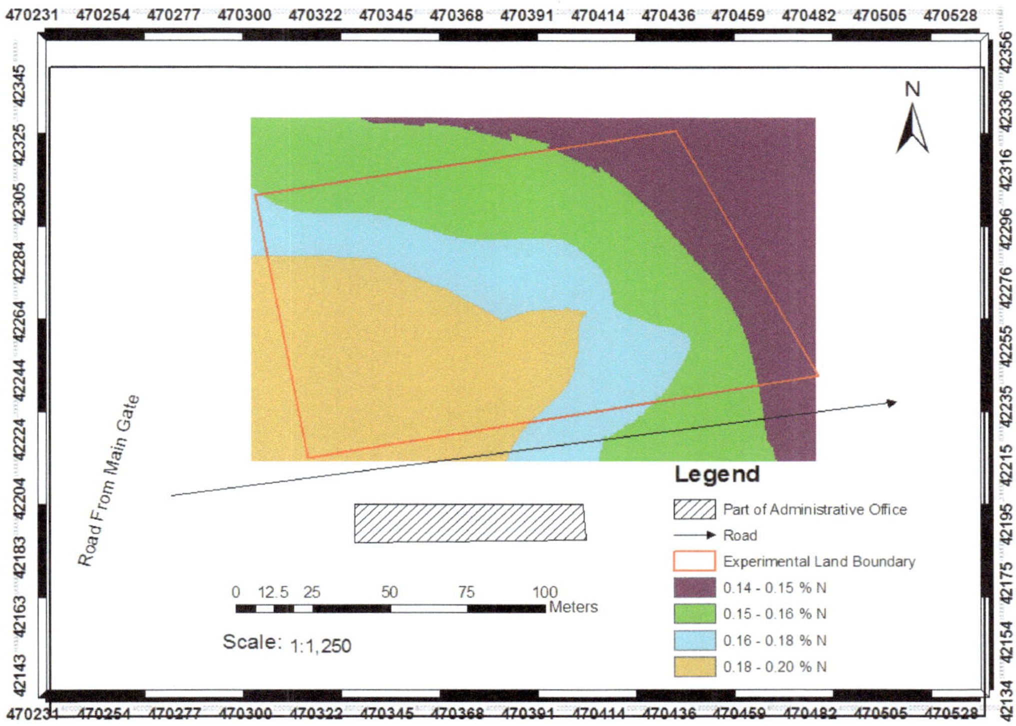

Figure 2. Nitrogen spatial concentration.

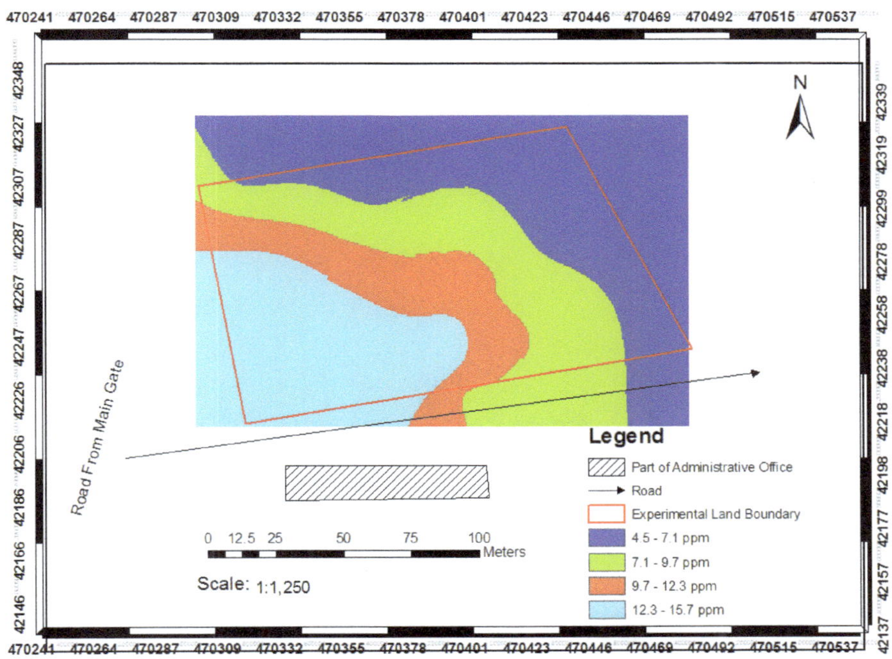

Figure 3. Phosphorus spatial concentration.

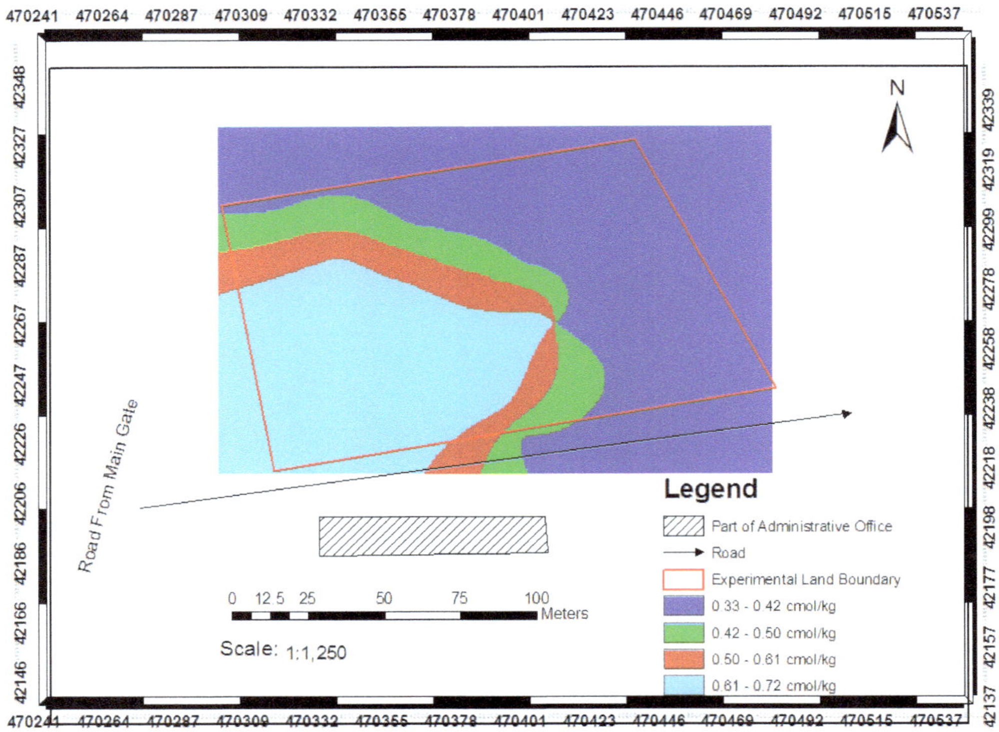

Figure 4. Potassium spatial concentration.

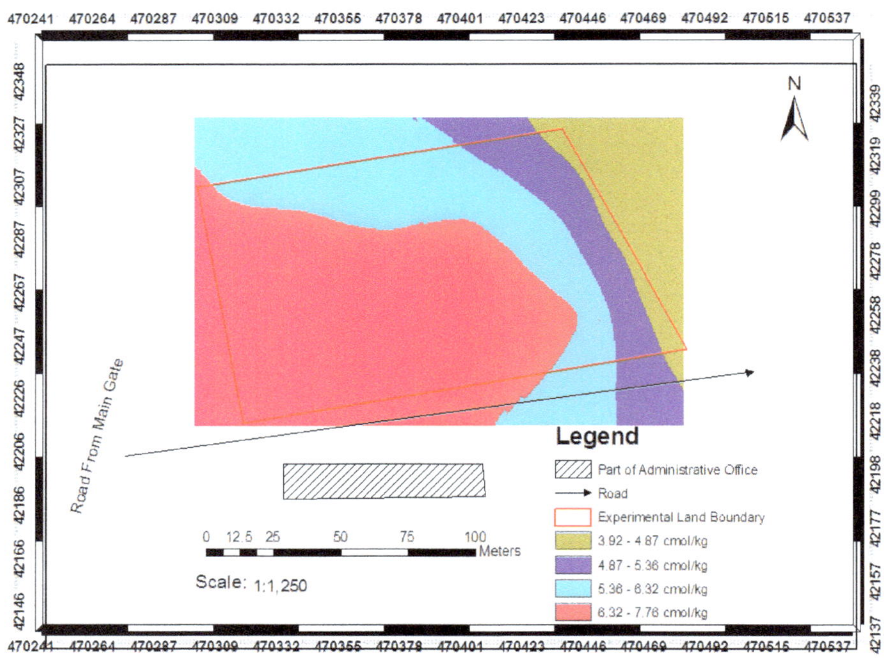

Figure 5. Calcium spatial concentration.

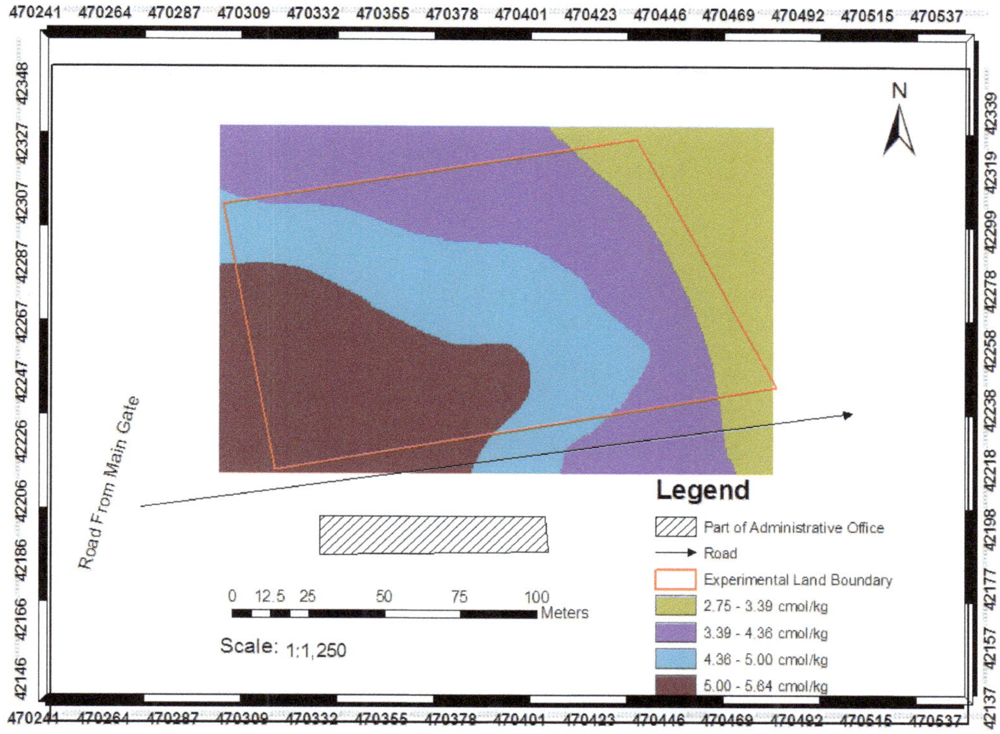

Figure 6. Magnessium spatial concentration.

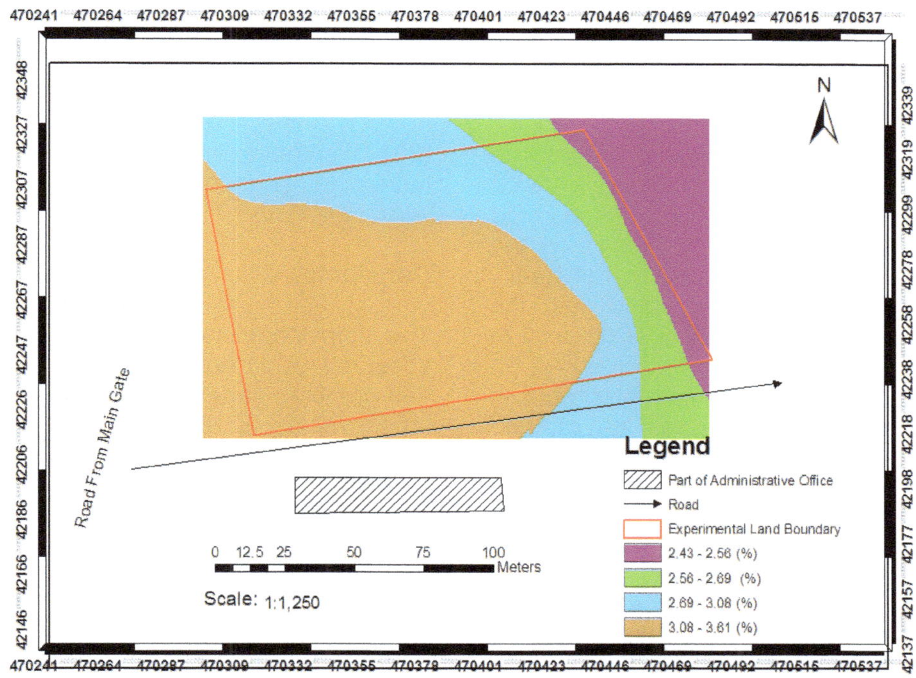

Figure 7. Organic matter spatial distribution.

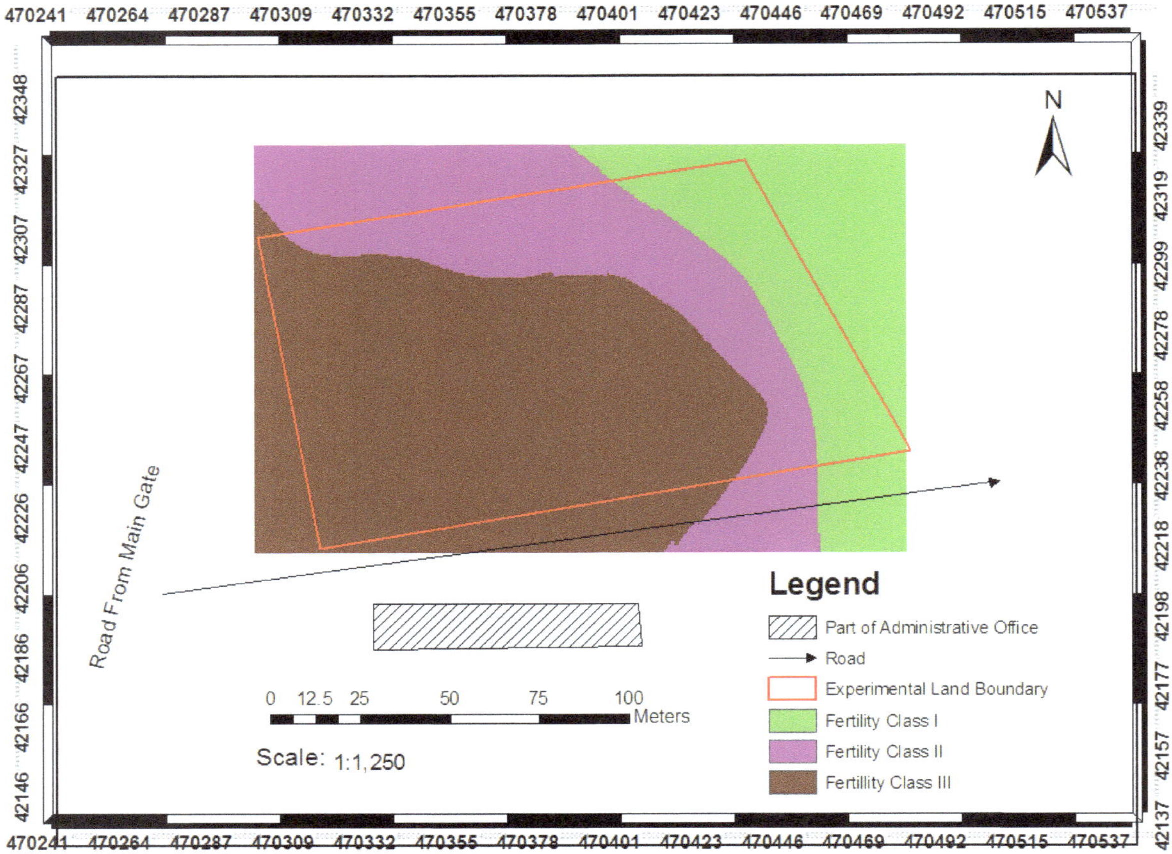

Figure 8. Soil fertility map.

the Moran's 1 Index of 0.37, P with clustered pattern with Moran's Index of 0.5 and K which was neither clustered nor dispersed with Moran's Index of 0.28 corroborated earlier observation of Huo et al. (2010, 2011) on the improvement of spatial interpolation accuracy of heavy metals concentration in soils and also in the autocorrelation analysis of soil pollution data on soils in Taiwan (Chu and Chang, 2011).

The spatial nutrient concentration maps integrated in an overlay operation to generate the soil fertility map corroborated the overlay procedure adopted by Onunkwo-Akunne et al. (2012) for the production of industrial, residential and waste disposal maps that were further superimposed to produce a composite land use map useful for regional and urban planning.

The spatial distribution of the soil fertility status of the experimental land as shown in Figure 8 reflected the spatial concentration of nutrients in Figures 2 to 7 which corroborated the previous observation of Salehi et al. (2013) on the determination of soil fertility status from the nutrient concentration.

Conclusion

The somewhat clustered nutrient concentration pattern observed could be adduced to past fertilizer application and cropping pattern on the experimental land. The high mean values of 0.44 and 5.35 cmol/kg for K and Ca, respectively could be attributed to past continuous application of CAN and NPK. A situation map of nutrient distribution was generated with the use of GIS Kriging technique.

Conflict of Interests

The author(s) have not declared any conflict of interests.

REFERENCES

Andrienko G, Andrienko N (2001). Exploring spatial data with dominant attribute map and parallel coordinates. Computers, Environment and Urban Systems, 25:5-15.

Carter MR (1993). Soil Sampling and Methods of Soil Analysis. Canadian Society of Soil Science. Lewis Publishers, London. 823pp.

Chen F, Kissel DE, West LT, Adkins W (2000). Field-Scale Mapping of Surface Soil Organic Carbon Using Remotely Sensed Imagery. Soil Sci. Soc. Am. J. 64(2):746-753.

Chu H, Lin Y, Chang T (2011). Spatial Autocorrelation Analysis of Soil Pollution Data in Central Taiwan. International Conference on Computational Science and its Applications held in Spain. pp. 219-222.

Craine JM, Dybzinski R (2013). Mechanisms of plant competition for nutrients, water and light. Functional Ecol. 27(4): 833-840.

Gebrelibanos T, Assen M (2013). Effects of slope aspect and vegetation types on selected soil properties in a dryland Hirmi watershed and adjacent agro-ecosystem, northern highlands of Ethiopia. Afr. J. Ecol. doi: 10.1111/aje.12118.

GISTUTOR (2001). GIS Data: the link between spatial and attribute data. http://www.gistutor.com/esri-arcgis/1-beginner-arcgis-tutorials/4-gis-data- the-link-between-spatial-and-attribute-data.html

Huo XN, Li H, Sun DF, Zhou LD, Li BG (2010). Multiscale spatial structure of heavy metals in agricultural soils in Beijing. Environmental Monitoring Assessment, 164:605-616.

Huo XN, Zhang WW, Sun DF, Li H, Zhou LD, Li BG (2011). Spatial pattern analysis of heavy metals in Beijing agricultural soils based on spatial autocorrelation statistics. Int. J. Environ. Res. Public Health 8:2074-2089.

Ismail MH, Junusi RM (2009). Determining and Mapping Soil Nutrient Content Using Geostatistical Technique in a Durian Orchard in Malaysia. J. Agric. Sci. 1(1):86-91.

Jobbagy EG, Jackson RB (2000). The vertical distribution of soil organic carbon and its relation to climate and vegetation. Ecol. Applic. 10:423-436.

Jobbagy EG, Jackson RB (2001). The distribution of soil nutrients with depth: Global patterns and the imprint of plants. Biogeochemistry 53:51-77.

Laiho R, Penttila T, Laine J (2004). Variation in soil nutrient concentrations and bulk density within peatland forest sites. Silva Fennica, 38(1):29-41.

Liu Z P, Shao MA, Wang YQ (2013). Large-scale spatial interpolation of soil pH across the Loess Plateau, China. Environmental Earth Sciences, 69(8):2731-2741.

Murray AT, Shyy T (2000). Integrating attribute and space characteristics in choropleth display and spatial data mining. Int. J. Geog. Inform. Sci. 14(7):649-667.

Oliver MA, Webster R (1999). Kriging: A method of interpolation for geographical information systems. Int. J. Geog. Information Systems, 4(3):313-332.

Omran E (2012). Improving the Prediction Accuracy of Soil Mapping through Geostatistics. Int. J. Geosci. 3(3):574-590.

Onunkwo-Akunne C, Onyekuru SO, Nwankwu GI (2012). Land Capability Index Mapping for Waste Disposal Landuse Option Using Geographic Information System (GIS) in Enugu Area, South Eastern Nigeria. J. Geog. Information System, 4:444-461.

Periaswamy SP, Ashaye TI (1982). Updated Classification of some Southwestern Nigeria Soils. Ife J. Agric. 4(1&2):25-41.

Roslee R, Jamaluddin TA, Talip MA (2012). Integration of GIS Using GEOSTAtistical INterpolation Techniques (Kriging) (GEOSTAINT-K) in Deterministic Models for Landslide Susceptibility Analysis (LSA) at Kota Kinabalu, Sabah, Malaysia. J. Geog. Geology, 4(1):18-32.

Sadeghi A, Graff CD, Starr J, Mc Carty G, Codling E, Seffon K (2006). Spatial Variability of Soil Phosphorus Levels Before and After Poultry Litter Application. Soil Sci. 171(11):850-857.

Salehi A, Ghorbanzadeh N, Salehi M (2013). Soil nutrient status, nutrient return and retranslocation in poplar species and clones in Northern Iran. J. Biogeosci. Forestry 6:341-346.

Shah Z, Malik W, Bhatti A, Rahman H (2013). Spatial Variability of Nutrients in Wheat Plants in Semi-arid Regions of Northwestern Pakistan. Communications in Soil Sci. and Plant Analysis, 44(16):2472-2487.

She DL, Shao MA (2009). Spatial variability of soil organic C and total N in a small catchment of the Loess Plateau, China. Acta Agriculturae Scandinavica, Section B. Soil and Plant Sci. 59(6):514-524.

Smyth AJ, Montgomery RF (1962). Soils and Land use in Central Western Nigeria, Ibadan Government Printer, pp 265.

Song X, Kou C, Chen J (2011). Soil nutrient distribution and its relations with topography in Huangshui River drainage basin. Chinese J. Appl. Ecol. 22(12):3163-3168.

Wollenhaupt NC, Wolkowski RP, Clayton MK (2013). Mapping soil test phosphorus and potassium for variable-rate fertilizer application. J. Production Agric. 7(4):441-448.

Species composition and biomass of annelids of Wular Lake, a Ramsar site in Kashmir, India

Rafia Rashid, Ashok Kumar Pandit and Sami Ullah Bhat

Department of Environmental Science/Centre of Research for Development, University of Kashmir, Srinagar-190 006, Kashmir India.

During the present investigation of Wular lake in terms of species composition and biomass of annelids, 10 taxa were recorded which belonged to two major classes namely Oligochaeta (7) and Hirudineae (3). The class Oligochaeta included *Limnodrilus hoffmeisteri*, *Tubifex tubifex*, *Branchiura sowerbyii, Nais* sp., *Aelosoma* sp., *Pristina* sp. and an unidentified taxion. Similarly, the class Hirudinaea was comprised by *Erpobdella* sp., *Placobdella* sp. and *Glossiphonia* sp. The seasonal mean value for biomass of annelids fluctuated between 0.31 g/m^2 at site I in winter to 14.92 g/m^2 at site III in summer. The annual mean biomass was highest at site IV (10.82±2.02 g/m^2), followed by site III (10.47±2.07 g/m^2), site II (8.94±1.90 g/m^2), site I (1.85±0.94 g/m^2) and site V (1.71±0.50 g/m^2).

Key words: Biomass, species composition, annelid, Lake and Ramsar site.

INTRODUCTION

In an aquatic ecosystem the life of aquatic biota is closely dependent on the physical, chemical and biological characteristics of water that directly acts as a controlling factor (Yaqoob and Pandit, 2009). Macrozoobenthos is an important constituent of aquatic ecosystem and has functional importance in assessing the trophic status. Thus as the abundance of benthic fauna mainly depends on physical and chemical properties of the substratum, the benthic communities are known to respond to changes in the quality of water or habitat. The benthic macroinvertebrates are associated with bottom or any solid liquid interface, which includes a heterogeneous assemblage of organisms belonging to various phyla like Arthropoda, Annelida, Mollusca and others. The benthos occupies an important position in the lake ecosystem, serving as a link between primary producers, decomposers and higher trophic levels (Pandit, 1980). They also play an important role in the detrital food web which in turn affects the cycling of minerals (Gardner et al., 1981). Macroinvertebrates are used as indicators of pollution as their communities change in response to changes in physicochemical factors and available habitats (Sharma and Chowdhary, 2011). According to Jumppanen (1976) the first signs of eutrophication and pollution in a lake are reflected in the benthic flora and fauna as the suspended waste immediately sink to the bottom to decompose and thus cause a change in the benthic composition and abundance. The lakes and wetlands having soft bottom sediments are characterized by annelids either as the dominant or one of the most abundant group. Of the fresh water annelids,

Table 1. Geographical coordinates of the five study sites

Sites	Latitude	Longitude	Altitude (m) (a.m.s.l)
Vintage	34° 21' 56.9" N	74° 39' 42.0" E	1583
Ashtung	34° 24' 3.8" N	74° 32' 41.7" E	1580
Watlab	34° 21' 29.4" N	74° 31' 48.2" E	1581
Makdomyari	34° 20' 39. 2" N	74° 34' 52.2" E	1579
Ningle	34° 17' 74.31"N	74° 31' 29.8" E	1578

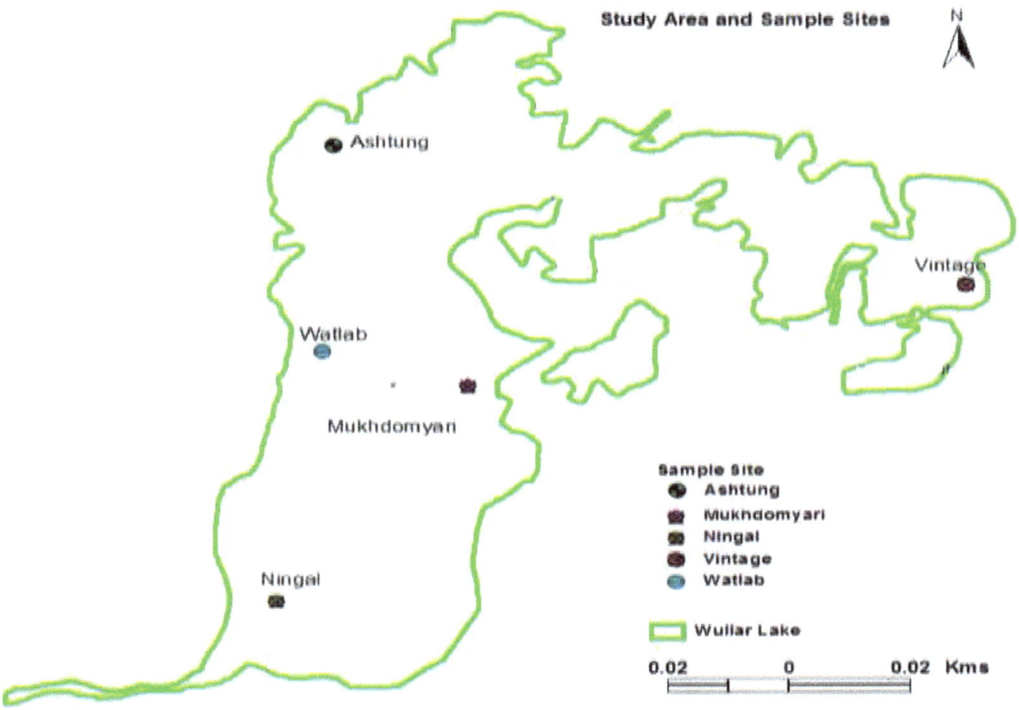

Figure 1. Map showing the study sites of Wular lake.

the oligochaetes display the greatest diversity and have the greatest indicator value. Oligochaetes worms are diverse and occur in a wide spectrum of freshwaters from unproductive to extremely eutrophic lakes and rivers. Leeches are found in warm water of shallow standing sites (Peckarsky et al., 1990) and are generally pollution tolerant. Oligochaeta, especially the Tubificidae family, have been universally applied on bioassessment assays, as bioindicators to reflect the organic pollution (Lin and Yo, 2008). This is because their capacity to increase in number with increasing organic matter, replacing other benthic macroinvertebrates, less tolerant for this condition (Schenkova and Helesic, 2006). It is in the backdrop of paucity of researches on Wular Lake, the largest freshwater body of Indian subcontinent, being recognised as a Ramsar Site, that the present study on community structure and biomass of annelids has been undertaken during 2012.

MATERIALS AND METHODS

Study area and study sites

Wular lake is the largest freshwater lake of Indian subcontinent, located in the flood plains of Jhelum river with an open water area of 24 km^2 (Pandit, 2002). The rural valley lake in the north-west of Kashmir extends from Bandipore to Sopore and is at a distance of about 54 km from Srinagar city and is situated at an altitude of 1,580 m (amsl), lying between 34°16′-34°20′N latitudes and 74°33′-74°44′E longitudes. The lake is mono-basined, elliptical in shape and is of fluviatile origin, formed by the meandering of Jhelum River. Its depth on an average is 3.6 m throughout length, reaching 5.8 m at its deepest point. The major inflows to Wular Lake are Jhelum, Madumati and Erin. The lake plays a significant role in the hydrographic system of Kashmir valley by acting as a large reservoir and by absorbing high annual flood of the Jhelum River. In 1990, this shallow lake was designated as a Ramsar Site, a Wetland of International Importance. For the present investigation, five sampling sites were selected which are described in Table 1 and Figure 1.

Table 2. Species composition of annelids at five sites in Wular lake (+ indicate taxa presence and – indicates taxa absence).

Annelids	Sites				
	Site I	Site II	Site III	Site IV	Site V
Oligochaetea					
Limnodrilus hoffmeisteri	+	+	+	+	+
Tubifex tubifex	+	+	+	+	+
Branchiuria sowerbyii	-	+	+	+	-
Aelosoma sp.	-	+	+	+	-
Nais sp.	-	+	+	+	+
Pristina sp.	-	+	-	+	-
Unidentified taxion	+	-	-	-	-
Total	3	6	5	6	3
Hirudinea					
Erpobdella sp.	+	+	+	+	+
Placobdella sp.	+	+	+	+	+
Glossiphonia sp.	-	+	+	-	-
Total	2	3	3	2	2
Annelids total	5	9	8	8	5

Collection and preservation

The benthic fauna encompassing annelids was collected from all five sites of the lake, with an Ekman Dredge (15 x 15 cm). The sites differered in water depth, vegetation and bottom sediments. The samples were taken in triplicate. The sediment samples collected were sieved carefully in order to remove fine sediments and other extraneous material without damaging the fragile organisms, using sieves of 1mm and 0.5 mm mesh size for checking annelids. The macroscopic organisms were collected with the help of forceps and brushes. The organisms collected were preserved in 70% alcohol for detailed examination.

Identification

Preserved annelids were identified by observing them under a microscope and identification was done with the help of standard taxanomical works of Edmondson (1992), APHA (1998), Pennak (1978), Adoni (1985), Brinkhurst (1971) and Kiemm (1995).

Biomass

Biomass of annelids was determined as fresh weight. The organisms were kept on filter paper for some time until the samples dry and weight became constant. Then organisms were weighed by using electronic balance with the accuracy of 1 mg. The biomass was calculated by using the formula:

$$\text{Biomass (g m}^{-2}) = \frac{n}{a \times s} \times 10{,}000$$

Where, n = weight of organisms counted; a = extraction area of dredge; s = number of samples

RESULTS

Species composition

During the present investigation of Wular lake, 10 taxa of annelids were recorded which belonged to two major classes namely Oligochaeta (7) and Hirudinaea (3). The class Oligochaeta included *Limnodrilus hoffmeisteri*, *Tubifex tubifex*, *Branchiura sowerbyii, Nais* sp., *Aelosoma* sp., *Pristina* sp. and an unidentified taxion. Similarly, the class Hirudinea was comprised by *Erpobdella* sp., *Placobdella* sp. and *Glossiphonia* sp. Amongst the 10 species listed, the highest taxa richness was obtained at site II (9), followed by site III (8), site IV (8), and decreasing to the minimum of 3 species each at sites I and V (Table 2). The most common taxa encountered across all the sites were *L. hoffmeisteri*, *T. tubifex*, *Erpobdella* sp. and *Placobdella* sp. The unidentified annelid taxion was found only at site I. Among the two classes, Oligochaeta dominated both qualitatively and quantitatively at each site and was represented by a maximum number of 6 species each at sites II and IV, followed by site III (5), till it reached a minimum number of 3 species each at sites I and V. The class Hirudinea was represented by 3 taxa each at sites II and III and 2 taxa each at sites I, IV and V.

Biomass

The annelid biomass showed a considerable variation between the study sites. Depending upon the species composition biomass of annelids varied seasonally. The seasonal mean value for biomass of annelids fluctuated between a lowest of 0.31 g/m^2 at site I in winter to a highest of 14.92 g/m^2 at site III in summer. However, the annual mean biomass was revealed to be maximum at site IV (10.82±2.02 g/m^2), followed by site III (10.47±2.07 g/m^2), site II (8.94±1.90 g/m^2), site I (1.85±0.94 g/m^2) and decreasing to the lowest ebb (1.71±0.50 g/m^2) at site V.

At site I, the annual mean biomass of Hirudineae was highest (1.18±0.61 g/m^2) as compared to class Oligochaeta

Table 3. Seasonal variation in biomass of annelids (g/m^2) at five different sites of Wular lake.

S/N	Class/ Taxa	Sites	Spring	Summer	Autumn	Winter	Mean
	Oligochaeta						
		Vintage	0.57	0.73	0.28	0.04	0.41
		Ashtung	4.04	4.81	3.51	1.74	3.53
1	*Limnodrilus hoffmeisteri*	Watlab	5.05	6.1	4.02	2.48	4.41
		Makdomyari	4.99	6.18	4.4	2.66	4.56
		Ningle	0.82	0.96	0.55	0.19	0.63
		Vintage	0.03	0.53	0.08	0.03	0.17
		Ashtung	3.99	4.5	3.37	1.72	3.39
2	*Tubifex tubifex*	Watlab	4.89	5.7	4.02	2.39	4.25
		Makdomyari	4.56	5.43	3.99	2.52	4.13
		Ningle	0.67	0.88	0.39	0.09	0.51
		Vintage	0	0	0	0	0
		Ashtung	0.12	0.19	0.11	0	0.11
3	*Branchiuria sowerbyii*	Watlab	0.12	0.19	0.11	0	0.11
		Makdomyari	0.17	0.19	0.11	0	0.12
		Ningle	0	0	0	0	0
		Vintage	0	0	0	0	0
		Ashtung	0.44	0.56	0.35	0.13	0.37
4	*Aelosoma* sp.	Watlab	0.49	0.59	0.37	0.07	0.38
		Makdomyari	0.58	0.61	0.41	0.11	0.43
		Ningle	0	0	0	0	0
		Vintage	0	0	0	0	0
		Ashtung	0.47	0.66	0.34	0.13	0.40
5	*Nais* sp.	Watlab	0.55	0.71	0.35	0.09	0.43
		Makdomyari	0.56	0.71	0.45	0.11	0.46
		Ningle	0.25	0.07	0.06	0	0.09
		Vintage	0	0	0	0	0
		Ashtung	0.05	0.12	0.04	0	0.05
6	*Pristina* sp.	Watlab	0	0	0	0	0
		Makdomyari	0.18	0.23	0.07	0	0.12
		Ningle	0	0	0	0	0
		Vintage	0.26	0	0.08	0	0.09
		Ashtung	0	0	0	0	0
7	Unidentified taxion	Watlab	0	0	0	0	0
		Makdomyari	0	0	0	0	0
		Ningle	0	0	0	0	0
	Total		33.85	40.65	27.46	14.5	29.15
	Hirudinea						
		Vintage	0.77	0.59	0.36	0.18	0.48
		Ashtung	0.58	0.82	0.47	0.11	0.49
8	*Erpobdella* sp.	Watlab	0.51	0.88	0.41	0.12	0.48
		Makdomyari	1.06	0.77	0.53	0.12	0.62
		Ningle	0.35	0.28	0.29	0.06	0.25
9	*Placobdella* sp.	Vintage	2.13	0.47	0.17	0.06	0.71

Table 3. Condt.

S/N	Class/ Taxa	Sites	Spring	Summer	Autumn	Winter	Mean
	Placobdella sp.	Ashtung	0.48	0.53	0.24	0.06	0.33
		Watlab	0.35	0.53	0.29	0.06	0.31
		Makdomyari	0.71	0.54	0.22	0.05	0.38
		Ningle	0.29	0.3	0.17	0.11	0.22
10	*Glossiphonia* sp.	Vintage	0	0	0	0	0
		Ashtung	0.11	0.29	0.59	0	0.25
		Watlab	0.11	0.22	0.05	0	0.10
		Makdomyari	0	0	0	0	0
		Ningle	0	0	0	0	0
Total			7.45	6.22	3.79	0.93	4.62
Grand total			41.3	46.87	31.25	15.43	33.77

(0.67±0.33 g/m^2). Among Oligochaetes, the *L. hoffmeisteri* acquired maximum mean biomass (0.41±0.15 g/m^2) while the unidentified taxion made a minimum mean biomass (0.09±0.06 g/m^2). Among Hirudineae, only *Placobdella* sp. maintained the greatest biomass (0.71±0.48 g/m^2). On a seasonal basis, the oligochaetes registered the maximum biomass during summer (1.26 g/m^2) and the minimum in winter (0.07 g/m^2) whereas the Hirudineae registered the highest biomass in spring (2.90 g/m^2) and the lowest in winter (0.24 g/m^2) (Table 3).

At site II, the annual mean biomass of Oligochaeta was highest (7.86±1.51 g/m^2) as compared to class Hirudinea (1.08±0.39 g/m^2). Further, among Oligochaeta, *L. hoffmeisteri* attained the maximum mean biomass (3.53± 0.65 g/m^2) whereas *Pristina* sp. had the minimum mean biomass (0.05±0.02 g/m^2). On a seasonal basis, the annelids showed maximum biomass in summer (12.48 g/m^2) and lowest in winter (3.89 g/m^2). *Erpobdella* sp. had the highest mean biomass (0.50±0.15 g/m^2) whereas *Glossophonia* sp. had the lowest mean biomass (0.25± 0.13g/m^2) in class Hirudinae (Table 3).

At site III, the annual mean biomass of annelids varied between 9.58±1.76 g/m^2 (Oligochaeta) and 0.89±0.31 g/m^2 (Hirudinea). *L. hoffmeisteri* obtained maximum mean biomass (4.41± 0.77g/m^2) and *B. soewerbyii* registered the minimum mean biomass (0.11±0.04 g/m^2). *Erpobdella* sp. obtained the highest mean biomass (0.48±0.16 g/m^2) and *Glossophonia* sp. the lowest mean biomass (0.10± 0.05g/m^2). On a seasonal basis, the annelids showed a maximum biomass during summer (14.92 g/m^2) and the lowest in winter (5.21 g/m^2) as depicted in Table 3.

At site IV, the mean biomass of *L. hoffmeisteri* was maximum (4.56±0.73 g/m^2), followed by *T. tubifex* (4.13±0.61 g/m^2) while *B. soewerbyii* and *Pristina* sp. registered lower mean biomass (0.12±0.05 g/m^2 for each). *Erpobdella* sp. recorded its greatest biomass (0.62±0.20 g/m^2) while *Placobdella* sp. had its lowest biomass (0.38± 0.15g/m^2). The annual mean biomass of annelids varied between 9.82±1.67 g/m^2 for Oligochaeta and 1.0±0.35 g/m^2 for Hirudinea. In general, on a seasonal basis, the annelids showed maximum biomass (14.66 g/m^2) in summer and minimum (5.57 g/m^2) in winter (Table 3).

At site V, the annelids registered their maximum biomass (2.49 g/m^2) in summer and their minimum (0.45 g/m^2) in winter. The biomass of Oligochaeta was highest (1.24±0.39 g/m^2) than the biomass of Hirudinea (0.47±0.11 g/m^2). Among Oligochaetes, the *L. hoffmeisteri* attained greatest mean biomass (0.63±0.17 g/m^2) while the lowest mean biomass (0.10±0.05 g/m^2) was found for *Nais* sp. *Erpobdella* sp. was the dominant maintained leeches (0.25±0.06 g/m^2) and *Placobdella* sp. was the less abundant one (0.22± 0.05 g/m^2) (Table 3).

DISCUSSION

Assessment of species composition, distribution and biomass of macroinvertebrate community often gives an important clue to the functional status of a water body. Benthic macroinvertebrates are tools that enable rapid bioassessment which is cost effective and quick assessment strategy to determine the health of an ecosystem (Loeb and Spacie, 1994). According to Jumppanen (1996) the first sign of eutrophication and pollution is reflected by the benthic biota, because the pollutants are rapidly deposited into the sediments where they evolve an impact on the benthic organisms. Therefore, it is essential to study the benthic composition to help in the evaluation of the trophic status of the water bodies.

During the present study, 10 taxa of annelids belonging to two major classes namely Oligochaeta (7) and Hirudinea (3) were recorded. Amongst the 10 species listed, the maximum number of species was obtained at site II (9), followed by site III (8), site IV (8), and decreasing to the minimum of 3 species each at sites I and site V. The lowest species richness at site V may be due to the physical heterogeneity of substrate, being reflected in the number of taxa (Marshall and Winterbourn,

1979) as site V is chracterized by sandy sediment with small boulders and diminished macrophytic growth. Conversely the highest richness at sites II, III and IV may be due to the dense macrophytic cover and soft bottom sediments as macrophytes provide suitable habitat for macroinvertebrates. The above finding corroborates the fact that oligochaete community thrives well in soft depositing substrates rather than stony beds (Bhat and Pandit, 2010).

Biomass is a potential renewable source of energy, which is related to the ecological conditions of the habitat. Estimation of biomass is necessary for understanding the trophic dynamics and productivity of an ecosystem (Mir and Yousuf, 2003). During the course of the present study, biomass of annelids fluctuated between 0.31 g/m^2 at site I and 14.92 g/m^2 at site III. This reflected the higher productivity and trophic status of site III and inversely, the relatively lower productivity at site I. Biomass showed a seasonal trend with maximum in spring and summer and minimum in winter. Thus, biomass fluctuated in close relation with density of annelids, which also showed higher value during summer and lower in winter, indicating a positive correlation between the two parameters. Sunder and Subla (1986) and Mortensen and Simonson (1983) also noticed a positive relationship between density and biomass of zoobenthos.

Conclusion

During the present investigation 10 taxa of annelids belonging to two major classes namely Oligochaeta (7) and Hirudinea (3) were recorded. The most common taxa encountered across all the sites included *L. hoffmeisteri*, *T. tubifex*, *Erpobdella* sp. and *Placobdella* sp. However, the unidentified annelid taxion was found only at site I. Among the two classes, Oligochaeta dominated both qualitatively and quantitatively at each site and was represented by a maximum number of taxa at each site. The presence of *L. hoffmeisteri* and *T. tubifex* on an average in great numbers is indicative of organically rich sediments. The class Hirudinea was represented by three taxa each at sites II and III and two taxa each at sites I, IV and V. In general, Oligochaeta comprised 95% of the total annelid community and remaining 5% was made by Hirudinea. During the course of the present study, biomass of annelids fluctuated from the lowest of 0.31g/m^2 at site I to the highest of 14.92 g/m^2 at site III. However, the biomass of oligochaetes fluctuated from 0.67 to 9.82 g/m^2 and the biomass of Hirudinea varied from 0.47 to 1.18 g/m^2, indicating that overall biomass was dominated by oligochaetes and hence reflecting the moderate eutrophication of lake.

REFERENCES

APHA (1998). Standard Methods for Examination of Water and Waste Water. 20th ed. American Public Health Association, New York.

Adoni AD (1985). Work Book of Limnology, Pratiba Publication Sagar, (M.P), India.

Bhat S, Pandit AK (2010). Ecological study of benthic communities in three limnocrene fresh water springs in Kashmir Himalaya. J. Natural Sciences and Mathematics, 3(2):89-96.

Brinkhurst RO (1971). A Guide for the Identification of British Aquatic Oligochaeta. Freshwater Biological Association, Scientific Publication No.22.

Edmondson WT (1992). Ward and Whiple Freshwater Biology. 2nd ed. Intern. Books and Periodicals Supply Service, New Delhi.

Jumppanen K (1976). Effects of waste waters on a lake ecosystem. Ann. Zool. Fennici. 13:85-138.

Kiemm DJ (1995). Identification Guide to the Freshwater Leeches (Annelida: Hirudinea) of Florida and Other Southern States. Bioassesment and Ecotoxicology branch, Ecological Monitoring Research Division, Environmental Monitoring System Laboratory, U.S.

Environmental Protection Agency, Cincinnati, Ohio- 45268. Leob SL, Spacie A (1994). Biological Monitoring of Aquatic Systems. Lewis Publishers, Florida. 381p.

Lin KJ, Yo SP (2008). The effect of organic pollution on the abundance and distribution of aquatic oligochaetes in an urban water basin, Taiwan. Hydrobiologia, 596(1):213-223.

Marshall JW, Winterbourn MJ (1979). An ecological study of a small New Zealand stream with particular reference to the Oligochaeta. Hydrobiologia, 65:199-208.

Mir MF, Yousuf AR (2003). Oligochaete community of Dal lake, Kashmir. Oriental Sci. 8:83-87.

Mortensen E, Simonsen IJ (1983). Production estimates of the benthic invertebrate community in a small Danish stream. Hydrobiol. 102: 155-162.

Pandit AK (2002). Plankton as indicators of trophic status of wetlands. p. 341-360. In: Ecology and Ethology of Aquatic Biota. (Arvind Kumar, ed.). Daya Publishing House, New Delhi-110002.

Pandit AK (1980). Biotic factor and food chain structure in some typical wetlands of Kashmir. Ph.D. thesis, University of Kashmir, Srinagar-190006, J&K, India.

Peckarsky BL, Fraissinet PR, Penton MA, Conklin DJ (1990). Freshwater macroinvertebrates of northeastern North America. Cornell Univ. Press, U.S.A. pp.442.

Pennak RW (1978). Freshwater Invertebrates of United States of America. Wiley Interscience Pub., New York.

Schenkova J, Helesic J (2006). Habitat preferences of aquatic Oligochaeta (Annelida) in the

Rokttna river, Czech Republic- a small highland stream. Hydrobiologia, 564(1):117-126.

Sharma KK, Chowdhary S (2011). Macroinvertebrate assemblages as biological indicators of pollution in a Central Himalayan River, Tawi (J&K). Int. J. Biodivers. Conserv. 3(5):167-174.

Sunder S, Subla BA (1986). Macrobenthic fauna of a Himalayan river. Indian J. Ecol., 13(1):127-132.

Yaqoob KU, Pandit AK (2009). Distribution and abundance of macrozoobenthos in Dal lake of Kashmir Valley. J. Res. Dev. 9:20-29.

Survival of enteric bacteria in source-separated human urine used as fertiliser: Effects of temperature and ammonia

Joseph M. Makaya[1], Marius K. Somda[1], Aly Savadogo[1], Dayéri Dianou[2], Nicolas Barro[1] and Alfred S. Traoré[1]

[1]Research Center for Biological, Food and Nutritional Sciences, Research and Training Unit, Life and Earth Sciences, University of Ouagadougou, 03 BP 7021, Ouagadougou, Burkina Faso.
[2]National Center for Scientific and Technological Researches, Institute for Health Sciences Research, 03 BP. 7047, Ouagadougou, Burkina Faso.

To promote the use of eco-toilets is an alternative to the lack of sanitation and high cost of artificial fertilisers in developing countries. Human urine is the fraction of excreta containing most nitrogen, phosphorus and potassium which are essential nutrients for plant growth. The major concern related to the use of the urine as fertiliser is the possible presence of enteric pathogens. The reduction in the numbers of *Escherichia coli*, *Enterococcus faecalis*, *Salmonella* sp., *Staphylococcus aureus*, *Clostridium perfringens* spores during urine storage was studied in diluted and undiluted urine samples. These experimental samples were stored in ambient air and at temperatures of 25, 35 and 42°C. The initial concentration of ammoniacal nitrogen reduced from 8260 to 1070 mg/L in quarter-diluted urine (1:3), while the pH values remained stable at 8.6 (± 0.2). *E. faecalis* survived for 35 days in the quarter-diluted urine at 25°C. Survival times of the tested bacteria were more shortened in ammonia concentrated samples, on storage at 35, 42°C. Except *C. perfringens* spores, *E. faecalis* was the least sensitive to storage conditions as compared to *S. aureus* and the two Gram-negative bacteria tested. Therefore, temperature and ammonia acted synergistically to remove bacteria in the stored urine. Likewise, *C. perfringens* spores were the most persistent bacteria, with only 1-$\log_{10}$ reduction within 100 days storage at ambient temperature. This study shows that a period of at least 30 days of undiluted urine storage at temperatures ≥ 25°C is sufficient to sanitize urine, considering only vegetative bacteria. Since urine from eco-toilets may contain other enteric organisms (viruses, protozoa, etc) appearing more persistent in external media, storage time must be lengthened for a satisfactory sanitation.

Key words: Human urine, storage, enteric bacteria, ammonia, temperature, sanitation, fertilizer.

INTRODUCTION

Separating the human excreta at the source can contribute to effective management of wastewater. To do that, urine diversion dry toilets (UDDTs) are the best indicated systems, because they are designed to separately collect urine and faeces (Mnkeni and Austin, 2009; Münch and Winker, 2009; Dagerskog and Bonzi, 2010). Thus, to promote the use of these eco-toilets, UDDTs represents an alternative to the lack of sanitation and high cost of artificial fertilisers that affect many people in developing countries. To this end, since 2006, large-scale ecological systems were implemented in many areas in Burkina Faso (Dagerskog and Bonzi, 2010).

Human urine is the fraction of excreta which contains more nutrients, mainly, nitrogen, phosphorus and potassium (Heinonen-Tanski and van Wijk-Sijbesma, 2005), which are essential nutrients for plant growth. Also, fresh urine can contain microorganisms. In individuals suffering from typhoid fever, *Salmonella* can be found in their urine, only in cases of septicemia (Feachem et al., 1983). Recycling urine nutrients in agriculture therefore seems easier and more profitable as compared to faeces.

For health risks associated with the use of urine as fertiliser, the great concern is about enteric pathogens resulting from environmental samples or faecal cross-contamination during urine collection (Esrey et al., 2001; Schönning et al., 2002). In view of that, it is possible to assume that the source-separated human urine can contain enteric organisms. It is possible that applying in soil the untreated urine, may introduce another pathogens transmission route in the environment.

In order to sanitize urine from UDDTs, several treatment options have been proposed (Maurer et al., 2006; Pronk and Koné, 2010). The urine storage in closed containers and then exposed to the ambient air can remove microorganisms. The storage seems to be an appropriate method for the urine sanitation, as its realization is very simple. Apart from pathogens removal, the urine treatment by storage may preserve the useful nutrients for plant growth (Jönsson and Vinnerås, 2007; Pradhan et al., 2009).

The parameters which may influence the survival of microorganisms in urine during storage are mainly temperature, ammonia and pH. The temperature depends on the storage conditions and the climate. Because of high temperatures regularly reported in tropical regions especially in Sahel (Ouédraogo et al., 2007), urine storage time can be shortened, as compared to temperate regions, where storage times of 2-6 months were recommended to sanitize urine satisfactory (at 20°C or higher), including a wide range of bacteria, protozoan (oo) cysts and viruses (Höglund et al., 2002; WHO, 2006; Vinnerås et al., 2008; Chandran et al., 2009).

The pH of fresh urine is usually around 6 (Haneaus et al., 1996), but this value increases to approximately 9 in the collection and storage tanks due to rapid hydrolysis of urea (Jönsson and Vinnerås, 2007). In aqueous solution, the alkaline pH affects the equilibrium between the uncharged ammonia (NH_3) and ammonium (NH_4^+) in favor of the formation of the NH_3. The microbicidal effects of NH_3 were shown in several matrices (Jenkins et al., 1998; Nordin et al., 2009). The urine dilution can affect the ammoniacal nitrogen concentrations and therefore these uncharged ammonia (Pecson et al., 2007).

The survival capacity of enteric microorganisms in external environment differs according to microbial groups (Vinnerås et al., 2008; Chandran et al., 2009). Viruses, protozoan oocysts seem to be more persistent in the environment than bacteria. That could justify the choice of *Clostridium perfringens* spores as biological tracers to follow the fate of pathogenic viruses and protozoa in the wastewater (Payment et al., 2001).

This study investigated the effectiveness of the urine treatment by storage on determining the reduction numbers of five indicators or potentially pathogenic bacterial species: *Escherichia coli, Salmonella* sp., *Staphylococcus aureus*, *Enterococcus faecalis* and *C. perfringens* spores. The main objective was to evaluate the survival in urine of the above-mentioned bacteria, highlighting the effects of ammonia concentrations and temperatures during storage.

MATERIALS AND METHODS

Isolation of bacteria

Vegetative bacteria were respectively isolated from a polluted soil (*E. coli* and *E. faecalis*), wastewater (*Salmonella* sp.) and pathological exudate (*S. aureus*). To do that, standard methods were used (ISO, 1999, 2000). Isolated bacteria were used in the experiments, for a bacterial suspension. On the other hand, a stock of *C. perfringens* spores was prepared with a dried compost (faeces + soil) taken from an UDDT pit. 30 g of crushed and sieved compost were transferred into 270 mL of saline solution (0.9%). After homogenization, the supernatant was removed and used to prepare decimal dilutions (10^{-1}-10^{-3}), 250 mL each. Clostridia spore numbers in dilutions were determined as the described method (Anonymous, 2005), using Tryptone-Sulfite-Neomycin (TSN) Agar. The dilution containing 10^5 spores/mL was considered as the spore stock solution.

Urine samples preparation

Two-days-old urine from a urinal used by male staff at the Pan African Agency, Water and Sanitation for Africa (WSA) located in Ouagadougou (Burkina Faso) was used to prepare the urine samples. Three (03) matrices were prepared: undiluted urine (1:0), half-diluted urine (1:1) and quarter-diluted urine (1:3). The dilution was done with a sterile water (urine : water).

For the exposure to ambient conditions, each above-mentioned matrix was distributed in 5 L plastic jerry cans (white color) in a volume of 4 L per jerry can. On the other hand, three (03) sets of samples were prepared for storage in an incubator (Binder, Germany) under the following temperatures; 25, 35 or 42°C. Each set of samples was made up of twelve (12) bottles containing in triplicate the three urine matrices (1:0, 1:1, and 1:3) and sterile

Table 1. Temperatures of the urine during storage at ambient air.

Time* (days)	Temperature per day of measurement (°C)			Temperature in jerry cans (°C)	
				UD = 1:0	UD = 1:1
	Min.	Max.	Mean	Mean ± SD	Mean ± SD
0	22	31.2	25.7	30.0 ± 0.3	29.8 ± 0.2
5	24.2	34	27.7	30.3 ± 0.4	31.0 ± 0.2
10	25.8	36.6	30.8	30.1 ± 0.3	30.1 ± 0.2
15	24	36.5	29.4	30.2 ± 0.1	30.8 ± 0.3
20	24	35.6	29.6	30.4 ± 0.2	30.5 ± 0.0
25	23.2	36.2	29.5	29.5 ± 0.5	30.0 ± 0.1
30	26.7	36.8	30.6	31.0 ± 0.0	30.5 ± 0.2

*Time: Measurements were performed with a time interval of five days, between October 07 and Nov. 06, 2010; UD : Urine diluted.

water (control). These matrices were distributed in 100 mL sterile polyethylene bottles (dark brown color) in a volume of 80 mL each.

Survival tests of bacterial species

Samples stored at ambient

Each sample in jerry can was contaminated either with a bacterial suspension (in which initial concentrations were 10^8-10^9 CFU/mL for *E. coli*, *Salmonella* sp. and 10^7-10^8 CFU/mL for *S. aureus*, *E. faecalis*) or with a Clostridia spores solution (10^5 cells/mL). Inocula were distributed at a rate of 40 mL per jerry can (1% [vol/vol]). Experiments were performed in duplicate. Contaminated samples were immediately exposed to ambient air (completely out in the sun, at Ouagadougou University site), for 05 months (September 2010-January 2011). Physicochemical and bacteriological parameters in samples were analyzed at regular time intervals (5, 10 days for vegetative bacteria and spores, respectively). Simultaneously, ambient temperatures were recorded daily during the exposure period.

Samples stored in an incubator

Samples in bottles were contaminated with a bacterial suspension (as above-mentioned) at a rate of 1.6 mL per bottle (2% [vol/vol]). Each set of 12 bottles contaminated samples was immediately placed in the incubator under one of these temperatures 25, 35 or 42°C. Bacterial parameters in samples were analyzed within 45 days, for 4 h and 5 days as regular times, respectively for rods and cocci tested bacteria.

Laboratory analyses

Physicochemical analysis

The pH and temperature values in stored samples were measured *in situ* using a multifunctional Inolab 340i pH/Thermo-meter (WTW, Germany). The ammoniacal nitrogen (NH_4-N) was analysed by the Nessler method, using a DR 2400 spectrophotometer HACH (Loveland, Co. USA).

Bacteriological analysis

Stored samples taken at regular time intervals were analyzed for microbial quantification. Clostridia spores were activated by heat (80°C for 10 min) and then counted in Tryptone-Sulfite-Neomycin (TSN) Agar, according to the method previously described (Anonymous, 2005). *E. faecalis* and *S. aureus* were enumerated respectively in Slanetz & Bartley and Mannitol Salt Agar, after incubation at 37°C for 24 to 48 h (APHA, 2005). *E. coli* and *Salmonella* sp. were counted respectively in Eosin Methylene Blue (EMB) and Salmonella-Shigella (SS) Agar (ISO, 1993, 2001).

Data analysis

For drawing graphs and calculating geometric means, standard deviations, minimum and maximum values of data obtained, the Microsoft Excel package was used.

RESULTS

Storage at ambient

Temperature values

During the samples exposure in the sun, ambient temperatures were recorded. The results showed that minimal and maximal temperatures respectively ranged from 13.9-24.9 and 30.7-38.8°C. The high thermal amplitudes were recorded significantly from November (~20°C), indicating the gaps between daytime and nighttime temperatures in the study area at this period of the year.

Table 1 presents values of daily temperature (sampling days) and those measured in the stored urine, exposed to ambient air. The data show that the differences in temperature between the two matrices of urine are not important. Temperatures in the urine storage containers depend on ambient temperature.

Reduction of vegetative bacteria

The graphs of Figure 1A and B show the reduction of bacterial numbers during samples exposure at ambient

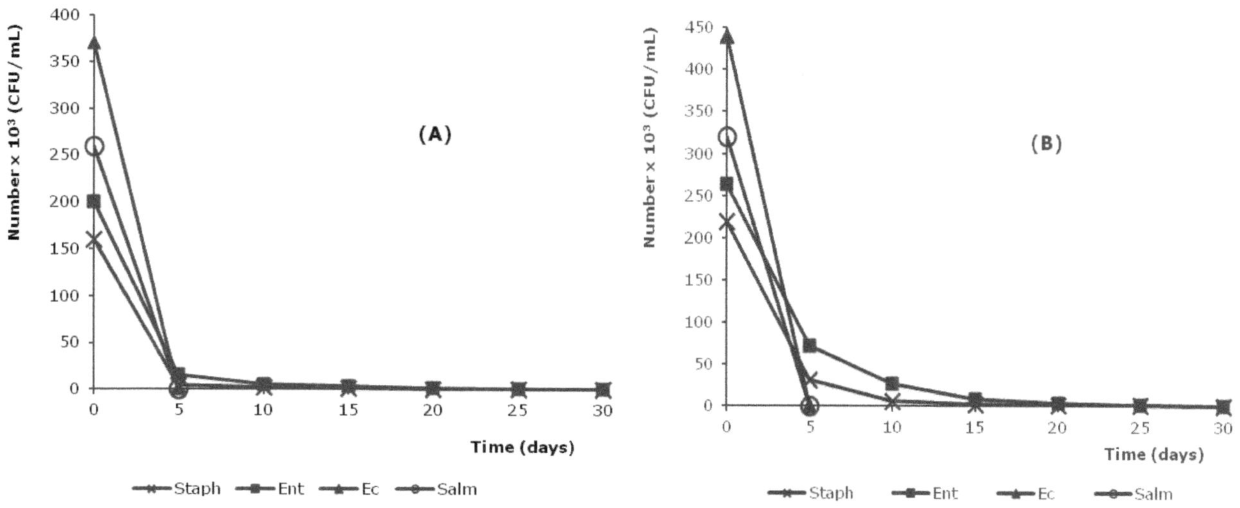

Figure 1. Reduction of bacterial numbers (*E. coli* [▲], *Salmonella* sp. [○], *S. aureus* [X], *E. faecalis* [■]) during samples exposure at ambient temperatures. The initial values were 8260 mg/L NH_4-N (pH 8.8) for the undiluted urine (A) and 4450 mg/L (pH 8.7) for the half-diluted urine (B).

temperatures.

The results show that after five (05) days of samples exposure, *E. coli* and *Salmonella* sp. were eliminated (0 CFU/mL) in both diluted and undiluted urine samples. In the undiluted urine samples, *S. aureus* and *E. faecalis* survived ~10 days (Figure 1A), while in the half-diluted urine, they survived around 15 and 20 days, respectively (Figure 1B).

The above data indicate that Gram-negative rods (*E. coli*, *Salmonella* sp.) were more sensitive in urine samples than Gram-positive cocci (*E. faecalis*, *S. aureus*).

Globally, the survival time for each of bacterial species tested was extended in the diluted samples as compared to the undiluted ones.

For the chemical analysis, a low decline of 0.1 pH units was observed in the corresponding pH on diluting the urine by half, while its ammoniacal nitrogen concentration decreased from 8260 to 4450 mg/L. Then, at the end of the experiments (after 30 days of storage), 7730 and 4060 mg/L were respectively the average concentrations of NH_4-N obtained in the undiluted urine and the half-diluted samples. These decreases of the ammonia concentration in the stored urine can be due to the nitrogen losses (N) occurring during the tests (measurements).

Reduction of Clostridia spores

The results of the reduction of *C. perfringens* spores in urine samples (and control) during the storage are shown in Figure 2. After one hundred (100) days of exposure, 0.85 and 1.07 $\log_{10}$ reductions of Clostridia spores were obtained respectively in the half-diluted urine samples (1:1) and undiluted ones (1:0). In the control (water), spore numbers did not decrease.

The reduced number of *C. perfringens* spores was slightly higher in the undiluted urine than in the half-diluted samples. It is well known that bacterial spores are substantially destroyed by heat (≥100°C) and by the tyndallisation process which requires moderate and intermittent temperatures (≤60°C). In fact, a moderate heat for 30 min is sufficient to remove the vegetative cells and to trigger the spores transforming in vegetative cells, and then may be destroyed at the next heating (Stanbury et al., 1995).

Storage in controlled environment

Physicochemical parameters

Means (± standard deviation) for the pH, ammoniacal nitrogen concentrations and temperature values in stored samples are indicated in Table 2.

Mean values of the initial concentrations of ammoniacal nitrogen were respectively 8260, 4450 and 1070 mg/L in the undiluted (1:0), half-diluted (1:1) and quarter-diluted (1:3) urine samples. At the end of the experiments, we found respectively 7980, 4295 and 1040 mg/L as average concentrations of ammonia. Between the start and the end of the experiments, small decreases in NH_4-N concentrations were observed. Apparently, nitrogen (N) losses in the stored urine were lower in the controlled environment (incubator), as compared to the ambient air. The pH was stable, with only a slight decrease (at the most 0.3 pH units) on diluting quarterly the urine, which

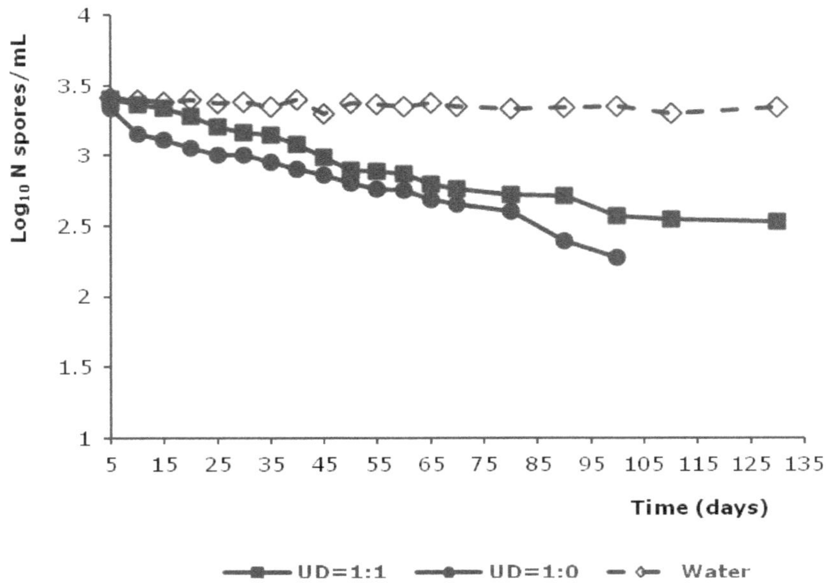

Figure 2. Reduction of *C. perfringens* spores during samples exposure at ambient temperatures. The initial values were 8260 mg/L NH_4-N (pH 8.8) for the undiluted urine (•) ; 4450 mg/L NH_4-N (pH 8.7) for the half-diluted urine (■) and pH 7.4 for the control (◊).

Table 2. Physicochemical parameters of the urine samples under incubating temperatures.

Sample	NH_4-N (mg/L)		Temperature (incubator)					
			25°C		35°C		42°C	
	Start	End	pH	T (°C)	pH	T (°C)	pH	T (°C)
UD=1:0	8260	7980	8.8±0.0	25.1±0.3	8.6±0.0	35.1±0.2	8.6±0.1	41.3±0.6
UD=1:1	4450	4295	8.7±0.0	25.2±0.1	8.4±0.0	35.2±0.2	8.4±0.0	41.7±0.3
UD=1:3	1070	1040	8.3±0.2	25.2±0.2	8.0±0.0	35.1±0.2	8.0±0.0	41.3±0.1
Control	ND	ND	7.4±0.1	25.1±0.2	7.4±0.0	35.1±0.3	7.4±0.2	41.5±0.0

ND: Not determined; Means ± SD; UD: urine diluted.

indicates its buffer property. The results (Table 2) show that temperatures in samples are similar to those shown on the incubator. Globally, a decrease of 0.2 pH units in urine samples was observed over time, under 25, 35 and 42°C.

Reduction of E. coli and Salmonella sp. number

Dark colonies with a metallic green sheen of *E. coli* were enumerated on EMB Agar. For *Salmonella* sp., colorless colonies with black centers were counted on SS Agar. The results in Table 3a and b show, regardless of incubation temperatures, 4 h of storage were sufficient to destroy in the undiluted urine, the two tested bacterial species (0 CFU/mL). The above data indicate that the survival of the two bacteria was better in the quarter-diluted urine (1:3) than in the undiluted (1:0). Similarly, we observed that survival of these bacteria was greater by incubating at 25°C than at 35 and 42°C, since in the case of the quarter-diluted urine samples, *E. coli* survival times were 24, 12 and 4 h respectively at 25, 35 and 42°C of incubation. For the *Salmonella* sp. 28, 16 and 8 h were the survival times obtained respectively in the same conditions as above mentioned. In the control, the reduction of bacterial numbers was particularly fast in function of the incubation temperatures.

Reduction of *E. faecalis and S. aureus* numbers

These bacteria were enumerated firstly, after two days of samples storage and then, at regular time intervals of 5 days. Red to maroon colonies of *E. faecalis* were counted

Table 3a. Reduction *of Escherichia coli* numbers in samples under storage conditions.

Time (h)	Number of *Escherichia coli* x 10^3 CFU/mL											
	25°C				35°C				42°C			
	UD=1:0	UD=1:1	UD=1:3	Water	UD=1:0	UD=1:1	UD=1:3	Water	UD=1:0	UD=1:1	UD=1:3	Water
0	1030±310	1120±670	1200±330	>300	980±211	1000±110	1070±111	>300	990±28	850±41	1010±38	>300
4	0	1.2±0.3	16.7±4.1	>300	0	1.0±0.1	5.9±1.2	>300	0	0.1±0.0	1.8±0.1	>100
8	0	0.1±0.0	13.6±2.5	>300	0	0	1.4±0.1	>100	0	0	0	<100
12	0	0	10.0±2.6	>300	0	0	0.3±0.2	>100	ND	ND	ND	ND
16	0	0	7.5±1.7	>300	0	0	0	>100	ND	ND	ND	ND
20	ND	ND	5.4±2.5	>100	0	0	0	>100	ND	ND	ND	ND
24	ND	ND	0.7±0.6	>100	ND	ND	ND	ND	ND	ND	ND	ND
28	ND	ND	0	>100	ND	ND	ND	ND	ND	ND	ND	ND
32	ND	ND	0	>100	ND	ND	ND	ND	ND	ND	ND	ND

Table 3b. Reduction of *Salmonella sp.* numbers in samples under storage conditions.

Time (h)	Number of *Salmonella sp.* x 10^3 CFU/mL											
	25°C				35°C				42°C			
	UD=1:0	UD=1:1	UD=1:3	Water	UD=1:0	UD=1:1	UD=1:3	Water	UD=1:0	UD=1:1	UD=1:3	Water
0	240±34	330±27	217±56	>300	280±85	341±71	490±141	>300	320±132	460±133	547±169	>300
4	0	1.4±0.4	22.0±5.0	>300	0	1.4±0.1	23.0±5.3	>300	0	0.2±0.0	17.6±4.7	>100
8	0	0.1±0.0	17.9±5.7	>300	0	0	5.0±0.7	>300	0	0	3.5±0.8	>100
12	0	0	11.0±3.8	>300	0	0	1.2±0.2	>300	0	0	0	>100
16	0	0	7.3±3.2	>100	0	0	0.2±0.0	>100	ND	ND	ND	ND
20	ND	ND	3.1±1.9	>100	ND	ND	0	>100	ND	ND	ND	ND
24	ND	ND	0.6±0.3	>100	ND	ND	ND	ND	ND	ND	ND	ND
28	ND	ND	0.1±0.0	>100	ND	ND	ND	ND	ND	ND	ND	ND
32	ND	ND	0	>100	ND	ND	ND	ND	ND	ND	ND	ND

on Slanetz & Bartley agar. For *S. aureus*, yellow colonies with yellow zones were enumerated on Mannitol Salt Agar. Data obtained (Table 3c and d) show that the two tested bacteria have survived at least 5 days in the undiluted urine samples (1:0) under 25°C of incubation, however they could not survive for up to two days at 42°C in the same samples as mentioned before. In the control, bacterial removal was faster at 42°C than at 25°C. Considering the incubation at 25°C, *E. faecalis* survival time in the quarter-diluted urine was 35 days approximatively, however it was less than 5 days in the undiluted urine samples. For the *S. aureus*, survival times were 15 and 5 days, respectively in the quarter-diluted and the undiluted urine samples. Extended times in the survival of the tested bacteria were observed by incubating in the diluted urine samples.

Table 3c. Reduction of *Enterococcus faecalis* numbers in samples under storage conditions.

Time (days)	Number of *Enterococcus faecalis* x 10^3 CFU/mL											
	25°C				35°C				42°C			
	UD=1:0	UD=1:1	UD=1:3	Water	UD=1:0	UD=1:1	UD=1:3	Water	UD=1:0	UD=1:1	UD=1:3	Water
0	96±17	117±14	105±11	>300	103±36	99±15	120±10	>300	86±12.7	97±18	102±40	>100
2	0.8±0.2	9.7±2.2	59±10	>300	0.3±0.1	5.4±0.9	22±7	>300	0	0.4±0.1	2.3±0.6	>100
5	0.2±0.0	4.5±0.4	28.0±1.3	>300	0	1.2±0.4	11.5±0.6	>300	0	0	0.9±0.2	>100
10	0	0.1±0.0	23.0±2.6	>300	0	0.2±0.0	1.5±0.4	<300	0	0	0	99
15	0	0.04±0.0	19.0±2.0	>300	0	0	0.5±0.0	244	ND	ND	ND	ND
20	0	0	5.1±3.2	>100	ND	ND	0	191	ND	ND	ND	ND
25	0	0	1.6±0.7	<100	ND	ND	ND	ND	ND	ND	ND	ND
30	ND	ND	0.2±0.0	<100	ND	ND	ND	ND	ND	ND	ND	ND
35	ND	ND	0.15±0.0	<100	ND	ND	ND	ND	ND	ND	ND	ND
40	ND	ND	0	<100	ND	ND	ND	ND	ND	ND	ND	ND
45	ND	ND	0	26	ND	ND	ND	ND	ND	ND	ND	ND

UD: Urine diluted.

Table 3d. Reduction *of Staphylococcus aureus* numbers in samples under storage conditions.

Time (days)	Number of *Staphylococcus aureus* x 10^3 CFU/mL											
	25°C				35°C				42°C			
	UD=1:0	UD=1:1	UD=1:3	Water	UD=1:0	UD=1:1	UD=1:3	Water	UD=1:0	UD=1:1	UD=1:3	Water
0	88±16	107±24	94±20	>100	102±31	100±15	94±30	<300	84±6	90±13	109±14	<100
2	0.6±0.2	09±0.2	3.6±0.9	>100	1.0±0.3	1.6±0.2	3.2±0.9	<300	0	0.2±0.0	0.9±0.2	<100
5	0.1±0.0	0.3±0.1	1.2±0.1	>100	0.2±0.1	0.5±0.1	1.3±0.2	<300	0	0	0.2±0.0	40
10	0	0	0.3±0.0	<100	0	0.1±0.0	0.4±0.0	<100	0	0	0	0
15	0	0	0.1±0.0	<100	0	0	0.1±0.2	<100	ND	ND	ND	ND
20	ND	ND	0	<100	ND	ND	0	<100	ND	ND	ND	ND

UD: Urine diluted.

Overall, Gram-positive cocci particularly *E. faecalis,* were less sensitive in the above described conditions (samples, incubation temperatures) as compared to Gram-negative rods (*E. coli*, *Salmonella* sp.).

DISCUSSION

The chemical results were that ammoniacal nitrogen (NH_4-N) concentrations in samples decreased in function of the urine dilution rate (Table 2), while pH values remained almost stable, even during the storage. Other previous studies indicated a biocidal effect in urine due to the uncharged ammonia (Jenkins et al., 1998; Nordin et al., 2009). Note that in aqueous solution, ammoniacal nitrogen is present in both uncharged (NH_3) and ionized (NH_4^+) forms. The uncharged ammonia concentration is proportional to the NH_4-N concentration and then, can be calculated, using chemical equations (Erickson, 1985).

Our findings on microbial behaviour show that the destruction of the tested bacteria was faster in the urine samples greatly concentrated in ammonia (Figure 1 and Table 3a to d), thus explaining the biocidal properties of this urine component. It is well known that uncharged ammonia can cross easily biological membranes as compared to ions (NH_4^+, H_3O^+, OH^-). Nordin et al. (2009) indicated that inside microorganism, the uncharged ammonia acts by increasing the internal pH to deleterious levels, so destroys cell by causing the dysfunction of the metabolism. Previous studies showed that microbicidal effects of alkaline pH in media (~8.8 for the urine) were less significant than these due to the uncharged ammonia (Jenkins et al., 1998; Vinnerås et al., 2008; Nordin et al., 2009). Moreover, a link has been found between the NH_3 membrane permeability and the external temperature (Pecson et al., 2007; Vinnerås et al., 2008). In fact, these authors observed the removal of the NH_3 microbicidal properties at low external temperatures (~4°C) because of the inhibition of its membrane permeability. However, inverse effect was observed by increasing temperature. Likewise, it was shown in this study that the reduction of the bacterial numbers was more rapid in samples incubated at 42°C than those placed at 25°C (Table 3a to d). Accordingly, these results show that temperature and urine intrinsic components, particularly the ammonia act synergistically to sanitize the urine samples.

Moreover, the sensitivity of bacteria during sample storage varied depending on microbial group since, Gram-positive cocci have survived more than Gram-negative rods. The presence or not of a large peptidoglycan layer in the wall of the tested bacteria can explain the difference in bacterial behaviour. In fact, Gram-negative bacteria possess in their wall, a slim peptidoglycan layer, that make them more sensitive to environmental factors like heat shock and ionic strength (Sinton et al., 1994; Madigan and Martinko, 2006).

Usually, to assess the environmental risks due to enteric pathogens, *E. coli* is the common faecal indicator (Finney et al., 2003). Because of its rapid removal in the urine, *E. coli* would not be a suitable faecal indicator for the urine from UDDTs. All the same, *E. faecalis* and *C. perfringens* spores were suggested as suitable indicators of the effectiveness of wastewater treatment (Payment and Franco, 1993; Ashbolt et al., 2001). We have found, for *E. faecalis,* a survival time of 10 days in undiluted urine at 26.5°C as the average of the ambient temperatures. And then, their reduction was faster when temperatures increased (Table 3c). As urine from UDDTs can contain enteric microorganisms such as viruses and protozoan oocysts which are very resistant to the treatment by storage (Höglund et al., 2002; Nordin et al., 2009), so, *E. faecalis* could not be considered as reliable indicator of the effectiveness of the urine treatment by storage.

With respect to our results on *C. perfringens* spores, overall no or small reduction was observed in samples (Figure 2) during the storage. These findings may confirm the statement according to a conservative effect of bacterial spores in alkaline media (Venczel et al., 1997; Gaillard et al., 1998), in so far as the urine pH is ~9.

Only ~ 1-$\log_{10}$ reduction of Clostridia spores was observed within 100 days of storage. These small reduction in spore numbers may be due in part to the low activation of spores. In fact, the occurrence by activating spores is the wall rip, which may exhibit these spore-forming bacteria which has deleterious effects on the environment. Note that spore activating agents can be mechanical (shock), physical (heat) or chemical (acidity). In our case, *C. perfringens* spores were placed in alkaline samples (urine) and then, exposed to ambient (26.5°C). Under these conditions, spore activation cannot occur physically or chemically. Therefore, the mechanically treatment of samples before analyses (homogenization), could be the main cause of spores activation. *C. perfringens* spores have been showed as suitable tracers for determining the fate of viruses and protozoan oocysts in wastewater treatment (Payment et al., 2001; Ashbolt et al., 2001). But, for the treatment of the urine by storage, more data on viruses and protozoan behaviour in this matrix are needed in order to confirm the use of Clostridia spores as tracers in the urine sanitation. Nevertheless, it had been shown that Rotavirus and *Cryptosporidium parvum* oocysts remain viable in urine after 6 months of storage, under very low temperatures (Höglund and Stenström, 1999; Höglund et al., 2002; WHO, 2006).

The ammonia concentrations used for experiments (Table 2), are the same found in the urine collected in fields, by considering large-scale systems (Jönsson and Vinnerås, 2007; Dagerskog and Bonzi, 2010). In these conditions, low NH_4-N concentrations observed, can be due to the urine-water mixture that occurred when the UDDTs are used for washing bodies.

On taking into account only the samples incubated at 25°C, the survival times for *E. faecalis* (less sensitive bacteria), were 5, 15 and 35 days respectively in undiluted, half-diluted and quarter-diluted urine. Based on these data, storage times of 30-45 days recommended to sanitize urine from large-scale systems in Ouagadougou (Dagerskog and Bonzi, 2010), seem to be sufficient since in this Sahelian zone, annual average temperatures are above 25°C. Unfortunately, urine collected via UDDTs can contain other enteric microorganisms more persistent in environment than vegetative bacteria.

Also, the volumetric capacities in liquid samples can influence the microorganisms survival. Previous studies had shown that the sensitivity of microbes to the external factors (UV-light, temperature) was more important in the small volumes than in the high volumes (Lemunier et al., 2005; Pecson et al., 2007; Niwagaba, 2009). Consequently, the urine volumes in containers must be considered for determining the storage time needed to sanitize urine. Although, the effects linked to the matrix volumetric were not studied here, however the ways for the urine collection via UDDTs and the climate variability must be integrated into the process of the urine sanitation by storage, before been applied in soils as fertiliser.

Conclusion

Experimental results have shown that the survival times of the tested bacteria increased during the storage with the urine dilution for which, the direct effect was the reduction of the ammonia concentration. For an effective treatment by storage, adding water into the urine collection containers must be avoided. Also, our findings showed that ammonia and temperature act synergistically to reduce or remove microorganisms in urine, during storage. That is why climatic factors are decisive to estimate the exposure time needed to sanitize urine. On top of that, analyses have shown that the survival capacities of bacteria in the urine, varied depending on bacterial group. Gram-positive cocci, particularly *E. faecalis* which were less sensitive in urine as compared to Gram-negative rods, because of the difference in peptiglycan layer size. The use of *E. faecalis* as indicator of the effectiveness of the urine treatment seems inappropriate since, *C. perfringens* spores are more resistant to the storage conditions. If the *C. perfringens* spores are suitable tracers for the fate of viruses and protozoan oocysts in wastewater treatment, their use as tracers for the urine treatment (storage) may be possible, in view of their stability. Nevertheless, it would be important to elucidate before the survival potential of viruses and protozoan oocysts in the urine during storage.

Conflict of Interests

The author(s) have not declared any conflict of interests.

ACKNOWLEDGEMENTS

The authors thank the staff of Pan African Agency, Water and Sanitation for Africa (WSA) at the headquarters of the Institution (Ouagadougou) for technical and financial support.

REFERENCES

Anonymous (2005). French standard NF V 08-061: Microbiology of food and animal feeding stuffs- Anaerobic enumeration of sulfito-reducing bacteria by colony count technique. AFNOR, Paris, France.

APHA (American Public Health Association) (2005). Standard Methods for the Examination of Water and Wastewater, 21st Edition, Washington DC, EUA.

Ashbolt NJ, Grabow WOK, Snozzi M (2001). Indicators of microbial water quality. In: Water Quality Guidelines, Standards and Health. Risk assessment and management for water- related infectious disease. Fewtrell L and Bartram J (eds) IWA Publishing, London. pp. 289-315.

Chandran A, Pradhan SK, Heinonen-Tanski H (2009). Survival of enteric bacteria and coliphage MS2 in pure human urine. J. Appl. Microbiol. 107: 1651-1657

Dagerskog L, Bonzi M (2010). Opening Minds and Closing Loops-Productive Sanitation Initiatives in Burkina Faso and Niger. Sustainable Sanitation Practice No. 3, 4-11.

Erickson RJ (1985). An evaluation of mathematical models for the effects of pH and temperature on ammonia toxicity to aquatic organisms. Water Res. 19(8): 1047-1058.

Esrey S, Andersson I, Hillers A, Sawyer R (2001). Closing the Loop Ecological Sanitation for Food Security. Stockholm (Sweden): Swedish International Development Agency.

Feachem RG, Bradley DJ, Garelick H, Mara DD (1983). Sanitation and Disease-Health Aspects of Excreta and Wastewater Management. Chichester, UK: John Wiley and Sons.

Finney M, Smullen J, Foster HA, Brokx S, Storey DM (2003). Evaluation of Chromocult coliform agar for the detection and enumeration of Enterobacteriaceae from faecal samples from healthy subjects. J. Microbiol. Methods 54(3): 353-358.

Gaillard S, Leguérinel I, Mafart P (1998). Model for combined effects of temperature, pH and water activity on thermal inactivation of Bacillus cereus spores. J. Food Sci. 63 (5):887-889.

Haneaus A, Hellström D, Johansson E (1996). Conversion of urea during storage of human urine. Vatten 52: 263-270.

Heinonen-Tanski H, van Wijk-Sijbesma C (2005). Human excreta for plant production. Bioresour. Technol. 96:403-411.

Höglund C, Stenström TA, Ashbolt N (2002). Microbial risk assessment of source-separated urine used in agriculture. Waste Manage Res. 20:150-161.

Höglund CE, Stenström TA (1999). Survival of Cryptosporidium parvum oocysts in source separated human urine. Can. J. Microbiol. 45:740-746.

ISO (1993). Microbiology - General Guidance on Methods for the Detection of Salmonella. ISO 6579, International Organization for Standardization, Geneva, Switzerland.

ISO (1999). Microbiology of food and animal feeding stuffs - Horizontal method for the enumeration of coagulase-positive staphylococci (Staphylococcus aureus and other species). ISO 6888-1, International Organization for Standardization, Geneva, Switzerland

ISO (2000). Water Quality - Detection and Enumeration of Intestinal Enterococci : Colony counting method on solid medium. ISO 7899-2, International Organization for Standardization, Geneva, Switzerland.

ISO (2001). Horizontal method for the enumeration of β-glucuronidase-positive *E. coli*. ISO 16649-2, International Organization for Standardization, Geneva, Switzerland

Jenkins MB, Bowman DD, Ghiorse WC (1998). Inactivation of Cryptosporidium parvum Oocysts by Ammonia. Appl. Environ. Microbiol. 64 (2):784-788.

Jönsson H, Vinnerås B (2007). Experiences and Suggestions for Collection Systems for Source-separated Urine and Faeces. Water Sci. Technol. 56(5):71-76.

Lemunier M, Francou C, Rousseaux S, Houot S, Dantigny P, Piveteau P, Guzzo J (2005). Long-term survival of pathogenic and sanitation indicator bacteria in experimental biowaste composts. Appl. Environ. Microbiol. 71(10):5779-5786.

Madigan TM, Martinko MJ (2006). Brock-Biology of microorganisms, 11th Edition. Published by Pearson Prentice Hall. Pearson Education Inc., Upper Saddle River, United States of America. ISBN 0-13-196893-9.

Maurer M, Pronk W, Larsen TA (2006). Treatment processes for source-separated urine. Water Resour. 40(17): 3151-3166.

Mnkeni PNS, Austin LM (2009). Fertiliser value of human manure from pilot urine-diversion toilets. Water SA 35(1):133-138.

Münch E, Winker M (2009). Overview of urine diversion components such as waterless urinals, urine diversion toilets, urine storage and reuse systems. Technol. Rev. 1(5): 1-32.

Niwagaba C (2009). Treatment Technologies for Human Faeces and Urine. Doctoral thesis. Swedish University of Agricultural Sciences, Uppsala, n°70, 91 p.

Nordin A, Nyberg K, Vinnerås B (2009). Inactivation of Ascaris eggs in source-separated urine and faeces by ammonia at ambient temperatures. Appl. Environ. Microbiol. 75(3), 662-667.

Ouédraogo E, Mando A, Brussaard L, Stroosnijder L (2007). Tillage and fertility management effects on soil organic matter and sorghum yield in semi-arid West Africa. Soil Tillage Res. 94:64-74.

Payment P, Franco E (1993). Clostridium perfringens and somatic coliphages as indicators of the efficiency of drinking water treatment for viruses and protozoan cysts. Appl. Environ. Microbiol. 59:2418-2424.

Payment P, Plante R, Cejka P (2001). Removal of indicator bacteria, human enteric viruses, Giardia cysts, and Cryptosporidium oocysts at a large wastewater primary treatment facility. Can. J. Microbiol. 47(3):188-193.

Pecson BM, Barrios JA, Jimenez BE, Nelson KL (2007). The effects of temperature, pH, and ammonia concentration on the inactivation of Ascaris eggs in sewage sludge. Water Res. 41:2893-2902.

Pradhan SK, Pitkänen S, Heinonen-Tanski H (2009). Fertilizer Value of Urine in Pumpkin (Cucurbita maxima) Cultivation. Agric. Food Sci. 18:57-68.

Pronk W, Koné D (2010). Options for urine treatment in developing countries. Desalination 251:360-368.

Schönning C, Leeming R, Stenström TA (2002). Faecal contamination of source-separated human urine based on the content of faecal sterols. Water Res. 36:1965-1972.

Sinton LW, Davies-Colley RJ, Bell RG (1994). Inactivation of enterococci and fecal coliforms from sewage and meatworks effluents in seawater chambers. Appl. Environ. Microbiol. 60(6): 2040-2048.

Stanbury PF, Whitacer A, Hall SJ (1995). Principles of Fermentation Technology. Second Edition. Oxford. Pergamon. pp. 357.

Venczel LV, Arrowood M, Hurd M, Sobsey MD (1997). Inactivation of Cryptosporidium parvum oocysts and Clostridium perfringens spores by a mixed-oxidant disinfectant and by free chlorine. Appl. Environ. Microbiol. 63(4):1598-1601.

Vinnerås B, Nordin A, Niwagaba C, Nyberg K (2008). Inactivation of bacteria and viruses in human urine depending on temperature and dilution rate. Water Res. 42:4067-4074.

World Health Organization (WHO) (2006). Guidelines for the Safe Use of Wastewater, Excreta and Greywater, Volume 4, Wastewater Use in Agriculture, 196.

27

The comparative insecticidal and residual efficacy of sniper and alpha cypermethrin in the control of *Blattella germanica* (L.) (Dictyoptera: Blattellidae) infestation

K. A. Kemabonta and Amadi, G.

Department of Zoology, Faculty of Science, University of Lagos, Akoka, Lagos, Nigeria.

Blattella germanica **(German cockroach) is an urban insect pest in Nigeria due to an increase in strains that are resistant to the commonly used insecticides. The purpose of this study was to determine which of the insecticides used by pest control professionals are still effective. The insecticidal efficacies of dichlorvos (organophosphate-called sniper©) and alpha cypermethrin (a pyrethroid called alpha action ©) were determined in laboratory bioassays against *B. germanica*. Efficacy and response were affected by application rate of insecticide and exposure period. Insecticides mixed with diesel were more effective than those mixed with water. Insects exposed to the insecticides in closed chambers gave higher mortalities than insects in open chambers. *B. germanica* was more susceptible to dichlorvos than to alphacypermethrin in both open and closed chambers. Mortality of *B. germanica* increased with increase in concentration of the insecticide used. Sniper mixed in diesel was fastest acting insecticide against *B. germanica* where 100% mortality was observed 5 min post exposure in closed and opened chambers. Both insecticides gave 100% mortality of all insects exposed within 4 h of the exposure period. Sniper mixed with diesel had the LT_{95} of 10 min while sniper mixed in water was 17 and 23 min, respectively in closed and opened chambers. The LT_{95} of alphacypermethrin (5% (v/v)) mixed with diesel and water in closed chambers were 10 and 125 min, respectively. The residual effect of sniper and alphacypermethrin was less than one week.**

Key words: Resistance, commercial insecticide, susceptibility.

INTRODUCTION

German cockroaches live in close association with people. They are tropical in origin but in the temperate zone most species live in parts of house and other building where warmth, moisture and food are adequate. They usually live in groups and are mostly active at night. In the daytime they hide in cracks and crevices in walls, door frames, and furniture, and in secure places in bathrooms, cupboards, steam tunnels, animal houses, basement, television, radio and other electronic devices, drains and sewer systems. If the light is turned on in an infested kitchen at night the cockroaches will run from dishes, utensil, working surface and the floor toward shelters (Rivault et al., 1993).

German cockroaches eat all food used for human

consumption. They however prefer starchy and sugary materials. They sip milk and nibble at cheese, meats, pastry, grain product, sugar and sweet chocolate. They also feed on cardboard, book binding, ceiling boards containing starch, the sized inner lining of shoe sole, their own cast skin, dead and crippled cockroaches, fresh and dried blood, excrement, sputum, fingernails and toenails of babies and sleeping or sick persons (Rauh et al., 2002). More importantly, roaches are sometimes implicated as disease vectors, as carriers of intestinal diseases, such as diarrhea, dysentery, typhoid fever and cholera (Appel and Tucker, 1986; Gradcolas, 1996).

The females unlike most other cockroaches carry ootheca that protrude from their abdomen until the eggs are ready to hatch or may hatch from ootheca while female still carries it (often in rear cases). The ootheca is then dropped in a secluded location, where the nymphs emerge within 24 h. A female may produce four to eight cases during her lifetime, each containing 30 to 48 eggs. Eggs hatch in about one month and nymph develop in 1 to 4 months. Adult females live for about six to seven months and males live slightly less. The German cockroach produce more eggs and has more generation per year (three to four) than other cockroaches thus troublesome infestation can develop from a few individuals (Rust et al., 1995).

Chemical control is still the main approach for urban pest control (Castle et al., 1999; Rozendaal, 1997; Marrs, 1993; Lee and Yap, 2003; Tidwell et al., 1994). The use of insecticides is seen as the most effective tool in cockroach control program (WHO, 1996; Chavasse and Yap, 1997; Lee and Yap, 2003; Tidwell et al., 1994). The major insecticides used against cock-roaches in Nigerian households include pyrethroids (alpha cypermethrin) and organophosphates (dichlorvos).

Resistance has been reported amongst cockroaches to various groups of insecticides such as organophos-phates, carbamates and synthetic pyrethroids, when applied directly, as residual spray and or in topical application (Diaz et al., 1994; WHO, 1996, Wei et al., 2001). With the increase in the number of German cockroach resistance to insecticides lately, there is a need to re-evaluate/ re-ascertain the insecticides that are being used so as to know those that have become less effective or the efficacious ones, in order to sustain the control of this deleterious organism.

This study therefore seeks to investigate the comparative insecticidal and residual efficacies of sniper and alpha cypermethrin against *B. germanica* in laboratory during fumigation (closed) or disinfestation (opened) exercises.

MATERIALS AND METHODS

All the experiments were carried out at the laboratory of the Department of Zoology, University of Lagos, Nigeria. Temperature and relative humidity measurements were recorded daily using the Thermo-Hygro Analog Yenaco.

Insect collection

A number of German cockroaches identified at insectary of the department, were collected from some old cupboards in some hostels in New Hall Area of the University of Lagos. The cupboards have been locked for over four months. Several male and female nymphs and adults, including female carrying ootheca were found in the cupboards.

Insect culture

These insects were carefully removed, kept in vials and taken to the Zoological laboratory and transferred to another plastic vial measuring 25 cm in length and 17 cm in width. A wire gauze and net were used to cover the plastic vials. The cover and body of the vials were strapped together using rubber band to prevent escape or cross infestation of the cockroaches. The cockroaches were fed with bread, and the slices of bread were changed three times a week. The drinking water placed in each container was also renewed weekly. The culture was kept in the laboratory for over six months, until the roaches in each vial were more than hundred. Each vial contains nymphs, adult males and females.

Insecticides used

The insecticides used were alpha cypermethrin and dichlorvos called sniper by SaroAgroSciences

Experimental design

Four experimental groups (two groups and two formulations) were included in the design. The first two groups contained filter paper treated with diesel formulation of sniper and alpha cypermethrin while the other two groups contained filter paper treated with water formulation of sniper and alpha cypermethrin. Each group was exposed to closed environment (fumigation) and opened environment (disinfestation). There were two controls (treatment with water and diesel), that were each run parallel to the treatments. The controls consisted of filter papers treated with water and diesel for both closed and opened environments

Toxicity test

All life stages of German cockroach except the first, second nymphal stages and the gravid females were used for the toxicity tests. Sniper and alpha cypermethrin insecticide formulation were used in accordance to the instruction on the labels. Four concentrations 2.5% (v/v) and 5% (v/v) of water; and 2.5% (v/v) 2.5% (v/v) and 5% (v/v) of diesel were used. Each formulation was sprayed on filter paper to run off. Excess liquid was drained off. Thereafter ten (10) German cockroaches were confined into each jar.

The upper surfaces of the jar were lightly greased with petroleum jelly to prevent escape of the insects. Control jars had only water sprayed on the filter paper. Each treatment was replicated three times. Cockroaches' knockdown or mortality was recorded at 5, 10, 15, 20, 30 min to the 4th h after spraying. The number of cockroaches dead and alive was counted and noted. The cockroaches were removed from the jar and kept in the freezer for 30 min and thereafter discarded. The jars containing treated and untreated filter papers were kept aside on laboratory bench and used in subsequent experiments.

Table 1. Cockroach mortality upon exposure to water (control) in filter paper in open and closed chambers.

Percentage mortality (%)	Treatment	Time									
		5 min	10 min	15 min	20 min	30 min	45 min	1 h	2 h	3 h	4 h
	Open	0	0	0	0	0	0	0	0	0	0
	Closed	0	0	0	0	0	0	0	0	0	0

Number of insects per replicate was 10 and each experiment was replicated three times.

Table 2. Cockroach mortality upon exposure to diesel (control) soaked in filter paper in open and closed chambers.

Percentage mortality (%)	Treatment	Time									
		5 min	10 min	15 min	20 min	30 min	45 min	1 h	2 h	3 h	4 h
	Open	0	0	0	0	10	13	13	27	27	37
	Closed	0	0	0	10	10	20	20	27	43	43

Number of insects per replicate was 10 and each experiment was replicated three times.

Quantal response

On assumption of death, the German cockroach is pricked with the strands of hair from a camel hair brush to ascertain death. Responses of cockroaches were noted and the parameters considered include time of knock down and mortality rate per minute

Residual effect of the insecitides

The jars used in the experiments above containing treated filter paper were used for the residual test on weekly bases for two weeks, 10 cockroaches were confined to each jar and mortality and or knockdown noted as in toxicity test above.

Statistical analysis

Statistical analysis of the results was also done after the results have been ascertained and its corresponding mean and standard deviation were recorded against time. This was done using the equations:

Mean, $\bar{x}= \Sigma FX / \Sigma F$

Standard deviation= $\sqrt{\Sigma(x-\bar{x})}/n-1$

The lethal time value (LT_{50}) and the regression slope for each treatment were obtained using probit analysis (SPSS 2000). Mean percentage of insect mortality value was subjected to arcsine transformation followed by comparison of mean using LSD test.

RESULTS

Mortality of *B. germanica* exposed to diesel and water

No insects died throughout the period of the experiment when exposed to water in both the closed and opened chambers (Table 1). No mortality was recorded in the insects exposed to diesel until after 20 and 30 min in closed (Io) and opened (Ic) chambers respectively with the highest mortality (43%) recorded in closed (I_o) chambers (Table 2).

Mortality in Insects exposed to sniper and alphacypermethrin mixed with water

Mortality of *B, germanica* increased with increase in concentration of the insecticide used. It was higher in closed than in opened chambers and higher in insecticide mixed with diesel than in water. Mortality of *B. germanica* was recorded from 5 min in the insects exposed to both 2.5% (v/v) and 5% (v/v) of aqueous sniper in both opened (Io) and closed (Ic) chambers. All insects placed in aqueous sniper (5% (v/v)) died within 20 min while 100% mortality was recorded in all concentration used in the closed and opened chambers within one hour (Table 3). On the other hand, no mortality was recorded until after 20 min (6 and 33%) in *B. germanica* exposed to 2.5% (v/v) and 5% (v/v) of aqueous alpha cypermethrin respectively. 100% mortality was recorded after 4 h of exposure in both open (Io) and closed (Ic) chambers (Table 4).

Mortality in insects exposed to sniper and alphacypermethrin mixed with diesel

100% mortality was recorded in less than 5 min in the insects exposed to sniper mixed with diesel in both concentrations (2.5% (v/v) and 5% (v/v)) used and in both open (Io) and closed (Ic) chambers (Table 5) On the

Table 3. Cockroach mortality upon exposure to sniper (mixed with water) in filter paper in an open(Lo) and closed (Lc) chambers.

	Treatment	Time									
		5 min	10 min	15 min	20 min	30 min	45 min	1 h	2 h	3 h	4 h
Percentage mortality (%)	2.5% (v/v) L o (I)	27	27	50	70	93	100	-	-	-	-
	2.5% (v/v) Lc (II)	13	27	33	37	73	86	100	-	-	-
	5% (v/v) Lo (I)	40	70	76	96	100					
	5% (v/v) Lc (II)	56	66	96	100	-	-	-	-	-	-

Number of insects per replicate was 10 and Each experiment was replicated three times.

Table 4. Cockroach mortality upon exposure to alpha cypermethrin (mixed with water) in filter paper in an open(Lo) and closed (Lc) chambers.

	Treatment	Time									
		5 min	10 min	15 min	20 min	30 min	45 min	1 h	2 h	3 h	4 h
Percentage mortality (%)	2.5% (v/v)L o (I)	0	0	0	6	6	16	53	90	96	100
	2.5% (v/v)Lc (II)	0	0	0	33	33	40	70	60	60	67
	5% (v/v)Lo (I)	0	0	0	26	33	33	56	83	93	100
	5% (v/v)Lc (II)	0	0	0	33	40	43	63	90	90	100

Number of insects per replicate was 10 and each experiment was replicated three times.

Table 5. Cockroach mortality upon exposure to sniper(mixed with Diesel) in filter paper in an open(Lo) and closed (Lc) chambers.

	Treatment	Time									
		2 min	5 min	8 min	10 min	20 min	30 min	45 min	1 h	2 h	3 h
Percentage mortality (%)	2.5% (v/v)L o (I)	100	-	-	-	-	-	-	-	-	-
	2.5% (v/v)Lc (II)	100	-	-	-	-	-	-	-	-	-
	5% (v/v)Lo (I)	100	-	-	-	-	-	-	-	-	-
	5% (v/v)Lc (II)	100	-	-	-	-	-	-	-	-	-

Number of insects per replicate was 10 and each experiment was replicated three times.

Table 6. Cockroach mortality upon exposure to alpha cypermethrin (mixed with diesel) in filter paper in an open (Lo) and closed (Lc) chambers.

	Treatment	Time									
		2 min	5 min	10 min	15 min	30 min	45 min	1 h	2 h	3 h	4 h
Percentage mortality (%)	2.5% (v/v)L o (I)	56	100	-	-	-	-	-	-	-	-
	2.5% (v/v)Lc (II)	46	93	100	-	-	-	-	-	-	-
	5% (v/v)Lo (I)	40	93	100	-	-	-	-	-	-	-
	5% (v/v)Lc (II)	46	100	-	-	-	-	-	-	-	-

Number of insects per replicate was 10 and each experiment was replicated three times.

other hand, 100% mortality was recorded after 10 min in insects exposed to Alpha cypermethrin mixed with diesel in both open and closed chambers (Table 6).

Residual effect of sniper and alpha cypermethrin on *B germanica* at *one week after treatment*

At one week after treatment 5% (v/v) of sniper mixed with water gave 10 and 50% mortality of *B. germanica* at 15 min and 4 h respectively of exposure in the closed chambers (Table 7) while sniper diluted with diesel gave 26 and 46% mortality respectively in closed and opened chambers under the same conditions (Table 8). On the other hand under same concentrations, 0 and 23% mortality respectively of *B. germanica* was recorded for Alphacypermethrin mixed in water (Table 9) and 16 and 40% mortality recorded for Alphacypermethrin mixed in

Table 7. Residual effect of sniper (mixed with water) in filter paper on cockroach morality in an open(Lo) and closed (Lc) chambers (after one week).

	Treatment	Time									
		5 min	10 min	15 min	20 min	30 min	45 min	1 h	2 h	3 h	4 h
Percentage mortality (%)	2.5% (v/v)L_O (I)	0	0	0	23	23	30	33	33	37	37
	2.5% (v/v)L_C (II)	0	0	23	26	26	33	40	40	46	46
	5% (v/v)L_O (I)	0	0	3	6	36	43	43	43	50	50
	5% (v/v)L_C (II)	0	0	10	13	26	33	40	40	50	50

Number of insects per replicate was 10 and each experiment was replicated three times.

Table 8. Residual effect of sniper (mixed with diesel) in filter paper on cockroach morality in an open (Lo) and closed (Lc) chambers (after one week).

	Treatment	Time									
		2 min	5 min	10 min	20 min	30 min	45 min	1 h	2 h	3 h	4 h
Percentage mortality (%)	2.5% (v/v)L_O (I)	0	10	16	16	20	23	26	26	26	26
	2.5% (v/v)L_C (II)	0	13	16	16	16	30	36	36	36	36
	5% (v/v)L_O (I)	0	16	23	23	26	33	43	43	43	43
	5% (v/v)L_C (II)	0	20	26	26	30	36	46	46	46	46

Number of insects per replicate was 10 and each experiment was replicated three times.

Table 9. Residual effect of alpha cypermethrin (mixed with water in filter paper) on cockroach morality in an open (Lo) and closed (Lc) chambers (after one week).

	Treatment	Time									
		5 min	10 min	15 min	20 min	30 min	45 min	1 h	2 h	3 h	4 h
Percentage mortality (%)	2.5% (v/v)L_O (I)	0	0	0	0	0	13	13	16	20	20
	2.5% (v/v)L_C (II)	0	0	0	0	0	13	16	16	20	20
	5% (v/v)L_O (I)	0	0	0	0	0	3	10	16	16	16
	5% (v/v)L_C (II)	0	0	0	0	0	20	20	23	23	23

Number of insects per replicate was 10 and each experiment was replicated three times.

Table 10. Residual effect of alpha cypermethrin (mixed with diesel in filter paper) on cockroach morality in an open (Lo) and closed (Lc) chambers (After one week).

	Treatment	Time									
		2 min	5 min	10 min	20 min	30 min	45 min	1 h	2 h	3 h	4 h
Percentage mortality (%)	2.5% (v/v)L_O (I)	0	0	0	6	10	10	16	26	26	26
	2.5% (v/v)L_C (II)	0	0	0	13	13	13	23	30	30	30
	5% (v/v)L_O (I)	0	0	0	6	10	23	23	33	33	33
	5% (v/v)L_C (II)	0	0	0	16	16	26	26	40	40	40

Number of insects per replicate was 10 and Each experiment was replicated three times.

diesel (Table 10).

Residual effect of sniper and alpha cypermethrin on *B germanica at two weeks after treatment*

No mortality was recorded at two weeks of treatment until after 3 and 4 h in *B. germanica* exposed to sniper mixed with diesel and water respectively and the highest mortality of 16 and 10% were recorded after 4 h with 5% (v/v) sniper in closed chambers (Tables 11 and 12). Same trend of 3 and 4 h after exposure was found in *B. germanica* exposed to alphacypermethrin mixed with diesel and water. Highest mortality of 20% was recorded

Table 11. Residual Effect of sniper (mixed with water in filter paper) on cockroach morality in an open(Lo) and closed (Lc) chambers (After two weeks).

	Treatment	Time									
		5 min	10 min	15 min	20 min	30 min	45 min	1 h	2 h	3 h	4 h
Percentage mortality (%)	7.5ML/300ML$_O$ (I)	0	0	0	0	0	0	0	0	0	3
	7.5ML/300ML$_C$ (II)	0	0	0	0	0	0	0	0	0	3
	5% (v/v)L$_O$ (I)	0	0	0	0	0	0	0	0	0	3
	5% (v/v)L$_C$ (II)	0	0	0	0	0	0	0	0	0	10

Number of insects per replicate was 10 and each experiment was replicated three times.

Table 12. Residual effect of sniper (mixed with diesel in filter paper) on cockroach morality in an open(Lo) and closed (Lc) chambers (after two weeks).

	Treatment	Time									
		2 min	5 min	10 min	20 min	30 min	45 min	1 h	2 h	3 h	4 h
Percentage mortality (%)	2.5% (v/v)L$_O$ (I)	0	0	0	0	0	0	0	0	6	6
	2.5% (v/v)L$_C$ (II)	0	0	0	0	0	0	0	0	10	10
	5% (v/v)L$_O$ (I)	0	0	0	0	0	0	0	0	13	13
	5% (v/v)L$_C$ (II)	0	0	0	0	0	0	0	0	16	16

Number of insects per replicate was 10 and each experiment was replicated three times.

Table 13. Residual effect of alpha cypermethrin (mixed with water in filter paper) on cockroach morality in an open (Lo) and closed (Lc) chambers (after two weeks).

	Treatment	Time									
		5 min	10 min	15 min	20 min	30 min	45 min	1 h	2 h	3 h	4 h
Percentage mortality (%)	2.5% (v/v)L$_O$ (I)	0	0	0	0	0	0	0	0	0	0
	2.5% (v/v)L$_C$ (II)	0	0	0	0	0	0	0	0	0	3
	5% (v/v)L$_O$ (I)	0	0	0	0	0	0	0	0	0	3
	5% (v/v)L$_C$ (II)	0	0	0	0	0	0	0	0	0	6

Number of insects per replicate was 10 and each experiment was replicated three times.

Table 14. Residual effect of alpha cypermethrin (mixed with diesel in filter paper) on cockroach morality in an open(Lo) and closed (Lc) Chambers (After two week).

	Treatment	Time									
		2 min	5 min	10 min	20 min	30 min	45 min	1 h	2 h	3 h	4 h
Percentage mortality (%)	2.5% (v/v)L$_O$ (I)	0	0	0	0	0	0	0	0	6	6
	2.5% (v/v)L$_C$ (II)	0	0	0	0	0	0	0	0	13	13
	5% (v/v)L$_O$ (I)	0	0	0	0	0	0	0	0	16	16
	5% (v/v)L$_C$ (II)	0	0	0	0	0	0	0	0	20	20

Number of insects per replicate was 10 and each experiment was replicated three times.

after 4 h in 5% (v/v) alphacypermethrin in closed chambers (Tables 13 and 14).

Lethal time of sniper to *B. germanica*

Table 15 shows the LT_5, LT_{50} and LT_{95} of 2.5% (v/v) and 5% (v/v) of sniper mixed in diesel and water. LT decreases with increase in the concentration of sniper. The LT_5 and LT_{50} could not be calculated because 95% mortality of *B. germanica* was recorded within 10 min of the experiment. Sniper mixed with diesel had the LT_{95} of 10 min in both closed and opened chambers while with sniper mixed in water was 17 and 23 min respectively in closed and opened chambers.

Table 15. Lethal time of sniper to mortality of *B. germanica*.

Lethal time (min)	Aqueous sniper				Sniper and diesel			
	2.5% (v/v)		5% (v/v)		2.5% (v/v)		5% (v/v)	
	Io	Ic	Io	Ic	Io	Ic	Io	Ic
LT_5 (min)	8	6	2	2	-	-	-	-
LT_{50} (min	**12**	**18**	**7**	**5**	**-**	**-**	**-**	**-**
LT_{95} (min)	45	72	23	17	10	10	10	10

Number of insects per replicate was 10 and each experiment was replicated three times

Table 16. Lethal time of alpha cypermethrin to mortality of *B. germanica*.

Lethal time	Aqueous alpha-cypermethrine				Alpha-cypermethrine mixed with diesel			
	2.5% (v/v)		5% (v/v)		2.5% (v/v)		5% (v/v)	
	Io	Ic	Io	Ic	Io	Ic	Io	Ic
LT_5(minutes)	25	12	13	11	2	-	3	-
LT_{50}(minutes)	62	40	44	30	5	-	6	-
LT_{95}(minutes)	183	175	161	124	10	10	10	10

Number of insects per replicate was 10 and each experiment was replicated three times

Lethal time of alphacypermetrin to *B. germanica*

Table 16 shows the LT_5, LT_{50} and LT_{95} of 2.5% (v/v) and 5% (v/v) of alpha cypermethrin mixed in diesel and water. LT also decreases as concentration of alphacypermethrin increases. The LT_5 and LT_{50} is unknown for alpha-cypermethrin and diesel solution in both 2.5% (v/v) and 15 ml/300 ml in closed (Ic) chambers because 95 mortality was recorded within 10 min. The LT_{95} of 5% (v/v) of Alphacypermethrin mixed with diesel and water closed chambers were 10 and 125 min respectively (Table 16).

Percentage mortality after four hours of exposure

At the onset of the experiment, both insecticides gave 100% mortality of all insects exposed within 4 h of the exposure period. However, the efficacy decreased with increase in residual time that by the second week after treatment, 17 and 20% mortality of *B. germanica* were recorded in sniper and alphacypermethrin respectively mixed in diesel in the closed chambers (Table 17).

DISCUSSION

For effective pest control of the German cockroach, closed (fumigation environment) and mixing the insecticide with diesel rather than water is being suggested. This is because *B. germanica* mortality was higher and time of death faster in closed chambers than in open chambers. Moreover, mixing each of the two insecticides with diesel gave better results than mixing with water.

The study revealed the effectiveness of two groups of insecticides, sniper (organophosphate) and alpha-cypermethrin (Pyrethroid) in the control of German cockroaches. Label-recommended doses of sniper and alphacypermethrin (5% v/v), were effective by having 100% mortality after four hours of exposure to *B. germanica*.

However, the 100% mortality recorded in 2 min in the insects exposed to sniper (Table 5) mixed with diesel in both concentrations used (2.5% (v/v) and 5% (v/v)) and in both open (Io) and closed(Ic) chambers suggest sniper to be more effective in controlling *B. germanica* than alphacypermethrin that had 100% mortality in 10 min under the same conditions (Table 6). Lots of studies have shown that cockroaches show high level of resistance to pyrethroid compared to organophosphates (Wei et al., 2001; Cochran, 1995).

This is not the case in this situation as both insecticides were able to control *B. germanica* in less than 10 min when mixed in diesel and less than 20 min and 4 h when sniper and alphacypermethrin respectively were mixed with water. This experiment conforms with that by Agrawal et al., (2005) who found that propoxur (an organophosphate) was more effective in controlling the German cockroach, *B. Germanica,* than deltametrin (a pyrethroid).

This study also showed that efficacy and time of response were affected by time of application of insecticide and time of exposure of insects after treatment. The insecticides (sniper and alpha- cypermethrin) were more

Table 17. Percentage mortality of *B. germanica* after four hours of exposure.

Parameter	Period	2.5% (v/v)		5% (v/v)	
		I_O	I_C	I_O	I_C
Diesel	Control	36	43		
Water	Control	0	0		
Sniper/water	Initial	100	100	100	100
	One week after treatment	37	47	50	50
	Two weeks after treatment	3	3	3	10
Sniper/diesel	Initial	100	100	100	100
	One week after treatment	37	40	43	53
	Two weeks after treatment	7	10	13	17
Alpha cypermethrine/water	Initial	100	100	100	100
	One week after treatment	20	20	13	23
	Two weeks after treatment	0	3	3	7
Alpha cypermathrine/diesel	Initial	100	100	100	100
	One week after treatment	27	30	33	40
	Two weeks after treatment	7	13	17	20

Number of insects per replicate was 10 and each experiment was replicated three times

effective when the insects were exposed immediately after application of insecticides as compared to when they were exposed one and two weeks after application. Thus the residual effect of sniper and alphacypermethrin can be said to be less than one week.

B. germanica collected from several places at one geographical location, University of Lagos, Nigeria with a total area of 8194.93 m^2 (consisting of both wet and dry land) showed high susceptibility to sniper and alpha-cypermethrin. There is a need for more research to be carried out to ascertain its susceptibility on these insecticides in other geographical locations in Lagos State, Nigeria.

Conclusion

From this study, closed chambers (fumigation) are more effective than open (disinfestation) exercise. In a situation where fumigation exercise is to take place, an environment that is well enclosed should be ensured as much as possible so as to have an effective and thorough pest control exercise. Sniper and alphacypermethrin were found to be effective in the control of *B. germanica*. Sniper however, was more effective than alpha cypermethrin. Four hours of no- entry after pest control exercise is recommended for fumigation exercise using sniper. However, an additional hour may be added if alphacypermethrin mixed with water is used. Diesel was found to increase the efficacy and can be a preferred choice over water in fumigation (closed) and disinfestation (opened) exercises. However caution should be exercised in the use of diesel as diluents due to its volatile nature and resultant toxic effect on the environment.

Conflict of Interests

The author(s) have not declared any conflict of interests.

REFERENCES

Agrawal VK, Tilak R, Gupta KK (2005). Efficacy of synthetic pyrethroid and propoxur aerosol in the control of German cockroaches (Dictyoptera: Blatellidae) in cookhouses. J. Vector Borne Dis. 42: 117-121.

Appel AG, Tucker JB (1986). Occurrence of the German cockroach *Blattella germanica* (Dictyoptera: Blattellidae) outdoors in Alabama and Texas. Florida Entomologist 69:422-423

Chavasse DC, Yap HH (1997). Chemical methods for the control of vector and pest of public health importance. WHO Geneva Switzerland 201pp

Cochran DG (1995). Insecticides resistance. In "Understanding and Controlling German Cockroach", (Rust, M.K., Owens, J.M. and Reierson, D.A., eds.). New York, Oxford University Press. 3: 171-192.

Diaz C, Bisset JA, Gonzalez T, Rodriguez MM (1994). Resistance to organophosphate, carbamate and pyrethroid insecticides in *Blatella germanica* (Dictyoptera: Blattellidae) in two municipalities of the city of Havana. Rev. Cubana Med. Trop. 46: 130-132.

Gradcolas P (1996). The phylogeny of coackroaches family, a cladistics apraisal of morpho-anatomical data. Can. J. Zool. 74: 508-527.

Lee CY, Yap HH (2003). Status of urban pest control in Malaysia. In, Lee CY, Yap HH, Chong N.L, and Jaal Z. (eds.), Urban Pest Control, A Malaysian Perspective. Universiti Sains Malaysia. pp.1-8.

Rauh VA, Chew GC, Garfinkel RS (2002). Deteriorated housing contributes to high cockroach allergen level in inner city household. Environ. Health Perspect. 110: 323-327.

Rivault C, Cloarec A, Guyader AL (1993). Bacterial load of Cockroaches in relation to urban environment. Epidermiol. Infect. 110 (2): 317-325

Rust MK, Owens JM, Reierson DA (1995). Understanding and Controlling the German Cockroach. Oxford University Press. 448pp.

Tidwell MA, Williams DC, Gwinn TA, Pena CJ, Tedders SH, Gonzalvez GE, Mekuria Y (1994). Emergency control of *Aedes aegypti* in the Dominican Republic using the Scorpion 20 ULV forced-air generator. J. Am. Mosq. Control Assoc. 10:403-406.

Wei Y, Appel AG, Moar WJ, Liu N (2001). Pyrethroid resistance and cross-resistance in the German cockroach, *Blatella germanica* (L). Pest Manage. Sci. 57 (11): 1055-1059.

WHO (World Health Organization) (1996). Report of the WHO informal consultation on the evaluation and testing of insecticides. WHO Geneva Switzaland. 69pp.

28

The benefits of the use of biogas energy in rural areas in Ethiopia: A case study from the Amhara National Regional State, Fogera District

Zerihun Yohannes AMARE

[1]Economic and Environmental Sustainability Affairs Advisor of FACT, P.O. Box: 3540, Bahir Dar, Ethiopia.

This study was conducted to assess socio-economic and environmental benefits of biogas energy and its challenges. The study used descriptive type of research design consisting of quantitative and qualitative research approach. The data collection instruments were household survey, key informants interview and self-observation to collect quantitative and qualitative data types. The study revealed that the biogas installation made each household to save on average 144 min per day from fuel wood collection, cooking, cleaning utensils/kitchen materials) and the reduction in the physical stress was also remarkable. Use of biogas provide an annual saving of 3833.28 Birr from fuel wood, Birr 1243 .20 from charcoal, Birr 128.50 from dung cake and Birr 266 and 717.65 from kerosene and chemical fertilizer, respectively, with net cash flow of Birr 1530 per HH/year. The households encountered lack of skills for adding raw materials to the biogas digester and fertilizer application to their farm land. Training for operation and liquid bio-slurry application is vital for the continuity of the project and revision of feeding material to the digester in different season needs consideration.

Key words: Biogas energy, environment benefits, biomass, alternative energy.

INTRODUCTION

Energy is undoubtedly a fundamental means for meeting the needs of life support system and development efforts now and for the future. The energy supply and use system has also many implications in the household economy, the indoor environment, women's activities, child safety, family nourishment and other aspects including local and global climate.

Energy plays a central role in national development process as a domestic necessity and major factor of production, whose cost directly affects price of other goods and services (Amigun et al., 2008). It affects all aspects of development, such as social, economic, political and environmental, including access to health, water, agricultural productivity, industrial productivity, education and other vital services that improve the quality of life. There is currently intense interest and strong policy direction to increase the proportion of energy derived from renewable sources

(Thomas et al., 2010).

Worldwide, energy consumption and demand are growing since the past 50 years (OpenStax College, 2012). Most of the resources used like petroleum, natural gas, coal are not sustainable sources of energy. Numbers of countries in the world including Ethiopia are currently passing through the critical phase of population explosion and the growing population demands more energy inputs.

Global experience shows that biogas technology is a simple and readily usable technology that does not require overtly sophisticated capacity to construct and manage. It has also been recognized as a simple, adaptable and locally acceptable technology for Africa (Taleghani and Kia, 2005).

Fuel wood consumption is often portrayed as a cause of environmental degradation, and may lead to energy insecurity for rural African households; especially where the resource is commercialized (Hiemstra-van et al., 2009). The high dependence on wood fuel in the sub-Saharan Africa has resulted in an alarming rate of tree felling and deforestation that causes faster depletion of biomass resources. According to the UNEP (2010), nearly half of the forest loss in Africa is due to removal of wood fuel. On the other hand, rural women are the ones who are directly affected by the rural energy crisis (Amigun et al., 2012; Smith et al., 2005).

The price increase of fossil fuels (an economic process) boosts the demand for bio fuels, which causes changes in land-use cover through deforestation, increases greenhouse gas emissions through the drainage of peat marshes, expands use of agrochemicals and raises the likelihood of establishment of invasive species. A long-term intervention could be to reduce the demand for fossil fuel by changing consumer and producer behavior (UNEP, 2010).

Currently, many African countries experience frequent blackouts and the cost of electricity blackouts is not known. The continent's energy consumption and demand is expected to continue to grow as development progresses at rates faster than those of developed countries (Amigun et al., 2008).

Ethiopia has a population of 89.6 million people, of which 82.4% live in the rural areas (United Nations, 2007). Through the Ethiopia Rural Energy Development and Promotion Centre (EREDPC), the National Biogas Program (NBP) was also launched. The aim of the programme is to establish 14000 biogas plants between 2008 and 2012, in four regions of Ethiopia (EREDPC, 2008). The NBP utilizes cattle manure as the feedstock for biogas production (EREDPC, 2008). In 2009, some households had already started experiencing the benefits of the project such as: use of clean cooking fuel; income savings made in terms of time and money to search for fuel and purchase other traditional fuels (wood, charcoal and kerosene) respectively; and income generation from the sale of biogas to the neighboring towns (Hivos, 2009b).

According to Central Statistical Authority Welfare Survey of Ethiopia (CSA, 2004), the major types of cooking fuels used by all households are firewood, leaves/dung cakes and kerosene. At country level, about 81.4% of the households use firewood, about 11.5% cook their food by using leaves/dung cakes and only 2.4% use kerosene for cooking.

In Ethiopia, the demand for household energy is far greater than the availability. This can lead to vulnerability to deforestation, health impacts and increasing climate change. According to Fogera district Agriculture and Rural Development Annual Report (2010), in 2002-2006, the coverage of the forest in the district was 10,240 ha, while in 2009 the coverage declined to 4,795.26 ha which is a reduction of 53.17% from 2002 - 2006. This is due to over exploitation of fuel wood by the poor households to fulfill their basic needs.

The objectives of the study are to assess the social, economic and environmental benefits of biogas energy technology. This paper was aimed to identify the actual benefits of biogas energy, perception of users in terms of level of satisfaction on the biogas technology, the level of understanding their living environment, challenges encountered during usage/application and dissemination, identifying the indigenous knowledge of rural biogas user households to solve their problems regarding biogas energy application. Finally, the findings are used as input for government and nongovernmental organizations (NGOs) for the dissemination of biogas energy technology for the other target households.

MATERIALS AND METHODS

Description of the study area

Fogera district is one of the 106 districts of Amhara National Regional State and found in South Gondar Zone (Figure 1). The district is located in North West of Addis Ababa with a distant of 625 and 55 km north east of regional capital, Bahir Dar. It has an altitude of 1774-2410 m above sea level and characterized by an average rain fall of 1216.3 mm, minimum and maximum temperature of 16 and 20°C respectively, and black clay soil type. It has a population 228, 449 (52,905 households) that lives in an area of 117405 ha with an estimated population density of 108 people per square kilometer (CSA, 2007). The main resource of the district could be taken as forest area, which is presently encroached due to high population density and urbanization processes; in fact the forest is consumed in various purposes especially for fire wood, furniture and for house construction. It is observed that the forest will be surely destroyed in few years, if proper solutions are not taken in consideration (FDFED, 2009).

Methods of data collection

This study uses both qualitative and quantitative approaches. As per Kothari (1990), for this study the quantitative approach was used for the analysis of household data collected through survey questionnaire; whereas, the qualitative approach is concerned with subjective assessment of attitudes, opinions and behaviors of

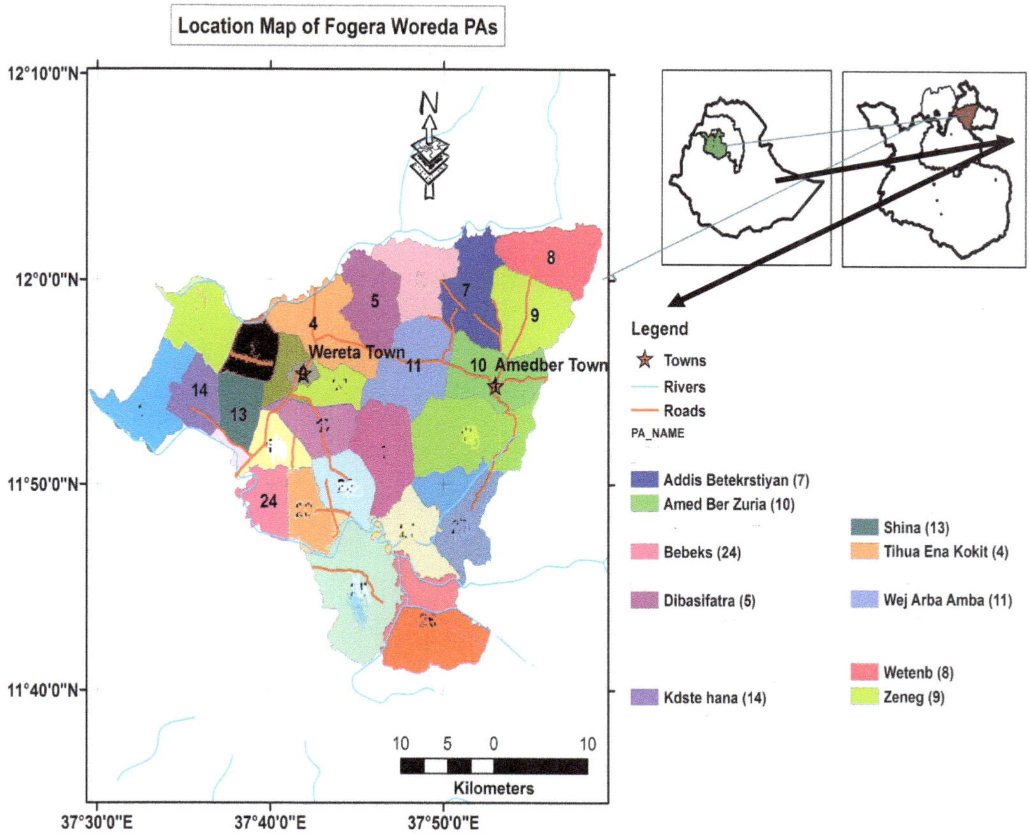

Figure 1. Map of the study area. Source: Fogera district Anti-Malaria Association Office, 2011.

biogas user households. Since the target population was manageable, by considering these parameters, this study used census study design. The data collection instruments were household survey questionnaire.

The biogas energy on which the study was conducted was selected purposefully and all the biogas user households were sampled. In addition to biogas users who work on biogas energy dissemination, the municipality sanitation and beatification process owners in the study area, agricultural and rural development officials were included purposively.

Primary data sources

Household data: Household data were collected through semi structured questionnaire. This survey was intended to gather information on demographic characteristics of households (household size, land size, education level and livestock number), household fuel consumption before and after biogas plant installation (consumption of charcoal, dung cake, fuel wood and kerosene).

Key informants interview: In addition to the household survey data, the interview was carried out on government and non-government partners and stakeholders who work on biogas energy technology dissemination. The key informants were NGO representatives from SNV-Netherlands/Ethiopia Field Office, government officials from the municipality, agriculture and rural development (department of alternative energy technology and animal science process owners).

Self-observation: The actual biogas plant status, toilet availabilities/toilet attached biogas plant/ and usage conditions, market value of household fuel at the local market (charcoal, fuel wood, kerosene and dung cake), state of waste management, utilization of spare time due to reduction of workload as a result of biogas plant were observed and snapped by camera, and voice of the users when they explained about the benefits of biogas were recorded in video to have evidenced data to the findings at their respective peasant associations (PAs).

Secondary data sources

Secondary data sources such as books, policy and published and unpublished documents, journals, and websites that were relevant and strengthened the study were reviewed and studied. Moreover, to have a deeper insight about biogas energy benefits and constraints of Fogera district, different organizations were visited and related documents about biogas energy were extensively used.

RESULTS AND DISCUSIION

Demographic data

The demographic characteristics of the biogas user households are presented in data presenting tools such as Figures and tables. The average household size of the surveyed biogas user households was 6.9 (Figure 2).

Table 1. Possession of animals by biogas user households.

Types of animals	Average holding size /household	Remark
Cattle	12	
Sheep	2	
Goats	5	
Poultry	5	
Horses	1-2	28 Households have no horse
Donkeys	2	8 Households have no donkey

Source: Field survey, 2011.

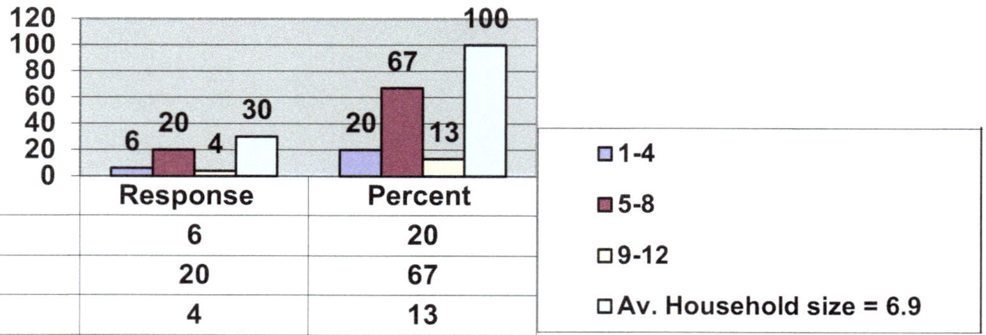

	Response	Percent
1-4	6	20
5-8	20	67
9-12	4	13
Av. Household size = 6.9	30	100

Figure 2. Average household size of the biogas household. Source: Field Survey, 2011.

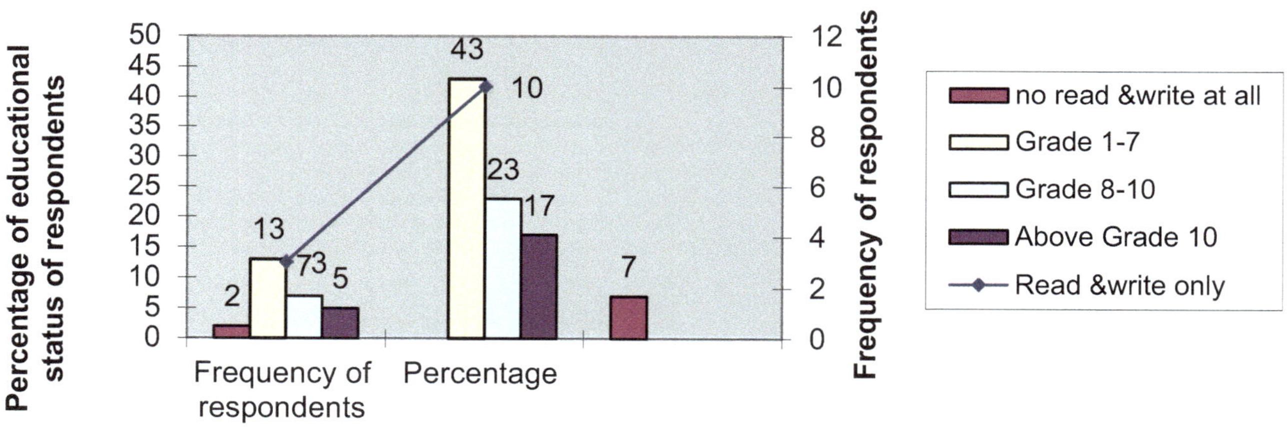

Figure 3. Educational status of biogas households. Source: Field Survey, 2011.

The surveyed biogas user households possess on average 12 cattle per household which is more than the minimum requirement of 4 cows for establishing biogas plant. The details are shown in Table 1.

From the total respondents, majority (83%) are well educated which are from grade 1 to 7, 8 to 10 and from 11 to 12 complete (Figure 3)

Environmental benefits of household biogas investment in the study area

Perception of households on impacts of traditional fuel on forest

All the respondents were asked whether using of Trade-

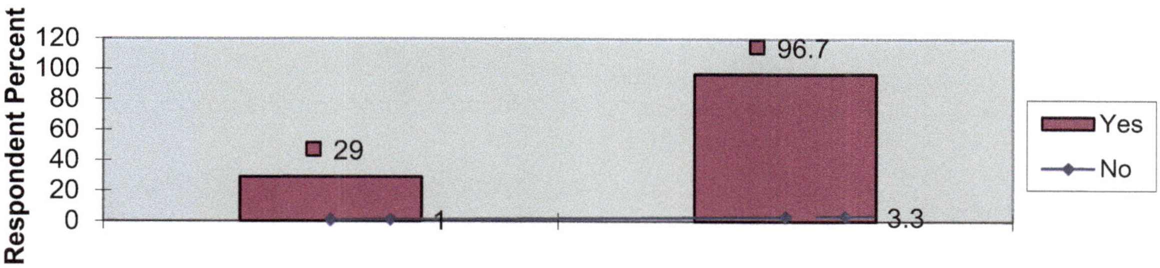

Figure 4. Respondents by knowledge of the impacts of traditional fuel on forest. Source: Field Survey, 2011.

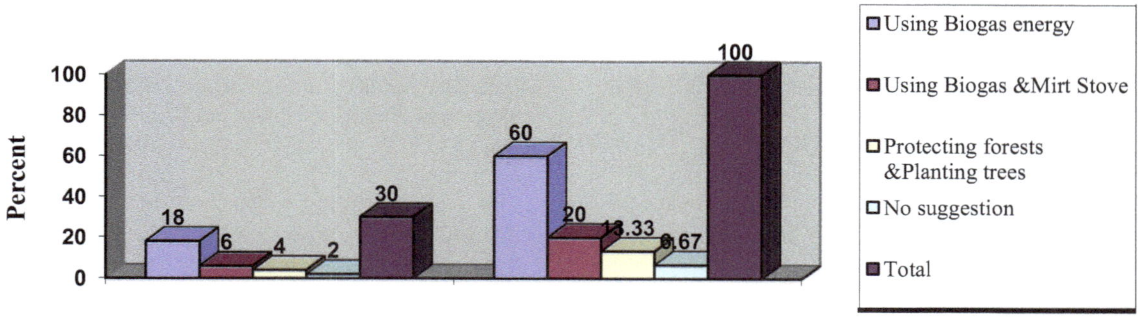

Figure 5. Mitigation measures of respondents' suggestion to minimize the problem of traditional fuel. Source: Field Survey, 2011.

tional fuel have direct impact on forest or not. Accordingly, the majority of the respondents (96.7%) perceived the negative impact of traditional fuel for the existing forest destruction and the rest (3.3%) did not perceive the impacts of traditional fuel for deforestation (Figure 4).

Regarding the solutions for deforestation, (60%) respondents suggested the use of biogas energy (20%) the use of biogas and Mirt stove, (13.3%) protecting forests and planting trees as a solution to control forest depletion and the rest (6.67%) did not show suggestion to minimize loss of forest due to use of traditional fuel (Figure 5).

It can be concluded that majority (96.7%) of the respondents perceived the negative impact of traditional household fuel on deforestation, and this could be attributed to their relatively better educational background. Even, all the solutions suggested by respondents are true, the researcher recommended the households to use biogas energy due to its clean and smokeless technology and ability to replace and reduce the use of firewood, charcoal, dung cake and other imported household energy sources such as kerosene.

Contribution of biogas as cleaner production mechanism

Biogas potential in the study area of Fogera district is in favorable condition in respect of the climatic and availability of raw material for biogas production. The potentials are: Municipal waste, livestock and human population. Thus, potentials needs to be recycled as cleaner production such as biogas energy, to get dual benefits from getting energy and making the environment clean.

Livestock population in Fogera district: Documents gained from Fogera district Agriculture and rural development office, out of 302,800 livestock; 182,699 are cattle, 15,575 sheep, 25,956 goats, 64,227 poultry, 571 horses and 13,772 donkeys found in the district. Moreover, the annual dung production is about 666, 851,350 - 1,000,277,025 kg, which has a potential production from 24,006,648.6 to 36,009,972.9 m^3 of biogas annually. However, it is estimated that only 90% of the theoretical potential, that is, 21,605,983.74 to 32,408,975.6 m^3 (Av.27, 0074, 79.67 m^3) of biogas would be practically available since the number of animals also include the households with only one cattle or goat. This has a potential for saving fuel wood from 118,832,910.6 to 178,249,365.8 kg, charcoal from 34569573.98 to 51854360.96 kg, kerosene from 16226, 093.79 L to 24339,140.68 L and electricity from 34569,573.98 to 51854360.96 Kwh annually (Table 3).

Human population in Fogera district: In Fogera district, 228,449 people reside there (CSA, 2007), which have a potential to produce 228449*0.3*365 = 2,501,5165.5 kg of human waste annually (Nijaguna, 2002). Assuming that all people have pit latrines and if they properly utilized their excreta, this would have a potential of producing 25,015165.5 kg x 0.046 m^3 = 1,150,697.613 m^3 of biogas, which saves 6,328,836.872 kg of firewood,1,841,116.181 kg of charcoal, 864,173.9 L of kerosene and 1,726,046.42 to 1956,185.942 Kwh annually (Filed survey, 2011; Nijaguna, 2002).

Municipal waste of the town administration: Fogera district town administration generates approximately 34,500 kg wastes (34.5 tons) of solid waste and 40,000 L (40tons) liquid waste was generated per day. Among these, the municipality collects and disposed only on average 32,000 kg of solid waste and 20,365 L of liquid waste per day, which is 92.5 and 50.91% of the total solid and liquid waste, respectively. The main sources of waste are from residential and commercial activities in the town. These wastes are collected and disposed in open space except small amount of liquid waste used for urban agriculture as fertilizer. Due to this small amount of utilization, the disposed waste creates bad smell to town and its surroundings that will create health problems.

COMEC (2006) stated that, in Brazil the biggest part of municipal waste generation is deposited without any methodology/without technological aid like Fogera district municipality, but Brazil uses high amount of waste for biogas energy production as energy source and waste treatment mechanism. This is also contended by UNESCO (1992), biogas technology have attracted considerable attention in waste recycling, pollution control and improvement of sanitary condition in addition to fuel and fertilizer.

On the contrary, the municipality of Fogera district have no future plan to use the potential waste as energy source officially except personal motivation and promise of experts after interview. As can be seen from Table 4, the town administration was collected and disposed 52.365 tons of waste per day and 19,113.225 tons of waste annually.

Assuming that all wastes are properly utilized, this has a potential of 19,113,225 kg × 0.03 m^3 = 573,396.75 m^3 of biogas, which saves 105,122,737.5 kg of firewood, 30,581,160 kg of charcoal,14,354,031.98 L of kerosene and 28,669,837.5 to 32,492,482.5 Kwh electricity annually. Besides, all this potentiality presented above such as wastes from livestock population, human population and municipal are dangerous unless it is recycled as cleaner production such as biogas to have dual purpose, killing two birds with one stone principle like Brazil (as source of energy and as environmental sanitation).

*Pair of oxen ploughs, 1 hectare per day: this is called "Temad", which is the name of local measurement.

Social benefits of household biogas investment in the study area

Benefits from replacement of traditional household fuel

As the household traditional fuel consumption decrease, the contact hour of the family member to the traditional stove also decreased. As a result, the type of adverse effects of biomass combustion on human health stated by WHO (1991) could be decreased.

Household fuel wood consumption: The study showed that in the surveyed area, the biogas user households used Cordia Africana (Wanza) and acacia for cooking among others, *Injera*[1]; Sauce (Wot), porridge and coffee/tea, before and after installation of biogas plant. However, after biogas plant installation, the traditional household fuel consumption was limited to baking *Injera*. As can be seen in Table 7, before installation of biogas plant, households used 3,596.4 kg of fuel wood/HH annually, but after installation of biogas plant each household uses on average 1062 kg of fuel wood/HH/year which is reduction of 2,534.4 kg (70.47%)/HH/ annually.

Household charcoal consumption: The average charcoal consumption in the surveyed households used 1 sack of quintal (27 kg) per month and 12 sack of quintal (324 kg) per year. As can be seen from Table 8, in the study area of biogas households, all of them used charcoal in different amount before installation of biogas plant. However, after installation of biogas plant, all the biogas households have been fully replaced with biogas energy.

Household dung cake consumption: All the surveyed households, before installation of biogas, the majority (43.3%), used 6-10 sacks of quintal (138-230 kg), 36.67%, used 1-5 sacks of quintal dung cake (23-115 kg), 13.3% used 11-15 sacks of quintal (253-345 kg), 6.67% used above 15 sacks of quintal (460 kg) per month with on an average consumption of 8 sack of quintal(184 kg)/HH per month and 96 sacks of quintal (2208 kg) used annually. After installation of biogas plant, as presented in Table 9, majority (60%) used 0.5 to 2 sacks of quintal (11.5-46 kg), 33% used 3-5 sacks of quintal(69-115kg), and the rest 7% used 6-8 sacks of quintal dung cake(138-184 kg) and on average of 3 sacks of quintal (63kg)/HH per month which is reduction of 5 sacks (115 kg)/HH per month, and 60 sacks (1380 kg) dung cake annually which is 62.5% from the total. This traditional household fuel consumption could contribute in

[1] Injera is Ethiopian Traditional food made from Teff (crop).

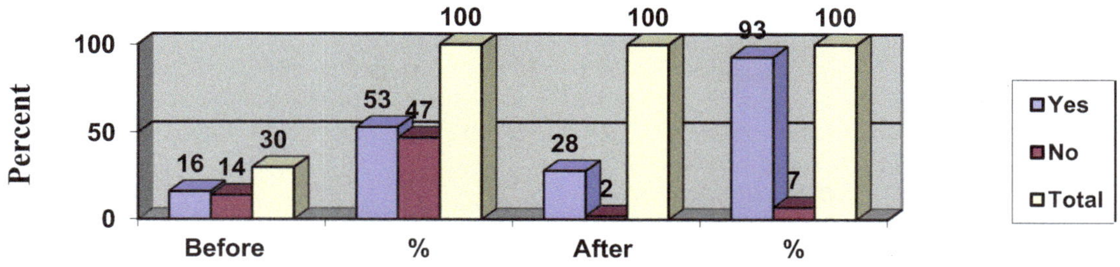

Figure 6. Possession of toilet before and after biogas installation. Source: Field Survey, 2011.

reducing the health impacts of households that appeared due to indoor smoke. Bajgain and Shakya (2005) stated that, the burning of fire wood, dung cake and agricultural residue creates many hazardous particles. Cooking is usually done indoors; this can lead to severe health problems. The particles from the smoke can give rise to acute respiratory infections among the people who are in contact with the smoke.

Benefits of biogas on health and sanitation of biogas user households

The study showed that in the surveyed area, after installation of biogas plant, the construction of toilet is increased which can be compared to IEIA (2002), study that was carried out by SNV/BSP of Nepal, showing that the record of toilet construction is higher among biogas users. Before installation of biogas plant, toilet possession of the surveyed HHs were 53%, while after installation of biogas, the possession of toilet reached 93% (Figure 6). From these toilet owners, 13% of them have attached their toilet to their biogas plant and the majority (80%) of respondents ready to attached their toilet to the biogas and the system is already installed which shows promising to use the integrated biogas plant for the near future. Only few (6.65%) respondents have no toilet at all. As data gained from the household survey, these two respondents cannot read and write at all. It seems that education has its own impact on dissemination of technology.

The biogas user households were asked to express their opinion about the difference in using traditional fuel and biogas energy use in relation to health impact. They replied by remembering what the local nurses and doctors recommend about negative impacts of traditional fuel on health and by adding their indigenous knowledge gained from experience. A comparison of the status of household in terms of frequently complained health problems, from the total of respondents, 21 (70%) were complain in eye illness, respiratory disease (such as common cold) and headache which accounts for 56.6, 6.6 and 6.6% respectively (Table 5).

Bajgain and Shakya (2005) stated that, the particles from the smoke can give rise to acute respiratory infections and eye ailments among the people who are in contact with smoke. These peoples are mainly women, children and infants, while in this study, the major victims were women, children and old people due to access to kitchen activities frequently and stay at home during food preparation. It has the same connotation with the study of Bajgain and Shakya (2005).

After biogas plant installation, the households reduce the use of traditional fuel such as firewood and dung cake, and limited these fuel only for *Injera* baking purpose which replaces the use of traditional fuel for cooking sauce (Wot), porridge, tea/coffee and at the same time the households replace kerosene and charcoal by lighting biogas lamp, and by using biogas stove for cooking sauce (Wot), porridge, tea/coffee etc. Due to this reduction of traditional fuel, after installation of biogas only 8 (26.6%) respondents complained about eye illness and head ache. The major victims were females due to baking of *Injera* by using firewood and dung cake but the frequency is less when compared with previous (before use of biogas energy). The direct effects of biogas plant on health and sanitation were found to be more visible than indirect ones; since, the study revealed that smokeless biogas had greatly benefited the plant owners by contributing to a significant reduction in eye related problems and respiratory diseases.

Biogas technology reduces contact hours to open stove fire, that is, after biogas installation the contact time is on average twice a week only for baking *Injera* and the rest activities are accomplished by biogas energy which is clean and smokeless.

Time saving and workload reduction

In the study area of the surveyed households, biogas installation makes each biogas household to save on average 144 min per day from fuel wood collection, cooking, cleaning utensils/kitchen materials, and the reduction in the physical stress was also remarkable. Women and children are the groups that collect fire wood traditionally. Table 6 depicts the time comparison that has been estimated to save up to 51 min on average every day due to installation of biogas energy. Gautam et al. (2005) stated that today's use of biomass in list developed countries does not provide any use of sufficient lighting during the dark hours. The time consumed in collecting fire wood and other bio fuel carried out by women and children results in less time available for education. This means, as the study revealed that, children have time to go to school and they could use the biogas light for studying overnight unlike the previous time that they were used to kerosene for lighting purpose only from 2 to 3 h, and it has dangerous smoke that could affect respiratory organ and cannot be afforded by the poor. On the other hand, biogas gains equity of work among family members in accomplishing works such as slurry mixing which was accomplished by household members. Biogas households use their spare time/saved time in better care of family, in keeping household sanitation, education and other social affair. Due to absence of managing firewood and charcoal during cooking of sauce, porridge and tea/coffee, women could accomplish additional works at the same time.

Economic benefits of household biogas energy investment

Saving from fire wood purchasing expenditure

Due to the installation of biogas plant, there is an annual reduction of fuel wood consumption approximately 79 bundle of fire wood (2528 kg) per year per household and provides each biogas households an equivalent saving of 3833.28 Birr per year at the local rate of Birr 48.40 per 32 kg of fire wood (Table 7).

Saving from charcoal purchasing expenditure

In the surveyed area, after household biogas investment, purchasing of 12 sacks of quintal (324 kg) of charcoal is fully replaced by biogas stove. This amount of charcoal provides each biogas household an equivalent saving of 1,243.20 Birr per year at the local rate of Birr 103.60 per 27 kg of charcoal (Table 8).

Saving from dung cake purchasing expenditure

Due to household biogas investment, 60 sacks of quintal (1,380 kg) of dung cake are saved. This has an equivalent saving of Birr 1,542 annually at local rate of Birr 25.70/sack of quintal dung cake.

Saving from kerosene purchasing expenditure

Besides the above household fuel, all the surveyed households used kerosene for lighting on average of 1.78 L per month per household and 21.36 L per year. Each household spent Birr 22.1789 and Birr 266.1456 per month and per year, respectively. After installation of biogas plant, all the surveyed HHs substituted their kerosene consumption fully by biogas lamp. This shows that Birr 266.1456 was saved annually at the local rate of Birr 12.46/L of kerosene per household (Table 10).

Savings from chemical fertilizer purchasing expenditure

Further, reducing chemical fertilizer has an effect on households' expense, from the total surveyed households, in the demographic characteristics of respondents, 27 of them have an average 1.453 ha/HH agricultural land and the rest have no agricultural land rather they have backyard farm used for managing the biogas slurry and cropping of cash crops such as khat (Chat). Before installation of biogas plant, the surveyed households used an average of 153.26 kg of chemical fertilizer per household annually/crop season/ in an equal proportion of dap and urea in their agricultural land to grow well the crop and increase the production but with an annual expenditure of Birr 1065.16/HH. After installation of biogas, due to use of bio-slurry fertilizer, the average household chemical fertilizer consumption in the crop season was reduced to 50 kg/HH which is saving of 103.26 kg of chemical fertilizer per household per crop season. This has an equivalent saving of Birr 717.657/HH/ crop season with an average local rate of Birr 695/100 kg of chemical fertilizer (Table 2).

Besides the above listed economic benefits of biogas, improving the hygiene and thereby reducing diseases also has an economic value. If people can avoid diseases it also means their working time will not be reduced as a result. The study also revealed that, household family illness due to use of traditional fuel also have an implication on health expense for treatment (Table 11).

In Fogera district, the cost of investment per plant varies due to personal contribution made by the respondents during construction work in the form of labor, variation in the year of construction, size of plants and access to the delivery of construction materials such as stone and sand/gravel. The total investment cost of the plant was an average of Birr 8,762.48 and Birr 9,813.46 for 6 and 8 m^3 respectively and Birr 8,832.55 for an average plant size of 6.14 m^3 for ease to calculate the approximate payback period of the plant. For a plant with total investment cost of Birr 8,832.55, the payback period was 5.77 years to recover investment of biogas plant per household without subsidy, whereas with subsidy of Birr 4,199.25, it could take only 2.7 years, which seem affordable as compared to study conducted by Li et al. (2005) and Woods et al. (2006). The calculation was based on saving from fuel wood, charcoal, dung cake, kerosene and chemical fertilizer (Table 12).

The shorter payback period makes biogas plant affordable for most peri-urban and rural households, even in poor areas (Li et al., 2005). The payback period for

Table 2. Land holdings of the surveyed households.

Land holding type	Average land holding /household in *Temad and ha	Number of respondents who have	Percentage of land holder respondent
Agriculture land	5.8 Temad = 1.453 ha	27	90
Grazing land	1.56 Temad = 0.39ha	16	53
	Common grazing land	14	47
Forest land	1.2 Temad = 0.3ha	5	17

Total land holding: 187.86 Temad = 46.971ha; Av. landholding/HH: 3.91 Temad = 0.978ha. Source: Field Survey, 2011.

Table 3. Total number of livestock and biogas produced per kg of animal dung.

Type of animals	Total number of animals	Daily produced dung/animal in kg	Total dung available per day in /kg/	Gas produced per day/ m^3/
Cattle	182,699	10-15	1,826,990-2,740,485	65,771.6-98,657.5
Sheep	15,575	0.75-1	11,681.25-15,575	420.525-560.7
Goats	25,956	0.75-1	19,467-25,956	408.807-545.076
Poultry	64,227	0.06-0.2	3853.62-12845.4	1,965.35-6,551.15
Horses	571	14-16	7,994-9,136	7,274.54-8313.76
Donkeys	13,772	12-15	165,246-206,580	134,675.49-168,362.7
Total	302,800		2035231.87-3010577.4	210,516.312-282,990.886

Source: Fogera District Agricultural and Rural Development Office, 2011 and Nijaguna, 2002.

Table 4. The annual collected and disposed waste in the town administration.

Type of waste	Unit	Daily generated	Daily collected and disposed waste	Remark
Solid	kg	34,500	32,000	
Liquid	Liter	40,000	20365	
Total		74,500	52,365	

Source: Fogera district municipality Office, 2011. NB: 1 kg of solid waste equal to 1 L of liquid waste (Nijaguna, 2002

Table 5. Analysis of health benefits before and after installation of biogas.

Health problems	Before			After		
	Number of respondent	Percent	Major victim	Number of respondent	Percent	Major victim
Eye illness	17	56.7	Females, child and old	6	20	Females
Respiratory disease	2	6.6	Females and child	-	-	-
Headache	2	6.6	Females and old	2	6.6	Females
No complain	9	30	-	22	73	-
Total	30	100		8	100	

Source: Field Survey, 2011

Table 6. Analysis of average time for daily works before and after biogas installation.

Daily works	Average time in minutes per day		Av. time saved per day in min/HH	Implementers
	Before	After		
Fuel wood collection	76	25	+51	Women and Children
Cooking	240	164	+77	Women and Children
Fetching water	30	56	-26	Women and Children
Cleaning utensils	54	35	+19	Women and Children
Livestock caring	35	35	-	Men and children
Dung cake collection and moulding	57	19	+38	Women and Children
Slurry mixing	-	15	-15	All Family members
Total	**492**	**349**	**+144**	

Source: Field Survey, 2011. +shows saved time due to household biogas investment.

Table 7. The amount of fuel wood consumption before and after installation of biogas plant.

Before installation of biogas plant			After installation of biogas plant		
Amount of fuel wood /HH/month in bundle/	Number of respondent	percent	Amount of fuel wood /HH/month in bundle/	Number of respondent	Percent
4-7	10	33	1-2	16	53
8-11	12	40	3-4	12	40
12-20	8	27	5-6	2	7
Total	30	100		30	100
Av. Bundle/HH/Month = 9.366(299.7kg)			Av. Bundle/HH/Month = 2.766(88.5kg) Difference = 6.6bundle (211.2kg)		

1 Bundle of fuel wood weighs on average 32 kg; price of 1 bundle of wood = on average Birr 48.40 (local market, February 14, 2011); 1 US$ = Birr 16.66 (National Bank of Ethiopia) (NBE, March 10, 2011) Source: Field Survey, 2011.

Table 8: Amount of charcoal consumption per household per month before Installation of Biogas Plant

Amount of charcoal in sack of quintal/HH	Number of Respondent	Percent	Remark
0.5	13	43	1 Sack of quintal Charcoal weighs on average 27 kg
1	7	6.67	
1.5	10	34	
Total	30	100	

Average used /HH/month/ in sack of quintal = 1; Price of 1sack of quintal of charcoal at local market = Birr 103.60 Average used/HH/year in sack of quintal = 12, that is, 324 kg; 1 US$ = Birr 16.66 (National Bank of Ethiopia) (NBE, March 10, 2011) Source: Field Survey, 2011.

Table 9. Amount of dung cake consumption per household per month before and after installation of biogas plant.

Before installation of biogas plant			After installation of biogas plant		
Amount of dung cake /HH/month in kg	Number of respondent	Percent	Amount of dung cake /HH/month in kg	Number of respondent	Percent
1-5	11	36.67	.5 -2	18	60
6-10	13	43.3	3-5	10	33
11-15	4	13.3	6-8	2	7
≥ 16	2	6.67	-	-	-
Total	30	100	Total	30	100

Average sack of quintal dung cake /HH/Month = 8; Average sack of quintal dung cake /HH/Month = 3 (5 sack of quintal saved). 1 quintal sack of dung cake weighs on Average 23 kg. Price of 1 quintal sack of dung cake = on Average Birr 25.70 (Local market, February 14,2011). 1 US$ = Birr 16.66 (National Bank of Ethiopia) (NBE, March 10, 2011). Source: Field Survey, 2011.

Table 10. Amount of kerosene consumption per HH before installation of biogas plant.

Amount of kerosene in liter/HH/ month		Number of respondent	Saving of kerosene after installation of the plant/ in liter
	1 to 2	26	
	>2	4	
Average household consumption in liter			
Per month	1.78	30	1.78
Per year	21.4	30	21.4

Price of 1 liter of kerosene = on average Birr 12.46 (Trade transport Office of town Adm., February 1, 2011); 1 US$ = Birr 16.66 (National Bank of Ethiopia) (NBE, March 10, 2011); Source: Field Survey, 2011.

Table 11. Fertilizer consumption before and after biogas Installation.

Particulars	Fertilizer used	Number of respondent	Percent
Before biogas installation	Farmyard Manure and Chemical fertilizer	26	100
Total		26	100
After biogas installation	Farmyard Manure and Bio-slurry	23	88.46
	Farmyard Manure, Chemical fertilizer, Bio-slurry	3	11.54
Total		26	100

Source: Field Survey, 2011.

Table 12. Saving of chemical fertilizer due to household biogas investment.

Amount of Average chemical fertilizer used before installation of biogas/HH/crop season in kg	Amount of chemical fertilizer before installation of biogas/HH/crop season in kg	Difference/saved due to installation of biogas plant	Percent
153.26	50	103.26	67.37

100 kg of chemical fertilizer on Average = Birr 695 (Fogera District Agricultural and Rural Development (2011), 1US$ = Birr 16.66 (National Bank of Ethiopia) (NBE, March 10, 2011). Source: Field Survey, 2011.

Table 13. Households' level of satisfaction.

Particulars	Number of respondent	Percent
Fully Satisfied	26	87
Moderately Satisfied	4	13
Not Satisfied	-	-
Total	30	100

Source: *Field Survey, 2011.*

Chinese type fixed dome biogas digester depends on how the biogas digester is used, what substrates, size, price on fuel wood, etc and without any subsidies would be around 3.6 to 5.8 years of payback period (Woods et al., 2006). The regional biogas coordination office and SNV/Ethiopia, estimated the cost of annual maintenance and miscellaneous expenses to be Birr 200 and 100 per plant, respectively. These costs are reserved in bank in the name of the plant owners dedicated from the subsidy.

Perception of users on benefits of household biogas energy

All the surveyed biogas plants were operational except temporary problems of lamp and other accessories. Due to this, the responses were quite satisfactory (Table 13, plate 1 and 2, box 1). There was significant satisfaction in terms of reduction of household fuel consumption, cease to expend for chemical fertilizer and HH traditional fuel, improvement in health and sanitation, time saving and studying for students among others.

Plate 1. User while lights biogas stove

Plate 2. User while cooking with biogas stove.
Source: *Field Survey, 2011*

Mrs. Debre Tsega one of biogas user, lights her stove, emitting a full blue clean fire. *"Did you see how fast it sets fire?" This is one of the main benefits". It used to take a long time to get an appropriate fire with charcoal, fuel wood and dung cake. "It makes life a lot easy".*

With full Charm, Waga Alem leads us to her kitchen, which is equipped with a stove and a lamp that both run on biogas plant. With a big smile she speaks about the changes this new technology brought to her daily life. *"Look at my kitchen it has never been so clean and see the time to cook Shero (local Sauce) it's a matter of less than 5minutes. Before I got biogas, I used to cook with charcoal, wood, dung cake and agricultural residue which produced a lot of smoke and takes time to give fire and it brought a bad smell. This now belongs to the past and its history".*

Box 1. Opinion of the users.

Conclusions

The use of biogas has a potential to reduce the demand for wood and charcoal use, hence reducing greenhouse gas emissions. In addition, the slurry and waste from the biogas plants provides a high quality fertiliser that can be used to improve the soil fertility and increase productivity in agriculture dependent rural communities of the study area.

In the study area, the biogas user households benefit from reduced indoor smoke, improved sanitation and better lighting. The biogas installation make each household to save on average 144 min per day from fuel wood collection, cooking, cleaning utensils/kitchen materials),and there is also reduction in the physical stress and health improvement.

The economy of a biogas plant is characterized by initial high investments costs, some operation and maintenance costs, mostly free raw materials (animal dung, weeds, plants, sewage sludge, human wastes, municipal wastes, etc.) and income from replacement of purchasing traditional fuels. In addition, the slurry and waste from the biogas plants provides a high quality fertilizer that can be used to improve the soil fertility and increase productivity in agriculture dependent rural communities in the study area. Use of biogas provide an annual saving of 3833.28 Birr from fuel wood, Birr 1243.20 from charcoal, Birr 128.50 from dung cake and Birr 266 and Birr 717.65 from kerosene and chemical fertilizer, respectively with Net cash flow of Birr 1530 per HH/year.

There are some challenges that must be tackled in order to ensure sustainable future of biogas technology. These challenges were seen from two angles: from the users' side and on the disseminators' side.

Attitudinal problems

According to the field survey, due to unavailability of demonstration areas for biogas plant in the nearby peasant associations (PAs), from the total surveyed users of biogas, 4 of them did not believe on gaining of light and cooking fuel from dung and human excreta until they saw it on the due date.

Lack of awareness on lamp care

Due to awareness problems on the technology, and lack of responsible technology disseminator technician, from the total respondent, in the majority (40%) their biogas lamp was broken due to improper lamp care (4) and feeding amount (4).

Standard problems in feeding amount of the raw material to biogas digester

The users complained about fixed standards of the raw material amount of the digester that were recommended by the disseminator to feed the biogas digester. Since the cellulose content of the grass that the cow eats is not equal throughout the year, the chemical content of the grass makes the biogas lamp to break. From 40% stated above, (4) respondents complained the constant standard of feeding amount recommended by technicians of biogas disseminators, Surprisingly, 4 respondents stated that, "the power/cellulose/content of dung that we add per day throughout the year is not equal, due to this, the power (in their expression) of the grass that the cows eat in the summer season makes the lamp to break". Currently, these households use their own indigenous knowledge and make correction about the feeding amount without any externality advice on the side of the responsible bodies.

Bio-slurry management problems

Due to awareness problems, 2 respondents have not reserved bio slurry storage tank. As a result, they faced problems in using efficiently the bio slurry for their farm.

Missed application of liquid bio-slurry in their back yard farm

Unknowingly, 4 respondents' cash crop such as Khat (Chat) had got dry due to missed application of liquid bio-slurry. After the problem appeared, one user uses his indigenous knowledge by understanding the cause through experience without external technical assistance about liquid bio-slurry application. As stated by this respondent, "the chemical content of the liquid bio-slurry is dangerous unless it mix with water during application".

Fear of the future due to unavailability of accessories /appliances

The biogas accessories are lamp, stove, connecters, gas pipes, pressure meter. Not less than 3 users have no slurry mixture and this doubles the time to add dung to the digester. Such problems make the user to fear the future about getting these appliances due to absence of the appliances in the local market as they need. Currently, 12 respondents have no lamp either one of the two (8 m^3 owners) or from one of the one (6 m^3 owners) but some of the user uses local materials to gain light from biogas and 1 user encounters problems of stove, 3 users problems of connectors and 1 user problems of pressure meter. Even lamps are guarantee of up to two years by the disseminator to be replaced when broken, for sustainability of the technology; users fear the cost of lamp which is Birr 85 per lamp.

The challenges encountered by biogas disseminator organizations, non-governmental organizations (SNV/Ethiopia) and the local government organizations raised their problems for not disseminated as expected.

Variation of market price and cost of installation

The criteria for disseminating the biogas plant to the community in the study area was; purchasing power and willingness of the user, having at least 4 cows, accessibility of water and construction material, area for slurry management. However, according to regional biogas coordination office and SNV/Ethiopia regional branch office, "due to construction material cost variation (such as cement) by not less than 50% in 2010/11, cost are born by the user and absence of loan to the poor inhibits the dissemination of household biogas investment as planned and or expected".

Lack of users' promotion

Even the promotion and awareness of users were the responsibility of the local government bodies such as district agricultural and rural development, higher government bodies and energy officials, the higher government bodies are not aware of biogas energy benefits and there is no sufficient experts at district level, and in the PAs there is no expert at all that follows biogas energy affairs.

Recommendations

Mainstreaming and promotion of biogas technology in different development activities

Government commitments to the development and promotion of renewable energy sources are advisable. It could be helpful to learn from the experiences gained in the developed world but should adapt to the needs and situation in the study area and in Ethiopia in general.

Develop training programs

Develop training programs for engineers, artesian, users, and all professionals involved in biogas dissemination.

Good understanding of the relation between capital costs and plant size

It can provide useful information in assessing economic viability of biogas plants, and providing means whereby decisions are taken on developmental of a new project.

REFERENCES

Amigun B, Sigamoney R, von Blottnitz H (2008). Commercialization of biofuel industry in Africa: A review. Renewable and Sustainable Energy Reviews, 12:690-711. http://dx.doi.org/10.1016/j.rser.2006.10.019

Amigun W, Parawira JK, Musango AO, Aboyade, Badmos AS (2012).Anaerobic Biogas Generation for Rural Area Energy Provision in Africa, Biogas, ISBN: 978-953-51-0204-5, In Tech. China.

Bajgain S, Shakyal (2005).The Nepal Biogas Support: A successful model of public private partnership for rural household energy supply, Ministry of Foreign Affairs. The Netherlands, Netherland.

Central Statistical Agency (2007). Summary and Statistical Report of population and Housing Census, Addis Ababa, Ethiopia.

Central Statistical Authority (2004). Welfare Moni

Toring Survey: Indicator on Living Standards, Accessibility, Household assets, Food security and HIV/AIDS, Addis Ababa, Ethiopia.

[COMEC] Conference of Mechanical Engineering Company (2006).Landfill biogas and its use for energy generation, Lasvillas, Cuba. Available at: <http/.www.globalmethane.org/.../events_land_20090127_subcom_brazil.pdf> [accessed 12 April 2011].

EREDPC (2008). National Biogas Programme Ethiopia. Programme Implementation Document. http://www.snvworld.org/en/Documents/NBP_implementation_document_Ethiopia_2008.pdf [accessed23Feb2015].

Fogera district Agriculture and Rural Development Office (2010). Annual report, Woreta town, Ethiopia.

Fogera district Finance and Economic Development Office (2009). Annual report, Woreta, Ethiopia.

Fogera district Anti- Malaria Association Office (2011), Working site document, Woreta, Ethiopia.

GautamR, Barallb S, and Heart S (2005). Biogas as Sustainable energy source in Nepal: Present status and future challenges; Renewable and sustainable Energy Reviews. Lalipur, Nepal.

Hiemstra-van derHorst G, HovorkaAJ (2009). Fuel wood: The "other" renewable energy source for Africa? Biomass and Bioenergy, 33, 1605-1616. http://dx.doi.org/10.1016/j.biombioe.2009.08.007

Hivos (2009b). Ethiopian households now enjoy the benefit of domestic biogas. http://www.hivos.nl/dut/Users/Members/jmsika/Jean-Marc-Sika/Ethiopian- Households-now-enjoy-the-benefit-of-domestic-biogas [accessed23Feb2015].

Kothari CR (1990). Research methodology: methods and techniques, 2nd edition, V.S. Johri, New Delhi, India.

Nijaguna BT (2002). Biogas technology, new age international (P) limited, New Delhi. Banglore, Mumbai.

UNEP (2010). Ecosystem Management programme. A new approach to sustainability, Nairobi, Kenya.

United Nations (2010). Number of city dwellers grows amidst continuing disparities in level and pace of urbanization worldwide. available at: http://esa.un.org/unpd/wup/Documents/WUP2009_Press-Release_Final_Rev1.pdf [accessed 23 Feb2015].

United Nations Economic and Social Council 2007. Africa review report on drought and desertification.http://www.un.org/esa/sustdev/csd/csd16/rim/eca_bg3.pdf [accessed23Feb2015]

United Nations Education, Science and Cultural Organization (1992). International Directory of New and Renewable sources of energy. Paris, France. available at: <http://unescodoc.unesco.org/ulis/.>[accessed12 December2010]. National Bank of Ethiopia (2011).Daily Market information, Addis Ababa, Eth

iopia. Netherlands Biogas Support Programme (2002). An Integrated Environmental Impact Assessment. Biogas Support Programme, Lalitpur, Nepal. OpenStax College, World Energy Use, College Physics (2012) Available online: https://www.inkling.com/read/college-physics-openstax-college-1st/chapter-7/7-9-world-energy-use [accessed 26 February 2014]. Li Z, Tang R, Xia C, Luo H, Zhong H(2005).Towards green rural energy in Yunnan, China. Renewable Energy 30: 99 - 108 pp. Yunnan. Smith KR, Rogers J, Cowlin SC (2005). Household fuels and ill-health in developing countries: what improvements can be brought by LP gas (LPG)? Paris, France, World LP Gas Association and Intermediate Technology Development Group.

Taleghani G, Kia AS (2005).Technical–economical analysis of the Saveh biogas power plant. Renewable Energy. 30:441–6. http://dx.doi.org/10.1016/j.renene.2004.06.004

Thomas Elmqvist, Edward Maltby, Tom Barker, Martin Mortimer, Charles Perrings James Aronson, Rudolf De Groot, Alastair Fitter, Georgina Mace, Jon Norberg, Isabel Sousa Pinto, Irene Ring(2010). Biodiversity, ecosystems and ecosystem services, China.

Woods J, Hemstock S, Burnyeat W(2006).Bio- energy systems at the community level in the South Pacific: Impacts & Monitoring; Mitigation and Adaptation Strategies for Global Change11: 469- 500 pp.

WHO (1991). Indoor air pollution from biomass fuel. Geneva: World Health Organization. available at:http://www.google.com.et/search?hl=en&q=WHO+1991+%3Aindoor+air+pollution+from +biomass+fuel+in+ethiopia> [accessed 10 december2010].

Permissions

All chapters in this book were first published in AJEST, by Academic Journals; hereby published with permission under the Creative Commons Attribution License or equivalent. Every chapter published in this book has been scrutinized by our experts. Their significance has been extensively debated. The topics covered herein carry significant findings which will fuel the growth of the discipline. They may even be implemented as practical applications or may be referred to as a beginning point for another development.

The contributors of this book come from diverse backgrounds, making this book a truly international effort. This book will bring forth new frontiers with its revolutionizing research information and detailed analysis of the nascent developments around the world.

We would like to thank all the contributing authors for lending their expertise to make the book truly unique. They have played a crucial role in the development of this book. Without their invaluable contributions this book wouldn't have been possible. They have made vital efforts to compile up to date information on the varied aspects of this subject to make this book a valuable addition to the collection of many professionals and students.

This book was conceptualized with the vision of imparting up-to-date information and advanced data in this field. To ensure the same, a matchless editorial board was set up. Every individual on the board went through rigorous rounds of assessment to prove their worth. After which they invested a large part of their time researching and compiling the most relevant data for our readers.

The editorial board has been involved in producing this book since its inception. They have spent rigorous hours researching and exploring the diverse topics which have resulted in the successful publishing of this book. They have passed on their knowledge of decades through this book. To expedite this challenging task, the publisher supported the team at every step. A small team of assistant editors was also appointed to further simplify the editing procedure and attain best results for the readers.

Apart from the editorial board, the designing team has also invested a significant amount of their time in understanding the subject and creating the most relevant covers. They scrutinized every image to scout for the most suitable representation of the subject and create an appropriate cover for the book.

The publishing team has been an ardent support to the editorial, designing and production team. Their endless efforts to recruit the best for this project, has resulted in the accomplishment of this book. They are a veteran in the field of academics and their pool of knowledge is as vast as their experience in printing. Their expertise and guidance has proved useful at every step. Their uncompromising quality standards have made this book an exceptional effort. Their encouragement from time to time has been an inspiration for everyone.

The publisher and the editorial board hope that this book will prove to be a valuable piece of knowledge for researchers, students, practitioners and scholars across the globe.

List of Contributors

Wolde Mekuria
International Water Management Institute (IWMI), P. O. Box 4199, Ban Nongviengkham, Xaythany District, Vientiane, Laos People's Democratic Republic

Mastewal Yami
International Water Management Institute (IWMI), P. O. Box 5689, Addis Ababa, Ethiopia

Jacobus Engelbrecht
Department of Environmental Health, Faculty of Science, Tshwane University of Technology, Private Bag X680, PRETORIA 0001, South Africa

Jacobus Joubert
Department of Environmental Health, Faculty of Science, Tshwane University of Technology, Private Bag X680, PRETORIA 0001, South Africa

Johannes Harmse
Department of Environmental Health, Faculty of Science, Tshwane University of Technology, Private Bag X680, PRETORIA 0001, South Africa

Charles Hongoro
Department of Environmental Health, Faculty of Science, Tshwane University of Technology, Private Bag X680, PRETORIA 0001, South Africa
Health Systems Research Unit, South African Medical Research Council, South Africa

Mathieu Nsenga Kumwimba
Institute of Environment and Sustainable Development in Agriculture, Chinese Academy of Agricultural Sciences / Key
Laboratory of Agro-Environment, Ministry of Agriculture, Beijing 100081, P.R. China

Zeng Xibai
Institute of Environment and Sustainable Development in Agriculture, Chinese Academy of Agricultural Sciences / Key
Laboratory of Agro-Environment, Ministry of Agriculture, Beijing 100081, P.R. China

Bai Lingyu
Institute of Environment and Sustainable Development in Agriculture, Chinese Academy of Agricultural Sciences / Key
Laboratory of Agro-Environment, Ministry of Agriculture, Beijing 100081, P.R. China

S. C. Jhansi
Gujarat Research Society's P.N. Shah Centre for Lifelong Learning, Dr Madhuri Shah Campus, Ramkrishna Mission Road, Corner of 16th & 17th Road, Khar (W), Mumbai, Maharashtra, 400052, India
Population Education Resource Centre (PERC), Department of Continuing and Adult Education and Extension Work, S. N. D. T
Women's University, Patkar Hall Buil, First Floor, Room. Number: 03, New Marine Lines, Mumbai - 400020, Maharashtra, India

Santosh Kumar Mishra
Gujarat Research Society's P.N. Shah Centre for Lifelong Learning, Dr Madhuri Shah Campus, Ramkrishna Mission Road, Corner of 16th & 17th Road, Khar (W), Mumbai, Maharashtra, 400052, India
Population Education Resource Centre (PERC), Department of Continuing and Adult Education and Extension Work, S. N. D. T
Women's University, Patkar Hall Buil, First Floor, Room. Number: 03, New Marine Lines, Mumbai - 400020, Maharashtra, India

Shihab A. SALMAN
The Central Environmental Laboratory, College of Science, University of Baghdad, Iraq

Mohammed J. AL-OBAIDI
Biological Researches Unit for Tropical Diseases, College of Science, University of Baghdad, Iraq

Fadhaa O. SAMEER
Biological Researches Unit for Tropical Diseases, College of Science, University of Baghdad, Iraq

R. K. Aggarwal
Department of Environmental Science, Dr Y S Parmar University of Horticulture and Forestry, Nauni (Solan) 173 230 India

Almas Kashindye
Forestry Training Institute, Olmotonyi, P.O Box 943, Arusha, Tanzania

Elifuraha Mtalo
College of Science, Informatics and Built Environment, University of Bagamoyo, Dar es Salaam, Tanzania

Mathew M. Mpanda
World Agroforestry Centre, Tanzania Country Program, Dar es Salaam, Tanzania

Evaristo Liwa
School of Geospatial Sciences and Technology, Ardhi University, Dar es Salaam, Tanzania

Richard Giliba
Forestry Training Institute, Olmotonyi, P.O Box 943, Arusha, Tanzania

Gregory M. Dowo
Tropical Resource Ecology Programme, Department of Biological Sciences, University of Zimbabwe, P. O. Box MP167, Mount Pleasant, Harare, Zimbabwe

Shakkie Kativu
Tropical Resource Ecology Programme, Department of Biological Sciences, University of Zimbabwe, P. O. Box MP167, Mount Pleasant, Harare, Zimbabwe

Jacob Godfrey AGEA
Department of Extension and Innovation Studies, College of Agriculture and Environmental Sciences, Makerere University, P.O. Box 7062 Kampala, Uganda

Clement Akais OKIA
Department of Extension and Innovation Studies, College of Agriculture and Environmental Sciences, Makerere University, P.O. Box 7062 Kampala, Uganda

Bernard Bonton OBAA
Department of Extension and Innovation Studies, College of Agriculture and Environmental Sciences, Makerere University, P.O. Box 7062 Kampala, Uganda

James Munga KIMONDO
Kenya Forestry Research Institute, P.O Box 20412-00200, Nairobi, Kenya

Prossy ISUBIKALU
Department of Extension and Innovation Studies, College of Agriculture and Environmental Sciences, Makerere University, P.O. Box 7062 Kampala, Uganda

Dino Andrew WOISO
Department of Biological Science, Sokoine University of Agriculture, P.O. Box 3038, Morogoro, Tanzania

Joseph OBUA
Inter-University Council of East Africa, P.O Box 7110, Kampala Uganda

Zewge TEKLEHAIMANOT
School of Environment, Natural Resources and Geography, Bangor University, Bangor-Gwynedd, LL57 2UW, United Kingdom

Mohammed El-Bekkay
Department of Biological Sciences, Faculty of Sciences, AGADIR, Morocco

Abedellatif Moukrim
Department of Biological Sciences, Faculty of Sciences, AGADIR, Morocco

Faiçal Benchakroun
High Commissary of Waters and Forests and Fighting Against Desertification, RABAT, Morocco

Jean Lagarde BETTI
Department of Botany, Faculty of Sciences, University of Douala, BP 24 157 Cameroon

Joseph AMBARA
Ministry of Environment and Nature Protection, Cameroon

E. E. Ezenwaji
Department of Geography and Meteorology, Nnamdi Azikiwe University, Awka, Anambra State, Nigeria

A. C. Okoye
Department of Environmental Management, Nnamdi Azikiwe University, Awka, Anambra State, Nigeria

A. K. Awopeju
Department of Statistics, Nnamdi Azikiwe University, Awka, Anambra State, Nigeria

Nimusiima Alex
Department of Geography, Geo-Informatics and Climatic Sciences, Makerere University, Uganda

C. P. K. Basalirwa
Department of Geography, Geo-Informatics and Climatic Sciences, Makerere University, Uganda

J.G.M Majaliwa
Department of Geography, Geo-Informatics and Climatic Sciences, Makerere University, Uganda

W. Otim-Nape
Africa Innovations Institute, Kampala, Uganda

J. Okello-Onen
Department of Biology, Gulu University, Uganda

C. Rubaire-Akiiki
College of Veterinary Medicine, Animal Resources and Bio-security, Makerere University, Uganda

J. Konde-Lule
College of Health Sciences, Makerere University, Uganda

S. Ogwal-Byenek
Africa Innovations Institute, Kampala, Uganda

A. AbdulKadir
Department of Geography, Federal University of Technology, PMB 65, Minna, Bosso Campus 920003, Niger State, Nigeria

M. T. Usman
Department of Geography, Federal University of Technology, PMB 65, Minna, Bosso Campus 920003, Niger State, Nigeria

A. H. Shaba
Strategic Space Applications, National Space Research and Development Agency (NASRDA), P.M.B. 437, Garki, Abuja, Nigeria

S. Saidu
Department of Geography, Federal University of Technology, PMB 65, Minna, Bosso Campus 920003, Niger State, Nigeria

K. Vasanthi
Department of Zoology, Sri Parasakthi College for Women, Courtallam, M. S. University, Tirunelveli, Tamilnadu- 627 802, India

K. Chairman
Department of Zoology, Sri Paramakalyani College, Alwarkurichi, M. S. University, Tirunelveli, Tamilnadu-627 412, India

A. J. A. Ranjit Singh
Department of Zoology, Sri Paramakalyani College, Alwarkurichi, M. S. University, Tirunelveli, Tamilnadu-627 412, India

G. A. Ayolagha
Department of Crop/Soil Science, Faculty of Agriculture, Rivers State University of Science and Technology, Port Harcourt, Nigeria

K. D. Peter
Department of Crop/Soil Science, Faculty of Agriculture, Rivers State University of Science and Technology, Port Harcourt, Nigeria

Dejene A. Tessema
Chemistry Department, College of Natural Sciences, Jimma University, P.O.Box 387, Jimma, Ethiopia

Dessalew D. Alemayehu
Chemistry Department, College of Natural Sciences, Jimma University, P.O.Box 387, Jimma, Ethiopia

C. S. Mesike
Rubber Research Institute of Nigeria, P.M.B. 1049, Benin City, Edo State, Nigeria

T. U. Esekhade
Rubber Research Institute of Nigeria, P.M.B. 1049, Benin City, Edo State, Nigeria

Darshan Marjadi
Department of Biotechnology, Shree Ramkrishna Institute of Computer Education and Applied Sciences, Surat, Gujarat India

Nishith Dharaiya
Department of Life Sciences, Hemchnandracharya North Gujarat University, Patan, Gujarat, India

Dedeke Oyinade Aderoju
Department of Botany, Federal University, Lafia, Nasarawa State, Nigeria

Akomolafe Gbenga Festus
Department of Botany, Federal University, Lafia, Nasarawa State, Nigeria

Alemayehu Dugassa
Jimma University, Collage of Natural Sciences, Ethiopia

Ketema Bacha
Jimma University, Collage of Natural Sciences, Ethiopia

Tisge Ketema
Jimma University, Collage of Natural Sciences, Ethiopia

A. C. C. Ezeabasili
School of the Built Environment, Heriot-Watt University, UK

O. L. Anike
Department of Geological Sciences, Nnamdi Azikiwe University, Anambra state, Nigeria

B. U. Okoro
Department of Civil Engineering, Nnamdi Azikiwe University, Anambra State, Nigeria

C. M. U-Dominic
Department of Industrial Engineering, University of Ibadan, Nigeria

Girmaye Benti
Department of Biology, Dilla University College of Natural and Computational Science, P. O. Box.419 Dilla, Ethiopia

F. O. Adekayode
Department of Crop, Soil and Pest Management, Federal University of Technology, Akure, Ondo State, Nigeria

T. Lutaaya
National Agricultural Research Organization, Mukono Zonal Agricultural Research and Development Institute (NARO MuZARDI)

M. O. Ogunkoya
Federal College of Agriculture, Akure, Ondo State, Nigeria
Adekayode Farms and Agricultural Consultancy, Fanibi Road, Akure, Ondo State, Nigeria

P. Lusembo
National Agricultural Research Organization, Mukono Zonal Agricultural Research and Development Institute (NARO MuZARDI)

P. O. Adekayode
Adekayode Farms and Agricultural Consultancy, Fanibi Road, Akure, Ondo State, Nigeria

Rafia Rashid
Department of Environmental Science/Centre of Research for Development, University of Kashmir, Srinagar-190 006, Kashmir India

Ashok Kumar Pandit
Department of Environmental Science/Centre of Research for Development, University of Kashmir, Srinagar-190 006, Kashmir India

Sami Ullah Bhat
Department of Environmental Science/Centre of Research for Development, University of Kashmir, Srinagar-190 006, Kashmir India

Joseph M. Makaya
Research Center for Biological, Food and Nutritional Sciences, Research and Training Unit, Life and Earth Sciences, University of Ouagadougou, 03 BP 7021, Ouagadougou, Burkina Faso

Marius K. Somda
Research Center for Biological, Food and Nutritional Sciences, Research and Training Unit, Life and Earth Sciences, University of Ouagadougou, 03 BP 7021, Ouagadougou, Burkina Faso

Aly Savadogo
Research Center for Biological, Food and Nutritional Sciences, Research and Training Unit, Life and Earth Sciences, University of Ouagadougou, 03 BP 7021, Ouagadougou, Burkina Faso

Dayéri Dianou
National Center for Scientific and Technological Researches, Institute for Health Sciences Research, 03 BP. 7047, Ouagadougou, Burkina Faso

Nicolas Barro
Research Center for Biological, Food and Nutritional Sciences, Research and Training Unit, Life and Earth Sciences, University of Ouagadougou, 03 BP 7021, Ouagadougou, Burkina Faso

Alfred S. Traoré
Research Center for Biological, Food and Nutritional Sciences, Research and Training Unit, Life and Earth Sciences, University of Ouagadougou, 03 BP 7021, Ouagadougou, Burkina Faso

K. A. Kemabonta
Department of Zoology, Faculty of Science, University of Lagos, Akoka, Lagos, Nigeria

G. Amadi
Department of Zoology, Faculty of Science, University of Lagos, Akoka, Lagos, Nigeria

Zerihun Yohannes AMARE
Economic and Environmental Sustainability Affairs Advisor of FACT, P.O. Box: 3540, Bahir Dar, Ethiopia